NANOBIOMATERIALS

Applications in Drug Delivery

NANOBIOMATERIALS

Applications in Drug Delivery

Edited By
Anil K. Sharma, MPharm
Raj K. Keservani, MPharm
Rajesh K. Kesharwani, PhD

CRC Press
Taylor & Francis Group
Boca Raton London New York

CRC Press is an imprint of the
Taylor & Francis Group, an **informa** business

First published 2018 by Apple Academic Press, Inc.

Published 2019 by CRC Press
Taylor & Francis Group
6000 Broken Sound Parkway NW, Suite 300
Boca Raton, FL 33487-2742

© 2018 by Taylor & Francis Group, LLC
First issued in paperback 2021

CRC Press is an imprint of the Taylor & Francis Group, an informa business

No claim to original U.S. Government works

ISBN 13: 978-1-77-463644-2 (pbk)
ISBN 13: 978-1-77-188591-1 (hbk)

Library of Congress Control Number: 2017952472

Library and Archives Canada Cataloguing in Publication

Nanobiomaterials : applications in drug delivery / edited by Anil K. Sharma, MPharm, Raj K. Keservani, MPharm, Rajesh K. Kesharwani, PhD.

Includes bibliographical references and index.

Issued in print and electronic formats.

ISBN 978-1-77188-591-1 (hardcover).--ISBN 978-1-315-20491-8 (PDF)

1. Drug delivery systems. 2. Nanobiotechnology. 3. Nanostructured materials. I. Sharma, Anil K., 1980-, editor II. Keservani, Raj K., 1981-, editor III. Kesharwani, Rajesh Kumar, 1978-, editor

| RS199.5.N36 2017 | 615'.6 | C2017-905673-5 | C2017-905674-3 |

CIP data on file with US Library of Congress

Visit the Taylor & Francis Web site at
http://www.taylorandfrancis.com

and the CRC Press Web site at
http://www.crcpress.com

CONTENTS

DEDICATION

The Present Book is Dedicated to Our Beloved

Aashna, Atharva, and Vihaan

CONTRIBUTORS

Mohd Fasih Ahmad
Barrocyte Pvt. Ltd., 105, First Floor, Block No-CHBS, Sukhdev Vihar CSC, Near Fortis-Escorts Hospital, New Delhi, 110025, India

Mirza Ehtesham Baig
Barrocyte Pvt. Ltd., 105, First Floor, Block No-CHBS, Sukhdev Vihar CSC, Near Fortis-Escorts Hospital, New Delhi, 110025, India

Mirza Sarwar Baig
Department of Biosciences, Jamia Millia Islamia (A Central University), New Delhi, 110025, India

Mahmood Alaei-Beirami
Drug Applied Research Center and Students' Research Committee, Tabriz University of Medical Science, Tabriz, Iran/Faculty of Pharmacy, Tabriz University of Medical Sciences, Tabriz, Iran

Christopher Clark
The Movement Group, Inc., Atlanta, Georgia, The United States of America

Cecilia Cristea
Analytical Chemistry Department, Faculty of Pharmacy, Iuliu Hațieganu University of Medicine and Pharmacy, 4 Pasteur St., 400012, Cluj-Napoca, Romania, E-mail: ccristea@umfcluj.ro

Paul Dalhaimer
Department of Chemical and Biomolecular Engineering, University of Tennessee, Knoxville, TN, 37996, USA, E-mail: pdalhaim@utk.edu/Department of Biochemistry, Cellular and Molecular Biology, University of Tennessee, Knoxville, TN 37996, USA

Mila A. Emerald
Phytoceuticals International™ and Novotek Global Solutions™, Ontario, Canada, E-mail: drmilaemerald@gmail.com

Ramona Gălătuș
Basis of Electronics Department, Faculty of Electronics, Telecommunication and Information Technology, Technical University of Cluj-Napoca, 28 Memorandumului St., 400114 Cluj-Napoca, Romania

Florin Graur
Surgical Department, Faculty of Medicine, Iuliu Hațieganu University of Medicine and Pharmacy, 8 Victor Babes St., 400012, Cluj-Napoca, Romania

Can Huang
Hunan Province Cooperative Innovation Center for Molecular Target New Drug Study, University of South China, Hengyang, 421001, China, E-mail: yucuiyunusc@hotmail.com

Wen Huang
Learning Key Laboratory for Pharmacoproteomics of Hunan Province, Institute of Pharmacy and Pharmacology, University of South China, Hengyang, 421001, China

Josef Jampílek
Department of Pharmaceutical Chemistry, Faculty of Pharmacy, Comenius University, Odbojárov 10, 83232, Bratislava, Slovakia, E-mail: josef.jampilek@gmail.com

Urmila Jarouliya
School of Studies in Biotechnology, Jiwaji University, Gwalior (M.P.), 474011, India

Pragatii Karna
Department of Pharmacy, Galgotia University, Gautam Budh Nagar, U.P., India

O. P. Katare
University Institute of Pharmaceutical Sciences, UGC Center of Advanced Studies, Punjab University, Chandigarh, 160–014, India

Raj K. Keservani
Faculty of Pharmaceutics, Sagar Institute of Research and Technology-Pharmacy, Bhopal–462041, India, E-mail: rajksops@gmail.com

Rajesh K. Kesharwani
Department of Biotechnology, NIET, Nims University, Jaipur-303121, India

Katarína Kráľová
Institute of Chemistry, Faculty of Natural Sciences, Comenius University, Ilkovičova 6, 84215, Bratislava, Slovakia

Zhi-Ping Li
Hunan Province Cooperative Innovation Center for Molecular Target New Drug Study, University of South China, Hengyang, 421001, China, E-mail: yucuiyunusc@hotmail.com

Garima Mathur
Department of Pharmacy, Galgotia University, Gautam Budh Nagar, U.P., India

Sumitra Nain
Department of Pharmacy, Banasthali University, Banasthali, Rajasthan, 304022, India, E-mail: nainsumitra@gmail.com

Sanju Nanda
Department of Pharmaceutical Sciences, Maharshi Dayanand University, Rohtak, 124001, India, E-mail: sn_mdu@rediffmail.com

Lalit Mohan Negi
Department of Pharmaceutics, Faculty of Pharmacy, Jamia Hamdard University, New Delhi, 110062, India, E-mail: stalegaonkar@gmail.com

Qian Ning
Hunan Province Cooperative Innovation Center for Molecular Target New Drug Study, University of South China, Hengyang, 421001, China, E-mail: yucuiyunusc@hotmail.com

Hari Om Singh
National AIDS Research Institute, Pune, India

Anamaria Orza
Department of Radiology and Imaging Sciences/Center for Systems Imaging, Emory University School of Medicine Atlanta, Georgia, The United States of America

Doina Pîslă
Research Center for Industrial Robots Simulation and Testing, Technical University of Cluj-Napoca, 28 Memorandumului St., 400114 Cluj-Napoca, Romania

Kevin J. Quigley
Department of Chemical and Biomolecular Engineering, University of Tennessee, Knoxville, TN, 37996, USA, E-mail: pdalhaim@utk.edu

Contributors

Rekha Rao
Department of Pharmaceutical Sciences, Guru Jambheshwar University of Science and Technology, Hisar–125001, India

Rahimeh Rasouli
Department of Medical Nanotechnology, International Campus, Tehran University of Medical Sciences, Tehran, Iran, E-mail: r-rasouli@razi.tums.ac.ir

Robert Săndulescu
Analytical Chemistry Department, Faculty of Pharmacy, Iuliu Hațieganu University of Medicine and Pharmacy, 4 Pasteur St., 400012, Cluj-Napoca, Romania, E-mail: ccristea@umfcluj.ro

Kirti Rani Sharma
Amity Institute of Biotechnology, Amity University Uttar Pradesh, Noida, Sector 125, Noida–201303 (UP), India, E-mail: krsharma@amity.edu, Kirtisharma2k@rediffmail.com

Anil K. Sharma
Delhi Institute of Pharmaceutical Sciences and Research, University of Delhi, New Delhi, 110017, India

Harshita Sharma
Department of Pharmaceutics, Faculty of Pharmacy, Jamia Hamdard University, New Delhi, 110062, India, E-mail: stalegaonkar@gmail.com

Sushama Talegaonkar
Department of Pharmaceutics, Faculty of Pharmacy, Jamia Hamdard University, New Delhi, 110062, India, E-mail: stalegaonkar@gmail.com

Călin Vaida
Research Center for Industrial Robots Simulation and Testing, Technical University of Cluj-Napoca, 28 Memorandumului St., 400114 Cluj-Napoca, Romania

Wen-Qin Wang
Hunan Province Cooperative Innovation Center for Molecular Target New Drug Study University of South China, Hengyang, 421001, China

N. K. Yadav
Department of Pharmaceutical Sciences, Maharshi Dayanand University, Rohtak, 124001, India, E-mail: sn_mdu@rediffmail.com

Sa Yang
Hunan Province Cooperative Innovation Center for Molecular Target New Drug Study, University of South China, Hengyang, 421001, China, E-mail: yucuiyunusc@hotmail.com/Changsha Medical University Affiliated People's Hospital of Xiangxiang City, Hunan Province, 411400, *China*

Cui-Yun Yu
Hunan Province Cooperative Innovation Center for Molecular Target New Drug Study, University of South China, Hengyang, 421001, China, E-mail: yucuiyunusc@hotmail.com/Learning Key Laboratory for Pharmacoproteomics of Hunan Province, Institute of Pharmacy and Pharmacology, University of South China, Hengyang, 421001, China

Farzaneh Zaaeri
Department of Pharmaceutics, Faculty of Pharmacy, Tehran University of Medical Sciences, Tehran, Iran

Sobiya Zafar
Department of Pharmaceutics, Faculty of Pharmacy, Jamia Hamdard University, New Delhi, 110062, India, E-mail: stalegaonkar@gmail.com

ABBREVIATIONS

AC NP	amorphous chititn nanoparticles
ASGPR	asialoglycoprotein receptor
BAC	bisacryloylcystamine
BBB	brain blood barrier
BC	bacterial cellulose
CA	cellulose acetate
CCNT	collagen/carbon nanotube
CD	cyclodextrins
CFD	computational fluid dynamics
CFNTs	covalently functionalized nanotubes
CLIJ	confined liquid impinging jets
CLS	cellulose
CM	carboxymethyl
CMC	carboxymethyl cellulose
CMCS	carboxymethyl chitosan
CMCT	carboxymethyl chitin
CNH	carbon nanohorns
CNS	central nervous system
CNTs	carbon nanotubes
CPP	cell-penetrating peptides
CRM	controlled-release mechanisms
CRS	controlled release systems
CRT	controlled-release technology
CTH	camptothecin
DDS	drug delivery systems
DNR	daunorubicin
DOX-HA-SPION	doxorubicin-hyaluronan conjugated super-paramagnetic iron oxide nanoparticles
DTT	dithiothreitol
ECs	endothelial cells
EGCG	epigallocatechin gallate
EGF	epidermal growth factor
EGFR	epidermal growth factor receptor

ELPs	elastin-like polypeptides
EMA	European Medicine Agency
EPR	enhanced permeability and retention
FA	folic acid
FB	flurbiprofen
FDA	Food and Drug Administration
Fe	iron
FNP	flash nanoprecipitation
FR	folate receptor
FRET	fluorescence resonance energy transfer
GA	glycyrrhetinic acid
GA	gum arabic
GC	galactosylated chitosan
GDL	gadoteridol
GEP	gastroenteropancreatic
GG	gellan gum
GPCR	G-protein coupled receptors
GRAS	generally recognized as safe
GRDDS	gastro retentive drug delivery system
GTC	galactosylated trimethyl chitosan-cysteine
HA	hyaluronic acid
HES	hydroxyethyl starch
HPMC	hydroxypropyl methylcellulose
IEC	intestinal epithelial cells
IND	indomethacin
IONPs	iron oxide nanoparticles
LBL	layer-by-layer
LCST	lower critical solution temperature
LDC	lipid drug conjugates
LLC	lewis lung carcinoma
LPD	liposome polycation pDNA
MB	methylene blue
MDDS	multiple drug delivery systems
MMT	montmorillonite
MN	manganese
MNP	magnetic nanoparticles
MRI	magnetic resonance imaging
MSN	mesoporous silica nanoparticles

MTP	membrane transduction peptides
MTX	methotrexate
NCS	nanotoxicological classification system
NIPAAm	nisopropylacrylamide
NIR	near infrared
NLC	nanostructure lipid carriers
NM	nanobiomaterial
NNI	National Nanotechnology Initiative
NP	nanoparticles
NRPs	neuropilins
PACA	polyalkyl cyanoacrylates
PCL	polycaprolactone
PDT	photo-dynamic therapy
PEG	poly ethylene glycol
PEI	polyethylenimine
PET	positron emission tomography
PGA	polyglycolide
PGLA	poly(lactic-co-glycolic) acid
PHDFDA	poly(heptadecafluorodecylacrylate)
PLA	polylactic acid
PLGA	poly lactic-co-glycolic acid
PNIPAM	poly(N-isopropylacrylamide)
PPy	polypyrrole
PR	plasmon resonance
PS	polysacchride
PTX	paclitaxel
PVMMA	poly(vinyl methyl ether-co-maleic anhydride)
PVP	poly(vinylpyrrolidone)
QD	quantum dots
QSAR	quantitative structure–activity relationship
RES	reticuloendothelial system
RESS	rapid expansion of supercritical solutions
RIF	rifampicin
ROS	reactive oxygen species
SAS	solvent to antisolvent
SCF	supercritical fluid
SCO_2	carbon dioxide
SELPs	silk-elastin-like protein polymers

SERS	surface enhanced Raman scattering
SF	silk fibroin
SLN	solid lipid nanoparticles
SPION	super paramagnetic iron oxide nanoparticles
SPR	surface plasmon resonance
SPs	surface plasmons
SSTA	synthetic somatostatin analogs
SWCNTs	single-walled carbon nanotubes
TC	trimethyl chitosan-cysteine
TDD	transdermal drug delivery
TDS	transdermal delivery systems
TEG	tetraethylene glycol
TfR	transferrin receptor
TOAB	tetraoctylammonium bromide
TOPO	trioctylphosphine oxide
TRS	The Royal Society
UCST	upper critical solution temperature
VEGF	vascular endothelial growth factor
VEGFRs	vascular endothelial growth factor receptors
XG	xanthan gum

PREFACE

The field of nanotechnology has been a constantly evolving arena of science and technology. In recent times diversified applications of nanotechnology-based products have definitely helped human beings. Nanotechnology has been observed to uplift the level of sophistication through a variety of ways. The uses of nanotechnology embrace materials science, engineering, medical, dentistry, drug delivery, etc. The utilities relevant to drug delivery are of interest to pharmaceutical manufacturers, healthcare personnel, and researchers. The last few decades have witnessed enormous developments with respect to the delivery of active pharmaceutical ingredients to the target sites, thereby sparing the normal functioning biological systems. The nanomaterials purported to be used in biological systems are termed 'nanobiomaterials.'

Keeping the drug delivery aspect of nanobiomaterials under consideration, this volume is an endeavor to furnish the requisite information to readers. The content of this book is written by highly skilled, experienced and renowned scientists and researchers from all over the world. They provide updated knowledge to provide drug delivery information to readers, researchers, academician, scientists and industrialists around the globe.

The book *Nanobiomaterials: Applications in Drug Delivery* is comprised of 14 chapters that provide an introduction to nanobiomaterials, their physicochemical features, their generalized and specific applications dealing with drug delivery in particular. The materials used as well as formulation and characterization have been discussed in detail. The emphasis of certain chapters is to provide specific input regarding treatment of a disease/disorder causing the highest mortality.

Chapter 1, *Nanoparticulate Nanocarriers in Drug Delivery,* written by Raj K. Keservani and colleagues, introduces the terminology prevailing in nanoscience. Further, it provides an overview of various nano-systems used for the delivery of drug molecules. The authors have summarized the nanocarriers with their key characteristics.

The details of general principles, classification, and methods of preparation of nanoparticles have been presented in Chapter 2, *Nanoparticles: Formulation Aspects and Applications,* written by Sushama Talegaonkar and associates. The authors have discussed metallic and non-metallic nanoparticles with suitable instances. In addition, they provide a thorough coverage of polymeric nanoparticles. The polymers of natural origin as well as synthetic ones are adequately described.

Chapter 3, *Physical and Chemical Properties of Nanobiomaterials,* written by Sumitra Nain and colleagues, gives a brief account of physicochemical attributes of nanoparticles. The last section of chapter summarizes applications of nanobiomaterials in tabular form.

Different aspects of drug delivery are described in Chapter 4, *Introduction to Drug Delivery,* written by Mila A. Emerald. The author has provided information regarding controlled release drug delivery systems with examples of polymers exploited for this purpose. The various routes by which the drug could be delivered have been discussed. The biosafety issue has also been addressed in this chapter.

The description of biodegradable polymers is provided by Chapter 5, *Application of Nanobioformulations for Controlled Release and Targeted Bio-distribution of Drugs,* written by Josef Jampílek and Katarína Kráľová. Nanoparticles-based preparations made up of biodegradable natural and synthetic polymers have been discussed following a brief introduction of nanotechnology-based products and their pros and cons. The studies having mention of bio-distribution of drug molecules have been chosen in order to offer the reader accurate information regarding targeting aspects of nanotechnology.

The uses of nanoparticles for drug delivery and other fields of medicine have been discussed in Chapter 6, *Targeted Delivery Systems-Based on Polymeric Nanoparticles for Biomedical Applications,* written by Sa Yang and associates. The authors have provided information relevant to tumor targeting. The drugs as well as genes that have been incorporated in polymeric nanoparticles have been prominently described. The details of polymers covered both of natural origin and synthetic ones.

The customized applications of nanobiomaterials-based on specific requirements are given by Chapter 7, *Smart Delivery of Nanobiomaterials in Drug Delivery,* written by Mirza Sarwar Baig and associates. The authors have strived to provide an overall preview of nanobiomaterials (NMs) or nanoparticles (NPs) and their applications in life sciences and medicine, insight into the drug delivery system (DDS) plus most recent developments in this field. Finally they discuss the limitations and future of NMs.

An overview of nanoscience-based preparation for the treatment of cancer has been presented in Chapter 8, *Advances of Nanotechnology in Cancer Therapy,* written by Urmila Jarouliya and colleagues. The authors have discussed a number of preparations having relevance in tumor treatment. The targeting to the cancerous tissues the core of this chapter. Moreover, the clinical and preclinical status of nanocarriers has been described covering future prospects.

Chapter 9, *Nanobiomaterials for Cancer Diagnosis and Therapy,* written by Cecilia Cristea and associates, deals with testing protocols and treatment approaches in tumor therapeutics. The authors have provided a brief preamble to different nanocarriers for cancer diagnosis; thereafter, the mechanism of tumor targeting have been discussed in detail. In addition, novel nanotechnology-based drug carriers have been described with respect to tumor therapy.

Chapter 10, *Nanobiomaterials for Cancer Therapy,* written by Kevin J. Quigley and Paul Dalhaimer, gives a focused view of nanocarriers explored for drug delivery in cancer and that have demonstrated improved drug delivery to tumor sites. The authors have availed the current status of nanotechnology-based drug formulations. Subsequently the hurdles posed by biological milieu have been discussed. Attempts to overcome barriers have also been described.

Chapter 11, *Nanotechnology in Hyperthermia-Based Therapy and Controlled Drug Delivery,* written by Anamaria Orza and Christopher Clark, describes the drug delivery approaches based on phenomenon of hyperthermia. The in-depth details regarding the mechanism of such systems have been provided in this chapter. Moreover, characterization of nanotechnology products has been described with incorporating recent studies. An outlook of obstacles and future course of hyperthermia-based formulations is also given.

Chapter 12, *Nano Carrier Systems of Ubidecarenone (Coenzyme Q10) for Cosmetic Applications,* written by N. K. Yadav and colleagues, has addressed the eternal desire of men and women to look beautiful. However, females appear more concerned toward beauty versus their counterparts. The cosmetic applications of antioxidant enzymes have been elaborated on with suitable instances wherever applicable. The variety of nano preparations incorporating Coenzyme Q10 has been discussed. A summary of cosmetic products is provided in tabulated form.

Chapter 13, *Nanobased CNS Delivery Systems,* written by Rahimeh Rasouli and Mahmood Alaei-Beirami, deals in drug delivery to the brain by different nanotechnology-based products. The authors have provided details with respect of the brain's physiology, inherent barriers in the way of drug delivery and emergence of nanotechnology. Further, the formulations are discussed with a focus on targeting aspect to organs of central nervous system (CNS).

The details of nanocarriers employed for delivery of genes have been provided in Chapter 14, *Nanobiomaterial for Non-Viral Gene Delivery,* written by Kirti Rani Sharma. As we are familiar with the fact that the delivery of genes is carried out predominantly by viral vectors, the author has complied information regarding use of variety of nanosystems to deliver genes in treatment of diverse ailments.

ABOUT THE EDITORS

Anil K. Sharma, MPharm
*Delhi Institute of Pharmaceutical Sciences and Research,
University of Delhi, India*

Anil K. Sharma, MPharm, is a lecturer at the Delhi Institute of Pharmaceutical Sciences and Research, University of Delhi, India. He has published 25 peer-reviewed papers in the field of pharmaceutical sciences in national and international reputed journals as well as 10 book chapters. His research interests include nutraceutical and functional foods, novel drug delivery systems (NDDs), drug delivery, nanotechnology, health science/life science, and biology/cancer biology/neurobiology. He graduated with a degree in pharmacy from the University of Rajasthan, Jaipur, India, and received a Master of Pharmacy (MPharm) from the School of Pharmaceutical Sciences, Rajiv Gandhi Proudyogiki Vishwavidyalaya, Bhopal, India, with a specialization in pharmaceutics.

Raj K. Keservani, MPharm
Faculty of Pharmaceutics, Sagar Institute of Research and Technology-Pharmacy, Bhopal, India

Raj K. Keservani, MPharm, has more than seven years of academic teaching experience at various institutes in India in pharmaceutical science. He has published 25 peer-reviewed papers in the field of pharmaceutical sciences in national and international reputed journals, 10 book chapters, and two co-authored books. He is also active as a reviewer for several international scientific journals. His research interests encompass nutraceutical and functional foods, novel drug delivery systems (NDDs), transdermal drug delivery, health science/life science, and biology/cancer biology/neurobiology. Mr. Keservani earned his BPharm at the Department of Pharmacy at Kumaun University, Nainital, India, and received his MPharm from the School of Pharmaceutical Sciences at Rajiv Gandhi Proudyogiki Vishwavidyalaya, Bhopal, India, with a specialization in pharmaceutics.

Rajesh K. Kesharwani, PhD

Department of Biotechnology, NIET, Nims University, Jaipur-303121, India

Rajesh K. Kesharwani, PhD, has more than seven years of research and two years of teaching experience at various institutes in India in bioinformatics and biotechnology education. He has authored over 29 peer-reviewed articles and 15 book chapters and has co-edited one book. He has been a member of many scientific communities and is a reviewer for many international journals. His research fields of interest are medical informatics, protein structure and function prediction, computer-aided drug designing, structural biology, drug delivery, cancer biology, and next-generation sequence analysis. Dr. Kesharwani is a recipient of the Swarna Jayanti Puraskar Award from the National Academy of Sciences, India, for best paper presentation. Dr. Kesharwani earned a BSc in Biology from the Ewing Christian College, Allahabad, an autonomous college of the University of Allahabad, and received his MSc in Biochemistry from Singh University, Rewa, Madhya Pradesh, India. He earned a PhD from the Indian Institute of Information Technology, Allahabad and received a Ministry of Human Resource Development/India fellowship and a Senior Research Fellowship from the Indian Council of Medical Research, India.

PART I

AN OVERVIEW OF NANOBIOMATERIALS

CHAPTER 1

NANOPARTICULATE NANOCARRIERS IN DRUG DELIVERY

RAJ K. KESERVANI,[1] ANIL K. SHARMA,[2] and
RAJESH K. KESHARWANI[3]

[1]*Faculty of Pharmaceutics, Sagar Institute of Research and Technology-Pharmacy, Bhopal–462041, India, E-mail: rajksops@gmail.com*

[2]*Delhi Institute of Pharmaceutical Sciences and Research, University of Delhi, New Delhi, 110017, India*

[3]*Department of Biotechnology, NIET, Nims University, Jaipur-303121, India*

CONTENTS

1.1   INTRODUCTION

The word *nano* is derived from the Greek language means *dwarf*. The range of nano is typically from 0.2 nm to 100 nm and first time the term "nanotechnology" was first introduced by Japanese scientist at the University of Tokyo, Norio Taniguchi in 1974. The nano particles are in tiny in size and can have different and enhanced properties compared with the same material at a larger size (TRS, 2004). Nanotechnology is the branch of science which deals with the study of manipulation of any matter at the molecular level or in nano size with the application of technology. Nanotechnology-based drug delivery is an ideal approach to treat many diseases. Nanoparticles are having long circulating time, therapeutic activity and reaches to the active site with appropriate concentration are called as "low-hanging fruit" of nanotechnology-based drug delivery system (Duncan, 2005; Joseph and Moore, 2008; Vasir et al., 2005). The chemotherapeutic effect of many important drugs are not use full in case of neurological disorder for the treatment of various deadly brain related disease or disorders due to the presence of blood–brain barrier (Matsumura and Maeda, 1986). The blood–brain barrier is a unique safeguard that tightly separate the brain from body blood circulatory system. Thus crossing of blood brain barrier for a drug is a biggest challenge and to develop a delivery system.

Thus, drug delivery to this organ is a challenge, because the brain benefits from very efficient protection. Nanotechnology offers a solution for using the numerous chemical entities for treating brain disorders that are not clinically useful because of the presence of the blood–brain barrier. Nanoparticles can be effectively used to deliver relevant drugs to the brain (Pardridge, 1999). The protein albumin has been modified to create novel nanostructures for applications in drug delivery. The surface of albumin has several groups available for covalent conjugation of biomolecules and drugs. Albumin-DNA-polyethylenimine (PEI) conjugates have been used for gene delivery, with reduced irritation, damage and toxicity (Li, 2008).

Many biomaterials, primarily polymer- or lipid-based, can be used to this end, offering extensive chemical diversity and the potential for further modification using nanoparticles. The particularly large surface area on the nanoparticles presents diverse opportunities to place functional groups on the surface. Particles can be created by expanding or contracting with changes in temperature or pH, or interact with anti-bodies in special ways to provide rapid ex-vivo medical diagnostic tests. Nanopharmaceuticals offer the potential to create "magic bullets" to improve and innovate the therapeutic armory to diagnose, prevent, and treat human diseases, both emerging diseases and those that have not yet been appropriately addressed. In this chapter, we discussed about the smart delivery of nanoparticulate nanocarriers such as nanoparticle, nanoemulsion, nanocapsule, nanogels, nanomedicines, dendrimers, liposome, grapheme, carbon nanotube, etc., and its applications in drug delivery.

1.2 NANOPARTICLES

Nanoparticles are in the range of 1 to 100 nm in size or structures of particle with several hundred nanometers in size. The living cells are easily absorbed the drugs which are delivered with the help of nanocarriers with optimized biological and physicochemical properties as compared to larger drug molecule alone, so they can be successfully used as delivery tools for currently available bioactive compounds (Keservani and Sharma, 2016; Singh and Keservani, 2016; Suri et al., 2007).

The development of conjugated drug opened new way to treat disease effectively and in this process drug are attached with nano sized career molecule in targeted therapy. A drug may be covalently attached or adsorbed to the nano sized carrier molecular surface or else it can be encapsulated into it (Nevozhay et al., 2007). Once the conjugates of drug-nanocarrier reach the targeted site of diseased organs or tissues, the chemotherapeutic agents are released. A controlled release of drugs at target site is a very important by using nanocarriers and it can be only achieved with the help of optimized physiological environment such as temperature, pH, osmolality, or *via* an enzymatic activity (Ai et al., 2011). Solid lipid nanoparticles (SLN), nanostructured lipid carriers (NLC) and lipid drug conjugates (LDC) are types of carrier systems-based on solid lipid matrix, for example, lipids solid at the body temperature (Wissing et al., 2004). Currently nano-based drug delivery system are advanced method and has been used for the drug delivery in the region of ocular (Attama et al., 2007), dermal (Abdel-Mottaleb et al., 2011), parenteral (Nayak et al., 2010), per oral (Muchow et al., 2008), rectal (Sznitowska et al., 2001) and pulmonary (Liu et al., 2008). SLNs are used for the process of in vitro transfection of DNA in to mammalian cells and in this process DNA are in the form of condense nanometric colloidal particles (Tabatt et al., 2004).

The use of SLN for DNA transfection offers many technological advantages as compared to standard DNA carriers (cationic lipids or cationic polymers) are as (Rudolph et al., 2004):

1. Relative ease of production without organic solvents;
2. The possibility of large-scale production with qualified production lines;
3. Good storage stabilities;
4. Possibility of steam sterilization;
5. Lyophilization.

The bioavailability of drug at their targeted site is biggest challenge and now-a-days SLN is very popular drug delivery system for ophthalmic application and also gaining prominence as promising method to improve the poor bioavailability of biomolecules in eye regions. The SLN and NLC are regarded as the first and second generation of lipid nanoparticles and are currently being applied in drug delivery system (Joshi and Muller, 2009; Muller, 2007). Apart from the advantages of SLN, it have some limitation like low drug loading capacity, poor long-term stability and early drug expulsion caused by lipid polymorphism. Nanostructured lipid carriers was designed and developed, and have advantages over of SLN.

The development of gene delivery system is biggest challenge and the development of a multifunctional envelope-type nanodevice has been 1st time reported by Kogure et al. (2004) the designed envelop are less cytotoxic contains membrane-permeable peptide R8. This lipid-based nanocarrier device can be a useful tool for gene delivery, gene therapy and biochemical research. This system can incorporate different functional units or devices such as a specific ligand to a specific cell, intracellular sorting devices that permit endosomal escape, and nuclear localization. Carbon is used as nanocarriers in drug delivery system are differentiated into carbon nanotubes (CNTs) and carbon nanohorns (CNH).

Carbon nanotubes are specialized structures and formed by rolling of single (SWNCTs – single-walled carbon nanotubes) or multi (MWCNTs

– multi-walled carbon nanotubes) layers of graphite. These carbon nano-carriers are having an enormous surface area and are good electrical and thermal conductivity (Beg et al., 2011). The release of drug using carbon nanocarriers can be controlled by chemically and electrically. The open ends of carbon nanotubes were sealed by using polypyrrole (PPy) films which prevents undesired release of drugs (Luo et al., 2011). Homing devices, for example, folic acid (Dhar et al., 2008) and epidermal growth factor (Bhirde et al., 2009), were attached to improve selectivity of such drug delivery systems.

In vivo studies suggested that intra tracheal administration of nano-tubes in rats facilitate the change in lung structures, such as granuloma formation and interstitial fibrosis (Lam et al., 2004; Muller et al., 2005). Magnetic nanoparticles show a wide variety of characteristics, which make them highly important drug carriers for delivery at the specific tar-get site.

Magnetic nanoparticles can be easily handled with the help of external magnetic field, the possibility of using passive and active drug delivery strat-egies. These MNPs are also having ability and used for visualization purpose in case of MRI. MNPs are enhanced the uptake of drugs and which in treat-ment effectively at the therapeutically optimal doses (Arruebo et al., 2007). MNPs have been tested as multitasking device simultaneously for the treat-ment and diagnosis as drug delivery system and biosensors, respectively. The combine use of targeted therapy and magnetic resonance or magneto fluorescent imaging and targeted therapy (via conjugation of targeting moi-eties) can increase the effectiveness for the killing of cancer cells in cancer chemotherapy (Chomoucka et al., 2010).

The sources of magnetic material used in combination therapy usu-ally Iron oxides with core/shell structure (Drbohlavova et al., 2009). Iron oxides have several crystalline phases or polymorphs viz., Fe3O4 (magne-tite), α-Fe$_2$O$_3$ (hematite), γ-Fe$_2$O$_3$ (maghemite), β-Fe$_2$O$_3$, ϵ-Fe$_2$O$_3$, and some others, form due to high pressure (Zboril et al., 2002). Among all the Iron oxides, only two (magnetite and maghemite) found the greatest importance in bioapplications (Tucek et al., 2006). The well-known small size material, carbonyl iron with a unique form of elemental iron was also used as mag-netic core (Reshmi et al., 2009).

Polymeric micelles have also attracted more attention in case drug deliv-ery system due to their small particle size, ability to solubilize hydropho-bic molecules, good thermodynamic solution stability, prevention of rapid clearance by the reticuloendothelial system (RES), extend release of various drugs and good thermodynamic solution stability (Kabanov et al., 2002).

Gao et al. (2005) described a new modality of drug targeting tumors is-based on drug encapsulation in polymeric micelles followed by a localized triggering of the drug intracellular uptake induced by ultrasound, which is focused into the tumor volume. The idea behind the use of polymeric micelles for drug encapsulation, to decreases a systemic concentration of free drug, slowdown the intracellular drug uptake by normal cells, and which provides passive drug targeting of tumors via enhanced penetration and retention effect and finally as a result of abnormal permeability of tumor blood vessels (Wu et al., 2001). The polymeric micelles for drug encapsulation were first developed and prepared by method of solubilization of the outer shell material in organic solvent (Couvreur et al., 2002).

Interesting chemotherapeutic application of drugs encapsulated in polymeric nanoparticles have been reported for the parenteral (Barrat et al., 1994; Hubert et al., 1991), oral (Dalencon et al., 1998), and the ocular routes (Calvo et al., 1996). However, the industrial constraints like handling of solvent, limited scale, and particular efforts needed to lowers the residual solvent up to few parts per million induced high manufacturing costs.

1.3 NANOMEDICINE

Nanotechnology have many sub branch and their importance in medical field and among that Nanomedicine is important and used in the process of diagnosis of disease, treatments and preventing disease and traumatic injury, of relieving pain and of preserving and improving human health, using molecular tools and molecular knowledge of the human body.

It is the application of nanotechnology (engineering of tiny machines) to the prevention and treatment of disease in the human body. The most elementary of nanomedical devices will be used in the diagnosis of illnesses (Keservani and Sharma, 2016).

In making advancement of nanotechnology in near future, Nanomedicine develops a set of valuable tools/medical devices/units for research, medical application, especially useful for clinicians (Freitas, 2009; Wagner et al., 2006). The designed tools should have ability to improve the solubility of drug molecules, controlled circulation, biodistribution, bioavailability and release rate or altogether enhancing the efficacy of drugs (Duncan, 2003; Langer, 1998; Moghimi, et al., 2005; Panyam and Labhasetwar, 2003).

The use of biodegradable nano-systems for the designing and development of nanomedicines is one of the most successful ideas. Nanocarriers are made up of biodegradable compounds or polymers and after reaches inside the body, it undergo hydrolysis and producing biodegradable monomeric metabolites, such as lactic acid and glycolic acid. The production of such monomeric causes less systemic toxicity associated with using of PLGA for drug delivery or biomaterial applications. Such nanoparticles are biocompatible with human tissue and cells (Panyam and Labhasetwar, 2003). Nanoparticles must perform a complex and series of tasks, it becomes obvious that one way to accomplish this is to contain each step in a single layer of a multilayered nanoparticle (Prow et al., 2004). Nanomaterials are inherently self-assembled atom-by-atom, or layer-by-layer (LBL) (Ai et al., 2003; De Smedt et al., 2000; Luov and Caruso, 2001).

Cancer is an uncontrolled division of cells or progressive disease. The early stage of transformation of cancer to normal cell might be possible with the use of more sophisticated strategy to repair tissues and organs at the single cell level using targeted nanoparticles as compared to simply killing the cancer cells. Hence the most profound use of nanomedicine may eventually be in the areas of regenerative medicine for treatment of many diseases including heart disease and infectious diseases. Prew first time show some very preliminary data in regenerative medicine involving hepatitis virus infected cells (Prow, 2005).

1.4 NANOCOMPOSITE MATERIALS

Nanocomposite is also known as multiphase solid material with one of the phases has one, two or three dimensions of less than 100 nm. Bionanocomposites form an advance multidisciplinary area that are combination of biology, mathematics, materials science, and nanotechnology. New bionanocomposites are impacting diverse areas, in particular, biomedical science. Generally, polymer nanocomposites are the result of the combination of polymers and inorganic/organic fillers at the nanometer scale. The extraordinary versatility of these new materials springs from the large selection of biopolymers and fillers available to researchers. Existing biopolymers include, but are not limited to, polysaccharides, aliphatic polyesters, polypeptides and proteins, and polynucleic acids, whereas fillers include clays, hydroxyapatite, and metal nanoparticles (Rhima et al., 2013). To date, primarily polysaccharide and polypeptidic matrices have been used with HAP nanoparticles for composite formation. Yamaguchi and co-workers have synthesized and studied flexible chitosan–HAP nanocomposites (Yamaguchi et al., 2001). The matrix used for this study, chitosan (a cationic, biodegradable polysaccharide), is flexible and has a high resistance upon heating because of intramolecular hydrogen bonds formed between the hydroxyl and amino groups (Trivedi et al., 2016).

1.5   NANOEMULSIONS

The role of nanotechnology in drug delivery systems have shown remarkable advance in the present pharmaceutical research. Among these, we placed our focus on nanoemulsions which could be defined as isotropic, thermodynamically stable, transparent or translucent with mean droplet diameters in the range between 50 to 1000 nm (Aboofazeli, 2010). The advantages of nanoemulsions are site-specific delivery of drugs, ability to protect drugs from degradation with long-term stability, capacity to dissolve large quantities of hydrophobics making them ideal drug delivery system. During the therapy frequency and dosage drugs injections can be reduced as the release pattern of drugs takes place in a sustained and controlled mode over long periods of time (Lovelyn and Attama, 2011). There are several routes for delivery of drugs of nanoemulsion as they are well tolerated orally, on the skin and mucous membranes for delivery of tropical active drugs. Nanoemulsion are good when used as vehicles for drugs active against herpes labialis, fungal infections, bacterial infections, vaginitis, etc. (Kitches et al., 1977). Interaction of nanoemulsion globules to membrane facilitate the penetration for easily transfer. Less quantity of surfactant is required in nanoemulsions compared to other emulsion systems. The above properties increases the bioavailability of poorly soluble drugs. However, extra small size provides large surface area which in turn eases the solubilization and penetration through the skin or epithelial layer (Sutradhar and Amin, 2013).

Nanoemulsions are primarily seen as vehicles for insoluble drugs administration. Recently, they have received increasing attention as colloidal carriers in reference to targeted delivery of various anticancer drugs, photosensitizers, genes, or diagnostic agents. Scientific research with perflurochemical nanoemulsions has explained their promising results for enhanced sonography imaging and the treatment of cancer in conjugation with other treatment modalities or by directed delivery to the neovasculature (Tiwari and Amiji, 2007). Nanoemulsion of antimicrobial containing water and soybean oil of ubiquitous droplets of 200 to 400 nm range, effectively destroy the microbes without side effects (Hamouda et al., 2001). Nanoemulsions could be topical antimicrobial activity and achieved by systemic antibiotics previously. Parachlorometaxylenol is relevant at concentration ranges 1–2 times lower than other disinfectants, which reflects no to□iceffects on animals, human or the environment (Charles and Attama, 2011; Rutvij et al., 2011; Subhashis et al., 2011; Yashpal et al., 2013). It is well known that the use of eye drops as conventional ophthalmic delivery systems reflects a poor bioavailability and therapeutic response because of lacrimal secretion and nasolacrimal drainage in the eye (Patton and Robinson, 1976; Sieg and Robinson, 1977).

1.6 NANOGELS

Nanogels are nanoscalar polymer networks which imbibe water when placed into an aqueous environment. Their affinity to aqueous solutions, inertness in the blood stream, superior colloidal stability, internal aqueous environment and suitable for bulky drugs incorporation make nanogels ideal candidates for uptake and delivery of peptides, proteins and several biological compounds. In respect to the classic nanoparticles, advantage of these hydrogel nanoparticles is the possibility of obtaining an increased degree of encapsulation and provides an ideal tridimensional microenvironment for several macromolecule types. Due to their molecular shape and size, ranging between 100–700 nm, nanogels could escape renal clearance with prolonged serum half-life period (Wilk et al., 2009).

1.7 NANOCAPSULES

Nanocapsules are vesicular or reservoir system of oil/water, essentially confines to cavity which is surrounded by tiny polymeric membrane. Nanocapsules are drug delivery systems for different drugs by various routes of administrations such including oral and parental significantly reported to reduce the toxicity of drugs and improve the durability/stability of the drug in the biological fluid as well as in the formulation (Amaral et al., 2007). Nanocapsules are active vectors due to their good capacity to release drugs; their sub cellular size induce relatively higher intracellular uptake than other particulate systems. They could increase the stability of active substances (Ourique et al., 2008). The ultra structure of the nanocapsules make it as finer drug carrier are apt to be absorbed more easily through biological systems, pronounced as enhanced permeation and retention (EPR effect-the diffusion and flow of nanoparticles through narrow channels), and can effectively manipulate affected diseased cells by releasing drug at sub cellular level (Maeda, 1992).

More recently nanoparticles and nanocapsules for biomedical purposes are synthesized-based on polyvinylpyrrolidone, polylactic- and polyglycolic acids, poly-ε-caprolactone and polyalkyl cyanoacrylates (PACA) (Kreuter, 1994). In the present scenario there are countless research groups working to synthesize polymeric nanocapsules systems for controlled drug release. The founder of nanoparticles creation for controlled release purposes was Prof. Peter Speiser (ETH, Zurich) (Kreuter and Speiser, 1976).

1.8 NANOFIBERS

These are a nanomaterial with one dimension of <100 nm. Wide range of polymers such as polyvinyl alcohol, collagen, gelatin, chitosan and carboxymethylcelulose are subjected to electro spinning technique for the synthesis of nanofibers. Nanofibers have unique properties of large specific surface area with small pore size reflecting the opportunities in management of wound care applications (Hung et al., 2003). Electrospun PLGA-based nano-fibers is developed by Xie et al., as implants for the sustained delivery of anticancer drug to treat C6 glioma cells in vitro (Gupta et al., 2010; Xie and Wang, 2006). Zeng et al., studied the methods of encapsulation and release kinetics of the lipophilic drug paclitaxel and the hydrophilic drug doxorubicin hydrochloride in the electro spun PLLA fiber mats. Paclitaxel encapsulation was found to be with good compatibility with PLLA and solubility in chloroform/acetone solvent, however, doxorubicin hydrochloride was observed on or near the surfaces of PLLA fibers. The releasing behavior of these drugs evidenced that the release of paclitaxel from electrospun PLLA fiber samples due to the degradation of the fibers followed nearly zero-order kinetics. In addition, a burst release was found for doxorubicin hydrochloride because the diffusion of the drug on or near the surfaces of the fiber sample (Zeng et al., 2005). Drug-loaded Eudragit RS100 (ERS) and Eudragit S100 (ES) nanofibers could be prepared by selecting the range of variables like type of drug: polymer ratio, solvent and solution viscosity. In addition, the optimized formulations could be helpful for colonic drug delivery (Akhgari et al., 2013).

1.9 LIPOSOMES

Liposomes are generally vesicles that consist phospholipid bilayers in which polar character of the liposomal core enables polar drug molecules to be encapsulated. Lipophilic/ amphiphilic molecules solubilized in the phospholipid bilayer according to their affinity towards the phospholipids (Ahmad et al., 1993; Keservani and Sharma, 2016; Sharma et al., 2010). Bangham and co-workers defined liposomes as small, spherical shape vesicle produced from non-toxic surfactants, cholesterols, glycolipids, sphingolipids, long chain fatty acids and even membrane protein (Bangham et al., 1965).

Liposomes have been well reported to use to enhance the therapeutic potential of new or established drugs by modifying drug absorption, decreasing their metabolism, prolonging the biological half-life and reducing toxicity. Drug distribution is controlled by nature and properties of the carrier, no longer by the physico-chemical characteristics of the drug substance (Gregoriadis and Florene, 1993). Liposomes potential is explained by Rogers et al. in oral drug delivery methods. Polymerized, microencapsulated, and polymer coated liposomes have been reported to increased potential of oral liposomes (Rogers and Anderson, 1998). Targeted liposomes uses and the understanding of their cellular processing will increased the lead to effective therapies from oral liposomes. Krauze et al. (2006) use Gadoteridol (GDL) liposomes as a tracer that allows us to trace infusions in real-time on magnetic resonance imaging (MRI). Taxanes are semi synthetic analogs as complexes of diterpenoid natural products. Currently, these drugs are prominent anticancer agents used in combined chemotherapy (Michaud et al., 2000). Taxane in liposomes have shown slower elimination with higher antitumor activity against various murine and human tumors, have lower systemic toxic effect as compared to Taxol® (Sharma et al., 1997). They are also reported to have antitumor effect in Taxol-resistant tumor models (Sharma et al., 1993).

1.10 GRAPHENE

In the present scenario, graphene and graphene-based materials are outstanding in function due to their fantabulous electronic (Berger et al., 2006), mechanical properties (Geim and Novoselov, 2007) by high surface area, higher aqueous solubility, useful non-covalent interactions with aromatic drugs with comparatively gamier stability in physiological solution as like human serum. Graphene and derivatives have been reported to have greater potentials in fields of nanoelectronics (Xuan et al., 2008), energy technology (for examples, fuel cell, super capacitor, hydrogen storage) (Liu et al., 2010), composite materials (Yang et al., 2010), sensors (Lu et al., 2009), and catalysis (Qu et al., 2010), which are summarized in several review articles by emeritus authors (Feng and Liu, 2011).

Genetic disorders and associated complications are treated by gene therapy, a novel and promising approach for cystic fibrosis, Parkinson's disease, and cancer (Yang et al., 2007). Process of gene therapy requires a gene vector protecting the DNA from nuclease degradation and facilitates the cellular uptake of DNA with high transfection efficiency (Naldini et al., 1996). Cui and his co-workers (Huang et al., 2011) reported the uses of folic acid and sulfonic acid conjugated GO loaded with porphyrin photosensitizers for targeting photo-dynamic therapy (PDT) are good method. Graphene-based materials demonstrate excellent capability to adsorb a variety of aromatic bio-molecules through a π–π stacking interaction and/or electrostatic interaction, which make them ideal materials for constructing biosensors and loading drugs (Chaudhary, 2015).

1.11 QUANTUM DOTS

Quantum dot (QD) is semiconductor crystal of 1–10 nm. Due to specific size range, QD exhibits quantum phenomena which in turn yield significant benefits in optical properties. It has been established that on excitation, smaller the size of QD results a higher energy and intensity of emitted light. QDs can be obtained from semiconductors (e.g., selenium, cadmium, etc.), metals (gold), or carbon-based materials (carbon dots and graphene) (Novoselov et al., 2004; Novoselov et al., 2005). Cadmium selenide nano particles in the form of quantum dots are used in detection of cancer tumors because when exposed to ultraviolet light, they glow. The surgeon injects these quantum dots into cancer tumors and can see the glowing tumor, thus the tumor can easily be removed (Nahar et al., 2006; Owen et al., 2008). Quantum dots have been used for molecular imaging and tracing of stem cells, for delivery of gene or drugs into stem cells, nano materials such as carbon nanotubes, fluorescent CNTs and fluorescent MNPs have been used (Wang et al., 2009; Zhao et al., 2015). Bhatia et al., recently reported on siRNA delivery using QDs as delivery vehicles (Derfus et al., 2007). Targeting peptides and siRNAs were conjugated to QDs in a 'parallel' manner. That is, the targeting peptide and siRNA were synthesized separately and simultaneously linked to the QD surface. Quantum dots represent a new tool of significant potential in neuroscience research. In addition to offering an alternative to traditional immunocytochemistry, they are particularly valuable for studies of neurons. Quantum dots can be used to visualize, measure, and track individual molecular events using fluorescence microscopy, and they provide the ability to visualize and track dynamic molecular processes over extended periods (e.g., from seconds to many minutes) (Larson et al., 2003).

1.12 DENDRIMERS

Dendrimer chemistry was first introduced in 1978 by Fritz Vogtle and co-workers. He synthesized the first "cascade molecules." In 1985, Donald A. Tomalia, synthesized the first family of dendrimers. The word *"dendrimer"* originated from two words, the Greek word *dendron*, meaning tree, and *meros*, meaning part (Bai, 2006; Padilla et al., 2002). Many surface modified PAMAM dendrimers are non-immunogenic, water-soluble and possess terminal-modifiable amine functional groups for binding various targeting or guest molecules (Sougata et al., 2012). Jevprasesphant et al. investigated effect of dendrimer generation and conjugation on the cytotoxicity, permeation and transport mechanism of PAMAM dendrimer and surface-modified cationic PAMAM dendrimer using mono layers of the human colon adenocarcinoma cell line (Jevprasesphant et al., 2003). The first in vivo diagnostic imaging applications using dendrimer-based MRI contrast agents were demonstrated in the early 1990s by Lauterbur and colleagues (Bakalova et al., 2007).

1.13   CONCLUSIONS

Medical fields always need to optimized the healthcare systems specially drug delivery. Nanotechnology provides nanocarriers for drug delivery systems and are designed to improve the pharmacological and therapeutic properties over the conventional medicinal system and drugs. The incorporation of drug molecules into nanocarrier facilitates the controlled distribution of drugs at their targeted sites and it can protect a drug against degradation.

KEYWORDS

- **biocompatibility**
- **dendrimers**
- **nanocomposites**
- **nanomedicine**
- **polymeric nanoparticles**
- **vesicular carriers**

REFERENCES

Abdel-Mottaleb, M. M. A., Neumann, D., & Lamprecht, A., (2011). Lipid nanocapsules for dermal application: A comparative study of lipid-based versus polymer-based nanocarriers. *Eur J Pharm Biopharm, 79*, 36–42.

Aboofazeli, R., (2010). Nanometric-scaled emulsions (nanoemulsions), *Iran. J. Pharm. Res., 9*, 325–326.

Ahmad, et al., (1993). Antibody-Targeted Delivery of Doxorubicin Entrapped in Sterically Stabilized Liposomes Can Eradicate Lung Cancer in Mice, *Cancer Res., 53*, 1484–1488.

Ai, H., Jones, S. A., & Lvov, Y. M., (2003). Biomedical applications of electrostatic layer-by-layer nano-assembly of polymers, enzymes, and nanoparticles. *Cell Biochem Biophys, 39*.

Ai, J., Biazar, E., Montazeri, M., Majdi, A., Aminifard, S., Safari, M., et al., (2011). Nanotoxicology and nanoparticle safety in biomedical designs. *Int J Nanomedicine, 6*, 1117–1127.

Akhgari, A., Heshmati, Z., & Makhmalzadeh, B. S., (2013). Indomethacin Electrospun Nano-fibers for Colonic Drug Delivery: Preparation and Characterization, *Adv Pharm Bull,* *3*(1), 85–90.

Amaral, E., Grabe-Guimaraes, A., Nogueira, H., Machado, G. L., Barratt, G., & Mosqueira, V., (2007). Cardioto□icity reduction induced by halofantrine entrapped in nanocapsule devices. *Life Sci, 80*, 1327–1334.

Arruebo, M., Fernández-Pacheco, R., Ibarra, M. R., & Santamaría, J., (2007). Magnetic nanoparticles for drug delivery, *Nano Today, 2*, 22–32.

Attama, A. A., Schicke, B. C., Paepenmüller, T., & Müller-Goymann, C. C., (2007). Solid lipid nanodispersions containing mixed lipid core and a polar heterolipid: characterization. *Eur J Pharm Biopharm, 67*, 48–57.

Bai, S., Thomas, C., & Rawat, A., (2006). Recent progress in dendrimer-based nanocarriers, *Crit. Rev. Ther. Drug Carrier Syst., 23*, 437–495.

Bakalova, R., Zhelev, Z., Aoki, I., & Kanno, I., (2007). Designing quantum dot probes. *Nat Photonics, 1*, 487–489.

Bangham, A. D., Standish, M. M., & Watkins, J. C., (1965). Diffusion of univalent ions across the lamellae of swollen phospholipids, *J Mol Biol, 13*, 238–252.

Barrat, G., Puisieux, F., Yu, W. P., Foucher, C., Fessi, H., & Devissaguet, J. P., (1994). Anti meta-static activity of MDP-l-alanyl cholesterol incorporated into various types of nanocapsules, *Int J Immuno-Pharmacol., 16*, 457.

Beg, S., Rizwan, M., Sheikh, A. M., Hasnain, M. S., Anwer, K., & Kohli, K., (2011). Advance-ment in carbon nanotubes: basics, biomedical applications and toxicity, *J Pharm Phar-macol, 63*, 141–163.

Berger, C., Song, Z. M., Li, X. B., Wu, X. S., Brown, N., Naud, C., et al., (2006). First PN, deHeer WA. Electronic Confinement and Coherence in Patterned Epitaxial Graphene. *Science, 312*, 1191–1196.

Bhirde, A. A., Patel, V., Gavard, J., Zhang, G., Sousa, A. A., Masedunskas, A., et al., (2009). Targeted killing of cancer cells in vivo and in vitro with EGF-directed carbon nanotube-based drug delivery. *ACS Nano, 3*, 307–316.

Calvo, P., Alonso, M. J., VilaJato, J, L., & Robinson, J. R., (1996). Improved ocular bioavail-ability of Indomethacin by novel ocular drug carriers. *J Pharm Pharmacol, 48*, 1147.

Charles, L., & Attama, A. A., (2011). Current state of nanoemulsions in drug delivery. *J. Biomat. Nanobiotech., 2*, 626–639.

Chomoucka, J., Drbohlavova, J., Huska, D., Adam, V., Kizek, R., & Hubalem, J., (2010). Magnetic nanoparticles and targeted drug delivering. *Pharm Res, 62*, 144–149.

Couvreur. P., Barratt, G., Fattal, E., Legrand, P., & Vauthier, C., (2002). Nanocapsule technol-ogy: A review. *Crit Rev Ther Drug Carrier Syst., 19*, 99.

Dalencon, F., Amjaud, Y., Lafforgue, C. Derouin, F., & Fessi, H., (1998). Atovaquone and rifabutine loaded nano capsules: formulation studies. *Int J Pharm., 153*, 127.

Derfus, A. M., Chen, A. A., Bhatia, S. N., et al., (2007). Targeted quantum dot conjugates for siRNA delivery. *Bioconjug Chem, 18*(5), 1391–1396.

DeSmedt, S. C., Demeester, J., & Hennink, W. E., (2000). Cationic polymer-based gene delivery systems, *Pharm Res, 17*, 113–26.

Dhar, S., Liu, Z., Thomale, J., Dai, H., & Lippard, S. J., (2008). Targeted single-wall carbon nanotube-mediated Pt (IV) prodrug delivery using folate as a homing device. *J Am Chem Soc, 27*(130), 11467–11476.

Drbohlavova, J., Hrdy, R., Adam, V., Kizek, R., Schneeweiss, O., & Hubalek, J., (2009). Preparation and properties of various magnetic nanoparticles. *Sensors, 9*, 2352–2362.

Duncan, R., (2003). The dawning era of polymer therapeutics. *Nat Rev Drug Discov, 2*(5), 347–360.

Duncan, R., (2005). Polymer therapeutics: Nanomedicines in routine clinical use. *Proc Euronanoforum.* Session 2B, 48.

Feng, L. Z., & Liu, Z., (2011). Graphene in biomedicine: Opportunities and challenges. *Nanomedicine, 6*(2), 317–324.

Freitas, R. A., (2005). What is Nanomedicine?: Nanomedicine. *Nanotech. Biol. Med., 1*, 2–9.

Gao, Z. G., Fain, H. D., & Rapoport, N., (2005). Controlled and targeted tumor chemotherapy by micellarencapsulated drug and ultrasound. *J Control Release, 102*, 203.

Geim, A. K., & Novoselov, K. S., (2007). The rise of graphene. *Nat Mater, 6*, 183–191.

Gregoriadis, G., & Florene, A. T., (1993). Liposomes in drug delivery: Clinical, diagnostic and opthalmic potential. *Drugs, 45*, 15–28.

Gupta, S., Yadav, B. S., Kesharwani, R., Mishra, K. P., & Singh, N. K., (2010). The role of nanodrugs for targeted drug delivery in cancer treatment. *Arc. of App. Sci. Res., 2*(1), 37–51.

Hamouda, T., Myc, A., Donovan, B., et al., (2001). A novel surfactant nanoemulsion with a unique non-irritant topical antimicrobial activity against bacteria, enveloped viruses and fungi. *Microbiol. Res., 156*, 1–7.

Huang, P., Xu, C., Lin, J., et al., (2011). Folic acid-conjugated graphene oxide loaded with photosensitizers for targeting photodynamic therapy. *Theranostics, 1*, 240–250.

Hubert, B., Atkinson, J., Guerret, M., Hoffman, M., Devissaguet, J. P., & Maincent, P., (1991). The preparation and acute antihypertensive effects of a nanoparticular form of darodipine, a dihydropyridine calcium entry blocker. *Pharm Res., 8*, 734.

Hung, Z. M., Zhang, Y. Z., Kotakic, M., & Ramakrishna, S., (2003). A review on polymer nanofibers by electro spinning and their applications in nanocomposites. *Composites Science and Technology, 63*, 2223–2253.

Jevprasesphant, R., Penny, J., Attwood, D., McKeown, N. B., & Demanuele, A., (2003). Engineering of dendrimer surface to enhance transepithelial transport and reduce cytotoxicity. *Pharm Res., 20*, 1543–1550.

Jong-Whan Rhima., Hwan-Man Parkb., & Chang-Sik Hac., (2013). Progress in Bionanocomposites: from green plastics to biomedical applications. *Progress in Polymer Science, 38*(10–11), 1629–1652.

Joseph, T., & Moore, R., (2008). Drug delivery using nanotechnology technologies: Markets and competitive environment. *Report-Institute of Nanotechnology.*

Joshi, M., & Müller, R. H., (2009). Lipid nanoparticles for parenteral delivery of actives. *Eur J Pharm Biopharm, 71*, 161–72.

Kabanov, A. V., Batrakova, E. V., Alakhov, V. Y., (2002). Pluronic block copolymers as novel polymer therapeutics for drug and gene delivery. *J Control Release, 82*, 189.

Keservani, R. K., & Sharma, A. K., (2016). Nanoarchitectured Biomaterials: Present Status and Future Prospects in Drug Delivery. In: Nanoarchitectonics for Smart Delivery and Drug Targeting, Alina Holban, M., & Grumezescu, A., (Eds.) William Andrews, Elsevier: USA, *2*, 35–66.

Kitces, E. N., Morahan, P. S., Tew, J. G., & Murray, B. K., (1977). Protection from oral herpes simplex virus infection by a nucleic acid-free virus vaccine, *Infect Immun, 16*(3), 955–960.

Kogure, K., Moriguchi, R., Sasaki, K., Ueno, M., Futaki, S., & Harashima, H., (2004). Development of a nonviral multifunctional envelop-type nanodevice by a novel lipid film hydration method. *J Control Release, 98*, 317–318.

Krauze, M. T., Forsayeth, J., Park, J. W., & Bankiewicz, K. S., (2006). Real-time imaging and quantification of brain delivery of liposomes. *Pharm Res.*, *23*(11), 2493–2504.

Kreuter, J., (1994). Colloidal Drug Delivery Systems, Marcel Dekker: New York, 344.

Lam, C. W., James, J. T., McCluskey, R., & Hunter, R. L., (2004). Pulmonary toxicity of single-wall carbon nanotubes in mice 7 and 90 days after intratracheal instillation. *Toxicol Sci.*, *77*, 126–134.

Langer, R., (1998). Drug delivery and targeting. *Nature*, *392*(6679), 5–10.

Larson, D. R., Zipfel, W. R., & Williams, R. M., (2003). Water-soluble quantum dots for multiphoton fluorescence imaging in vivo. *Science*, *300*, 1434–1436.

Li, F. Q., Su, H., Wang, J., Liu, J. Y., Zhu, Q. G., Fei, Y. B., et al., (2008). Preparation and characterization of sodium ferulate entrapped bovine serum albumin nanoparticles for liver targeting. *Int J Pharm.*, *349*(1–2), 274–282.

Liu, C., Alwarappan, S., Chen, Z. F., et al., (2010). Membrane less enzymatic biofuel cells-based on graphene nanosheets. *Biosens Bioelectron.*, 25, 1829–1833.

Liu, J., Gong, T., Fu, H., Wang, C., Wang, X., Chen, Q., et al., (2008). Solid lipid nanoparticles for pulmonary delivery of insulin, *Int J Pharm*, *356*, 333–344.

Lovelyn, C., & Attama, A. A., (2011). Current State of Nanoemulsions in Drug Delivery. *J Biomater Nanobiotechnol.*, *2*, 626–639.

Lu, C. H., Yang, H. H., Zhu, C. L., et al., (2009). A graphene platform for sensing biomolecules. *Angew Chem Int Ed.*, *121*(26), 4879–4881.

Luo, X., Matranga, C., Tan, S., Alba, N., & Cui, X. T., (2011). Carbon nanotube nanoreservior for controlled release of anti-inflammatory dexamethasone. *Biomaterials*, *32*, 6316–6323.

Lvov, Y., & Caruso, F., (2001). Biocolloids with ordered unease multilayer shells as enzymatic reactors, *Anal Chem*, *73*, 4212–4217.

Maeda, H., (1992). Nanoparticle technology handbook, *J. Control. Release*, *19*, 315–324.

Matsumura, Y., & Maeda, H., (1986). A new concept for macromolecular therapeutics in cancer chemotherapy: Mechanism of tumoritropic accumulation of proteins and the antitumor agent smancs. *Cancer Res.*, *46*, 6387–6392.

Mei-Xia Zhao., & Er-Zao Zeng., (2015). Application of functional quantum dot nanoparticles as fluorescence probes in cell labeling and tumor diagnostic imaging, *Nanoscale Res Lett*, *10*, 171.

Michaud, L. B., Valero, V., & Hortobagyi, G., (2000). Risks and benefits of taⵏanes in breast and ovarian cancer. *Drug Safety*, *23*, 401–428.

Moghimi, S. M., Hunter, A. C., & Murray, J. C., (2005). Nanomedicine: Current status and future prospects. *FASEB J*, *19*(3), 311–330.

Muchow, M., Maincent, P., & Müller, R. H., (2008). Lipid nanoparticles with a solid matrix (SLN, NLC LDC) for oral drug delivery. *Drug Dev Ind Pharm*, *34*, 1394–1405.

Muller, J., Huaux, F., Moreau, N., Misson, P., Heilier, J. F., Delos, M., et al., (2005). Respiratory toxicity of multi-wall carbon nanotubes. *Toxicol Appl Pharmacol*, *207*, 221–231.

Muller, R. H., (2007). Lipid nanoparticles: Recent advances. *Adv Drug Deliv Rev.*, *59*, 522–530.

Nahar, M., Dutta, T., Murugesan, S., Asthana, A., Mishra, D., et al., (2006). Functional polymeric nanoparticles: An efficient and promising tool for active delivery of bioactives. *Crit Rev Ther Drug Carrier Syst.*, *23*, 259–318.

Naldini, L., Blömer, U., Gallay, P., et al., (1996). *In vivo* gene delivery and stable transduction of nondividing cells by a lentiviral vector. *Science*, *272*, 263–267.

Nayak, A. P., Tiyaboonchai, W., Patankar, S., Madhusudhan, B., & Souto, E. B., (2010). Curcuminoids-loaded lipid nanoparticles: Novel approach towards malaria treatment. *Colloids Surf B Biointerfaces*, *81*, 263–273.

Nevozhay, D., Kañska, U., Budzyñska, R., & Boratyñski, J., (2007). Current status of research on conjugates and related drug delivery systems in the treatment of cancer and other diseases (Polish). *Postepy Hig. Med Dosw, 61*, 350–360.

Novoselov, K. S., Geim, A. K., Morozov, S. V., Jiang, D., Katsnelson, M. I., Grigorieva, I. V., et al., (2005). Two-dimensional gas of mass less Dirac fermions in graphene. *Nature, 438*, 197–200.

Novoselov, K. S., Geim, A. K., Morozov, S. V., Jiang, D., Zhang, Y., Dubonos, S. V., et al., (2004). Electric field effect in atomically thin carbon films. *Science, 306*, 666–669.

Ourique, A. F., Pohlman, A. R., Guterres, S. S., & Beck, R. C. R., (2008). Tretionoin-loaded nanocapsules: Preparation, physicochemical characterization, and photo stability study. *Int J Pharm., 352*, 1–4.

Owen, J. S., Park, J., Trudeau, P. E., & Alivisatos, A. P., (2008). "Reaction chemistry and ligand exchange at cadmium-selenide nanocrystal surfaces". *J. Am. Chem. Soc, 130*, 12279–12281.

Padilla, O. L., Ihre, H. R., Gagne, L., Fréchet, J. M., & Szoka, F. C., (2002). Polyester dendritic systems for drug delivery applications: in vitro and in vivo evaluation, *Bioconjug. Chem., 13*, 453–461.

Panyam, J., & Labhasetwar, V., (2003). Biodegradable nanoparticles for drug and gene delivery to cells and tissue. *Adv Drug Deliv Rev, 55*(3), 329–347.

Pardridge, W. M., (1999). Vector-mediated drug delivery to the brain. *Adv Drug Deliv Rev., 36*, 299–321.

Patton, T. F., & Robinson, J. R., (1976). Quantitative precorneal disposition of topically applied pilocarpine nitrate in rabbit eyes. *J. Pharm. Sci., 65*, 1295–301.

Prow, T. W., Rose, W. A., Wang, N., Reece, L. M., Lvov, Y., & Leary, J. F., (2005). Biosensor-controlled Gene Therapy/Drug Delivery with Nanoparticles for Nanomedicine. *Proc. of SPIE, (5692)*, 199–208.

Prow, T., Kotov, N. A., Yuri, M., Lvov, R. R., & Leary, J. F., (2004). Nanoparticles, molecular biosensors, and multispectral confocal microscopy, *J Mol Histol, 35*, 555–564.

Qu, L. T., Liu, Y., Baek, J. B., et al., (2010). Nitrogen-doped graphene as efficient metal-free electro catalyst for oxygen reduction in fuel cells. *ACS Nano, 4*(3), 1321–1326.

Reshmi, G., Kumar, P. M., & Malathi, M., (2009). Preparation, characterization and dielectric studies on carbonyl iron/cellulose acetate hydrogen phthalate core/shell nanoparticles for drug delivery applications. *Int J Pharm, 365*, 131–135.

Rogers, J. A., & Anderson, K. E., (1998). The potential of liposomes in oral drug delivery. *Crit Rev Ther Drug Carrier Syst, 15*(5), 421–480.

Rudolph, C., Schillinger, U., Ortiz, A., Tabatt, K., Plank, C., Muller, R. H., et al., (2004). Application of novel solid–lipid (SLN) gene vector formulations-based on a dimeric HIV-1 TAT peptide in vitro and in vivo. *Pharm Res., 21*, 1662–1663.

Rutvij, J. P., Gunjan, J. P., Bharadia, P. D., Pandya, V. M., & Modi, D. A., (2011). Nanoemulsion: An advanced concept of dosage form. *Int. J. Pharm. Cosmetol., 1*(5), 122–133.

Sharma, A., Mayhew, E., & Straubinger, R. M., (1993). Antitumor effect of taxol-containing liposomes in a Taxol-resistant murine tumor model. *Cancer Res, 53*, 5877–5881.

Sharma, A., Mayhew, E., Bolcsak, L., Cavanaugh, C., Harmon, P., Janoff, A., et al., (1997). Activity of paclitaxel liposome formulations against human ovarian tumor xenografts. *Int. J. Cancer, 71*, 103–107.

Sharma, V. K., Mishra, D. N., Sharma, A. K., Keservani, R. K., & Dadarwal, S. C., (2010). Development and optimization of vitamin E acetate loaded liposomes. *Pharmacology-online Newsletter, 2*, 677–691.

Sieg, J. W., & Robinson, J. R., (1977). Vehicle effects on ocular drug bioavailability II: Evaluation of pilocarpine. *J. Pharm. Sci., 66,* 1222–1228.

Singh, V. K., & Keservani, R. K., (2017). Application of Nanoparticles as a Drug Delivery System. In I. Management Association (Ed.), *Materials Science and Engineering: Concepts, Methodologies, Tools, and Applications.* Hershey, PA: IGI Global. pp. 1358-1383.

Sougata, J., et al., (2012). Dendrimers: Synthesis, Properties, Biomedical and Drug Delivery Applications. *Am. J. PharmTech res.,* 2(1), 32–55.

Subhashis, D., Satayanarayana, J., & Gampa, V. K., (2011). Nanoemulsion a method to improve the solubility of lipophilic drugs. *Pharmanest. Int. J. Adv. Pharm. Sci.,* 2(2–3), 72–83.

Suri, S. S., Fenniri, H., & Singh, B., (2007). Nanotechnology-based drug delivery systems. *J Occup Med Toxicol, 2,* 16.

Sutradhar, K. B., & Amin, M. L., (2013). Nanoemulsions: Increasing possibilities in drug delivery. *Eur. J. Nanomed.,* 5(2), 97–110.

Sznitowska, M., Gajewska, M., Janicki, S., Radwanska, A., & Lukowski, G., (2001). Bioavailability of diazepam from aqueous-Organic solution, submicron emulsion and solid lipid nanoparticles after rectal administration in rabbits. *Eur J Pharm Biopharm, 52,* 159–163.

Tabatt, K., Sameti, M., Olbrich, C., Muller, R. H., & Lehr, C. M., (2004). Effect of cationic lipid and matrix lipid composition on solid–lipid nanoparticles mediated gene transfer. *Eur J Pharm Biopharm., 57,* 155–156.

The Royal Society (TRS)., (2004). Nanoscience and nanotechnologies: Opportunities and uncertainties, Royal Society: London, *4.*

Tiwari, S. B., & Amiji, M. M., (2007). Nanoemulsion formulations for tumor targeted delivery In: Amiji M. M., Nanothechnology for cancer therapy, CRC press: New York, USA, pp. 723–40.

Tucek, J., Zboril, R., & Petridis, D., (2006). Maghemite nanoparticles by view of Mossbauer spectroscopy. *J Nanosci Nanotechnol., 6,* 926–947.

Vasir, J. K., Reddy, M. K., & Labhasetwar, V., (2005). Nanosystems in drug targeting: Opportunities and challenges. *Curr Nanosci., 1,* 47–64.

Wagner, V., Dullaart, A., Bock, A. K., & Zweck, A., (2006). The emerging nanomedicine landscape. *Nat. Biotechnol., 24,* 1211–1217.

Wang, Z., Ruan, J., & Cui, D., (2009). Advances and prospect of nanotechnology in stem cells. *Nanoscale Res Lett, 4,* 593–605.

Wilk, K. A., Zielinska, K., Pietkiewicz, J., & Saczko, J., (2009). Loaded nanoparticles with cyanine-tipephotosensizers: preparation, characterization and encapsulation. *Chem Eng Trans, 17,* 987–992.

Wissing, S. A., Kayser, O., & Müller, R. H., (2004). Solid lipid nanoparticles for parenteral drug delivery. *Adv Drug Deliv Rev, 56,* 1257–1272.

Wu, J., Akaike, T., Hayashida, K., Okamoto, T., Okuyama, A., & Maeda, H., (2001). Enhanced vascular permeability in solid tumor involving peroxynitrite and matrix metalloproteinases. *Jpn J Cancer Res., 92,* 439.

Xie, J., & Wang, C. H., (2006). Electrospun micro- and nanofibers for sustained delivery of paclitaxel to treat C6 glioma *in vitro, Pharm. Res., 23,* 1817–1826.

Xuan, Y., Wu, Y. Q., Shen, T., et al., (2008). Atomic-layer-deposited nanostructures for graphene-based nanoelectronics. *Appl Phys Lett, 92*(1), 13101–13103.

Yamaguchi, I., Tokuchi, K., Fukuzaki, H., Koyama, Y., Takakuda, K., Monma, H., & et al., (2001). Preparation and microstructure analysis of chitosan/hydroxyapatite nanocomposites. *J Biomed Mater Res., 55*(1), 20–27.

Yang, X. M., Tu, Y. F., Li, L., et al., (2010). Well-dispersed chitosan/graphene oxide nano-composites. *ACS Appl Mater. Interfaces, 2*(6), 1707–1713.

Yang, Z. R., Wang, H. F., Zhao, J., et al., (2007). Recent developments in the use of adeno viruses and immunotoxins in cancer gene therapy. *Cancer Gene Ther, 14*, 599–615.

Yashpal, S., Tanuj, H., & Harsh, K., (2013), Nanoemulsions: A pharmaceutical review. *Int. J. Pharma. Prof. Res., 4*(2), 928–935.

Zboril, R., Mashlan, M., & Petridis, D., (2002). Iron (iii) oxides from thermal processes synthesis, structural and magnetic properties, Mossbauer spectroscopy characterization, and applications. *Chem Mater, 14*, 969–982.

Zeng, J., Yang, L., Liang, Q., Zhang, X., Guan, H., Xu, X., et al., (2005). Influence of the drug compatibility with polymer solution on the release kinetics of electro spun fiber formulation, *J. Control. Release, 105*, 43–51.

CHAPTER 2

NANOPARTICLES: FORMULATION ASPECTS AND APPLICATIONS

SUSHAMA TALEGAONKAR, LALIT MOHAN NEGI, HARSHITA SHARMA, and SOBIYA ZAFAR

Department of Pharmaceutics, Faculty of Pharmacy, Jamia Hamdard University, New Delhi, 110062, India, E-mail: stalegaonkar@gmail.com

CONTENTS

2.1   INTRODUCTION

Over the last few decades, nanoparticles have attracted immense interest in the field of therapeutics. Various types of nanoparticles have been disseminated in development of several formulations resulting in increased efficacy and fewer side effects and of incorporated drugs (Cooper et al., 2014).

Nanoparticles can be defined as the particles with at least one dimension sized from 1 to 100 nanometers (Moreno-Vega et al., 2012; Shinde et al.,

2012). One nanometer (nm) is 1 billionth part of a meter, or 10^{-9} of a meter. The comparative size of a nanometer to a meter can be estimated as the size of a marble to the earth's size. It was discovered by scientists and researchers that materials at nano scale size exhibit properties that considerably differ from the same materials at macro scale, such as altered chemical and physical reactivity, increased strength, altered conductivity, lighter weight, etc. Nanoparticles hold the potential of improving diagnosis and therapy of diseases through efficient delivery of drugs and imaging moieties to specific sub cellular targets (Kafshgari et al., 2014). Nanoparticles can be developed using various materials such as polymers, lipids and metals, depending on the use, and can incorporate wide range of therapeutics such as drugs, contrast agents, proteins and nucleic acid (Elsabahy et al., 2012). The drug can be incorporated into the nanoparticles via several approaches, such as, adsorption onto the nanoparticles matrix or may be entrapped or encapsulated within the nanoparticles thus giving rise to the terminologies like nanospheres and nanocapsules (Nagarwal et al., 2009). Nanoparticles have dimensions much smaller than the living cells which thus presents several innovatory opportunities in the development of formulations against all types of diseases such as cancer, neurodegenerative disorders, etc. (Shinde et al., 2012). Nanoparticles are used for therapeutic as well as diagnostic purposes (Jain et al., 2014). Nanoparticles-based formulations offer better alternatives to conventional therapies resulting in improved safety profiles of encapsulated drugs. Studies on nanoparticles have shown promising results by increasing the therapeutic inde☐of several drugs, mainly by organ specific targeting and thus reducing unwanted toxicity. Nanoparticles may also help in lessening the distribution of encapsulated drug to the kidneys and thus attenuating drug-associated adverse renal complications (Cooper et al., 2014). Since last couple of decades, a huge number of nanoformulations have received regulatory approval and many more are on their way to future clinical translation (Wolfram et al., 2014). Nanoparticles show great promise in the treatment of a wide range of diseases, which provides advantages and offers a new prospect for disease detection, prevention and treatment (Wang et al., 2014). In the design of commercial nanoparticles, size and surface characteristics have been studied and e☐ploited to achieve efficacious delivery of the drug in the desired manner (Truong et al., 2014). Nanoparticles possess many other advantages too, for example, retention of drug at the active site, increased bioavailability, faster dissolution of active agents in an aqueous environment, prolonged half-life, improved biodistribution, increased circulation time of the drug, smaller drug doses, reduction in fed/fasted variability, passive and active drug targeting, reduction in side effects, and versatility in the choice of route of drug administration (Shinde et al., 2012; Mohanraj et al., 2006; Jain et al., 2014).

However, along with the advantages, it is also important to consider the limitations of nanoparticles in order to achieve biocompatibility and desired activity. Major concern associated with nanoparticles is the toxicity and safety issue. Other limitations include particle-aggregation due to small particle size, chances of burst release and stability problem. Therefore, before the commercial applications of nanoparticles these problems need to be overcome (Mohanraj et al., 2006).

Several nanoparticles that are currently approved, such as Doxil and Abraxane, exhibit fewer side effects than their small molecule counterparts, while other nanoparticles (e.g., metallic and carbon-based particles) tend to display toxicity. However, the hazardous nature of certain nanoparticles could be exploited for the ablation of diseased tissue, if selective targeting can be achieved (Wolfram et al., 2014). Other issues to be considered are pulmonary inflammation, pulmonary carcinogenicity, alveolar inflammation and disturbance of autonomic imbalance by nanoparticles having direct effect on heart and vascular function (Shinde et al., 2012).

2.1.1 ADVANTAGES OF NANOPARTICLES

Nanoparticles offer several advantages as a drug delivery agent:
- Nanoparticles help in targeting the therapeutic agents to the diseased tissue, thus reducing the unwanted side effects.
- More than one drug can be administered through same formulation.
- The drug is protected from degradation when encapsulated inside the nanoparticle.
- Nanoparticles can help in delivering the proteins, peptides and genes for gene therapy.
- Imaging agent can also be attached to the nanoparticles for the simultaneous imaging of the diseased area.
- Administration of antibodies can also be achieved through nanoparticles.

Figure 2.1 describes the basic structure of a nanoparticle and also depicts its potential applications.

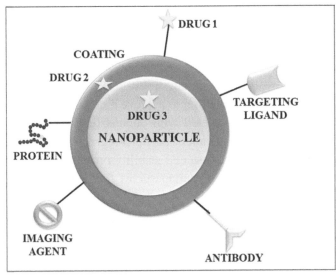

FIGURE 2.1   Structure and Applications of a Nanoparticle.

2.2   CLASSIFICATION OF NANOPARTICLES

On the basis of what the nanoparticles are composed of, they can be divided as:

1. Metallic Nanoparticles: These nanoparticles are made up of metals, such as iron, zinc, Silver, Gold, etc.
2. Non-metallic nanoparticles: These are composed of non-metallic materials, which can be further divided as:
   i. **Ceramics nanoparticles:** These nanoparticles are made up of ceramic materials such as hydroxyapatite, silica, alumina, etc.
   ii. **Polymeric nanoparticles:** These nanoparticles are made up of polymers such as PLGA, albumin, eudrajit, etc.
   iii. **Lipid nanoparticles:** These nanoparticles are made up of lipids. Figure 2.2.

In this chapter, different types of nanoparticles and an overview of their methods of preparation are discussed.

2.3   METALLIC NANOPARTICLES

Metallic nanoparticles have become an area of great interest in scientific research and medical applications, which is actually a result of their unique

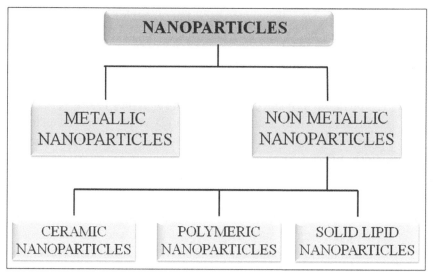

FIGURE 2.2 Types of Nanoparticles.

properties owing to their large surface-to-volume ratios and quantum effects. The unique properties of metallic nanoparticles include physical (e.g., plasmonic resonance, fluorescent enhancement) and chemical (e.g., catalytic activity enhancement) properties which make them suitable for drug delivery, targeting and other applications. These properties are exhibited by metallic properties because of the larger surface area and greater area/volume ratio which in turn is a result of their nanosized structure (Galdiero et al., 2011). Thus, at nanosized level, metals exhibit many such properties which are not exhibited by bulk forms. Here, few metals which are mainly used are described along with their unique properties.

2.3.1 METHODS OF PREPARATION OF METALLIC NANOPARTICLES

Metal Nanoparticles can be synthesized thorough any of the following methods:

1. **By chemical reduction of metal salts:** In chemical synthesis, metallic nanoparticles are produced by reducing a metal salt to its corresponding metal atoms. These atoms act as nucleation centers and form atomic clusters. These clusters are finally surrounded by a stabilizing

agent which prevents atoms from agglomerating (Toshima, 2000). Metal salts can be reduced by any of the several reducing agents available. The most commonly used reducing agents are:

a. **Alcohol**: Alcohols are the very commonly used reducing agents employed for metal nanoparticles synthesis. The alcohol gets converted into aldehydes and ketones thereby reducing the metal precursors into metal atoms. Obreja et al., prepared platinum nanoparticles by carrying out the reduction of chloroplatinic acid (H2PtCl6) with methanol, ethanol and propanol using three different polymers as capping agents. Following this method, they obtained nanoparticles as small as having a size of about 3 nm (Obreja et al., 2008).

b. **Sugars**: Sugars such as glucose, fructose and sucrose are good reducing agents and can be used to transform metal precursors into metal nanoparticles via their reducing abilities. Panigrahi and co-workers synthesized metal nanoparticles of gold, silver, platinum and palladium using these sugars. The nanoparticles were successfully prepared and also, they achieved control over particle size by just changing the sugars and evaporating the precursor solution on glass support (Panigrahi et al., 2004).

c. **Polyolsystem**: Metal nanoparticles can also be prepared using polyol system, for example, using polyols as reducing agent as well as capping agent. Polyol method offers several advantages such as good reproducibility, convenient operation, and easy synthesis involving only one step. A good size control can also be achieved by varying the ratio of polyol and metal precursor. This is attributed to the well-controlled precursor reduction kinetics in the nucleation and growth stages. Hei et al. (2012) prepared monodispersed, small sized and well-shaped nanoparticles of platinum, rhodium, and palladium. They used ethylene glycol solution as reducing system and poly (vinylpyrrolidone) (PVP) as capping agent. After the characterization of nanoparticles, it was revealed that the particles had spherical morphology and the particle size increased with the increase in metal precursor concentration as well as the amount of PVP.

d. **Biologicalagents**: Usually, reducing agents that are used for metal nanoparticles synthesis are expensive, toxic, and are not environment friendly. Biological agents thus provide a safer and more cost effective alternative to those toxic chemicals. It has been demonstrated that metabolites and proteins that are contained in the plant cells have the ability to reduce inorganic metal ions into metal nanoparticles. The main metabolites that are capable of reducing metal ions include terpenoids (such as eugenol), flavanoids (such as quercetin) and amino acids (such as tyrosine) (Makarov et al., 2014). Nazeruddin et al. successfully synthesized silver nanoparticles using the leaf extract of *Azadirachta indica* (Nazerrudin et al., 2014). Nanoparticles synthesized by chemical synthesis offer advantages such as reproducibility, narrow size distribution and lower costs of production (Toshima, 2000).

ii. **By microemulsion method:** Water in oil microemulsions are efficient platform for synthesising metal nanoparticles with good size control. As a common fact, water in oil microemulsions contain oil as continuous phase having dispersed nanosized water droplets which are covered with a thin film of surfactant. These nanosized water droplets act as nanoreactors for the synthesis of metal nanoparticles, where the reaction between metal precursor and reducing agent takes place. For synthesizing metal nanoparticles, two different microemulsions are prepared, one having metal precursor in the water droplets and second having a suitable reducing agents entrapped in the water droplets. In the next step, these two microemulsions are mixed. As a result of mixing, the water droplets of the two microemulsions collide and coalesce with each other, thus causing the mixing of the metal precursor and reducing agent. Finally the reaction takes place between two reactants inside the nano sized water droplets. In this method, where on one hand water droplets act as nano reactors and control the size of the nanoparticles, on the other hand, the surfactant adsorbed at the surface of the droplets helps in preventing the agglomeration of the resultant nanoparticles. Thus, the nanoparticles produced are obtained in a very fine size range and in monodispersed state (Capek, 2004).

iii. **By electrochemical decomposition:** In this method, a voltage is applied to induce a chemical reaction in the electrolyte solution to carry out the preparation of nanoparticles. The bulk metal is made as cathode in the process of electroplating. Then the metal ions are electrochemically reduced and deposited in the pores of the template membranes. Finally, the nanoparticles are released from the template by physicochemical means. Agglomeration of the particles can be prevented by the addition of ammonium stabilizer. Using electro-chemical decomposition method, highly pure nanoparticles can be obtained by using fast and simple procedures. Moreover, the size and morphology of the nanoparticles can be easily controlled by vary-ing the electrode position parameters, such as potential, current den-sity, temperature, deposition time, appropriate surfactants, or soluble polymers, etc., during the deposition process. Also, this method is environmental friendly because it does not involve the use of toxic reducing agents (Roldan et al., 2013).

2.3.2 IRON NANOPARTICLES

Iron nanoparticles exhibit super paramagnetic properties, owing to the fact that the orbital contribution to the total magnetization is much more in low dimensional systems (Amtenbrink et al., 2009). This implies that they respond and get attracted to magnetic field applied externally. Owing to this property, iron nanoparticles have attracted interest as therapeutic as well as diagnostic agent. They have already been used clinically for a num-ber of applications, such as for magnetic separation techniques, as contrast agents in magnetic resonance imaging, for local hyperthermia and as mag-netic targetable carriers for several drug delivery systems (Jurgons et al., 2006). Iron is a ferromagnetic substance and thus it exhibits a certain value of remnant magnetization. Remnant magnetization is the residual magne-tism when external field value becomes zero. Activation over an energy bar-rier is responsible for causing changes in magnetization of a material. If the number of atoms per particle decreases, the exchange energy might reach a value as low as the thermal energy at room temperature. This will result in a spontaneous random orientation of the magnetic spin inside the particles. This means, the remnant magnetization will become zero. This results in no hysteresis and therefore paramagnetic behavior is exhibited. Relaxation time (τ) is the characteristic of the super paramagnetic behavior which is defined as the time that system needs to achieve zero magnetization after the value of external magnetic field becomes zero. Another effect which is exhibited by super paramagnetic particles is the energy absorption due to Neel relaxation, in which, if the particles are exposed to an alternating external magnetic field, the magnetic dipole moments of the nanoparticles will reorient quickly according to the frequency and strength of the magnetic field. In this process, power loss of the particles takes place which is used for heating of the surroundings, for example, tumor (Kleibert et al., 2011; Sharma et al., 2015).

2.3.3 ZINC NANOPARTICLES

Zinc oxide (ZnO) is a conventional wide band gap semiconductor and its unique properties have made it useful in multiple areas of science and medical technology, such as cancer therapy (Rasmussen et al., 2010). FDA has recognized bulk ZnO as a GRAS substance and ZnO nanoparticles larger than 100 nm are found to be relatively biocompatible (Zhou et al., 2006). ZnO possess inherent photoluminescence properties which find role in diagnostic applications, and wide band-gap semiconductor properties of ZnO are useful in generation of reactive oxygen species which result in cytotoxicity if the antioxidative capacity of the cell is exceeded (Wang et al., 2007; Ryter et al., 2007). The ability of ZnO to generate reactive oxygen species is due to its semiconductor properties. The electrons in semiconductors have energies within certain bands. The band gap for crystalline ZnO is ~3.3 eV. As a result, UV rays contain sufficient energy to promote electrons (e–) to the conduction band to leave behind electron holes (h+). Electrons and holes can than migrate to the surface of the nanoparticles to react with oxygen and hydroxyl ions, respectively. This results in the formation of superoxide and hydroxyl radicals (Rasmussen et al., 2010). For ZnO nanoparticles, large number of valence band holes and/or conduction band electrons is available even in the absence of UV light. The reason being that nanocrystal quality decreases with the size of the ZnO nanoparticles which results in increased interstitial zinc ions and oxygen vacancies, and donor/acceptor impurities. These crystal defects can lead to a large number of electron-hole pairs (Sharma et al., 2009). Another useful property of ZnO nanoparticles is their electrostatic properties. ZnO nanoparticles are found to show a different type of surface charge behavior, which is due to the neutral hydroxyl groups chemisorbed on to their surface. Protons (H+) move out from the particle surface in aqueous medium at high pH, leaving a negatively charged surface with partially bonded oxygen atoms (ZnO–). At lower pH, protons from the environment get transferred to the particle surface, resulting in a positively charged surface (ZnOH2+) (Qu et al., 2001; Rasmussen et al., 2010). The isoelectric point of ZnO nanoparticles is found to be 9–10. Accordingly, ZnO nanoparticles will have a strong positive surface charge under physiological conditions (Degen et al., 2000). It is also found that cancer cells contain a high concentration of anionic phospholipids on their outer membrane and large membrane potentials (Abercrombie et al., 1962). Thus, interactions of cancer cells with positively charged ZnO nanoparticles will be driven by electrostatic interactions, and will thereby promote cellular uptake, phago-cytosis and cytotoxicity (Rasmussen et al., 2010; Sharma et al., 2015).

2.3.4 GOLD NANOPARTICLES

Gold nanoparticles have been widely studied in recent years for their appli-cations in the field of cancer treatment.-based on their facile synthesis and surface modification, enhanced and tunable optical properties and excellent biocompatibility, they have been used both for diagnosis and therapy of can-cer. The nanoparticles of various shapes like spheres, shells, rods or cages of gold can be bound to a wide variety of biochemically functional groups and made to target specific types of cells. Apart from their inherent imaging properties, they can also be conjugated with some external energy source such as infrared laser, to cause killing of cancer cells (Marsh et al., 2009).

Gold nanoparticles have unique optical properties which make them extremely useful for cancer diagnosis and photo thermal therapy. These properties are conferred by the interaction of light with electrons on the surface of gold nanoparticles. Many metals (e.g., alkali metals, Mg, Al, and noble metals such as Au, Ag, and Cu) are free-electron systems which contain equal numbers of positive ions (fiᐧed in position) and con-duction electrons. Such metals are called plasma. When the metal surface is irradiated with an electromagnetic wave, oscillating electric fields of the wave interact with the free electrons causing a concerted oscillation of electron charge that is in resonance with the frequency of visible light. The free electrons are driven by the electric field to oscillate coherently. Quantized plasma oscillations of the free electrons are called plasmons. An electromagnetic wave has a certain penetration depth (<50 nm for Ag and Au), thus the electrons on the surface of the metal are the most significant and their collective oscillations are termed surface plasmons (SPs). These plasmons interact with visible light in a phenomenon called surface plasmon resonance (SPR) (Bahadory). The surface plasmon resonance leads to strong electromagnetic fields on the particle surface and consequently enhances all the radiative and non-radiative properties of the nanoparticles (Huang et al., 2010). SPR results from confinement of photon to a small particle size, and is a significant spectroscopic feature of gold nanoparticles, which gives rise to a sharp and intense absorption band in the visible range. The response of a spherical nanoparticle to the oscillating electric field can be described by the so-called dipole approximation of Mie theory (Plech et al., 2007). The electron oscillation around the particle surface causes a charge separation, forming a dipole oscillation along the direction of the electric field of the light. The amplitude of the oscillation reaches maⅢimum at a specific frequency which is called SPR. The SPR band intensity and wavelength depends on the factors affecting the electron charge density on the particle surface such as the metal type, particle size, shape, structure, composition and the dielectric constant of the surrounding medium. Au nanoparticles show the SPR band around 520 nm in the visible region (Papavassiliou, 1979, Sharma et al., 2015).

2.3.5 SILVER NANOPARTICLES

Silver nanoparticles have been tremendously used in the areas of catalysis, optoelectronics, detection and diagnostic, antimicrobials and therapeutics. Attempts have also been made to use silver nanoparticles as an anti-cancer agent and they have all given new hopes in the field of anti-cancer therapy.

The silver nanoparticles are plasmonic structures and they absorb and scatter a portion of the impinging light. While scattered light can be used for imaging, the absorbed light can be used for thermal killing of cancer cells after their selective uptake into the cancer cells. Like gold (and copper), silver is also a free electron system, containing equal numbers of positive ions (fiⅢed in position) and conduction electrons (free and highly mobile). Under the irradiation of an electromagnetic wave, the free electrons are driven by the electric field to oscillate coherently at a plasma frequency of ωp relative to positive ions. Quantized plasma oscillations or collective oscillations of the free electrons are called plasmons. These plasmons can interact, under certain conditions, with visible light, exhibiting SPR (Honary et al., 2012). For this phenomenon to happen, the particle must be much smaller than the wavelength of incident light. The electric field of incident light can induce an electric dipole in the metal particle by displacing many of the delocalized electrons in one direction away from the rest of the metal particle and thus producing a net negative charge on one side. The side opposite the negative charge has a net positive charge (nuclei). Of the three metals (Ag, Au and Cu) that display plasmon resonance in the visible spectrum, Ag exhibits the highest efficiency of plasmon eⅡcitation. The light-interaction cross-section for Ag can be about ten times that of the geometric cross-section, which indicates that the particles capture much more light than is physically incident on them. Silver is also the only material whose plasmon resonance can be tuned to any wavelength in the visible spectrum (Evanoff et al., 2005).

Silver nanoparticles may also affect proteins and enzymes with thiol groups like glutathione, thioredoxin, SOD and thioredoxin peroxidise by interacting with mammalian cells. These proteins and enzymes are responsible to neutralize the oxidative stress of ROS which is generated by mitochondrial energy metabolism. Silver nanoparticles, thus causes ROS accumulation by depleting the antioxidant defense mechanism of the cell. ROS accumulation initiates inflammatory response and perturbation and leads to destruction of the mitochondria. Apoptogenic factors like cytochrome C are released afterwards which finally results in programmed cell death (Mohammadzadeh, 2012, Sharma et al., 2015). Table 2.1 enlists few recent applications of metallic nanoparticles.

2.4   NON-METALLIC NANOPARTICLES

Non-metallic nanoparticles are the nanoparticles composed of ceramic materials, polymers or lipids. These nanoparticles offer the advantages of being comparatively non-toxic, easier and simpler methods of preparation and a better control over size and other characteristics.

2.4.1   METHODS OF PREPARATION OF NON-METALLIC NANOPARTICLES

The two basic approaches for the preparation of nanoparticles are 'the top down approach' and 'the bottom up approach.' The top–down approach involves the particle size-reduction to the nanometer range using larger,

TABLE 2.1 Applications of Metallic Nanoparticles

Sl. no.	Metal	DrugDeliverd	Inference
1.	Iron (II,III) oxide	Paclitaxel	Folic acid was conjugated to the surface of Fe3O4 for targeted delivery. Rhodamine B isothiocyanate was conjugated for biological imaging applications. Paclitaxel was conjugated through hydrophobic interactions. These folic acid-conjugated magnetic nanoparticles showed stronger T2-weighted MRI contrast towards the cancer cells, proving their potential in drug delivery and imaging.
2.	Iron Oxide	Bortezomib	Results demonstrated that bortezomib loaded magnetic nanofiber matrix was highly beneficial due to the higher therapeutic efficacy and low toxicity towards normal cells and also due to the availability of magnetic nanoparticles for repeated hyperthermia application and tumor-triggered controlled drug release.
3.	Iron Oxide	Paclitaxel	Iron oxide nanoparticles were developed which self-assembled in magnetic nanoclusters in aqueous environment. Significant enhancement in magnetic resonance contrast was observed due to the large cluster size and dense iron oxide core. Tethering a tumor-targeting peptide to nanoparticles enhanced their uptake into tumor cells. Paclitaxel was efficiently loaded into β-CDs and was released in a controlled manner.
4.	Zinc oxide	Doxorubicin	Doxorubicin and zinc oxide nanoparticles, compared with free doxorubicin, effectively enhanced the intracellular drug concentration by simultaneously increasing cell uptake and decreasing cell efflux in MDR cancer cells. The acidic environment-triggered release of drug could be tracked real-time by the doxorubicin fluorescence recovery from its quenched state.
5.	Zinc oxide	Rose Bengal	It was observed that the ROS generation under green light illumination was greater at low concentrations of RB-ZnO nanohybrids compared with free RB. Substantial photodynamic activity of the nanohybrids in bacterial and fungal cell lines validated the in vitro toxicity results. The cytotoxic effect of these nanoparticles in HeLa cells also showed promising results.
6.	Zinc oxide	Doxorubicin	Zinc oxide nanoaparticles were synthesized and doxorubicin was passively loaded. Nanoparticles released the drug in a controlled manner. Very high cytoto□icity towards cancer cells was achieved due to better targeting and retention of the nanoparticles by cancer cells, whereas the formulation was non-toxic for normal cells.

TABLE 2.1 (Continued)

Sl. no.	Metal	DrugDeliverd	Inference
7.	Gold	Oxaliplatin	Gold nanoparticles encapsulating oxaliplatin and conjugated with anti-DR5 antibody were synthesized. Oxaliplatin immuno-nanoparticles showed sustained release of drug. MTT assay exhibited 3-fold decrease in cell viability of nanoparticles as compared to oxaliplatin. Reduction in tumor size and volume in xenograft tumor models justified its in vitro activity.
8.	Gold	Doxorubicin	Administration of doxorubicin-loaded gold nanoparticles resulted in a significant inhibition of the proliferation of cancer cells (A549, B16F10) in vitro and of tumor growth in an in vivo model compared to doxorubicin alone. Cellular uptake and release of conjugated Dox was faster than the uptake and release of unconjugated Dox.
9.	Silver	Methotrexate	Methotrexate loaded silver nanoparticles exhibited improved anticancer activity against MCF-7 cell line. Hemolytic activity of these particles was significantly less than methotrexate.
10.	Iron Oxide	Doxorubicin	Results demonstrated enhanced cytoplasmic uptake, more apoptosis, and down-regulation of both IL-6, a pro-inflammatory cytokine, and NF-κβ activity with these nanoparticles in comparison to treatment with free DOX in the human MDA-MB-231 invasive breast cancer cells.
11.	Zinc oxide	Paclitaxel	Energy dispersion X-ray mapping and whole animal fluorescent imaging studies revealed that combined pH and folate-receptor targeting reduces off-target accumulation of the nanocarriers. A dual cell-specific and pH-sensitive nanocarrier greatly improved the efficacy of paclitaxel to regress subcutaneous tumors in vivo.
12.	Silver	Plumbagin	Enhanced internalization of drug into the HeLa cells was observed. Treatment with nanoparticles inhibited proliferation of cells with IC50 value of about 18 ± 0.6 μM and blocked the cells at mitosis in a concentration-dependent manner. They also inhibited the post-drug exposure clonogenic survival of cells and induced apoptosis. The antiproliferative, antimitotic and apoptotic activities were also found to be increased.

externally-controlled devices such as milling or high pressure homogeni-zation, to direct their assembly, while the bottom–up technique generates nanoparticles by starting at the molecular level where the molecular compo-nents arrange themselves into some useful conformation using the concept of molecular self-assembly. Controlled precipitation (or crystallization) and evaporation are used to achieve the nanometer size particles using the bot-tom up approach. These processes can occur in the bulk solution or in drop-lets, depending on the technique (Iqbal et al., 2012). The two techniques, though widely used for nanoparticles synthesis, suffer from some limita-tions. The top down approach leads to the production of nanoparticles which are mostly crystalline in nature but the high energy or pressure is required to achieve nano-range which can also contaminate the sample if milling medium is used. The bottom up technique, in contrast, generates nanopar-ticles which may be crystalline or amorphous, depending on the process conditions. The crystal growth of the resultant crystalline particles must be controlled to limit the particle size (Chan and Kwok, 2012). The different bottom–up methods are categorized as-hydrosol production by anti-solvent precipitation, flash nanoprecipitation, and supercritical fluid (SCF) technol-ogy while the top down approaches include wet media milling and high-pressure homogenization.

i. **Hydrosol production by anti-solvent precipitation**: Pharmaceutical hydrosol refers to the aqueous nanosuspension of a hydrophobic drug molecule. The organic solution of the drug is mixed with an excess of vigorously stirred water, thus forming a supersaturated solution which results in the crystallization of the API in the nanosized form. The organic solvent used to solubilize the drug should be miscible with water, this may include solvents such as ethanol, isopropanol, or acetone. Rapid mixing results in the formation of large number of nanosized particles with a very narrow particle size distribution. An electrostatic stabilizer is also added to prevent particle agglomera-tion by adsorbing at the surface of nanoparticles (Chan and Kwok, 2012). Kakran et al., developed cucrumin nanoparticles with a mean diameter of 330 nm using the antisolvent precipitation method. The process parameters such as flow rate, solvent to antisolvent (SAS) ratio, stirring speed and drug concentration were investigated for their effects on the super saturation, nucleation, and growth rate to obtain the smallest particle size. The decrease in drug concentration and the increase in the flow rate, stirring rate, and antisolvent amount resulted in the production of nanoparticles. A higher solubility and dissolution rates with greater antioxidant activity was observed for the curcumin nanoparticles than the original curcumin (Kakran et al., 2012).

ii. **By flash nanoprecipitation (FNP)**: FNP is just a modification of controlled precipitation technique which produces smaller particles with higher drug loading. FNP involves the use of confined liquid impinging jets (CLIJ) where a jet of drug solution impinges a jet of anti-solvent, both coming from two opposing nozzles which is mounted in a small chamber and hence creates a region of extreme turbulence and intense mixing leading to the drug precipitation as fine particles. Precise control over the velocity of the two jets is important to prevent any unbalanced flow or mixing. The process parameters include speed of liquid jets, drug concentration and the volume ratio of drug solution to anti-solvent (Johnson et al., 2003). The two streams in the CIJ mixer should operate at near equal momentum (1:1 volume flow rate), so as to prevent any loss in mixing efficiency. Furthermore, programmed syringe pumps are required to introduce liquid in this design. Thus, Han et al., developed a modified CIJ design, called as CIJ-D (CIJ with dilution), which overcomes the limitation associated with the design of Johnson and Prudhomme. The new mixing process eliminated the need for syringe pump and employed a second antisolvent dilution stage, which increased the super saturation and causes a rapid quenching which enhanced the stability of nanoparticles (Han et al., 2012). Margulis et al. (2014) developed curcumin nanoparticles using this approach. The method employed a turbulent co-mixing of water with curcumin-loaded emulsion using hand-operated confined impingement jets mixer forming a clear stable dispersion of nanoparticles of mean size of 40 nm. The dispersion could be converted to dry, easily water-dispersible powder by spray drying.

iii. **By supercritical fluid (SCF) technology:** A distinct advantage of the SCF technology over the other process of solvent precipitation is the rapid removal of SCF and the solvent without need of any additional drying step. Conventional methods, for the preparation of nanoparticles, employ organic solvents which are hazardous to both the physiological system and the environment. Thus supercritical fluids can be used as safer green alternatives to produce the nanoparticles of high purity. A supercritical fluid is defined as a solvent at a temperature and pressure above its critical point, at which the fluid remains a single phase. The two main processes under SCF technology are rapid expansion of supercritical solution (RESS) and supercritical anti-solvent (SAS) (Chan and Kwok, 2012). The SCF with the unique physical properties of low density and viscosity with a high diffusivity attains rapid micro-mixing for precipitation. Supercritical carbon dioxide (SCO_2) having a critical temperature of 31.1°C and pressure of 72.9 atm is used as supercritical fluid for the processing of pharmaceuticals. In the RESS technique, the solute is dissolved in the supercritical fluid such as CO_2 at ambient temperature. The solution is then pumped into the pre-expansion unit using a syringe pump and is heated isobarically to the pre-expansion temperature. The solution is then allowed to expand through the orifice or the capillary nozzle with subsequent reduction in the pressure, resulting in the homogenous nucleation and hence the formation of nanoparticles. The process variable affecting the particle size and the other morphological characteristics of the particles prepared using RESS is the amount of polymer added, for example, the degree of saturation of the polymer (Rao and Geckeler, 2011). In the SAS process, SCO_2 is used as an anti-solvent to cause the precipitation of drugs from their polar solution. SAS process is used for the drugs which do not have sufficient solubility in SCO_2. The solvent used for solubilizing the drug should have miscibility with SCO_2. SAS process is achieved by either introducing the drug solution into the SCF, or passing the SCF into the drug solution to expand the solvent to cause precipitation (Fages et al., 2004). Liu et al., reported production of 10-hydroxycamptothecin proliposmes using SAS process. The effects of different variables, such as, mass ratio of HCPT to liposomal components, mass ratio of soy lecithin to cholesterol, temperature, pressure and HCPT solution flow rate were investigated using an OA_{25} (5^5) orthogonal experimental design. Under the optimized conditions, the liposomal formulation achieved the particle size of 209.8 ± 38.4 nm, with a drug loading of 5.33% and percentage encapsulation efficiency of 85.28% (Liu et al., 2014).

Lim et al., developed semiconducting polymeric particles of poly (2-(3-thienyl)acetyl 3,3,4,4,5,5,6,6,7,7,8,8,8-tridecafluorooctanoate) (PSFTE) using RESS process The nanoparticles with a size range of 50–500 nm were developed using 0.1–0.5 wt% PSFTE solutions in CO_2 with a pre-expansion temperature and pressure of 40°C and 276 bar. The particle size was found to be influenced by processing conditions, concentration and the degree of saturation, molecular mass of the polymer and several other factors (Lim et al., 2005). Chernyak et al. (2001) also used this technique to prepare droplets of perfluoropolyethers. They also studied the influence of different RESS processing conditions such as temperature, concentration of solute and nozzle configuration on droplets and spray characteristics. Meziani et al., prepared polymeric nanoparticles of poly (heptadeca-fluorodecylacrylate) (PHDFDA) with the particle size of 50 nm. The polymer was found to be highly soluble in SCO_2 at a temperature of 35°C and a pressure of 130–210 bar, but was insoluble in aqueous medium, thus the expansion was done into water. They also observed that a basic aqueous solution containing a surfactant, sodium dodecyl sulfate, at the expansion end could prevent the agglomeration of nanoparticles and was able to confer stability to the nanoparticles for several days (Meziani et al., 2004).

iv. **By milling:** It is a top down approach to produce nanoparticles. It is basically a unit operation where mechanical energy is applied to physically break down the coarser particles to finer ones. Milled drug particles are further developed as dosage form. Milled products are cohesive in nature and exhibit poor flow properties, due to their higher surface energies and thus inert pharmaceutical excipients or fillers such as, calcium phosphate, mannitol or lactose, is added to alleviate this problem. The drug particles may also be granulated with these fillers to form granules which exhibit improved flow properties and content uniformities. Milling reduces the particle size which leads to higher surface free energies and thinner diffusion boundary layers which enhances the dissolution rate of the milled particles. The particle size reduction, during milling process, ceases at certain limit beyond which the material cannot be comminute even though the milling time is prolonged. The continued transfer of mechanical energy from the mill to the drug substance results in the accumulation of defects on the drug crystal which leads to disordering of the crystal structure, eventually causing the disappearance of the order in the positions of atoms or molecules in the crystal. This may result in complete amorphization of the drug or a thin, amorphous (disordered) layer can be formed around the crystalline (ordered) core.

The drug is thus said to be "mechanochemically-transformed" or "activated" by the milling process. This improves the aqueous solubility and dissolution characteristics of the drug (Loh et al., 2015). The nanoparticles can be produced using wet media milling technique. Wet media milling is employed for potent drugs and drugs which possess high residual moisture contents. It involves size reduction of the drug particles suspended in liquid medium which may be aqueous or non-aqueous in nature. A sufficiently concentrated dispersion of drug particles in the aqueous or non-aqueous liquid medium is subjected to ball milling operation. The yield of nanoparticles is improved as the liquid medium prevents adhesion and consequent compaction of the milled drug particles to the wall of the vessel and/or the surfaces of the milling balls. The major limitation of the media milling process is the erosion of balls which may arise from the intense mixing forces in the vessel which may lead to product contamination and further chemical destabilization of the newly-formed particle surface (Merisko-Liversidge and Liversidge, 2011).

v. **By high pressure homogenization**: this is a high energy process in which the particle size reduction is achieved by repeatedly cycling the drug nanosuspension, with the help of piston, through a very thin gap at a high velocity and pressure (1000–1500 bars). The drug substance has to be pre-micronized prior to homogenization so as to minimize clogging of the homogenization gap and also reduce the milling time. The drug suspension, when forced through the gap at high flow rate, causes the static pressure exerted on the liquid to fall below the vapor pressure of the liquid, at the prevailing temperature. This results in the boiling of liquid and formation of gas bubbles which collapse when the liquid exits from the gap and the normal pressure is resumed. The cavitation forces coupled with the shearing effect, causes the nanonization of the drug microparticles (Mohr, 1987 Keck and Muller, 2006). Gonzalez-Mira et al., developed Flurbiprofen (FB)-loaded nanostructured lipid carriers (NLCs) for anti-inflammatory ocular therapy, using the process of high-pressure homogenization. The NLCs-based on Compritol® 888 ATO were prepared using factorial design and exhibited low particle size (<199 nm) with a high entrapment efficiency (90%), and long-term physical stability (Gonzalez-Mira et al., 2012).

2.4.2 CERAMIC NANOPARTICLES

Ceramic nanoparticles are inorganic systems with porous characteristics. In recent years, there has been a growing interest in using ceramic nanoparticles as drug vehicles, owing to their property of being easy to synthesize with desired size and porosity. Researchers have explored typical biocompatible ceramic materials such as silica, titania, alumina, etc. (Yih et al., 2006; Singh et al., 2011). This newly emerging area of using ceramics nanoparticles to entrap biomolecules/drug molecules has found wide applications in the field of drug delivery system, as they offer a number of advantages, including: easy preparation, desired size, shape and porosity, and no effect on swelling or porosity with no change in pH (Shinde et al., 2012; Thomas et al., 2015). Ceramics nanoparticles are made up of ceramic materials such as calcium phosphates (CaP), silica (SiO_2), alumina (Al_2O_3), titanium oxide (TiO_2), and zirconia (ZrO_2). These have been synthesized using new synthetic methods in order to improve their physical and chemical properties and reducing cytotoxicity in biological systems. Ceramic nanoparticles are most widely exploited for controlled release of drugs, for which their dose and size are given special consideration (Moreno-Vega et al., 2012; Fadeel et al., 2010).

2.4.3 PREPARATION OF CERAMICS NANOPARTICLES

Traditionally, ceramics nanoparticles are prepared by solid state methods, in which hydroxide, oxide, carbonate, nitrate, or sulphate raw materials are physically mixed, and then treated at high temperature (approximately at 1100°C) for long durations, to enable the formation of the nanoparticles. By this method, coarse sized agglomerated particles with low surface area are formed. Use of high temperatures for the formation of solid-state compound probably causes abnormal grain growth and lack of control of stoichiometry (Zhang, 2004). Various chemical synthesis techniques have been developed to obtain nanoscale ceramics particles with a proper morphology at a low temperature. Usually, the techniques start from the preparation of a precursor solution, in which the ions are well mixed on a molecular scale. Solid precursor compositions are then formed by co-precipitation, hydrothermal treatment (Zawadzki et al., 2000), sol–gel method (Zayat et al., 2000; Areán et al., 1999) and spray roasting (Okuyama et al., 2003). The solid precursors may be amorphous or crystalline single phases with a homogeneous or inhomogeneous composition or physical mixtures of such phases. The precursors are heated to cause decomposition and chemical reaction to produce the desired multi-component oxide phase. The nature of the multi-component oxide, and in particular, its morphology will critically depend on solid state morphology developing during the entire synthesis route (Zhang, 2004).

2.4.4 CALCIUM NANOPARTICLES

Calcium mainly forms calcium phosphate nanoparticles and calcium carbonate nanoparticles. Calcium nanoparticles provide a number of advantages, such as, they can deliver drugs in minimally invasive manner just as polymeric nanoparticles, easy to fabricate and inexpensive, usually have longer biodegradation time, do not swell or change porosity, are more stable when variation in temperature and pH are observed, can process same chemistry, crystalline structure, size as the constituents of targeted tissues, their fabrication enhances their bioavailability and biocompatibility even before releasing drugs (Ferraz et al., 2004; Chenab et al., 2012).

Calcium phosphates (CaP), class of materials are suitable for use as a carrier for drugs, non-viral gene delivery, antigens, enzymes, and proteins Calcium phosphates are abundant in nature, especially in calcified tissue of vertebrate (Gassmann et al., 1913). The constituting building blocks of bone are composites of biological apatite and molecules of collagen (Fratzl et al., 2004; Tzaphlidou et al., 2008). Hydroxyapatite $(Ca_{10}(PO_4)_6(OH)_2)$ is a type of calcium phosphate that has a similar chemical structure to bone mineral, and hence has excellent biocompatibility, bioactivity, and high affinity to proteins, chemotherapy drugs, and antigens. Apart from hydroxyapatite, another types of calcium phosphates are also used.

- Monocalcium phosphate – $Ca (H_2PO_4)_2$
- Dicalcium phosphate – $CaHPO_4$
- Calcium pyrophosphate – $Ca_2P_2O_7$
- Amorphous calcium phosphate – $Ca_3(PO_4)_2 \cdot nH_2O$
- Tricalcium phosphate – $Ca_3(PO_4)_2$
- Tetracalcium phosphate – $Ca_4(PO_4)_2O$
- Octacalcium phosphate – $Ca_8H_2(PO_4)_6 \cdot 5H_2O$

Types of calcium phosphate differ chemically-based on their calcium to phosphate ratio and hence their physical properties are also different (Uskokovic et al., 2013).

Like calcium phosphate, calcium carbonate nanoparticles are also found abundantly in nature, for, for example, in bones and egg shells. They also possess the similar advantages such as biocompatibility, easy fabrication into nanoparticles, inexpensive, slow biodegradability, etc. There are usually three polymorphic forms of calcium carbonate, ranked in order of decreasing thermodynamic stability: calcite, aragonite and vaterite. The needle-like aragonite has drawn much attention of research interests among material scientists due to their unique properties and potential applications (Nagaraja et al., 2014; Kammoe et al., 2012; Thomas et al., 2015).

2.4.5   TITANIUM OXIDE NANOPARTICLES

Titanium dioxide (TiO_2) is an important n-type wide band-gap semiconductor with light absorbing, charge transport, and surface adsorption properties. Three different crystallite structures of brookite, anatase and rutile have been found for titanium dioxide. Due to the exclusive properties like photoactivity, photostability, chemical and biological inertness, and high stability, titanium dioxide has found wide variety of applications in various fields (Ramimoghadam et al., 2014). Titanium dioxide nanoparticles have also been widely used as white pigment in paint, UV blocker in cosmetics, welding rod-coating material, disinfectant in environment and wastewater and photosensitizer for the photodynamic therapy (Choi et al., 2014). Titanium is widely used in biomedical applications due to its mechanical and photo catalytic properties, and its biocompatibility. It is a well established and proved fact that titanium dioxide, when photo excited, exhibits strong oxidizing and reducing ability and also affects cellular functions, thus leading to its applicability in treatment of cancer. According to recent studies, TiO2 induces death by apoptosis in different types of cells, such as mesenchymal stem cells, osteoblasts, and other cell types. Also, due to its photo catalytic properties, TiO_2 kills cancer cells upon irradiation with light of wavelength <390 nm via production of reactive oxygen species, which damage cancer cells (Lagopati et al., 2014; Thomas et al., 2015).

2.4.6   SILICA NANOPARTICLES

Silica is a chemical compound that is an oxide of silicon with the chemical formula SiO_2. Unlike metals, silica is a poor conductor of both heat and electricity. Silica is one of the most valuable materials for biomedical research. It has wide application in the field of drug delivery as nanoparticles. Mesoporous silica nanoparticles (MSNs) which are formed by polymerizing silica in the presence of surfactants offer several advantages such as large surface area and volumes, encapsulation of drugs and other therapeutic molecules, and tunable pore sizes. Furthermore, the surface of silica nanoparticles can be functionalized with a variety of functional groups or ligands which in turn will increase biocompatibility and modify other properties too, such as targeting ability of the nanoparticles. Silica nanoparticles have found application in controlled delivery systems as it is able to store and gradually release therapeutic agents. Out of several advantages that MSNs offer, the important ones are: automatic release of drugs, ease of dissolution, ease of availability, etc. Also, there are multimodal silica nanoparticles which are applicable as markers in cancer diagnosis and imaging. Silica nanoparticles are also found to significantly reduce cell viability by induction of apoptosis (Moreno-Vega et al., 2012). During last couple of decades, solid silica nanoparticles (SiNPs) and mesoporous silica nanoparticles (MSNs) have been extensively studied and characterized for use in biomedical science. Though SiNPs have been widely used in drug delivery systems and for imaging, the functionalization of SiNPs is limited by their surface. On the other hand, MSNs exhibit a higher surface area and a tunable pore volume, allowing for a higher loading capacity of therapeutic agents.(Chen et al., 2013; Thomas et al., 2015). Table 2.2 enlists few recent applications of ceramic nanoparticles.

2.4.7 POLYMERIC NANOPARTICLES

The polymeric nanoparticles can be surface functionalized with various groups which can make it responsive to certain stimulus in the biological environment such as, pH, temperature, enzymatic, etc. The chemotherapeutics can also be released from the polymeric nanoparticles in response to certain external stimuli such as ultrasound, magnetically modulated, photo responsive or when irradiated with near IR or UV-vis light. Depending on the route of administration, therapeutic application and the desired target site, for example, organs, tissues or at the cellular level, the polymeric nanoparticles of appropriate type could be designed. The chemistry and the payloads of the polymeric nanoparticles influence

TABLE 2.2 Applications of Ceramic Nanoparticles

S. No.	Ceramic material	Drug delivered	Results	Reference
1.	Calcium Phosphate	siRNA	In vitro experiments demonstrated that the negatively charged siRNA loaded CaP nanoparticles could effectively deliver EGFR-targeted siRNA into A549 cells and significantly down-regulate the level of EGFR expression. Further, internalized nanoparticles exhibited a pH-responsive release of siRNA. In vivo tumor therapy demonstrated significant tumor growth inhibition with a specific EGFR gene silencing effect.	Qiu et al., 2016
2.	Calcium Phosphate (Hydroxyapatite)	Risedronate	Pharmacokinetic studies revealed 6-fold and 4-fold increase in the relative bioavailability after intravenous and oral administration of nanoparticles, respectively as compared to marketed formulation confirming better effective drug transport. Biochemical investigations also showed a significant change in biomarker level which ultimately lead to bone formation/resorption.	Rawat et al., 2016
3.	Calcium Phosphate (Hydroxyapatite)	Doxorubicin	DOX-loaded nanoparticles exhibited satisfactory drug loading and could markedly enhance mitochondrial cytochrome C leakage thereby activate apoptotic cascade associated with it. In vivo anti-tumor efficacy and toxicity evaluation indicated that these nanoparticles were more effective and less harmful.	Xiong et al., 2016
4.	Silica	Puerarin	Release of puerarin was pH dependent, and the release rate was much faster at lower pH than that at higher pH. Also, nanoparticles exhibited improved blood compatibility in terms of low hemolysis, and it could also reduce the side effect of hemolysis induced by puerarin. Further, nanoparticles showed a 2.3-fold increase in half-life of puerarin and a 1.47-fold increase in bioavailability.	Liu et al., 2016

No.	Material	Drug	Description	Reference
5.	Silica	Ibuprofen	Nanoparticles exhibited superior properties such as good dispersion in aqueous medium, high drug loading efficiency, improved stability and high drug release rates. Nanocomposites presented a flexible control over drug release. Cumulative percent release of ibuprofen was much higher at pH 7.4 than at pH 2.0. Release rate was very slow in acidic medium and faster in a neutral medium, owing to hydrogen bonding in an acidic medium and electrostatic repulsion between negatively charged carboxyl groups in an alkaline medium.	Guo et al., 2016
6.	Silica	Doxorubixin and shRNA	In vitro antitumor activity assays revealed that HeLa cell growth was significantly inhibited. Intracellular accumulation study of DOX showed that FA targeted nanoparticles was more easily taken up than nontargeted ones. These nanoparticles induced a significant decrease in VEGF expression as compared to cells treated with either the control or other complexes.	Li et al., 2016
7.	Calcium Carbonate	Doxorubicin	Nanoparticles showed a high doxorubicin-loading capacity. They also exhibited sustainable releasing performance and inhibition of the initial burst release.	Zhang et al., 2015
8.	Titanium dioxide	Paclitaxel	Polyethylenimine (PEI) on the surface of multifunctional porous TiO_2 nanoparticles could effectively block the channel to prevent premature drug release. Following UV light radiation, PEI molecules on the surface were cut off by the free radicals (OH^- and O^{2-}) that TiO_2 produced, and then the drug loaded in the carrier was released rapidly into the cytoplasm. Amount of drug released from multifunctional porous TiO2 nanoparticles could be regulated by UV-light radiation time.	Wang et al., 2015
9.	Titanium dioxide	Enrofloxacin hydrochloride (ENRO)	Results indicated that the maximum loading content of drug reached to 33.28%. ENRO-TNTs performed a better release profile at low temperature than at high temperature in PBS solution.	Lai et al., 2013
10.	Titanium dioxide and Hydroxyapatite	Paclitaxel	Biochemical, hematological and histopathological analysis showed that the surface modified paclitaxel attached nanoparticles had a higher anticancer activity than the pure paclitaxel and surface modified nanoparticles without paclitaxel. This is due to the targeting of the drug to the folate receptor in the cancer cells.	Venkatasubbu et al., 2013

their stability, pharmacokinetics and biodistribution, biocompatibility and biodegradability. The full therapeutic application of the nanoparticles is achieved when they are built of the appropriate characteristics, thus the size, shape and the surface characteristics of the nanoparticles should be controlled to achieve the desired response (Elsabahy and Wooley, 2012).

2.4.8 PREPARATION OF POLYMERIC NANOPARTICLES

Preparing polymeric nanoparticles is an art of technology which requires choosing a suitable technique among the various possible methods, through the utility of homogenization process, appropriate surfactants, co-surfactants and suitable initiating system to obtain desired polymeric nanoparticles with optimum characteristics. Method like solvent evaporation, emulsion diffusion solvent displacement, etc. are environment friendly. Generally, two main strategies are employed for their preparation: the dispersion of preformed polymers and the polymerization of monomers. Various techniques can be used to produce polymer nanoparticles, such as solvent evaporation, salting-out, dialysis, supercritical fluid technology, micro-emulsion, mini-emulsion, surfactant-free emulsion, and interfacial polymerization. The choice of method depends on a number of factors, such as, particle size, particle size distribution, area of application, etc. (Rao et al., 2011).

A wide range of polymers are available to prepare polymeric nanoparticles are given in the following

2.4.8.1 Natural Polymers

The natural polymers used for biomedical application can be obtained from both the plants and animals. The polymers obtained from animals are chitosan, elastin, keratin, collagen, silk and chitin while those obtained from plants are starch, pectin, cellulose. These can be used to prepare nanoparticles with increased therapeutic efficacy (Sionkowska, 2011). Smitha et al., prepared amorphous chititn nanoparticles (AC NPs) for delivery of paclitaxel to colon cancer. The AC NPs with a particle size range of 200 ± 50 nm were obtained employing the ionic cross linking reaction with TPP. The nanoparticles were found to be hemocompatible with sustained release of paclitaxel from the nanoparticles. The enhanced anticancer efficacy was shown using human colorectal adenocarcinoma cells (HT-29), human colon carcinoma cells (COLO-205) and rat intestinal epithelial cells (IEC-6) (Smitha et al., 2013). Yang et al., developed reduction-sensitive disulfide cross linked starch nanoparticles N,N-bisacryloylcystamine (BAC) as cross linker. 5-aminosalicylic acid (5-ASA) could be efficiently loaded into the nanoparticles with a particle size of about 40 nm. It was observed with the in vitro drug release study that the drug release from nanoparticles release was accelerated in the presence of dithiothreitol (DTT) due to reductive cleavages of disulfide linkages. The biocompatibility of nanoparticles was observed using HeLa cells (Yang et al., 2014).

2.4.8.2 Synthetic Polymers

A wide range of synthetic polymers are currently in use. The nanoparticles prepared from synthetic polymers have emerged as a potential carrier system for the delivery of variety of cargoes such as drugs, peptides and proteins. The various polymers used for preparing polymeric nanoparticles are: aliphatic polyesters such as poly (lactic acid) (PLA), poly (ɛ-caprolactone) (PCL), poly (ethylene glycol), Poly (acrylic acid), poly (N-isopropylacrylamide) (PNIPAM), Poly (carboxybetaine methacrylate), polyethyelneimine, polyurethane and several others (Elsabhay and Wooley, 2012). Roussaki et al. (2014) prepared aureusidin loaded poly (lactic acid) nanoparticles with the particle size range of 231.1 ± 3.6–371.4 ± 12.9 nm and encapsulation efficiency of 68–82%. The blank PLA nanoparticles exhibited a zeta potential value of –2 to –18.7 mV, the negative charge attributed to the presence of carboxyl end groups on PLA while on encapsulating aureusidin into the nanoparticles the surface become slightly more charged due to the formation of H-bonds between the polar-OH groups of auresidin and the carboxyl group of PLA and/or due to the acidic character of the phenolic hydroxyl groups, thus enhancing the stability of nanoparticles. Bernabeu et al., developed poly (-caprolactone)–alpha tocopheryl polyethylene glycol 1000 succinate (PCL–TPGS) nanoparticles loaded with paclitaxel. Three different methods were employed to prepare the nanoparticles. The smallest particle size and highest entrapment was obtained using ultrasonication method. Cytotoxicity studies conducted using MCF-7 and MDA-MB-231 cell lines exhibited better anticancer activity of the prepared nanoparticles. It was evident from the

in vivo studies that the developed nanoparticles had longer systemic circulation time (Bernabeu et al., 2014). Chawla and Amiji, developed tamoxifen
loaded poly (ε-caprolactone) (PCL) nanoparticles for estrogen receptor (ER)
positive breast cancer. The nanoparticles with the particle size of 250–300
nm and a zeta potential of +25 mV were obtained. The biodegradation studies performed showed that the degradation of PCL was enhanced in the presence of pseudomonas lipase. The cell line studies performed using MCF-7
cells exhibited the enhanced uptake of the nanoparticles (Chawla and Amiji,
2002).

2.4.8.3   Biodegradable Polymers

The biodegradable and biocompatible nature of these types of nanoparticles
have made them a choice of carrier system for delivery of therapeutics, by
many researchers. The most widely used biodegradable polymers for preparing nanoparticles are Poly-d, l-lactide-co-glycolide (PLGA), Polylactic acid
(PLA), Poly-ε-caprolactone (PCL), Chitosan, Poly-alkyl-cyano-acrylates
(PAC), Gelatin (Kumari et al., 2010). Khuroo et al. (2014) developed dual
loaded PLGA nanoparticles simultaneously delivering topotecan hydrochloride (TOP) and tamoxifen citrate (TAM) for the treatment of breast cancer.
TAM acted as P-glycoprotein inhibitor and inhibited the efflux transporter
and thus enhanced the permeation of TOP within the cells, hence reducing
the dose and further the side effects of the individual drugs. The bioavailability of TOP was increased by 1.9 folds, as shown by the ex vivo gut
permeation study. Yordanov et al. (2013) developed and characterized etoposide loaded poly (butyl cyanoacrylate) nanoparticles. The two different
colloidal stabilizers were employed (pluronic F68 and polysorbate 80). The
nanoparticles with the size range of 170–260 nm and entrapment efficiency
of 63–68% were obtained. The pluronic coated nanoparticles exhibited zeta
potential of –4 mV while polysorbate 80 coated nanoparticles showed the
zeta potential of –12 mV. The nanoparticles exhibited the enhanced toxicity
towards adenocarcinoma human epithelial (A549) cells.

2.4.8.4   Mucoadhesive Polymers

The mucus layer covering the surface of various organs has led to the development of mucoadhesive dosage forms that result in the localization of

administered drugs at the site of action for a prolonged period of time and hence increasing the bioavailability of the drugs. Such type of nanoparticles prepared from mucoadhesive polymers are called mucoadhesive polymeric nanoparticles. The most commonly used mucoadhesive polymers for developing a mucoadhesive system are: Alginate, Chitosan and its derivatives, Carrageenan κ type II, Gelatin, Poly (ethylene glycol) and poly (ethylene oxide) and its copolymers, Poly (acrylic acid) and poly (methacrylic acid) derivatives, Poly (vinyl pyrrolidone), Poly (vinyl amine) (Sosnik et al., 2014). Anitha et al., developed curcumin (CRC) and 5-fluorouracil (5-FU) loaded thiolated chitosan nanoparticles for the treatment of colon cancer. Both the nanoparticles showed a sustained release for upto 4 days with increased release in acidic pH. The anticancer effects were enhanced by 2.5–3 folds, as shown by the MTT assay, mitochondrial membrane potential measurement and cell cycle analysis. The plasma concentrations of CRC and 5-FU were also found to be improved (Anitha et al., 2014). Haque et al., prepared alginate nanoparticles loaded with venlafaxine to be delivered by intranasal route for the treatment of depression. The improvement in behavioral analysis parameters was observed along with the improvement in loco motor activity. A greater brain/blood ratio was also observed during the pharmacokinetic studies indicating the superiority of the prepared nanoparticles over conventional treatment (Haque et al., 2014).

2.4.8.5   Bioerodible Polymers

The bioerodible polymeric system are those in which polymers breakdown into fragments through the enzymatic, hydrolytic or other chemical processes, simultaneously releasing drug (Smith et al., 1990). The polymer can undergo surface erosion or bulk erosion. Some of the bioerodible polymers use for preparing bioerodible polymeric nanoparticles are polyanhydrides, poly (vinyl methyl ether-co-maleic anhydride) (PVMMA) (Li and Lee, 2010), poly (ortho esters) (Schwach-Abdellaoui, 2001), poly (lactic acid) and poly (glycolic acid) (Smith et al., 1990). Palamoor and Jablonski, prepared celecoxib loaded poly (ortho ester) nanoparticles for the treatment of chronic eye diseases. The drug release was found to follow zero order release kinetics with a surface erosion-controlled mechanism. The minimal toxicity of the nanoparticles was observed towards HEK 293 cells. Since the nanoparticles were not found to be internalized by either Mueller or HEK

293 cells, thus the drug could be delivered for prolonged periods to the back of the eye (Palamoor and Jablonski, 2013).

2.4.8.6   pH Responsive Polymers

The pH sensitive polymeric nanoparticles are prepared using pH sensitive polymers, which contain an acidic or a basic group in their chain and thus undergo a change in their ionization state with the variation in environmental pH (James et al., 2013). Several pH sensitive polymers used for preparing nanoparticles are: poly (amino acids), aliphatic polyesters, polyanhydrides, poly (vinylpyridine), poly (vinylimidazole). Some pH sensitive polymers are biocompatible and biodegradable such as chitosan, poly (aspartic acid) (PASP) and its derivates, a Poly (amino acids) containing pendant carboxylic groups (Nemethy et al., 2013). A pH responsive cyclodextrin-containing star polymer was developed by Xiong et al., via host-guest interaction. The cyclodextrin polymer and pH-sensitive poly (2-(dimethylamino)ethyl methacrylate) serve as the core and poly (ethylene glycol) as the arm for the star polymer loaded with doxorubicin. A sustained release of doxorubicin was obtained. The in-vivo anti-tumor studies conducted on BALB/c mice bearing cervical tumor showed that the nanoparticles could effectively suppress the growth of tumor and hence reducing the significant side effects (Xiong et al., 2014). Zheng et al., prepared pH-sensitive poly (l-glutamic acid) grafted mesoporous silica nanoparticles encapsulating doxorubicin. The release of drug from the nanoparticles was found to be pH dependent. A higher drug release was observed at pH 5.5 than at pH 7.4. The Sulforhodamine B (SRB) colorimetric assay was performed on HeLa cells and the prepared nanoparticles were found to be biocompatible with enhanced Cytotoxicity (Zheng et al., 2013).

2.4.8.7   Temperature Responsive Polymers

Thermo responsive polymeric nanoparticles are prepared using thermo responsive polymers which undergo a sudden change in their solubility with response to a small variation in temperature. The rate of drug release from the thermo responsive system is controlled by the reversible sol–gel transition of the system at body temperature (James et al., 2013). Some of the temperature sensitive polymers are: poly (ethylene oxide), methylcellulose, poloxamers, gelatine, poly (N-acryloyl glycinamide), xyloglucan and poly (N-isopropylacrylamide) (Nemethy et al., 2013). Na et al., prepared thermo-senstive multi-block copolymers of dicarboxylated poly (ethylene glycol) with poly (l-lactic acid). The triblock (PLLA/PEG/PLLA) copolymer was synthesized and further used to prepare nanoparticles encapsulating doxoru-bicin. The thermo sensitivity was determined by the change in particle size, as on increasing the temperature from 37–42°C, the particle size decreased. Another measure of thermo sensitivity was change in internal structure that was measured by change in micro viscosity. Enhanced cytotoxicity was observed at higher temperature (42°C) as compared to lower (37°C) against Lewis Lung Carcinoma (LLC) cells (Na et al., 2006). Table 2.3 enlists few recent applications of polymeric nanoparticles.

2.4.9 SOLID LIPID NANOPARTICLES

A distinct category of nano drug delivery tools called solid lipid nanopar-ticles (SLN) can address most of these challenges in effective manner. SLN are nano to sub-micron range colloidal carriers (particle size 50 to 1000 nm) that are composed of physiological lipid, dispersed in aqueous surfactant solution (Muller et al., 2000). SLN offer unique advantages over other carri-ers such as (Mehnert and Mader, 2001):

i. Slow degradation kinetics and hence longer circulation in biologi-cal systems;

ii. Protection of drug payload from external environment;

iii. High drug encapsulation capacity;

iv. Encapsulation of both hydrophilic and lipophilic drugs due to mul-tiple encapsulation mechanisms like direct salvation and entrap-ment in imperfect crystal structures;

v. Lack of particle coalescence due to highly lipophilic nature;

vi. Prepared with generally recognized as safe (GRAS) materials;

vii. Avoidance of organic solvents;

viii. Easy industrial scale up; and

ix. Non-toxic biological clearance of the lipid excipients from the body.

Thus, SLN provides a versatile nano drug delivery system with lower toxicity concerns and better pharmaceutical acceptability.

Lipid nano carriers had an advent as lipid nanoemulsions which were initially introduced as parenteral nutrients (Lapillonne et al., 2011).

TABLE 2.3 Applications of Polymeric Nanoparticles

S. No.	Polymer	Drug delivered	Results	Reference
1.	Poly (aspartic acid)	Doxorubicin	These self-assembling nanoparticles increased the cellular uptake efficiency, bypassed P-glycoprotein-mediated drug efflux and improved intracellular drug retention. In nude mice bearing xenografted HCT8/ADR colon cancers, intravenous or peritumoral injection of these nanoparticles for 22 days effectively inhibited tumor growth.	Pan et al., 2016
2.	Eudragit E100	Efavirenz	In-vivo bio distribution profile of the nanoparticles showed considerably higher drug concentration in serum and major organs, especially in the brain compared to the free drug. Nanoparticles also demonstrated increase in dissolution, drug distribution, and bioavailability.	Hari et al., 2016
3.	Eudragit E100 (EE)-polyvinyl alcohol	7, 3', 4'-Trihydroxyisoflavone	Nanoparticles showed good drug loading, good encapsulation efficiencies, amorphous transformation and intermolecular hydrogen-bond formation. This resulted in increased water solubility and enhanced in vitro skin penetration, with no cytotoxicity toward HaCaT cells. In addition, they showed good antioxidant activity.	Huang et al., 2016
4.	Chitosan	Hydrocortisone and Hydroxytyrosol	Nanoparticles showed significantly improved drug penetration into the epidermal and dermal layers without saturation. Nanoparticles penetrated 2.46-fold deeper than the commercial formulation did, and had greater affinity at the skin target site without spreading to the surrounding tissues. Commercial formulation induced skin atrophy whereas prepared nanoparticles showed no evidence of toxicity.	Siddique et al., 2016

#				Reference
5.	Semitelechelic N-(2-hydroxypropyl) methacrylamide (HPMA)	Docetaxel	The amphiphilic star-shaped conjugates could self-assemble into nanoparticles and exhibited conspicuous drug-loading capacity. The stimuli-responsive drug release under acidic lysosomal and reducing cytoplasmic environments was observed, leading to enhanced cytotoxicity. Nanoparticles displayed uniform tumor distribution and suppressed tumor growth by 78.9% through enhanced depletion of cancer-associated fibroblasts and induction of apoptosis.	Yang et al., 2016
6.	Poly (lactic-co-glycolic acid) (PLGA)	Epirubicin	Epirubicin loaded nanoparticles (EPI-NP) was prepared for oral delivery. Nanoparticles showed uniform size, good entrapment and controlled release behaviour. Cytotoxicity studies conducted on human breast adenocarcinoma cell lines (MCF-7) revealed thesuperiority of epirubicin loaded poly-lactic-co-glycolic acid nanoparticles (EPI-NPs) over free epirubicinsolution (EPI-S). Further, flow cytometric analysis demonstrated improved drug uptake through nanoparticles and dominance of caveolae mediated endocytosis. Results from in vivo pharmacokinetic study revealed 3.9 fold increased in oral bioavailability of EPI through EPI-NPs	Md Tariq et al., 2016
7.	Polyethylene glycol-poly (lactic-co-glycolic acid)	Gemcitabine	Nanoparticles exhibited a controlled release pattern. Enhanced cellular uptake and cytotoxicity of aptamer-conjugated nanoparticles was observed in A549 cancer cells and verified nucleolin-mediated receptor-based active targeting.	Alibolandi et al., 2016
8.	Poly (amidoamine) dendrimers (PAMAM)	Doxorubicin	Drug release from these complexes followed a redox and acid-triggered manner and increased with increasing degree of PE Gylation. In vivo study in B16 tumor-bearing mice indicated that these nanoparticles could significantly improve antitumor efficiency with a good safety.	Hu et al., 2016

TABLE 2.3 (Continued)

S. No.	Polymer	Drug delivered	Results	Reference
9.	Polyacrylamide-grafted-xanthan gum (PAAm-g-XG)	Curcumin	Release of curcumin was comparatively faster at pH 7.2 than that observed with at 1.2 and 4.5. Excellent release was observed at pH 6.8. Highest curcumin release was observed when rat caecal contents were incorporated in pH 6.8 solution, indicating microflora-dependent drug release property. Curcumin was better absorbed systemically in nanoparticulate form with increased Cmaₓ (3 fold) and AUC (2.5 fold) than when delivered as free curcumin.	Mutalik et al., 2016
10.	Redox-sensitive polymer (PEG-S-S-PLA)	Gemcitabine and Doxorubicin	Nuclear-targeted polymersomes released more than 60% of encapsulated contents in response to 50 mM glutathione. Prepared formulation can increase drug's therapeutic index by delivering the drugs directly to the cells' nuclei.	Anajafi et al., 2016

However, later on their capability for delivering the drug molecules was recognized and they were employed as the bonafide drug carriers (Davis et al., 1987). Nevertheless, due to very small size and liquid state these carriers have low potential for controlled release of the drug molecules. Thus, SLN provided an attractive alternative to deliver the drug molecules with added advantages of nanoemulsion and controlled drug release (Muller et al., 2000).

Structurally SLN consists of a lipid core, which consists of solid lipids and stabilized by high HLB surfactants to provide an aqueous dispersion (Gastaldi et al., 2014) (Figure 2.3). Usually phospholipids are also used to stabilize the structure of the SLN particles (Siekmann and Westesen, 1996). The drug molecules are encapsulated in SLN via solubilization or entrapment in imperfections of the lipid crystals (Souto and Müller, 2006). The structural uniqueness and mode of drug encapsulation ensures a very high entrapment efficiency of drug molecules in SLN.

SLN are usually prepared from the lipids of GRAS category that are non-toxic and biodegradable in nature. Some of the commonly used lipids in SLN formulations are given in Table 2.4. It was observed that the SLN prepared with lower melting point lipids have smaller size as compared to the higher melting point lipids (Müller et al., 2002). The lower melting lipids get emulsified with lesser amount of surfactant

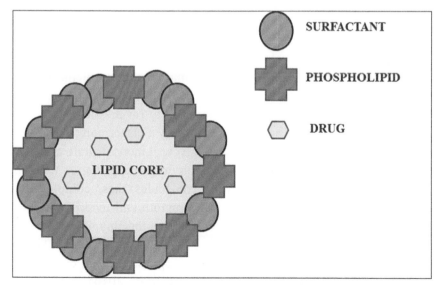

FIGURE 2.3 Structure of Solid Lipid Nanoparticles.

TABLE 2.4   Commonly Used Lipid in Solid Lipid Nanoparticles

S.No.	Lipids
Triglycerides	
1.	Tricaprin
2.	Trimyristin
3.	Tripalmitin
4.	Trilaurin
5.	Tristearin
Hard Fat type	
6.	Hydrogenated coco-glycerides
7.	Witepsol W 35, H35, E 85, H 42
Fatty acids	
8.	Stearic acid
9.	Palmitic acid
10.	Decanoic acid
11.	Behanic acid
Waxes	
12.	Glyceryl monostearate
13.	Glyceryl behnate
14.	Glyceryl palmitostearate
Steroids	
15.	Cholesterol

and produce smaller particles. However, the selected lipid should also have sufficiently high melting point to maintain the structural integrity. Selection of lipid can also influence the entrapment efficiency of the SLN. Selection of lipid with greater affinity for drug molecule may always lead to the high encapsulation efficiency. Furthermore, use of lipid molecules with complex structure (e.g., glyceryl monostearate and glyceryl behanate) also has a positive effect on drug encapsulation (Jenning and Gohla, 2000). Unlike, symmetrical molecules, these complex lipid molecules congeal into more imperfect structure and incorporate more drugs. Besides, selection of lipid according to the lipophilicity of the drug molecule was also found to affect the drug entrapment. The lipids with longer fatty acid chain are more lipophilic and can encapsulate more lipophilic drugs and vice versa (Sanna et al., 2010). Selection of an appropriate lipid is also required for the pharmaceutical acceptability of the formulation. The lipids which can be autoclaved and freeze dried in the aqueous suspension can provide distinctive advantage in case of parenteral preparation.

2.4.10 GLYCERIDES

Glycerides are substituted glycerols, and are classified as monoglycerides (mono-substituted), diglycerides (di-substituted) and triglycerides (tri-substituted). Out of these, the most well-known and most commonly used for the preparation of nanoparticles is the fatty acid trimesters of glycerol, for example, triglycerides. In triglycerides, the three hydroxyl groups of glycerol are all esterified, each by different fatty acids (Coleman et al., 2004).

Triglycerides are formed by combining glycerol with three fatty acid molecules. Alcohols have a hydroxyl (–HO) group. Organic acids have a carboxyl (–cOOH) group. Alcohols and organic acids join to form esters. The glycerol molecule has three hydroxyl (HO–) groups. Each fatty acid has a carboxyl group (–cOOH). In triglycerides, the hydroxyl groups of the glycerol join the carboxyl groups of the fatty acid to form ester bonds. The three fatty acids are usually different, but many kinds of triglycerides are known. The chain lengths of the fatty acids in naturally occurring triglycerides vary, but most contain 16, 18, or 20 carbon atoms (IUPAC–IUB). Examples of commonly used triglycerides are tristearin, trimyristin, etc. Examples of monoglycerides and diglycerides include glycerol monostearate and glycerol bahenate, respectively.

2.4.11 FATTY ACIDS

Fattyacids are carboxylic acids with a long aliphatic tail (chain), which is either saturated or unsaturated. Most naturally occurring fatty acids have a chain of an even number of carbon atoms, from 4 to 28. Fatty acids are usually derived from triglycerides or phospholipids. When they are not attached to other molecules, they are known as "free" fatty acids (IUPAC Compendium of Chemical Terminology, 2nd edn.). Fatty acids that have carbon–carbon double bonds are known as unsaturated. Fatty acids without double bonds are known as saturated. They differ in length as well.

Fatty acids are usually produced industrially by the hydrolysis of triglycerides, with the removal of glycerol. Phospholipids represent another source. Some fatty acids are produced synthetically by hydrocarboxylation of alkenes.

Examples of saturated fatty acids are stearic acid, myristic acid, etc. Examples of unsaturated fatty acids include oleic acid, elaidic acid, etc.

2.4.12   QUALITY ISSUES IN SLN

Apart from various advantages associated with SLN, some of the major quality issues provide the challenges for its reputation. These quality issues influence the physical stability of SLN and lead to inconsistency in their performance. The major quality issues associated with SLN are:

- **Gelation:** Gelation is a common problem with SLN, which leads to the formation of high viscosity gel from non-viscous particle suspension (Mehnert and Mader, 2001). Such physical changes may leads to the formation of large particles or gel like structure and thus compromising the whole purpose of preparing a nano system. Besides, it could prove fatal if develops during the intravenous administration as it can block the blood vessels. Gelation often occurs due to the insufficient surfactant concentration or other quality issues such as polymorphic changes. Thus, prevention of polymorphic changes and use of appropriate surfactant concentrations can reduce the gelation problem (Westesen and Siekmann, 1997).

- **Presence of many colloidal species:** The emulsifiers that are used in the formulation of nanoparticles not only occupy their surface interface but they can also orient themselves into different colloidal species like micelles and or liposomes (Mehnert and Mader, 2001, Negi et al., 2014). Thus, a single SLN formulation may contain other nano systems that may dissimilar characteristics and outcomes.

- **Super cool melts:** Supercooled melts may be described as the event in which the lipid may fail to crystallize even at a temperature below its melting point. Such formulations are not solid nanoparticles but nanoemulsions and hence loose the potential advantages like controlled drug release, which associated with the solid state of the lipids (Zimmermann et al., 2005).

- **Polymorphic transformation:** Solid lipids used in SLN have different polymorphic forms. During the preparation of SLN, these lipids crystallize into imperfect but unstable α polymorph. In due course of time during storage the unstable α polymorph get converted to more stable β' and finally to β polymorphs (Jores et al., 2003; Siekmann and Westesen, 1994). These stable polymorphs are more perfect and symmetrical in crystal structure as compared to unstable α polymorphs. Such polymorphic changes may leads to the expulsion of the drug incorporated in lipid imperfections and leads to unpredictable drug encapsulation. Besides, these structural changes may lead to other quality issues such as gelation (Westesen and Siekmann, 1997). However, polymorphic changes in SLN can be reduced by using mixed lipid. The structural differences of mixed lipids reduce the stabilization of polymorphic forms. Additionally, triglycerides are less prone to polymorphic changes (Bunjes et al., 1996). Furthermore, as these polymorphic conversions are promoted by the oxidation of lipids the prevention of oxidation by adequate storage or use of nitrogen environment may also reduce the polymorphic transformations (Mori, 1988). These polymorphic changes can be prevented by:

− Use of mixed lipids.
− Use of triglycerides.
− Storage under nitrogen.
− Switching to Nanostructured Lipid Carriers (NLCs.)

The NLCs consists of Solid Lipids and Liquid lipids (oils).

2.4.13 ADVANTAGES OF NLC OVER SLN

- Limitation of drug load by the solubility of the drug in the solid lipids is overcome.
- Drug expulsion phenomenon when lipid crystallizes to the stable β'-form is reduced.
- More flexible in texture and smaller size (better uptake).

2.4.14 TYPES OF NEW GENERATION NLC

- **Type I:** Liquid lipid increases imperfection in solid lipid matrix.
- **Type II:** Liquid lipid reduces crystallization of solid lipid and make amorphous matrix.

- **Type III:** liquid lipid have greater solubility of drug than solid lipid and hence increases overall solubility of drug

Table 2.5 enlists few recent applications of solid lipid nanoparticles.

2.5   LIMITATIONS OF NANOPARTICLES

When compared to conventional dosage forms such as capsules or drug solutions, synthesis of nanoparticles is more troublesome and longer time taking process. This also leads to higher costs of synthesis which makes the whole nano formulation much more expensive.

Apart from tedious synthesis methods and higher costs of synthesis, another major problem which is faced with nanoparticles is the toxicity associated with them. Their toxicity is mainly attributed to their extremely small size which enables them to easily cross living cell membranes of the body and thus causing damage to them. Another reason is the materials used in nanoparticles such as metals, certain polymers, etc. Yet another reason of their toxicity lies in their mode of action when they enter inside body. Nanoparticles, especially metallic ones, exert their action by producing reactive oxygen species, which leads to killing of foreign cells present inside the body, or the diseased cells of the body such as cancer cells. Normal cells usually trigger their protective mechanisms to escape the damage by ROS, but when oxidative stress is elevated further, an irreparable damage may be caused to the body cells.

Toxicity of the nanoparticles can be minimized by using non-toxic materials for the synthesis and by targeting them actively or passively to the diseased tissue, sparing the normal cells (Sharma et al., 2015).

2.6   CONCLUSION

Nanoparticles find a huge range of applications due to their unique physical and chemical properties which are not present in their bulk analogues. Nanoparticles are extensively researched and have shown promising results. Nanoparticles have been found to be more efficient, lower the need of dose of the drug, can be used for targeting, and also offer controlled or sustained release. Nanoparticles such as "Doxil" and "Abraxane" are already in market and many more are in clinical trials. Nanoparticles are broadly categorized into metallic and non-metallic

TABLE 2.5 Applications of Solid Lipid Nanoparticles

S. No.	Drug delivered	Results	Reference
1.	Voriconazole	In vitro release study of solid lipid nanoparticles demonstrated sustained release up to 12 hours. This study demonstrated that nanoparticles prepared by ultra-sonication method is more suitable than micro emulsion technique without causing any significant effect on corneal hydration level.	Khare et al., 2016
2.	Methotrexate	Ex-vivo results showed greater cellular uptake and better cytotoxicity at lower IC50 in breast cancer cells. Enhanced apoptosis with altered lysosomal membrane permeability and better rate of degradation of lysosomal membrane was also observed in-vitro. Further, in-vivo evaluation showed maximum bioavailability and tumor targeting efficiency when compared with free drug. Reduction in size of tumor was also observed.	Garg et al., 2016
3.	5-Fluorouracil	Entrapment efficiency of SLNs was $47.92 \pm 2.34\%$. The diffusion of the drug-loaded SLN was 2-fold higher when compared with the free drug when both diffused through a hydrophobic membrane. SLN treated mice exhibited reduced inflammatory reactions, with reduced degrees of keratosis, in addition to reduced symptoms of angiogenesis compared to 5-FU treated mice.	Khallaf et al., 2016
4.	Paclitaxel	These nanoparticles exhibited uniformly dispersed distribution, increased drug-encapsulation with a sustained-release pattern, as well as enhanced biostability during blood circulation. They induced efficient intracellular trafficking of drug in situ via multi-targeting mechanisms. Resulting anticancer treatment demonstrated a IC50 of 0.20 µg/mL, cell apoptosis of 18.04% 24h post-incubation mainly arresting G2/M cell cycle in vitro, and tumor weight inhibition of 70.51% in vivo.	Wang et al., 2016
5.	Epigallocatechin gallate (EGCG)	When delivered through SLNs, the cytotoxicity of EGCG loaded SLNs was found to be 8.1 times higher against MDA-MB 231 human breast cancer cells and 3.8 times higher against DU-145 human prostate cancer cells than that of the pure EGCG.	Radhakrishnan et al., 2016
6.	Paromomycin sulfate	The results of this study showed that the paromomycin sulfate loaded SLN formulation is a safe compound and is effective for treatment of leishmaniasis by improving the effectiveness of the drug in killing the parasite and switching towards Th1 response.	Heidari-Kharaji et al., 2016

S. No.	Drug delivered	Results	Reference
7.	Zidovudin	Cell culture studies using C6 glioma cells on the nanoparticles showed enhanced growth and proliferation of cells without exhibiting any toxicity. Normal cell morphology and improved uptake were observed by fluorescence microscopy images of rhodamine labeled modified solid lipid nanoparticles. Cellular uptake study suggested that these nanoparticles could be a promising drug delivery system to enhance the uptake of antiviral drug by brain cells and a suitable drug carrier system for the treatment of HIV.	Joshy et al., 2016
8.	Alendronate sodium	Results indicated highest entrapment efficiency value of 74.3%. Eudragit E100 coated nanoparticles released alendronate sodium only at an alkaline pH. Alendronate sodium bioavailability was enhanced by more than 7.4-fold in rabbits.	Hosny, 2016
9.	Celecoxib	Maximum entrapment efficiency of $92.46 \pm 0.07\%$ was achieved. Permeation across the cornea was also significantly better than aqueous suspension. SLN formulation demonstrated improved performance of entrapped celecoxib and exhibited prolonged retention over ocular surfaces.	Sharma et al., 2016
10.	Naringenin	SLNs showed an entrapment efficiency of 79.11%, and a cumulative drug release of 80% in 48 hours with a sustained profile. Pharmacokinetic study showed that the relative bioavailability of naringenin SLNs was 2.53-fold greater than that of naringenin suspension.	Ji et al., 2016

nanoparticles. Non- metallic nanoparticles are further of three types, namely, ceramics nanoparticles, polymeric nanoparticles and solid lipid nanoparticles. Each class of nanoparticles has its own unique advantages and properties. But, despite of the wide applicability and advantages of the nanoparticles, they also pose toxicity issues as a result of their extremely minute size. Studies are continuously performed to overcome the toxicity of the nanoparticles and improve their safety along with their efficiency.

KEYWORDS

- **ceramic nanoparticles**
- **lipid nanoparticles**
- **metallic nanoparticles**
- **nanoparticles**
- **polymeric nanoparticles**
- **synthesis**

REFERENCES

Abercrombie, M., & Ambrose, E. J., (1962). The surface properties of cancer cells: A review. *Cancer Res., 22*, 525–548.

Alibolandi, M., Ramezani, M., Abnous, K., & Hadizadeh, F., (2016). AS1411 aptamer-decorated biodegradable polyethylene glycol-poly (lactic-co-glycolic acid) nanopolymersomes for the targeted delivery of gemcitabine to non-small cell lung cancer in vitro. *J Pharm Sci., 105*, 1741–150.

Amtenbrink, M. H., Rechenberg, B., & Hofmann, H., (2009). Superparamagnetic nanoparticles for biomedical applications. In *Nanostructured Materials for Biomedical Applications*, Tan, M. C., Chow, G. M., Ren, L., (Eds.), Transworld Research Network: India, p 119.

Anajafi, T., Scott, M. D., You, S., Yang, X., Choi, Y., Qian, S. Y., et al., (2016). Acridine orange conjugated polymersomes for simultaneous nuclear delivery of gemcitabine and doxorubicin to pancreatic cancer cells. *Bioconjug Chem, 27*, 762–771.

Appadurai, P., & Rathinasamy, K., (2015). Plumbagin-silver nanoparticle formulations enhance the cellular uptake of plumbagin and its antiproliferative activities. *IET Nanobiotechnol., 9*, 264–272.

Areán, C. O., Mentruit, M. P., Platero, E. E., Xamena, F. X. L., & Parra, J. B., (1999). Sol–gel method for preparing high surface area $CoAl_2O_4$ and $Al_2O_3.CoAl_2O_4$ spinels. *Mater. Lett, 39*, 22–27.

Arias, J. L., Gallardo, V., Ruiz, M. A., & Delgado, A. V., (2007). Ftorafur loading and controlled release from poly (ethyl-2-cyanoacrylate) and poly (butylcyanoacrylate) nanospheres. *International Journal of Pharmaceutics, 337*, 282–290.

Bunjes, H., Westesen, K., & Koch, M. H., (1996). Crystallization tendency and polymorphic transitions in triglyceride nanoparticles. *International Journal of Pharmaceutics, 129*, 159–173.

Capek,, (2004). Preparation of metal nanoparticles in water-in-oil (w/o) microemulsions. *Advances in Colloid and Interface Science, 110*, 49–74.

Chan, H. K., & Kwok, P. C. L., (2011). Production methods for nanodrug particles using the bottom–up approach. *Adv Drug Deliv Rev., 63*, 406–416.

Chaudhuri, S., Sardar, S., Bagchi, D., Dutta, S., Debnath, S., Saha, P., et al., (2016). Photoinduced dynamics and toxicity of a cancer drug in proximity of inorganic nanoparticles under visible light. *Chemphyschem, 17*, 270–277.

Chawla, J. S., & Amiji, M. M., (2002). Biodegradable poly (o-caprolactone) nanoparticles for tumor targeted delivery of tamoxifen. *International Journal of Pharmaceutics, 249*, 127–138.

Chen, Y. C., Huang, X. C., Luo, Y. L., Chang, Y. C., Hsieh, Y. Z., & Hsu, H. Y., (2013). Nonmetallic nanomaterials in cancer theranostics: A review of silica and carbon based drug delivery systems. *Sci. Technol. Adv. Mater., 14*, 23 pages.

Chenab, F., Zhub, Y., Wub, J., Huanga, P., & Cuia, D., (2012). Nanostructured Calcium Phosphates: Preparation and Their Application in Biomedicine. *Nano Biomed. Eng., 4*, 41–49.

Chernyak, Y., Henon, F., Harris, R. B., Gould, R. D., Franklin, R. K., Edwards, J. R., et al., (2001). Formation of perfluoropolyether coatings by the rapid expansion of supercritical solutions (RESS) process. Part 1: experimental results. *Industrial and Engineering Chemistry Research., 40*, 6118–26.

Choi, J., Kim, H., Choi, J., Oh, S. M., Park, J., & Park, K., (2014). Skin corrosion and irritation test of sunscreen nanoparticles using reconstructed 3D human skin model. *Environ Health Toxicol., 29*, e2014004.

Cooper, D. L., Conder, C. M., & Harirforoosh, S., (2014). Nanoparticles in drug delivery: mechanism of action, formulation and clinical application towards reduction in drug-associated nephrotoxicity. *Expert Opin Drug Deliv., 11*, 1661–1680.

Davis, S.S., Washington, C., West, P., Illum, L., Liversidge, G., Sternson, L., et al., (1987). Lipid emulsions as drug delivery systems. *Annals of the New York Academy of Sciences, 507*, 75–88.

De Campos, A. M., Sanchez, A., & Alonso, M. J., (2001). Chitosan nanoparticles: A new vehicle for the improvement of the delivery of drugs to the ocular surface. Application to cyclosporin A. *International Journal of Pharmaceutics, 224*, 159–168.

Degen, A., & Kosec, M., (2000). Effect of pH and impurities on the surface charge of zinc oxide in aqueous Solution. *J Europena Ceramic Society, 20*, 667–73.

Eguía, L. P. H., Borrull, J. F., Macias, G., Pallarès, J., & Marsal, L. F., (2014). Engineering optical properties of gold-coated nanoporous anodic alumina for biosensing. *Nanoscale Res Lett., 9*, 414.

Elsabahy, M., & Wooley, K. L., (2012). Design of polymeric nanoparticles for biomedical delivery applications. *Chemical Society Reviews, 41*, 2545–2561.

Evanoff, D. D., & Chumanov, G., (2005). Synthesis and optical properties of silver nanoparticles and arrays. *Chem Phys Chem.*, *6*, 1221–1231.

Fadeel, B., & Bennett, A. E. G., (2010). Better safe than sorry: Understanding the toxicological properties of inorganic nanoparticles manufactured for biomedical applications. *Advanced Drug Delivery Reviews*, *62*, 362–374.

Ferraz, M. P., Monteiro, F. J., & Manuel, C. M., (2004). Hydroxyapatite nanoparticles: A review of preparation methodologies. *J. Appl. Biomater. Biomech*, *2*, 74–80.

Fratzl, P., Gupta, H. S., Paschalis, E. P., & Roschger, P., (2004). Structure and mechanical quality of the collagen-mineral nano-composite in bone. *J Mater Chem.*, *14*, 2115–2123.

Galdiero, S., Falanga, A., Vitiello, M., Cantisani, M., Marra, V., & Galdiero, M., (2011). Silver nanoparticles as potential antiviral agents. *Molecules*, *10*, 8894–8918.

Garg, N. K., Singh, B., Jain, A., Nirbhavane, P., Sharma, R., Tyagi, R. K., et al., (2016). Fucose decorated solid-lipid nanocarriers mediate efficient delivery of methotrexate in breast cancer therapeutics. *Colloids Surf B Biointerfaces*, *146*, 114–126.

Gassmann, T., (1913). The preparation of a complex salt corresponding to apatite-typus and its relations to the constitution of bones. *H-S Z Physiol Chem.*, *83*, 403–408.

Gastaldi, L., Battaglia, L., Peira, E., Chirio, D., Muntoni, E., Solazzi, I., et al., (2014). Solid lipid nanoparticles as vehicles of drugs to the brain: Current state of the art. *European journal of pharmaceutics and biopharmaceutics*, *87*, 433–444.

Gonzalez-Mira, E., Nikolić, S., Calpena, A. C., Egea, M. A., Souto, E. B, & García, M. L., (2012). Improved and safe transcorneal delivery of flurbiprofen by NLC and NLC-based hydrogels. *J Pharm Sci.*, *101*, 707–25.

Guo, Y., Sun, J., Bai, S., & Jin, X., (2016). Nanoassemblies constructed from bimodal mesoporous silica nanoparticles and surface-coated multilayer pH-responsive polymer for controlled delivery of ibuprofen. *J Biomater Appl. 31*(3), 411–420.

Han, J., Zhu, Z., Qian, H., Wohl, A. R., Beaman, C. J., Hoye, T. R., et al., (2012). A simple confined impingement jets mixer for flash nanoprecipitation. *J Pharm Sci.*, *101*, 10.

Hari, B. N., Narayanan, N., Dhevendaran, K., & Ramyadevi, D., (2016). Engineered nanoparticles of Efavirenz using methacrylate co-polymer (Eudragit-E100) and its biological effects in-vivo. *Mater Sci Eng C Mater Biol Appl.*, *67*, 522–532.

Hei, H., He, H., Wang, R., Liu, X., & Zhang, G., (2012). Controlled synthesis and characterization of noble metal nanoparticles. *Soft Nanoscience Letters*, *2*, 34–40.

Heidari-Kharaji, M., Taheri, T., Doroud, D., Habibzadeh, S., Badirzadeh, A., & Rafati, S., (2016). Enhanced paromomycin efficacy by Solid Lipid Nanoparticle formulation against Leishmania in mice model. *Parasite Immunol. 38* (10), 599–608.

Hom, C., Lu, J., & Tamanoi, F., (2009). Silica nanoparticles as a delivery system for nucleic acid-based reagents. *J Mater Chem.*, *19*, 6308–6316.

Honary, S., Barabadi, H., Gharaeifathabad, E., & Naghibi, F., (2012). Green synthesis of copper oxide nanoparticles using penicillium aurantiogriseum, penicillium citrinum and penicillium waksmanii. *Digest Journal of Nanomaterials and Biostructures*, *7*, 999–1005.

Hosny, K. M., (2016). Alendronate sodium as enteric coated solid lipid nanoparticles, preparation, optimization, and in vivo evaluation to enhance its oral bioavailability. *PLoS One*, *11*, e0154926.

Hu, W., Qiu, L., Cheng, L., Hu, Q., Liu, Y., Hu, Z., et al., (2016). Redox and pH dual responsive poly (amidoamine) dendrimer-poly (ethylene glycol) conjugates for intracellular delivery of doxorubicin. *Acta Biomater.*, *36*, 241–53.

Huang, P. H., Hu, S. C., Lee, C. W., Yeh, A. C., Tseng, C. H., & Yen, F. L., (2016). Design of acid-responsive polymeric nanoparticles for 7,3,'4'-trihydroxyisoflavone topical administration. *Int J Nanomedicine, 11*, 1615–27.

Huang, X., & El-Sayed, M. A., (2010). Gold nanoparticles: Optical properties and implementations in cancer diagnosis and photo thermal therapy. *Journal of Advanced Research, 1*, 13–28.

Iqbal, P., Preece, J. A., & Mendes, P. M., (2012). Nanotechnology: The "Top–Down" and "Bottom–Up" Approaches. In: Supramolecular Chemistry: From Molecules to Nanomaterials. John Wiley and Sons, Ltd. Steed, J.W., & Gale, P.A. (Eds.). United Kingdom.

Jain, V., Jain, S., & Mahajan, S. C., (2015). Nanomedicines-based drug delivery systems for anti-cancer targeting and treatment. *Curr Drug Deliv., 12*, 177–91.

Jenning, V., & Gohla, S., (2000). Comparison of wa□and glyceride solid lipid nanoparticles (SLN®). *International Journal of Pharmaceutics, 196*, 219–222.

Ji, P., Yu, T., Liu, Y., Jiang, J., Xu, J., Zhao, Y., et al., (2016). Naringenin-loaded solid lipid nanoparticles: preparation, controlled delivery, cellular uptake, and pulmonary pharmacokinetics. *Drug Des Devel Ther., 10*, 911–25.

Johnson, B. K., & Prud'homme, R. K., (2003). Chemical processing and micromixing in confined impinging jets. *AIChE Journal, 49*, 9.

Jores, K., Mehnert, W., & Mäder, K., (2003). Physicochemical investigations on solid lipid nanoparticles and on oil-loaded solid lipid nanoparticles: a nuclear magnetic resonance and electron spin resonance study. *Pharmaceutical Research, 20*, 1274–1283.

Jurgons, R., Seliger, C., Hilpert, A., Trahms, L., Odenbach, S., & Alexiou, C., (2006). Drug loaded magnetic nanoparticles for cancer therapy. *J. Phys.: Condens. Matter, 18*, S2893–S2902.

K. S. Joshy., Sharma, C. P., Kalarikkal, N., Sandeep, K., Thomas, S., & Pothen, L. A., (2016). Evaluation of *in-vitro* cytotoxicity and cellular uptake efficiency of zidovudine-loaded solid lipid nanoparticles modified with aloe vera in glioma cells. *Mater Sci Eng C Mater Biol Appl., 66*, 40–50.

Kafshgari, M. H., Harding, F. J., & Voelcker, N. H., (2015). Insights into cellular uptake of nanoparticles. *Curr Drug Deliv, 12*, 63–77.

Kakran, M., Sahoo, N. G., Tan, I. L., & Lin, L., (2012). Preparation of nanoparticles of poorly water-soluble antioxidant curcumin by antisolvent precipitation methods. *J Nanopart Res., 14*, 757.

Kammoe, R. B., Hamoudi, S., Larachi, F. A., & Belkacemi, K., (2012). Synthesis of CaCo₃ nanoparticles by controlled precipitation of saturated carbonate and calcium nitrate aqueous solutions. *Can. J. Chem. Eng., 90*, 26–33.

Katara, R., Majumdar, D. K., & Eudragit R. L., (2013). 100-based nanoparticulate system of aceclofenac for ocular delivery. *Colloids and Surfaces B: Biointerfaces, 103*, 455–462.

Keck, C. M., & Mu¨ller, R. H., (2006). Drug nanocrystals of poorly soluble drugs produced by high pressure homogenization. *Eur J Pharm Biopharm., 62*, 3–16.

Khallaf, R. A., Salem, H. F., & Abdelbary, A., (2016). 5-Fluorouracil Shell-enriched Solid Lipid Nanoparticles (SLN) for Effective Skin Carcinoma Treatment. *Drug Deliv., 1*–32.

Khare, A., Singh, I., Pawar, P., & Grover, K., (2016). Design and evaluation of voriconazole loaded solid lipid nanoparticles for ophthalmic application. *J Drug Deliv., 2016*, 6590361.

Kleibert, A., Rosellen, W., Getzlaff, M., & Bansmann, J., (2011). Structure, morphology, and magnetic properties of Fe nanoparticles deposited onto single-crystalline surfaces. *Beilstein J. Nanotechnol, 2*, 47–56.

Konan-Kouakoua, Y. N., Boch, R., Gurny, R., & Alle'mann, E., (2005). *In vitro* and *in vivo* activities of verteporfin-loaded nanoparticles. *Journal of Controlled Release, 103,* 83–91.

Kumari, A., Yadav, S. K., Pakade, Y. B., Singh, B., Yadav, S. C., (2010). Development of biodegradable nanoparticles for delivery of quercetin. *Colloids and Surfaces B: Biointerfaces, 80,* 184–192.

Lagopati, N., Tsilibary, E. P., Falaras, P., Papazafiri, P., Pavlatou, E. A., Kotsopoulou, E., et al., (2014). Effect of nanostructured TiO_2 crystal phase on photo induced apoptosis of breast cancer epithelial cells. *Int J Nanomedicine, 9,* 3219–3230.

Lai, S., Zhang, W., Liu, F., Wu, C., Zeng, D., Sun, Y., et al., (2013). TiO_2 nanotubes as animal drug delivery system and in vitro controlled release. *J Nanosci Nanotechnol., 13,* 91–97.

Lapillonne, A., Fellous, L., & Kermorvant-Duchemin, E., (2011). Use of parenteral lipid emulsions in French neonatal ICUs. *Nutrition in clinical practice: Official publication of the American Society for Parenteral and Enteral Nutrition., 26,* 672–680.

Li, T., Shen, X., Geng, Y., Chen, Z., Li, L., Li, S., et al., (2016). Folate-functionalized magnetic-mesoporous silica nanoparticles for drug/gene codelivery to potentiate the antitumor efficacy. *ACS Appl Mater Interfaces, 8,* 13748–13758.

Lim, W. T., Subban, G. H., Hwang, H. S., Kim, J. T., Ju, C. S., & Johnston, K. P., (2005). Novel semiconducting polymer particles by supercritical fluid process. *Macromolecular Rapid Communications, 26,* 1779–1783.

Liu, G., Wang, W., Wang, H., & Jiang, Y., (2014). Preparation of 10-hydroxycamptothecin proliposomes by the supercritical CO_2 anti-solvent process. *Chem Eng J., 243,* 289–296.

Liu, J., Ma, X., Jin, S., Xue, X., Zhang, C., Wei, T., et al., (2016). Zinc oxide nanoparticles as adjuvant to facilitate doxorubicin intracellular accumulation and visualize pH-responsive release for overcoming drug resistance. *Mol Pharm, 13,* 1723–1730.

Liu, X., Ding, Y., Zhao, B., Liu, Y., Luo, S., Wu, J., et al., (2016). *In vitro* and *in vivo* evaluation of puerarin-loaded PEGylated mesoporous silica nanoparticles. *Drug Dev Ind Pharm.,* 1–7.

Loh, Z. H., Samanta, A. K., Wan, P., & Heng, S., (2015). Overview of milling techniques for improving the solubility of poorly water-soluble drugs. *Asian J Pharmaceut Sci., 10,* 255–274.

Makarov, V. V., Love, A. J., Sinitsyna, O. V., Makarova, S. S., Yaminsky, I. V., Taliansky, M. E., et al., (2014). "Green" Nanotechnologies: Synthesis of Metal Nanoparticles Using Plants. *Acta Naturae, 6,* 35–44.

Margulis, K., Magdassi, S., Lee, H. S., & Macosko, C. W., (2014). Formation of curcumin nanoparticles by flash nanoprecipitation from emulsions. *J Colloid Interface Sci., 434,* 65–70.

Marsh, M., Schelew, E., Wolf, S., & Skippon, T., (2009). Gold Nanoparticles for Cancer Treatment. PHYS 483–Queen's University, Kingston.

Md, Tariq., Anu T Singh., Zeenat Iqbal., & Sushama Talegaonkar., (2015). Biodegradable polymeric nanoparticles for oral delivery of epirubicin: *in-vitro*, *ex-vivo*, and *in-vivo* investigations. *Colloids and Surfaces B: Biointerfaces, 128,* 448–456

Mehnert, W., & Mader, K., (2001). Solid lipid nanoparticles: production, characterization and applications. *Advanced Drug Delivery Reviews, 47,* 165–196.

Merisko-Liversidge, E., & Liversidge, G. G., (2011). Nanosizing for oral and parenteral drug delivery: A perspective on formulating poorly-water soluble compounds using wet media milling technology. *Adv Drug Deliv Rev., 63,* 427–440.

Meziani, M. J., Pathak, P., Hurezeanu, R., Thies, M. C., Enick, R. M., & Sun, Y. P., (2004). Supercritical-fluid processing technique for nanoscale polymer particles. *Angewandte Chemie International Edition, 43*, 704–7.

Mohammadzadeh, R., (2012). Hypothesis: Silver nanoparticles as an adjuvant for cancer therapy. *Advanced Pharmaceutical Bulletin, 2*, 133.

Mohanraj, V. J., & Chen, Y., (2006). Nanoparticles: A Review. *Tropical Journal of Pharmaceutical Research, 5*, 561–573.

Mohr, K. H., (1987). High-pressure homogenization. Part II. The influence of cavitation on liquid–liquid dispersion in turbulence fields of high energy density. J Food Eng, *6*, 311–324.

Moreno-Vega, A. I., Gomez-Quintero, T., Nuñez-Anita, R. E., Acosta-Torres, L. S., & Castano, V., (2012). Polymeric and Ceramic Nanoparticles in Biomedical Applications. *Journal of Nanotechnology, 936041*, 10.

Mori, H., (1988). Solidification problems in preparation of fats. *Crystallization and Polymorphism of Fats and Fatty Acids*, 423–442.

Morral-Ruíz, G., Melgar-Lesmes, P., García, M. L., Solans, C., & García-celma, M. J., (2012). Design of biocompatible surface-modified polyurethane and polyurea Nanoparticles. *Polymer, 53*, 6072–6080.

Muhammad, Z., Raza, A., Ghafoor, S., Naeem, A., Naz, S.S., Riaz, S., et al., (2016). PEG capped methotrexate silver nanoparticles for efficient anticancer activity and biocompatibility. *Eur J Pharm Sci. 91*, 251–255.

Mukherjee, S., Sau, S., Madhuri, D., Bollu, V. S., Madhusudana, K., Sreedhar, B., et al., (2016). Green synthesis and characterization of monodispersed gold nanoparticles: toxicity study, delivery of doxorubicin and its bio-distribution in mouse model. *J. Biomed Nanotechnol., 12*, 165–81.

Muller, R. H., Mader, K., & Gohla, S., (2000). Solid lipid nanoparticles (SLN) for controlled drug delivery: A review of the state of the art. *European Journal of Pharmaceutics and Biopharmaceutics, 50*, 161–177.

Müller, R., Radtke, M., & Wissing, S., (2002). Solid lipid nanoparticles (SLN) and nanostructured lipid carriers (NLC) in cosmetic and dermatological preparations. *Advanced Drug Delivery Reviews, 54*, S131–S155.

Mutalik, S., Suthar, N. A., Managuli, R. S., Shetty, P. K., Avadhani, K., Kalthur, G., et al., (2016). Development and performance evaluation of novel nanoparticles of a grafted copolymer loaded with curcumin. *Int J Biol Macromol., 86*, 709–20.

Nagaraja, A. T., Pradhan, S., & McShane, M. J., (2014). Poly (vinylsulfonic acid) assisted synthesis of aqueous solution stable vaterite calcium carbonate nanoparticles. *Journal of Colloid and Interface Science, 418*, 366–372.

Nagarwal, R. C., Kant, S., Singh, P. N., Maiti, P., & Pandit, J. K., (2009). Polymeric nanoparticulate system: A potential approach for ocular drug delivery. *Journal of Controlled Release, 136*, 2–13.

Nazeruddin, G. M., Prasadb, N. R., Waghmarec, S. R., Garadkarb, K. M., & Mullad, I. S., (2014). Extracellular biosynthesis of silver nanoparticle using Azadirachta indica leaf extract and its anti-microbial activity. *J. Alloys Compd., 583*, 272–277.

Negi, L. M., Jaggi, M., & Talegaonkar, S., (2014). Development of protocol for screening the formulation components and the assessment of common quality problems of nanostructured lipid carriers. *International Journal of Pharmaceutics., 461*, 403–410.

Nguyen, D. H., Lee, J. S., Choi, J. H., Park, K. M., Lee, Y., & Park, K. D., (2016). Hierarchical self-assembly of magnetic nanoclusters for theranostics: Tunable size, enhanced magnetic resonance imagability, and controlled and targeted drug delivery. *Acta Biomater.*, *35*, 109–17.

Niaz, T., Nasir, H., Shabbir, S., Rehman, A., & Imran, M., (2016). Polyionic hybrid nano-engineered systems comprising alginate and chitosan for antihypertensive therapeutics. *Int J Biol Macromol.*, *91*, 180–187.

Obreja, L., Foca, N., Popa, M. I., & Melnig, V., (2008). Alcoholic reduction platinum nanoparticles synthesis by sonochemical method. *Biomaterials in Biophysics, Medical Physics and Ecology*, 31–36. Available from http://www.phys.uaic.ro/labs/comb/analele%20 stintifice/2008/5 obreja%20laura.pdf

Okuyama, K., & Lenggoro, I. W., (2003). Preparation of nanoparticles via spray route. *Chem. Eng. Sci.*, *58*, 537–47.

Pakrashi, S., Dalai, S., Humayun, A., Chakravarty, S., Chandrasekaran, N., & Mukherjee, A., (2013). *Ceriodaphnia dubia* as a potential bio-indicator for assessing acute aluminum oxide nanoparticle toxicity in fresh water environment. *PLoS One, 8*, e74003.

Pan, Z. Z., Wang, H. Y., Zhang, M., Lin, T. T., Zhang, W. Y., Zhao, P. F., et al., (2016). Nuclear-targeting TAT-PEG-Asp8-doxorubicin polymeric nanoassembly to overcome drug-resistant colon cancer. *Acta Pharmacol Sin.* 37,1110–1120.

Panigrahi, S., Kundu, S., Ghosh, S. K., Nath, S., & Pal, T., (2004). General method of synthesis for metal nanoparticles. *Journal of Nanoparticle Research, 6*, 411–414.

Papavassiliou, G. C., (1979). Optical properties of small inorganic and organic metal particles. *Prog Solid State Chem, 12*, 185–271.

Plech, A., Cerna, R., Kotaidis, R. V., Hudert, F., Bartels, A., & Dekorsy, T., (2007). A Surface Phase Transition of Supported Gold Nanoparticles. *Nano Letters, 7*, 1026–1031.

Puvvada, N., Rajput, S., Kumar, B. N., Sarkar, S., Konar, S., Brunt, K. R., et al., (2015). Novel ZnO hollow-nanocarriers containing paclitaxel targeting folate-receptors in a malignant pH-microenvironment for effective monitoring and promoting breast tumor regression. *Sci Rep., 5*, 11760.

Qiu, C., Wei, W., Sun, J., Zhang, H. T., Ding, J. S., Wang, J. C., et al., (2016). Systemic delivery of siRNA by hyaluronan-functionalized calcium phosphate nanoparticles for tumor-targeted therapy. *Nanoscale.* 8, 13033–13044.

Qu, F., & Morais, P. C., (2001). The pH dependence of the surface charge density in oxide-based semiconductor nanoparticles immersed in aqueous solution. *IEEE Transactions on Magnetics, 37*, 2654–6.

Radhakrishnan, R., Kulhari, H., Pooja, D., Gudem, S., Bhargava, S., Shukla, R., et al., (2016). Encapsulation of biophenolic phytochemical EGCG within lipid nanoparticles enhances its stability and cytotoxicity against cancer. *Chem Phys Lipids, 198*, 51–60.

Ramimoghadam, D., Bagheri, S., & Hamid, S. B. A., (2014). Biotemplated synthesis of anatase titanium dioxide nanoparticles via lignocellulosic waste material. *Biomed Res Int., 2014*, 205636.

Rao, J. P., & Geckele, K. E., (2011). Polymer nanoparticles: Preparation techniques and size-control parameters. *Progress in Polymer Science, 36*, 887–913.

Rasmussen, J. W., Martinez, E., Louka, P., & Wingett, D. G., (2010). Zinc oxide nanoparticles for selective destruction of tumor cells and potential for drug delivery applications. *Expert Opin Drug Deliv., 7*, 1063–1077.

Rawat, P., Ahmad, I., Thomas, S. C., Pandey, S., Vohora, D., Gupta, S., et al., (2016). Revisiting bone targeting potential of novel hydroxyapatite-based surface modified PLGA nanoparticles of risedronate: Pharmacokinetic and biochemical assessment. *Int J Pharm., 506*, 253–261.

Ray Chowdhuri, A., Bhattacharya, D., & Sahu, S. K., (2016). Magnetic nanoscale metal organic frameworks for potential targeted anticancer drug delivery, imaging and as an MRI contrast agent. *Dalton Trans.*, *45*, 2963–2973.

Roldán, M. V., Pellegri, N., & Sanctis, O., (2013). Electrochemical Method for Ag-PEG Nanoparticles Synthesis. *Journal of Nanoparticles*, 7.

Ryter, S. W., Kim, H. P., Hoetzel, A., Park, J. W., Nakahira, K., Wang, X., et al., (2007). Mechanisms of cell death in oxidative stress. *Antioxid Redox Signal.*, *9*, 49–89.

Sanna, V., Caria, G., & Mariani, A., (2010). Effect of lipid nanoparticles containing fatty alcohols having different chain length on the *ex vivo* skin permeability of Econazole nitrate. *Powder Technology*, *201*, 32–36.

Sasikala, A. R., Unnithan, A. R., Yun, Y. H., Park, C. H., & Kim, C. S., (2016). An implantable smart magnetic nanofiber device for endoscopic hyperthermia treatment and tumor-triggered controlled drug release. *Acta Biomater.*, *31*, 122–133.

Sharma, A. K., Sahoo, P. K., Majumdar, D. K., Sharma, N., Sharma, R. K., & Kumar, A., (2016). Fabrication and evaluation of lipid nanoparticulates for ocular delivery of a COX-2 inhibitor. *Drug Deliv*, 1–10.

Sharma, H., Kumar, K., Choudhary, C., Mishra, P. K., & Vaidya, B., (2016). Development and characterization of metal oxide nanoparticles for the delivery of anticancer drug. *Artificial cells, Nanomedicine and Biotechnology*, 44, 672–679.

Sharma, H., Mishra, P. K., Talegaonkar, S., & Vaidya, B., (2015). Metal nanoparticles: A theranostic nanotool against cancer. *Drug Discovery Today*, *20*, 1143–1151.

Sharma, S. K., Pujari, P. K., Sudarshan, K., Duttaa, D., Mahapatraa, M., Godbolea, S. V., et al., (2009). Positron annihilation studies in ZnO nanoparticles. *Solid State Communications.*, *149*, 550–554.

Shenoy, D. B., & Amiji, M. M., (2005). Poly (ethylene oxide)-modified poly (epsiloncaprolactone) nanoparticles for targeted delivery of tamoxifen in breast cancer. *International Journal of Pharmaceutics*, *293*, 261–270.

Shinde, N. C., Keskar, N. J., & Argade, P. D., (2012). Nanoparticles: Advances in drug delivery systems. *Research Journal of Pharmaceutical, Biological and Chemical Sciences*, *3*, 922.

Siddique, M. I., Katas, H., Amin, M. C., Ng, S. F., Zulfakar, M. H., & Jamil, A., (2016). In-vivo dermal pharmacokinetics, efficacy, and safety of skin targeting nanoparticles for corticosteroid treatment of atopic dermatitis. *Int J Pharm*, *507*, 72–82.

Siekmann, B., & Westesen, K., (1994). Thermoanalysis of the recrystallization process of melt–HOmogenized glyceride nanoparticles. *Colloids and surfaces B: Biointerfaces*, *3*, 159–175.

Siekmann, B., & Westesen, K., (1996). Investigations on solid lipid nanoparticles prepared by precipitation in o/w emulsions. *European journal of pharmaceutics and biopharmaceutics*, *42*, 104–109.

Singh, S., Pandey, V. K., Tewari, R. P., & Agarwal, V., (2011). Nanoparticle-based drug delivery system: Advantages and applications. *Indian Journal of Science and Technology*, *4*, 177–180.

Sironmani, A., & Daniel, K., (2011). Silver nanoparticles–universal multifunctional nanoparticles for bio sensing, imaging for diagnostics and targeted drug delivery for therapeutic applications. In Drug Discovery and Development–Present and Future, Kapetanović, I., (Ed.), InTech, 2011.

Souto, E., & Müller, R., (2006). Investigation of the factors influencing the incorporation of clotrimazole in SLN and NLC prepared by hot high-pressure homogenization. *Journal of microencapsulation*, *23*, 377–388.

Thomas, S. C., Harshita, Mishra, P. K., Talegaonkar, S., (2015). Ceramic nanoparticles: Fabrication methods and applications in drug delivery. *Current Pharmaceutical Design, 21*, 6165–6188.

Truong, N. P., Whittaker, M. R., Mak, C. W., & Davis, T. P., (2015). The importance of nanoparticle shape in cancer drug delivery. *Expert Opin Drug Deliv, 12*, 129–42.

Tummala, S., M N, S. K., & Pindiprolu, S. K., (2016). Improved anti-tumor activity of Oxaliplatin by encapsulating in anti-DR5 targeted gold nanoparticles. *Drug Deliv*, 1–31.

Tzaphlidou, M., (2008). Bone architecture collagen structure and calcium/phosphorus maps. *J Biol Phys., 34*, 39–49.

Uskokovic, V., & Desai, T. A., (2013). Phase composition control of calcium phosphate nanoparticles for tunable drug delivery kinetics and treatment of osteomyelitis. I. Preparation and drug release. *J Biomed Mater Res Part A., 101A*, 1416–1426.

Venkatasubbu, G. D., Ramasamy, S., Reddy, G. P., & Kumar, J., (2013). *In vitro* and *in vivo* anticancer activity of surface modified paclitaxel attached hydroxyapatite and titanium dioxide nanoparticles. *Biomed Microdevices, 15*, 711–26.

Vyas, D., Lopez-Hisijos, N., Gandhi, S., El-Dakdouki, M., Basson, M. D., Walsh, M. F., et al., (2015). Doxorubicin-hyaluronan conjugated super-paramagnetic iron oxide nanoparticles (DOX-HA-SPION) enhanced cytoplasmic uptake of doxorubicin and modulated apoptosis, IL-6 Release and NF-kappaB activity in human MDA-MB-231 breast cancer cells. *J Nanosci Nanotechnol., 15*, 6413–22.

Wang, B., Yang, Q., Wang, Y., & Li, Z., (2014). The toolbox of designing nanoparticles for tumors. *Mini Rev Med Chem., 14*, 707–16.

Wang, R., Gu, X., Zhou, J., Shen, L., Yin, L., Hua, P., et al., (2016). Green design "bioinspired disassembly-reassembly strategy" applied for improved tumor-targeted anticancer drug delivery. *J Control Release, 235*, 134–146.

Wang, T., Jiang, H., Wan, L., Zhao, Q., Jiang, T., Wang, B., et al., (2015). Potential application of functional porous TiO2 nanoparticles in light-controlled drug release and targeted drug delivery. *Acta Biomater., 13*, 54–63.

Wang, X., Zhang, R., Wu, C., Dai, Y., Song, M., Gutmann, S., et al., (2007). The application of Fe_3O_4 nanoparticles in cancer research: a new strategy to inhibit drug resistance. *J Biomed Mater Res A., 80*, 852–60.

Westesen, K., & Siekmann, B., (1997). Investigation of the gel formation of phospholipid-stabilized solid lipid nanoparticles. *International Journal of Pharmaceutics, 151*, 35–45.

Wolfram, J., Zhu, M., Yang, Y., Shen, J., Gentile, E., Paolino, D., et al., (2015). Safety of Nanoparticles in Medicine. *Curr Drug Targets. 16*, 1671–1681.

Xiong, H., Du, S., Ni, J., Zhou, J., & Yao, J., (2016). Mitochondria and nuclei dual-targeted heterogeneous hydroxyapatite nanoparticles for enhancing therapeutic efficacy of doxorubicin. *Biomaterials, 94*, 70–83.

Yang, Q., Li, L., Sun, W., Zhou, Z., & Huang, Y., (2016). Dual stimuli-responsive hybrid polymeric nanoparticles self-assembled from poss-based star-like copolymer-drug conjugates for efficient intracellular delivery of hydrophobic drugs. *ACS Appl Mater Interfaces, 8*, 13251–13261.

Yih, T. C., & Al-Fandi, M., (2006). Engineered nanoparticles as precise drug delivery systems. *J. Cell. Biochem., 97*, 1184–1190.

Yildiz, U., & Landfester, K., (2008). Miniemulsion polymerization of styrene in the presence of macromonomeric initiators. *Polymer, 49*, 4930–4934.

Zawadzki, M., & Wrzyszcz, J., (2000). Hydrothermal synthesis of nanoporous zinc aluminate with high surface area. *Mater. Res. Bull., 35,* 109–114.

Zayat, M., & Levy, D., (2000). Blue CoAl$_2$O$_4$ Particles Prepared by the Sol–Gel and Citrate–Gel Methods. *Chem. Mater*, 12, 2763–2769.

Zhang, J., Li, Y., Xie, H., Su, B. L., Yao, B., Yin, Y., et al., (2015). Calcium Carbonate Nanoplate Assemblies with Directed High-Energy Facets: Additive-free Synthesis, High Drug Loading, and Sustainable Releasing. *ACS Appl Mater Interfaces*, 7, 15686–15691.

Zhang, L., (2004). Preparation of Multi-Component Ceramic Nanoparticles. Available from: http://api.ning.com/files/9ACOqW6CjadaR9MBpqWJnw1wHavp3UEvR9T9LseD7S j2NrQUm8lWPkeFRBqqJ0⊡4XSIFf6Zlsn*dS236dq1pUuVIwkdp9lie/Multio⊡ides_ preperation_Zhang.pdf

Zhou, J., Xu, N., & Wang, Z. L., (2006). Dissolving behavior and stability of ZnO wires in biofluids: A study on biodegradability and biocompatibility. *Advanced Materials, 18,* 2432–2435.

Zimmermann, E., Souto, E. B., & Muller, R. H., (2005). Physicochemical investigations on the structure of drug-free and drug-loaded solid lipid nanoparticles (SLN) by means of DSC and 1H NMR. *Die Pharmazie, 60,* 508–513.

CHAPTER 3

PHYSICAL AND CHEMICAL PROPERTIES OF NANOBIOMATERIALS

SUMITRA NAIN,[1] GARIMA MATHUR,[2] PRAGATII KARNA,[2] and HARI OM SINGH[3]

[1]Department of Pharmacy, Banasthali University, Banasthali, Rajasthan, 304022, India, E-mail: nainsumitra@gmail.com

[2]Department of Pharmacy, Galgotia University, Gautam Budh Nagar, U.P., India

[3]National AIDS Research Institute, Pune, India

CONTENTS

ABSTRACT

Nanobiomaterials are biocompounds that are nano in size and have several biological applications. The physical and chemical properties of these nanobiomaterial varies due to their nano size thus effect of these factors have been discussed below with example in detail. This chapter also comprises of different biological applications of nanobiomaterial with examples.

3.1 INTRODUCTION

The research and development in biomaterials is emerging rapidly after the discovery of nanomaterials. With the development of nanomaterials, nano-biomaterials have become topic of concern throughout the world these days. A talk titled "There's Plenty of Room at the Bottom" in 1959 by a physicist Richard P. Feynman expressed his vision on nanotechnology in a meeting of American Physics Society (Tahan, 2007). Although his idea was too revolutionary and unrealistic at that time, thus hard to believe that it will turn out to be multidisciplinary field of applied science and technology of today. In 1974 Japanese scientist Norio Taniguchi first coined the term nanotechnology at a conference (Taniguchi, 1974). In biomedical field nanaomaterial has gained attention due to the nano dimensionality (Li, et al., 2006, 2009 a,b). Since that time this new area of science grabbed interest of different scientist leading to development of different nanobiomaterial such as metallic nanobiomaterial, ceramic nanobiomaterial, semiconductor-based nanobiomaterial, lipoprotein-based nanobiomaterial, etc. (Sitharaman, 2011).

Biological activity of nanobiomaterials depends upon the nature and function of drug as well as physical and chemical properties of materials. Biomaterials with their unique physical properties and chemical properties can control the toxicity and may regenerate tissues. Today biomaterials are being used for various medical purposes like drug delivery to targeted site of action, tissue engineering, medical imaging, etc. (Peppas and Langer, 1994). Through blood nanobiomaterials can be transferred and start accumulating in secondary organs of body such as liver, placenta, the cardiovascular system, spleen, kidney, and central nervous system (CNS), where they may cause adverse effects (Dusinska et al., 2013; Li, et al., 2010, 2011). Nanomaterials can be organic or inorganic materials. Biomaterial can be obtained from natural sources or can be synthesized from laboratory practices using chemical components like ceramics, polymers, etc.

Nanobiomaterial act completely different in comparison to larger size particles thus it become difficult to predict their properties. Nanoparticles attract with each other differently as compared to the larger particles. So, depending on the different interactions of nanoparticles with different chemical entity it becomes complicated and result is unknown. Nanoparticles are mostly used in aerosols, suspensions and emulsions. Various field of researcher's are involved in this technology. The new physical and chemical entity in sample preparation is provoking the researchers to frame and introduce a set of new designed parameters in the field of nanoparticles. The outcomes from the development of the parameters have enhanced the study of surface effect, hygroscopic features, and non-linear susceptibility (Hartwig, 2013). Preparation of biomaterials may involve semiconductors, metals, insulators, polymers, dielectric or other organic compounds, etc. (Hartwig, 2013). In this chapter we are going to discuss about the physical and chemical properties of nanobiomaterials:

3.2 PHYSICAL AND CHEMICAL PROPERTIES

Development of physical and chemical properties in nanomaterials has completely changed the view of nanoscience. Various physical approaches are considered in nanoscience at dimensional level.

3.2.1 PARTICLE SIZE

For drug delivery at targeted site, particle size is curiously being studied for development in nanobiomaterials. Various methods are determined to synthesize micro and nano size uniformity of size, (Napier, et al., 2007; Tao, et al., 2006) micro fauna, (Moghimi et al., 2005) emulsion polymerization, (Clark, et al., 1999) and self-assembly (Berkland et al., 2001). The parameters like mixing method, capillary diameter and particles flow rate can be used to determine the physical properties of different polymer and surfactant.

Particle size is an important factor for in-vivo studies which is involves clearance, elimination, absorption, circulation, flow properties, degradation, metabolism, etc. (Gao, et al., 2005; Kohane, et al., 2007; Ravi, 2000, Singh, et al., 2007). Micro particles are imprisoned in capillary beds by kupffer cells present in the liver but nanoparticles lesser than 100 nm is permeable to endothelial surface and leave the blood vessels (Dusinska et al., 2013; Stolnik, et al., 1995; Tabata, et al., 2010). In the study of gold material, the individual particles are being studied and found that the particle diameter having 50 nm have high uptake rates in glycoviruses (Osaki. et al., 2004).

Study found particles having diameter of 40nm can active dendritic cells at lymph node. Particles can easily penetrate biological blockade by active transport mechanism and passive diffusion. This type of mechanisms affects the particle size distribution and clearance in living organism cells. The particle size having range between 2 and 3 µm are capable for drug delivery to pulmonic site (Champion, et al., 2008). Micro particles are not considered to be used in intravascular injection because it eliminates the drug rapidly from systemic circulation.

When formulating a specific formulation size distribution is another important factor to be analyzed parallel to particle diameter (Gratton, et al., 2007) as well as nanoparticles use also depend upon their stability against dilution, time and other environmental factor (Anton, et al., 2008).

3.2.2   SHAPE

In the field of biotechnology shape is also found to be another important regulatory property in having desirable cellular responses (Decuzzi, et al., 2008a,b; Decuzzi, et al., 2006). In drug delivery system shape regulates the phagocytosis by macrophages (Champion, et al., 2006). According to studies disc shape nanoparticles are preferred over sphere shaped for example: Muro et al. (2008) on comparison between the sphere shaped targeted accumulation in tissues of various diameters and elliptical disc shaped micro scale dimensions found better efficiency in targeting in case of disc than that of spheres. Long cylindrical micelles were found to have prolonged activity in some studies which can be occurred because of their shape (Geng, et al., 2007). Non-spherical particles enhance their activity by prolonging the half life as their removal from liver and spleen (Moghini, et al., 2001) is reduced as they get stack on increasing the flow.

3.2.3   SURFACE AREA

One of the important deciding factors in cell morphology, adhesion or motility (Dalby, et al., 2004) is found to be the surface area. Nanoparticles intercommunicate with various number of tissue–blood barriers. As the particle size decreases, the surface area of the particles increases which simultaneously results the drug to penetrate the encapsulated nanoparticle matrix and absorb faster at targeted site. The surface properties notify the broad area of optical properties which ultimately affect the biomedical applications. So, by reducing the particle size of nanobiomaterials we can, ultimately modify the surface area and drug absorption. Biomaterial with nanostructured surfaces proved to have versatile usage in biomedical applications such as carriers for high throughput screening (Pregibon, et al., 2007) and molecular imaging (Yoshida, et al., 2007). When compared between same sized surfaces patterned and non-patterned particles drastic difference was found (Verma, et al., 2008). Surface chemistry of particles also affects their optical properties which plays significant role in bio-imaging applications. Simple change in size of a particle may result in variation of surface area, these modifications can be obtained by two methods:

1. Controlled in situ modification: leads to formation of non-aggregated nanoparticles.
2. Post synthesis modification: helps in obtaining nanoparticle more easily. (Iijima, et al., 2009).

3.2.4 COMPARTMENTALIZATION

Compartmentalization is an important function of living organism cells which differentiate them from slightly developed form of lives. Organelles are formed by compartmentalization of internal eukaryotic cells which vary different sizes between 10–25 nm (ribosomes) to 100–500 nm (lysosomes, endosomes) which may interchange their nutrition from various transport processes. Due to the separation of biomolecules metabolic reactions are easily controlled for well functioning of cell (Mitragotri, et al., 2009). Examples of compartmentalization having multiple functions are:

Multicompartment particles can be entitled in the form of capsules having shell compartment which enhanced drug delivery, for example, core–shell component (Loscertales, et al., 2002; Roh, et al., 2005; Sengupta, et al., 2005). Emulsion polymerization and self-associative process for liposomes and polymerosomes are progressively been used in the core-shell compartment preparation for biomedical applications.

Multicompartment shows specific property that each compartment is self-associatively loaded with various biological components like growth factors, protein and anti-bacterial biomaterials (Ferrari, 2005). Studies on cell structure and biomedical applications can be useful in the field of drug delivery system or molecular imaging with the help of multicompartment particles.

3.2.5 COLOR

When the material is converted to nanoparticles it shows various colors. For example, gold particles show red color when converted from gold materials. Light strongly interacts with gold nanoparticles due to particle size of gold materials. Particle size ranging between 2~150 nm shows greater surface densities which is known as surface plasmons. At specific wavelength, light interact with surface plasmons and get excited due to continuous oscillation. This type of oscillation is produced by surface plasmons resonance (SPR). Uniform gold nanoparticles at scale bar ~30 nm shows rich red color when the blue-green portion is well absorbed and red light got reflected by surface plasmons resonance (SPR) (Lubick, et al., 2008; Vollath, et al., 2008).

3.2.6 MELTING POINT

When particle sizes ranges in nano scale it may leads to a stricking fall in the melting point of a material. A hundred degree fall in temperature is found while melting material of same element in nanoscale and in bulk sizes for example these variations are prominently found in case of nanowires, nanotubes and nanoparticles. This variation in melting point is observed as in case of nanobiomaterial the surface to volume ratio is much more than the bulk particles leads to change in their thermal properties and thermodynamics (Lubick, et al., 2008; Vollath, et al., 2008).

3.2.7 SOLUBILITY

Solubility is one of the major aspects in the field of nanobiomaterial. Approximately, 40% of pharmaceutical active compound detected via conjuctional screening programs were difficult to formulate due to water insolubility problems (Lipinski, et al., 2000; Lipinski, et al., 2002; Lipper, 1999).

Drugs which are hydrophobic or lipophilic in nature have ability to penetrate the biological membrane and drug molecules get readily absorbed. The main point of view for targeting these issues was to solve and develop various salts of appallingly water-soluble molecule to tweak solubility without affecting its biological activity. For improving solubility of prodrugs or analogs, screening methods are ultimately used. Formulation applications will be accepted in nanoparticle production if it succeeded.

The applications used in the formulation are not succeeded yet but the products are available with suboptimal properties including problems like:

- poor patient compliance.
- additional excipients limit the dose up swinging.
- inadequate optimal dosing.
- lack of gratify/fasted equivalence, etc.

When these types of problems are faced then generally nanoparticles development are commonly useful and is providing an opportunities to re-energize the marketed products with sub-optimal delivery.

There are various technologies which is being used for the formulation of nanoparticles for overcoming poorly water soluble drugs (Horn, et al., 2001; Muller, et al., 2001; Merisko-Liversidge, 2004) and they are as follows:

- a building-up approach via synthesis (Liu, et al., 1999).
- self-assembly association (Letchford, et al., 2007).
- precipitation of drug molecules (Horn, et al., 2001).
- drug-fragmentation processes such as homogenization (Liedtke, et al., 2000).

For enhancing the dissolution rate and for maintaining the pharmacokinetic parameters of dosage form, surface area and surface interactions should be elevated. This type of enormous increment in surface activity can result in nanometer particle size to extempore aggregation in a thermodynamically stable manner.

Despite of these chemical and physical properties it has been also found that nanobiomaterial are also affected by some electrical, mechanical and optical properties such as conduction, dielectric constant, quantum number, etc.

3.2.8 DRUG DELIVERY, TARGETED SITE, AND IMAGING

Nanomaterials are cross-linked structure of various molecules that are either physically or chemically assembled ranging from 1–1000 nm. Drug, biomolecules, etc. are compacted molecules which can be readily dissolved, absorbed or dissipated via forge (like nanospheres) or compacted to an oily phase or aqueous phase which is enclosed by shell like barrier (like nanocapsule). The enclosed molecules are carried to targeted site of infection without interfering the diffusion process of blood brain barrier of one's body. When the drug is liberated in matrix via dissipation, bulging or degenerated will show its therapeutic response in clinical background of nanoparticles. For increment of efficacious drug delivery, nanoparticles must unify to targeted molecules by orders of magnitude (Debbage, et al., 2010).

There are few limitations in the research field of nanomedicine such as:
- resolution limits for identification of biomolecules of microorganism.
- Mechanism of drug delivery.
- restrained defilement, etc.

The molecular level of biology and medicine must be understood to figure out the problems of limitations and others (Parveen, et al., 2011). Recently, according to studies the increment in therapeutic efficacy is seen in drug delivery system and imaging of nanoparticles which can be determined, identified, delivered at targeted site of action and monitored the ways and efficacy of treatment before, now, after treatment control and its development (Bhojani, et al., 2011; Janib, et al., 2010; Ma, et al., 2011; Yoo, et al., 2011).

For formulation of each dose therapy targeted techniques are commonly used. This technique is very useful in complex multi-compartment entity (Debbage, et al., 2010).

Examples of targeted delivery system of nanobiomaterials in anticancer therapy: nanobiomaterials are being engineered to prepare nanoplatforms having capacity to identify the receptors at cellular level which are having cancer-specific nature as well as to deliver anticancerous drug such as doxorubicin, paclitaxel, camptothecins, etc. at the site of abnormality. These nano platforms also play vital role in nanoimaging, real-time intraoperative imaging and bimodal imaging (Su-Eon Jin, et al., 2014).

3.2.9 SURFACE CHARGE

Surface Charge plays a significant role in determining the interaction of nanoparticles with the biological system. Surface charge is responsible for regulating a number of different aspects such as blood brain barrier integrity, transmembrane permeability, colloidal behavior, adsorption of nanoparticles and plasma protein binding. Relevant cellular uptake was shown by positively charged nanoparticles in comparison to negative and neutral nanoparticles whereas negatively charged nanoparticles permeate skin after dermal administration but no such effects were seen for positively charged and neutral particles irrespective of their sizes (Gatoo, et al, 2014).

3.3 APPLICATIONS OF NANOBIOMATERIALS

1. Used in dentistry: several nanobiomaterials has found their application in dentistry such as carbon nanotubes-based material, dental and skeletal applications of silica-based nanobiomaterial, nanobiomaterials are also used in preventive and restorative dentistry as well as in control of oral bio films.
2. Used in Orthodontics.
3. Used in Prosthodontics.
4. Used in Endodontics.
5. Used in orthopaedics.
6. Nanobiomaterials also found use in Periodontics, Implant Dentistry, and Dental Tissue Engineering (Table 3.1).

3.4 CONCLUSION

A number of studies have been made till date to understand the effect of physical and chemical properties on nanoparticles and biomaterials but its effect on nanobiomaterial is still a matter of curiosity. Now-a-days

TABLE 3.1 Applications of Nanobiomaterials Entities

S.No.	Nanobiomaterials	Application(s)
1.	Au@protein bio-nanoconjugates	Exploration of key chemistry of the biomolecule
2.	Chymotrypsin (CHT) NPs	Effectual host for organic dye Methylene Blue (MB) revealing the molecular recognition
3.	Nano–biocomposite	Photocatalytic activity
4.	Dexamethasone-loaded nanofibers	Drug carriers and scaffolds for tissue engineering
5.	Cu_{QC}@BSA bio-nano-conjugates	Toxic metal ion sensing
6.	Ni–CHT bio-nanoconjugates	Exploration of key chemistry of the biomolecule

it has been established that a profound impact is found on biomaterials due to the physical properties, such as surface texture, size, mechanical properties, shape and compartmentalization, when exposed to biological environment. These studies have opened the channel for further development of nanobiomaterials which help in biomedical applications. In past few years, a number of processes have been developed for the manufacturing of nanoparticles and to control the effect of physical and chemical properties. These designing of nanobiomaterial with controlled physical and chemical properties have enhanced the scope of future research in this field.

KEYWORDS

- color
- compartmentalization
- nanobiomaterial
- physical and chemical properties
- size
- surface area
- surface charge

REFERENCES

Anton, N., Benoit, J. P., & Saulnier, P., (2008). Design and production of nanoparticles formulated from nano-emulsion templates A review. *J. Controlled Release, 128*, 185–199.

Berkland, C., Kim, K., & Pack, D. W., (2001). Fabrication of PLG microspheres with precisely controlled and monodisperse size distributions. *J. Control. Release., 73*, 59–74.

Bhojani, M. S., Van Dort, M., & Rehemtulla, B. D., (2011). Targeted imaging and therapy of brain cancer using theranostic nanoparticles. *Ross. Mol. Pharm., 7*, 1921.

Champion, J. A., & Mitragotri, S., (2006). Role of target geometry in phagocytosis. *Proc. Natl Acad. Sci., 103*, 4930–4934.

Champion, J. A., Walker, A., & Mitragotri, S., (2008). Role of particle size in phagocytosis of polymeric microspheres. *Pharm. Res., 25*, 1815–1821.

Clark, H. A., Hoyer, M., Philbert, M. A., & Kopelman, R., (1999). Fabrication, characterization, and methods for intracellular delivery of pebble sensors. *Anal. Chem., 71*, 4831–4836.

Dalby, M., Riehle, M. O., Sutherland, D. S., Agheli, H., & Curtis, A. S., (2004). Use of nano-topography to study mechanotransduction in fibroblasts-methods and perspectives. *Eur. J. Cell Biol., 83*, 159–169.

Debbage, P., & Thurner, G. C., (2010). Nanomedicine Faces Barriers. *Pharmaceuticals, 3*, 3371.

Decuzzi, P., & Ferrari, M., (2006). The adhesive strength of non-spherical particles mediated by specific interactions. *Biomater., 27*, 5307–5314.

Decuzzi, P., & Ferrari, M., (2008a), The receptor-mediated endocytosis of non-spherical particles. *Biophys. J. 94*, 3790–3797.

Decuzzi, P., Pasqualini, R., Arap, W., & Ferrari, M., (2009). Intravascular delivery of particulate systems: Does geometry really matter?, *Pharm. Res.,* 2–6(1), 235–243.

Dusinska, M., Magdolenova, Z., & Fjellsbo, L., (2013). Toxicological aspects for nanomaterial in humans. *Methods in Molecular Biology. 948*, 1–12.

Ferrari, M., (2005). Cancer nanotechnology: Opportunities and challenges. *Nature Rev. Cancer, 5*, 161–171.

Gao, H., Shi, W., & Freund, L. B., (2005). Mechanics of receptor-mediated endocytosis. *Proc. Natl Acad. Sci., 102*, 9469–9474.

Gatoo, M. A., Naseem, S., Arfat, M. Y., Dar, A. M., Qasim, K., & Zubair, S., (2014). Physicochemical properties of nanomaterial: implication in associated toxic manifestations. *BioMed Research International, 2014.*

Geng, Y., Dalhaimer, P., Cai, S., Tsai, R., Tewari, M., & Minko, T., (2007). Shape effects of filaments versus spherical particles in flow and drug delivery. *Nature Nanotech, 2*, 249–255.

Giri, A., Goswami, N., Sarkar, S., & Pal, S. K. Bio-Nanomaterials: Understanding Key Biophysics and Their Applications. *Nanotechnology, 11*, 41–110.

Gratton, S. E. A., (2007). Nanofabricated particles for engineering drug therapies: A preliminary biodistribution study of nanoparticles. *J. Control. Rel., 121*, 10–18.

Horn, D., & Rieger, J., (2001). Organic nanoparticles in the aqueous phase-theory, experiment, and use. *Angew Chem Int., 40*, 4330–61.

Iijima, M., & Kamiya H., (2009). Surface modification for improving the stability of nanoparticles in liquid media. *KONA-Powder and Particle, 27* , 119–129.

Illum, L., (1982). Blood clearance and organ deposition of intravenously administered colloidal particles: the effects of particle-size, nature and shape. *Int., J. Pharm. 12*, 135–146.

Janib, S. M., Moses, A. S., MacKay, J. A., (2010). Imaging and drug delivery using theranostic nanoparticles. *Adv. Drug Deliv. Rev., 62*, 1052–1063.

Kohane, D. S., (2007). Microparticles and nanoparticles for drug delivery. *Biotechnol. Bioeng. 96*, 203–209.

Letchford, K., & Burt, H., (2007). A review of the formation and classification of amphiphilic block copolymer nanoparticulate structures: Micelles, nanospheres, nanocapsules and polymerosomes. *Eur J Pharm Biopharm., 65*, 259–69.

Li, X., Fan, Y., & Watari, F., (2010). Current investigations into carbon nanotubes for biomedical application. *Biomed. Mater, 5*(2), 12.

Li, X., Feng, Q., Liu, X., Dong, W., & Cui, F., (2006). Collagen-based implants reinforced by chitin fibers in a goat shank bone defect model. *Biomaterials, 27*(9), 1917–1923.

Li, X., Gao, H., & Uo, M., (2009a). Effect of carbon nanotubes cellular functions in vitro. *J. Biomed. Mater. Res. 91*(1), 132–139.

Li, X., Gao, H., & Uo, M., (2009b). Maturation of osteoblast-like SaoS$_2$ induced by carbon nanotubes. *Biomed. Mater. 4*, 1.

Li, X., Liu, X., Huang, J., Fan, Y., & Cui, F. Z., (2011). Biomedical investigation of CNT-based coatings. *Surf. Coat Technol*, *206*(4), 759–766.

Liedtke, S., Wissing, S., Muller, R. H., & Mader, K., (2000). Influence of high pressure homogenization equipment on nanodispersions characteristics. *Int J Pharm.*, *160*, 229–37.

Lipinski, C. A., (2000). Drug-like properties and the causes of poor solubility and poor permeability. *J Pharm Tox Meth.*, *44*, 235–49.

Lipinski, C., (2002). Poor aqueous solubility: An industry wide problem in drug discovery. *Am Pharm Rev.*, *5*, 82–5.

Lipper, R. A., (1999). E-pluribus product. *Mod. Drug Discovery, 2*, 55–60.

Liu, M., & Frechet, J. M. J., (1999). Designing dendrimers for drug delivery. *Pharmaceutical Science and Technology Today, 2*, 393–401.

Loscertales, I. G., Barrero, A., Guerrero, I., Cortijo, R., Marquez, M., & Gañán-Calvo, A. M., (2002). Micro/nano encapsulation via electrified coaxial liquid jets. *Sci.*, *295*, 1695–1698.

Lubick, N., & Betts, K., (2008). Silver socks have cloudy lining. *Environ Sci Technol.* *42*(11), 3910.

Ma, X., Zhao, Y., Liang, X. J., (2011). Theranostic nanoparticles engineered for clinic and pharmaceutics. *Acc. Chem. Res.*, *44*(10), 1114–22.

Md. Abidian, R., & Martin, D. C., (2009). Multifunctional Nanobiomaterials for Neural Interfaces. *Adv. Funct. Mater.*, *19*, 573–585.

Merisko-Liversidge, E., Liversidge, G. G., & Cooper, E. R., (2004). Nanosizing: a formulation approach for poorly water-soluble compounds. *Eur J Pharm Sci.*, *18*, 113–20.

Mitragotri, S., & Lahann, J., (2009). Physical approaches to biomaterial design. *Nature Mater.*, *8*, 15–23.

Moghimi, S. M., Hunter, A. C., & Murray, J. C., (2001). Long-circulating and target-specific nanoparticles: Theory to practice. *Pharmacol. Rev.*, *53*, 283–318.

Moghimi, S. M., Hunter, A. C., & Murray, J. C., (2005). Nanomedicine: Current status and future prospects. *Faseb, J. 19*, 311–330.

Muller, R. H., Jacobs, C., & Kayser, O., (2001). Nanosuspensions as particulate drug formulations in therapy: Rationale for development and what we can expect in the future. *Adv Drug Delivery Rev. 47*, 3–19.

Muro, S., (2008). Control of endothelial targeting and intracellular delivery of therapeutic enzymes by modulating the size and shape of ICAM-1-targeted carriers. *Mol. Ther*, *16*, 1450–1458.

Napier, M. E., & DeSimone, J. M., (2007). Nanoparticle drug delivery platform. *Polym. Rev. 47*, 321–327.

Osaki, F., Kanamori, T., Sando, S., Sera, T., & Aoyama, Y., (2004). A quantum dot conjugated sugar ball and its cellular uptake. On the size effects of endocytosis in the sub viral region. *J. Am. Chem. Soc.*, *126*, 6520–6521.

Parveen, S., Misra, R., & Sahoo, S. K., (2011). Nanoparticles: a boon to drug delivery, therapeutics, diagnostics and imaging. *Nanomedicine, 8*(2), 147–66.

Peppas, N. A., & Langer, R., (1994). New challenges in biomaterials. *Sci.*, *263*, 1715–1720.

Pregibon, D. C., Toner, M., & Doyle, P. S., (2007). Multifunctional encoded particles for high-throughput biomolecule analysis. *Sci.*, *315*, 1393–1396.

Prof. Dr. Andrea Hartwig., (2013). Commission for the Investigation of Health Hazards of Chemical Compounds in the Work Area, Wiley-VCH Verlag GmbH and Co. KGaA: Weinheim.

Ravi Kumar, M. N., (2000). Nano and micro particles as controlled drug delivery devices. *J. Pharm. Pharm. Sci.*, *3*, 234–258.

Roh, K. H., Martin, D. C., & Lahann, J., (2005). Biphasic Janus particles with nanoscale anisotropy. *Nature Mater.* *4*, 759–763.

Sengupta, S., Eavarone, D., Capila, I., Zhao, G., Watson, N., Kiziltepe, T., et al., (2005). Temporal targeting of tumor cells and neovasculature with a nanoscale delivery system. *Nature, 436*, 568–572.

Singh, M., Chakrapani, A., & O'Hagan, D., (2007). Nanoparticles and micro particles as vaccine-delivery systems. *Expert Rev. Vaccines 6*, 797–808.

Sitharaman, B., (2011). Nanobiomaterials: Current status and future prospects. In *Nanobiomaterials Handbook,* Ed., CRC Press, Taylor and Francis Group,. I-1–I-15.

Stolnik, S., Illum, L., & Davis, S. S., (1995). Long circulating micro particulate drug carriers. *Adv. Drug Delivery Rev., 16*, 195–214.

Su-Eon Jin., Hyo-Eon Jin., & Soon-Sun Hong., (2014). Targeted Delivery System of Nanobiomaterials in Anticancer Therapy: From Cells to Clinics. *BioMed Research International, 2014, 814208.*

Tabata, Y., & Ikada, Y., (1990). Phagocytosis of polymer microspheres by macrophages. *Adv. Polym. Sci., 94*, 107–141.

Tahan, C., (2007). Identifying Nanotechnology in Society. *Adv Comput, 71*, 271.

Taniguchi N., (1974). In Society of Precision Engineering, Proceedings of the International Conference on the basic concept of nano-technology, Tokyo, Japan.

Tao, S. L., Popat, K., & Desai, T. A., (2006). Off-water fabrication and surface modification of asymmetric 3D SU-8 microparticles. *Nat. Protocols. 1*, 3153–3158.

Verma, A., Uzun, O., Hu, Y., Hu, Y., Han, H., Watson, N., S., et al., (2008). Surface-structure-regulated cell-membrane penetration by monolayer-protected nanoparticles. *Nature Mater., 7*, 588–595.

Vollath, D., (2008). Nanomaterials: An introduction to synthesis, properties and application. *Environmental Engineering and Management Journal, 7*(6), 865–870.

Yoo, D., Lee, J. H., & Shin, T. H., Cheon (2011). Theranostic magnetic nanoparticles. *J. Acc. Chem. Res., 44*(10), 863–74.

Yoshida, M., Roh, K. H., & Lahann, J., (2007). Short-term biocompatibility of biphasic nano-colloids with potential use as anisotropic imaging probes. *Biomater., 28*, 2446–2456.

PART II

DRUG DELIVERY AND NANOTECHNOLOGY

CHAPTER 4

INTRODUCTION TO DRUG DELIVERY

MILA A. EMERALD

Phytoceuticals International™ and Novotek Global Solutions™, Ontario, Canada, E-mail: drmilaemerald@gmail.com

CONTENTS

4.1   INTRODUCTION

A visionary idea about future of science which is potentially leading to an operation of human life and health by machines varying in size and operations, and which can initiate the creation of a super tiny molecular and atomic size machines vital for further development and advance in human life, science and medicine has finally become a reality about 20 years later. Nanoscience, which is now targeting a multiple health, research and industrial applications, and especially nanomedicine has been progressively developing since 1970 (Caruso et al., 2012), and lead to creation and production of a large variety of molecular vehicles and super miniature biomarkers (Diehl et al., 2012; Ngoepe et al., 2013; Suhaimi and Jalaludin, 2015).

A variety of conventional drugs are powerful and tested pharmacological weapon against a different disease, which can be administered orally, intravenously, locally, transdermally, etc. But there are still many important obstacles which diminish the drugs therapeutic efficiency, such as drug solubility, stability, degradation (environmental or enzymatic), clearance rates, non-specific to□icity, inability to cross biological barriers, which are almost solved via creation and further development of the drug delivery systems (DDS) (Landau et al., 1994; Van der Pol et al., 2012; Soppimath and Aminabhavi, 2001; Friedman et al., 2013; Nitta and Numata, 2013; Ramishetti and Huang, 2012).

The DDS which represents an intelligent and advanced therapeutics, are designed to transport the molecules of drug to the specific and desirable biological target/s, with minimal or even no side effects. This process is-based on a total control over kinetics of therapeutic molecules release, and protecting of active ingredient with delivery and bio distribution which has minimal or even no side effects. In spite of the large variety of the DDS which are successfully developed and tested during the last decade and used for a different health conditions, there is still a lots of work need to be done to improve the precision of the target pharmaceutical and herbal drug delivery, their proper absorbance, effective distribution, creation of the selective defense mechanism against different types of pathological conditions, which are including but not limited to tumors, viral disease and parasitic infestations (Wang et al., 2014; Chen et al., 2013; Kanwar et al., 2012).

The design and development of DDS has been associated with a variety of problems, including the product development. There is a huge demand for a new type of non-toxic and non–immunogenic DDS candidates which are expected to be highly effective in vivo, which can be super small in size, be well designed to find and selectively target cellular signaling pathways, and can be used for proper diagnosis and assimilation in good standing with other pharmaceuticals and bio active molecules without any unwanted interaction and side effects. This chapter represents a brief review of the impressive variety of the DDS available in the market, their route of administration, challenges and a potential development.

4.2 CONTROLLED RELEASE SYSTEMS (CRS) AND DRUG DELIVERY METHODS

Bringing a novel drug to the International market including a complex process including drugs discovery, development, clinical testing, regulatory approval and a first production in average can cost pharmaceuticals companies from $4 to $11 billion (Herper, 2012). Administration techniques for the conventional drug, which have a variety of associated problems including a difficulty of maintain the safe and effective concentration of a drug in a blood, become much more simplified and more effective is-based on specific controlled-release delivery systems, such as encapsulated cells, biodegradable drugs loaded nanoparticles, lozenges, gels, aerosol sprays, polymers, solid implants which are able to sense and adopt to systemic and environmental changes and sustain continuous controlled drug release for a few month (Lee and Li, 2010). The drug delivery systems can be divided into three main groups: targeted, controlled and modulated drug delivery systems. The initial controlled drug delivery ideas which were implemented and developed in early 1950s, has been dedicated to a formulation and production of controlled drug release systems powered by fast development of the controlled-release technology (CRT). In general, drug release could be described as a process when drug migrates from the initial place where it has been delivered, to the delivery system outer surface and then to the release medium (Langer, 1990). CRS are able to maintain the drug exposure with constant rate release (zero order), and guarantee drug delivery through the physiological and cellular barriers, preventing them from the elimination. One of the examples could be a drug/s encapsulation with slowly dissolving biodegradable polymer coating which is effective in colon release and has credible stomach acid-liability. The most popular polymers used for creation of the CRS-based on their physical properties are: elastic polyurethanes, insulating polysiloxanes, strong and durable polymethyl methacrylate, hydrophilic swelling polyvinyl alcohol, forming a great suspension polyvinyl pyrrolidone, poly 2-hydroxy ethyl methacrylate, poly N-vinyl pyrrolidone, polyvinyl alcohol, polyacrylic acid, polyethylene glycol, polymethacrylic acid and others (Sambathkumar et al., 2011). The mechanisms involved into CRS are-based on the following stages: dissolution, partitioning, diffusion, osmosis, swelling, erosion, and targeting (Figure 4.1). Drugs, which can often have amphorous or crystal-like structure, during dissolution stage in the medium such as polymer, water, etc. (controlled by diffusion

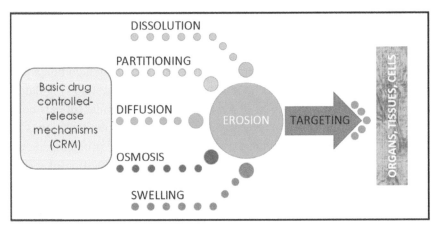

FIGURE 4.1 Basic drug controlled-release mechanisms (CRM) (Copyright @ Dr. M. Emerald, 2017).

and which increases with solubility and decreases with size of the drug molecule), transforms from its solid stage to the medium. Dissolution-controlled products usually sub-divided-based on encapsulation dissolution and matrix dissolution control (Welling and Dobrinska, 1978; Lakshmi, 2010). The coefficient of partitioning is depending on a drug affinity, solubility in the specific type of medium (hydrophobic or hydrophilic nature of the medium polymer micelles), and viscosity of the water medium. After drug partitioning, water-based osmotic force ruptures the polymer coating and release the drug. After that the swelling of the polymer coating occurs under the influence of water and slow release of drug is usually initiated. During the surface erosion which directly correlates with drug release, the polymers are going through the stages of drug release, degradation, and rapid release of the drug trapped. The target drug delivery using CRS could be localized on organ, tissue, cellular, sub cellular or organelle level, which will potentially guarantee the precision of the drug controlled release, will increase drug effectiveness. The main advantages perceived for CRS are: simplicity of formulation, reduction in drug blood level fluctuations, reduction in dosing frequency and side effects, reduction in health care costs, etc. A variety of the CRS mechanisms which lead to fulfillment of the advantages listed above, presented in Figure 4.2. Controlled drug release rate is also dependent on a few important factors: formulation aspects, polymer variables and drug characteristics. The formulation aspects which must be carefully considered when CRS is designed includes system geometry, processing techniques,

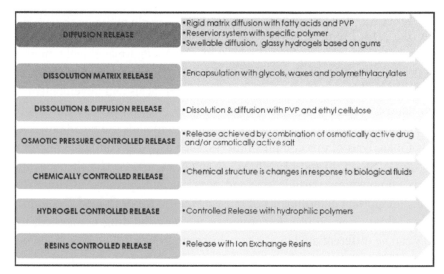

DIFFUSION RELEASE	•Rigid matrix diffusion with fatty acids and PVP •Reservior system with specific polymer •Swellable diffusion, glassy hydrogels based on gums
DISSOLUTION MATRIX RELEASE	•Encapsulation with glycols, waxes and polymethylacrylates
DISSOLUTION & DIFFUSION RELEASE	•Dissolution & diffusion with PVP and ethyl cellulose
OSMOTIC PRESSURE CONTROLLED RELEASE	•Release achieved by combination of osmotically active drug and/or osmotically active salt
CHEMICALLY CONTROLLED RELEASE	•Chemical structure is changes in response to biological fluids
HYDROGEL CONTROLLED RELEASE	•Controlled Release with hydrophilic polymers
RESINS CONTROLLED RELEASE	•Release with Ion Exchange Resins

FIGURE 4.2   Controlled Release Mechanisms for a different DDS. (Copyright @ Dr. M. Emerald, 2017).

nature of additives and surfactants, physical characteristics of dosage and the nature of excipients/additives. Incorporation of correct additive and surfactant are crucial for proper drug release, and concentrated surfactant will accelerate the drug release and reduce the inner particles adhesion.

According to their application routes, CRS can be divided onto: injectables (solutions, dispersions, microspheres, microcapsules, nanoparticles, niosomes, liposomes, pharmacosomes, resealed erythrocytes), implants, infusion devices, osmotic devices (pumps, vapor pressure powered pumps, battery powered), pulsatile miniaturized wearable delivery systems. Mathematical modeling has been engaged a lot to describe, examine and understand the complexity of nanoparticles-drug interaction, drug delivery mechanism, interactions between contributing components using computational simulations and quantitative analyses. A variety of mathematical models are used to describe mechanisms of drug release and their kinetics, such as Harland et al. (to predict the speed of dissolution process), Ritger-Peppas, Peppas and Sahlin model used by Chebli et al. (to analyze the diffusion and erosion mechanism of the CRS from substituted amylose matrices), Higuchi and Korsmeyer-Peppas, degradation and drug release of Poly (lactic-co-glycolic acid) (PLGA) microspheres, (Harland et al., 1988; Peppas and Sahlin, 1989; Chebli and Cartilier L, 1999), mathematical and simulation, as well as biophysical models used for description of drugs release process for a number of cancer drugs (Swierniak et al., 2009; Enderling and Rejniak, 2013; Liu et al., 2011) and release kinetics of semisolid drug molecules release from liposomes (Loew et al., 2011). Pharmacokinetic modeling is a major player in development of new CRS drug therapies (Brunton et al., 2010) and provides a great tool for the synthesis and development of new nanostructured delivery vesicles, such as NPs, nanocapsules, nanofibers, hollow nanofibers, etc. The release kinetics models use novel powerful tools for analysis of this complex process, such as X-ray computed microtomography (SR-μCT) and LC/MS/MS (Yang et al., 2014). The chemical and matrix composition, particle size and physical and chemical environment are also important determinants of release kinetics for drug-loaded nanoparticles (Mittal et al., 2007). According to the way the drug is released, the DDS can be differentiated onto a few different groups: (a) DDS with immediate release, when drug is released immediately after administration: (b) DDS with modified release, when drug release is insignificantly delayed after the drug administration which is also can be composed by drug delayed or extended release. Drug dose, diffusivity and solubility in polymer are important factors that determine drug proportion in the matrix use for DDS. There are a various approaches used, such as different surfactants, drug-cyclodextrin complexes and variable carrier systems are used to solve the solubility problem (Efentakis et al., 1991; Rao et al., 2001; Giunchedi et al., 1999). Poorly soluble drugs are released by osmosis and erosion mechanisms and pH sensitive drugs used for treatment of various gastrointestinal disease require use of buffer salts to change the environmental pH to help provide the proper drug release (Streubel et al., 2000; Martinez Gonzalez and Villafuerte Robles, 2003). Control release dosage form has a number of the commercial and clinical advantages and a few limitations as well (Figure 4.3). To qualify for a really effective controlled release, drug must have appropriate size and molecular weight (to be able to transport across a biologic barrier/membrane), proper solubility, high level of absolute bioavailability, great intrinsic absorption, low therapeutic concentration and great terminal disposition rates (Ansel et al., 1995; Genç et al., 1998; Vyas and Khar, 2002; Shargel, 1999).

Since the term "nanoparticle" was proposed in 1976, and the first research-based on nanoparticle systems has been published (Kreuter and Speiser, 1976; Gurny et al., 1985), the nanoparticle-based drug delivery systems went through an amazing transformation leading to a development of at least three generations of the revolutionary and more

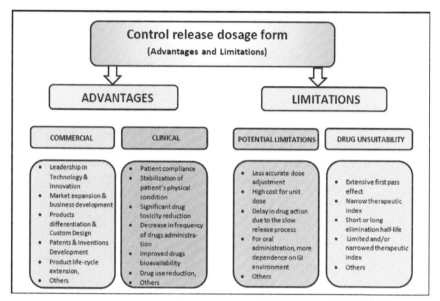

FIGURE 4.3 Main advantages and limitations of drug control release dosage form (Copyright @ Dr. M. Emerald, 2017).

efficient products. A large variety of formulated and developed intelligent therapeutic DDS-based on combination of pharmaceutics, polymer science, chemistry and molecular biology are currently available on the International market (Figure 4.4). The DDS vary in their type, indication for use and targeting a specific problem, as well as route of application and a wide number of parameters (such as drug release kinetics and detailed mechanism), which are hard to categorize. The main points which need to be addressed during the development of the best quality DDS is minimization of drug loss and degradation, proper controlled release mechanism, diminishing of the potential harmful side-effects, increased bioavailability, and many others. The most popular carriers for the drug delivery are soluble biodegradable natural and synthetic polymers, cell ghosts, lipoproteins, microcapsules, liposomes, micelles, inorganic nanoparticles, liquid crystals. Inorganic nanoparticles-based DDS (metallic nanoparticles, mesoporous silica), polymers (hyperbranched polymers, dendrimers and polymeric stimuli-responsive to PH, temperature and reductive environment micelles), carbon nanotubes and a variety of hybrid materials (carbon nanotubes and grafted hyperbranched polymers). A wide spectrum of nanoparticles used is DDS which are available

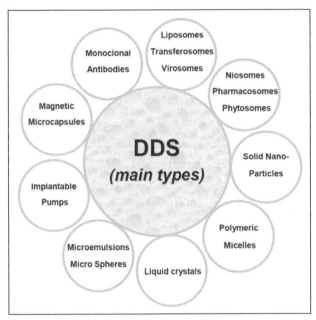

FIGURE 4.4 Current main types of the Drug Delivery Systems/DDS (Copyright @ Dr. M. Emerald, 2017).

in different sizes, shapes, and produced from a different material with various chemical and surface properties, have an important advantages, such as improved bioavailability, slow release via diffusion and dissolution, increased chemical and physical degradability resistance, targeting to specific location and safe delivery of to ic drugs with minimum side effects (Irving, 2007; Müller et al., 2000). The fullerenes which were discovered by Kroto, Curl and Smalley in 1985 (Kroto et al., 1985) are and which are similar to the graphite in structure, have a number of applications in biomedical science and DDS in particular, and they are used along with the targeting drugs for the treatment of leukemia and other types of cancer, HIV, liver disease and especially Hepatitis, Parkinson's, allergy, and many others (Dugan et al., 1997; Mashino et al., 2005; Thrash et al., 1999). Available in a variety of shapes ('buckybabies,' 'fuzzyballs,' gadofullerenes, metallofullerens, nanotubes, giant and C70 fullerenes) and attaching hydrophilic moieties, fullerenes are used to carry drugs and genes. They are an important DDS agent, but must be used with caution due to the potential opportunity to harm some living cells and accumulate in the living tissues (Darshana Nagda et al., 2010).

The natural polysaccharides such as gums, starch, pectin, alginates, chitosan, chondroitin sulfate, hyaluronic acid, bacteria, fungi and others, can be divided on linear, branched and long branched (Lemarchand et al., 2003; Fathima et al., 2011), and are usually safe, biocompatible, non-toxic, biodegradable, economical, eco-friendly and easily obtainable (Guo et al., 1998). They can be of plant (gums, starch, pectin), marine (alginates), animal (chitosan, chondroitin sulfate, hyaluronic acid) and microbial origin (from bacteria and fungi) and used in many medicinal applications (Lemarchand et al., 2003; Fathima et al., 2011). They are widely exployed for the delivery of antidiabetic, antibiotic, anti-inflammatory agents, proteins and enzymes (Patil, 2008; Chaurasia and Jain, 2006).

One of the popular drug delivery carrier is natural chitosan which is often used as a nanoparticle drug delivery vehicle in gels. The consistency of gel layer plays an important role in drug release and dissolution rate (Ford Versypt et al., 2013; Dash et al., 2010). As a natural polysaccharide with versatile chemical properties, which is soluble in acidic solvents only, chitosan has superior biocompatibility, can load hydrophobic agents and has been employed for delivery of anticancer chemical drugs, proteins, insulin, cisplatin, etc. (Park et al., 2010; Victor et al., 2013; Kean and Thanou, 2010; Ma et al., 2005; Janes et al., 2001; Kim et al., 2008). A chemical modified chitosan, N-imidazolyl-O-carbo□ymethyl chitosan (IOCMCS), improves solubility of chitosan in water and make its positively charged (Tables 4.1 and 4.2).

4.3 DDS AND ROUTES OF ADMINISTRATION

The drug delivery technology market is expected to reach USD 1, 504.7 Billion by 2020 from USD 1, 048.1 Billion in 2015 (9th World Drug Delivery Summit, June 30–July 02, 2016 New Orleans, Louisiana, USA), and creates constant demand for novel and more effective CRS and commercialized controlled release technologies with the most effective application routes. The research in the field of novel controlled release drug delivery systems lead to development of novel controlled release technologies, such as OROS, osmotic ally controlled release oral delivery system (Wang and Park, 2011), OCAS, oral controlled absorption system, TIMERx, controlled release tablets formulation with hydrophilic matrix:

TABLE 4.1 Lipid Particles as Drug Carriers

Type of the lipid particles	Advantages	Limitations
Liposomes	• Non-toxic, flexible, biocompatible, completely biodegradable • Nonimmunogenic • Used for variety of applications: for systemic and non-systemic administrations • Reduces tissues exposure to toxic drugs • Flexible coupling with specific ligands for precise targeting • Helps increase therapeutic index of drugs and stability via encapsulation	• Low solubility • Short half life • After same time can go through oxidation • Not very stable if physical conditions change • High production cost
Niosomes Composed of Cholesterol and non-ionic surfactants	• Have flexible composition, fluidity and size • Can accommodate the hydrophilic, lipophilic as well as amphiphilic drug moieties • Protect drug from the environmental changes and metabolic degradation • Release drug slowly and improve therapeutic performance • Stable and osmotically active • Have high skin penetration rate • Increase drugs oral, parenteral and topical permeability • Non-immunogenic, biocompatible, biodegradable • Have great level of surfactants storage • Provide easily controlled drug release • Reducing drug clearance time • Can be easily produced	• Aggregation • Physical instability • Partial leaking and hydrolysis of entrapped drug • Fusion
Transferosomes	• High deformability and penetration rate • High entrapment efficiency • Infrastructure which accommodate a different solubility drugs • Release drug slowly and improve therapeutic performance • Used for systemic and topical applications	• Expensive to produce • Made of phospholipids • Sometimes luck of purity • Chemically unstable

Phytosomes

Phosphatidylcholine as a main compound

- Protect drug from the environmental changes and metabolic degradation
- Do not need the additional pharmaceutical additives to improve the drug release
- Great stability and efficiency
- Excellent compatibility with botanical extracts
- Increased bioavailability and absorption require lover dose of drug
- Demonstrate fast absorption by Gastrointestinal tract and Liver protection
- Protect drug from the environmental changes and metabolic degradation
- Release drug slowly and improve therapeutic performance
- Nourishes skin vile demonstrate high transdermal penetration rate
- Can be used for delivery of large and specific group of drugs (peptides and proteins)
- Can be easily produced

- Slow drug delivery process: phytoconstituent is usually rapidly eliminated from phytosome

Pharmacosomes

Amphiphilic phospholipid complexes of drugs with phospholipids bind with active hydrogen

- Improves bioavailability of poorly soluble drugs
- Minimize drug side and toxic effects
- Sustaining drug concentration for long time and within an optimal Range
- Guarantee control and stability of whole drug release system
- Easily penetrate the cell membrane and tissues by endocytosis or exocytosis
- The drug entrapment efficiency of remain unaffected with time
- No drug leaking due to covalent binding of the drug to the carrier

- Potential aggregation and fusion, if stored
- To protect leakage of drugs, require covalent bounding
- Mainly used in encapsulation of water insoluble drugs

TABLE 4.2 Drug Delivery Systems and their Applications

Main types of the DDS	Size	Indication	Main Route of Application	Discovery	Reference
Liposomes (phospholipid vesicles loaded with drugs)	50–450 nm	Fungal infections, solid tumors, leukemia, ovarian cancer, breast cancer, colon cancer, osteosarcoma, rheumatoid arthritis, skin cancers, hepatitis A, influenza, ebola, cystic fibrosis, malaria, tuberculosis, leishmaniasis, neoplasms, others	Intravenous, Epidural, Intramuscular, Others	Discovered by Alec D. Bangham in the 1960s at the Babraham Institute, University of Cambridge	Bangham, A.D., Horne, R.W., 1964.
Transferosomes	200–300 nm	Pain relieve, immunization (tetanus, etc.), Parkinson's, cardiovascular, Alzheimer's diseases, depression, anxiety, attention deficit hyperactivity disorder (ADHD), skin cancer, female sexual dysfunction, post-menopausal bone loss, urinary incontinence, target delivery of plant bio-actives, hormones, insulin, anti-cancer drugs, others	Transdermal, Subcutaneous, Oral (proteins/peptides), Sublingual, Periocular, Intranasal, Rectal, Others	First introduced in early 1990s. Approved for use by the Swiss regulatory agency (SwissMedic) in 2007	Cevc, G., 1995; Schatzlein, A., Cevc, G., 1995.
Virosomes	110–200 nm	Vaccines for influenza, hepatitis A, hepatitis B, hepatitis C, antifungals, anti-parasitic agents, cancer chemotherapy	Inhalable, Intravenous, Intramuscular, Subcutaneous, Intra Arterial	Started to be used late 1990s. Licensed in Switzerland since 1997, Italy since 1998	Moser, C., Amacker, M., 2013.

Main types of the DDS	Size	Indication	Main Route of Application	Discovery	Reference
Niosomes (non-ionic surfactants vesicles)	120–250 nm 1000 nm and > (for cosmetic applications and some pharmaceutical products)	Mainly cholesterol and non-ionic surfactants-based. Used for tumors, gene delivery, leishmaniasis, haemoglobin carriers, antineoplastic drugs and proteins delivery, others	Drug targeting, Peptides delivery, Hemoglobin carriers, Transdermal drug delivery, Ophthalmic drug delivery, Dermal and Mucocutaneous infection, Anti-neoplastic Treatment (Cancer), Others	First niosomes were developed and technology has been patented by L'Oreal in 1975	Cosmetic and pharmaceutical compositions containing niosomes and a water-soluble polyamide, and a process for preparing these compositions. French Pat. No. 2,315,2,221,122, 2,315,991,83–04674; Khandare, J. N., Madhavi, G., Tamhankar, B. M., 1994.
Phytosomes (phytolipids)	1000 nm and >	Delivery of plant compounds-based drugs with a variety of therapeutic properties: cardioprotective, hepato-protective, anticancer, antihypertensive, anti-inflammatory, antimutogenic, antimicrobial, antioxidant, hypotensive, antiatherosclerotic, antiaging, hypocholesterolemic,	Oral, Sublingual, Intranasal, Transdermal, Inhalable, Others	Developed by Indena s.p.a of Italy	Bombardelli, E., Curri, S. B., Della, R. L., Del, N. P., Tubaro, A., Gariboldi, P., 1989; Bombardelli, E., Spelta, M., 1991; Vinod, K. R., Sandhya, S.,

TABLE 4.2 (Continued)

Main types of the DDS	Size	Indication	Main Route of Application	Discovery	Reference
		anticellulite, UV protection, anti-edema, Immunomodulatory, antiitching, anti-allergic, others			Chandrashekar, J., Swetha R., Rajeshwar, T., Banji, D., Anbuazhagan, S., 2010; Kalita, B., Das, M., Sharma, A., 2013.
Solid Nanoparticles Solid lipid nanoparticles (SLNs), Poly (ethylene glycol) coated nanospheres, Azidothymidin (AZT)/ dideoxycytidine (ddc) nanoparticles, Chitosan-poly (ethylene oxide) nanoparticles, Others	1 to 100 nm	Gene therapy, Antitumor, in MEL therapy	Oral, Topical, Parenteral, Intravenously, Intramuscularly/ Subcutaneously, Inhalable, Ocular, Rectal, Others	Introduced by Muller and Lucks in 1996	Muller, R., Lucks, J., 1996; Charcosset, C., El-Harati, A., Fessi, H., 2005; Heiati, H., Tawashi, R., Phillips, N.C., 1998; Hou, D. Z., Xie, C. S., Huang, K., et al., 2003.

Main types of the DDS	Size	Indication	Main Route of Application	Discovery	Reference
Polymeric Micelles	20-100 nm	Target delivery for: anticancer drugs, drugs for neurodegenerative, cardiovascular and other conditions, delivery of anti-fungal agents and polynucleotides, delivery of imaging agents	Oral, Oromucosal, Ocular (intracorneal)	Introduced by H. Ringsdorf in 1984, and developed by Kataoka in 1992	Kalal, J., Drobnik, J., Kopecek, J., Exner, J., 1978; Yokoyama, M., Kwon, G. S., Okano, T., Sakurai, Y., Seto, T., Kataoka, K., 1992
Liquid Crystals Lyotropic, Lamellar, Hexagonal, Thermotropic, Cholesteric, Smectic, Nematic, Cubic, Others	50 to 250 nm	Target delivery for: vitamins, aminoacids, proteins, peptides, target drugs, hormones, antibiotics, pain killers, etc.	Oral, Oromucosal, Transdermal, Intravenous injection, Subcutaneous injection, Ocular, Others	Introduced in late 1990s	Drummond, C.J., & Fong, C.,1999; Mueller-Goymann, C. C., 2002; Stevenson, C. L., Bennett, D. B., Ballesteros, D. L., 2005; Yaghmur, A., De Campo, L., Sagalowicz, L., Leser, M. E., Glatter, O., 2005; Stevenson, C. L., Bennett, D. B., Ballesteros, D. L, 2005.

TABLE 4.2 (Continued)

Main types of the DDS	Size	Indication	Main Route of Application	Discovery	Reference
Micro Emulsions (surfactant-based)	Up to 150 nm	Delivery of plant compounds-based drugs and essential oils with a variety of therapeutic properties, delivery of painkillers, antimicrobials and antifungals	Transdermal/ Topical, Intranasal, Oral, Ocular, Parenteral	The process was first recognized by Hoar and Schulman in 1943, and microemulsion term was Introduced by Schulman in 1959	Gasco, M.R., Trotta, M., 1986; Sjoblom, J., Lindberg, R., Friberg, S.E., 1996; Sainsbury, F., Zeng, B., Middelberg, A. P. J., 2014; Lawrence, M. J., Rees, G. D., 2000.
Others	Pharmacosomes, Genosomes. Photosomes. Cubosomes, Ufasomes, Enzymosomes, Colloidosomes, Hemosomes, Sphyngosomes, Aquasomes, Implantable Pumps, Magnetic Microcapsules, Monoclonal Antibodies, etc.				

SODAS, spheroidal oral drug absorption system, IPDAS, intestinal protective drug absorption system), CODAS (chronotherapeutic oral drug absorption system) (Kalantzi et al., 2009). The direct entry into the body through the oral and pulmonary administration, transdermal entry into the body, and entry into the body through the mucosal membranes are three major routes of administration typical for the DDS.

4.3.1 THE ORAL ROUTE OF DRUG DELIVERY

The oral route of drug delivery which is the most convenient, can be divided into sublingual (through the base of the mouth) and buccal delivery, but has some advantages and disadvantages. The main advantages of oral drug delivery are: simple way to use, painless, controlled drug dosing, patient-friendly, and painless. The main disadvantages are: drug degradation by intestinal enzymes, delayed mucosal absorption, and side effects due to waste amount of drug metabolized by liver causing hepatotoxicity. The oral consumption and following targeting release of the drug in a specific region of the gastrointestinal tract is especially useful for oral controlled release drugs with high absorption. Certain drugs are absorbed from the upper part of small intestine and stomach, while others are slowly absorbed by other parts of the gastrointestinal tract. Stomach acidity is an important factor compromises drug stability, target delivery and proper release, and represented one of the significant concerns for DDS research and development which lead to development of the gastro retentive drug delivery system (GRDDS) consisting of gelatin beads or hollow microspheres which are-based on a variety of polymers including chitosan, acrycoat, polyacrylates, gelatin, agar, polycarbonates, etc.

Buccal mucosa is quite effective route for controlled oral drug delivery, with main advantages are: drug delivery for extended periods of time, due to a great vascular and lymphatic drainage system, and avoidance of gastrointestinal and hepatic complications (de Vries et al., 1991). A variety of hydrogels such as xantham gum, guar gum, locust bean gum and others which has low mucoadhesion, can form the gel stable enough for the CRS when dissolved in neutral medium only (Watanabe et al., 1991). To increase the mucoadhesion and provide better control of drug release, Anders et al., designed the bilayer protirelin/ hydroxyethylcellulose patch. The mucoadhesive buccaldrug delivery systems (MBDDS) could be divided to: (a) drug delivery through the of the oral cavity lining, which could be used for local application and drugs effect (mucosal), and (b) systemic application and drugs effect (transmucosal), applied to a mucosal linings of the oral, nasal, ocular, rectal, vaginal cavity can cease unwanted side effects due to a significantly lower drug concentration use and decreased toxicity (Harris and Robinson, 1992; Puratchikody et al., 2011; Kianfar et al., 2011). Drugs administration via MBDDS route used to treat cardiovascular disease (Yamsani et al., 2007), migraine (Dechant et al., 1992), fungal and microbial infections (Yehia et al., 2008; Koks et al., 1996; Giunchedi et al., 2002; Mizrahi and Domb, 2007), inflammation (Heasman et al., 1993; Mura et al., 2010), diabetes (Muzib and Kumari, 2011), nausea and vomiting (post radiation effect in cancer patients) (Blake et al., 1993), smoking deterrent (Pongjanyakul and Suksri, 2009), used for pain management (Okamoto et al., 2001, Woolfson et al., 1998), allergies relive (Sekhar et al., 2008), etc. Drugs delivery using buccal polymer films produced from cellulose derivatives, copolymers and poly (acrylic acids), offer a significant comfort, fle⬚ibility and simple design then tablets (Anders and Merkle, 1989; Woolfson et al., 1995). The first usage of protein therapeutics via injection of purified bovine insulin for treatment of diabetes has been performed and documented by Banting et al. (1922).

In the present time, proteins and peptides therapeutics is widely used for treatment cancer, leukemia, diabetes, rheumatoid arthritis, hepatitis, and others chronic conditions (Tan et al., 2010; Dharap et al., 2005; Cai et al., 2011; Rigon et al., 2015). The first buccal delivery system for peptide/protein drugs was proposed in 1988 (Nagai, 1988), and since that time several technologies (such as encapsulation, spray drying, double emulsion, etc.) has been involved into the process of delivery of complex molecules (Hermann and Bodmeier, 1995; Takada et al., 1995; Hanson and Rouan, 1992). The most popular polymers used for proteins delivery are: Polyethylene Glycol (PG) (Torchilin and Papisov, 1994), a variety of Polyesters (Jeyanthi et al., 1997), natural polymers (gelatin, collagen, albumen and others, and polysaccharides (cellulose, dextran and others) and hydrogels (Peppas and Klier, 1991). The stimuli-responsive specialty materials, which can change their chemistry and conformation is one of the hot topics of modern protein delivery systems (Stuart et al., 2010). The external stimuli which can used for such type of proteins delivery are divided to: physiological stimuli (enzymatic activity, PH level, redox potential, glucose concentration, etc.) and e⬚ternal stimuli (electric field, magnetic field, light, temperature, mechanical forces, etc.) (Matsumura and Maeda, 1986; Gu et al., 2013; Stuart et al., 2010; Mo et al., 2012). Physiological stimuli triggered drugs delivery systems can be represented by: (a) pH-sensitive nanosystems, changing chemical structure, membrane fusion, charge, etc., for examplebio-degradable and pH-sensitive hydrogels (Mo et al., 2012; Gao et al., 2013), (b) redo☐responsive nanosystems-based on thiol-disulfide e☐change (Zhang et al., 2003), (c) enzyme-responsive nanosystems, for example PEG hydrogels for controlled protein delivery induced by photopolymerization which, or furin-triggered protein release (Biswas et al., 2011; Aimetti et al., 2009), (d) glucose oxidase-based Insulin delivery systems, for example, glucose-responsive gels (Tanna et al., 2002), (e) synthetic boronic acids-based Insulin delivery (Tanner et al., 2011; Zhao et al., 2009), (f) ultrasound initiated drug delivery (Schroeder et al., 2009; Rapoport et al., 2004), and others. A few examples of presently used drug delivery systems with stimulus-responsive capabilities are: Tablets coated with ion-sensitive PNIPAM layer for colon drug delivery (Eeckman et al., 2002), a bead system coated with ion-sensitive Eudragit RS (Narisawa et al., 1994), 'salting-out taste-masking systems' with a salting-out layer containing salts and polymers (Yoshida et al., 2008, 2009), sensitive to acidic PH polymeric mixed micelles (Lee et al., 2008). Over 95% of all the pathogens enter via mucosal routes, and local mucous immunization using peptides, bacterial toxoids, ovalbumin, DNA plasmids can effectively be used to stimulate systemic and local immune responses (Maloy et al., 1994; Challacombe et al., 1997; Kim et al., 1999; Esparza and Kissel, 1992; Tung et al., 2001). A buccal drug delivery leads to a novel types of vaccines delivery, which are being adhesive to mucous membranes can stimulate local immunity effectively and with minimal side effects (Eldridge et al., 1993).

4.3.2 THE INTRANASAL ROUTE OF DRUG DELIVERY

The intranasal delivery variety of drugs, as well as hormones, stem cells, insulin, neuropeptides, used to target and treat such chronic disease as Huntington's, depression, anxiety, Alzheimer, Parkinson's, obesity, ASD, seizures, drug addiction, brain tumors) is another quite effective option to target the CNS via effective crossing the blood–brain barrier. The drug is delivered from the nasal cavity via intracellular (olfactory sensory neurons, viaslow endocytosis and diffusion) and extracellular routes through the olfactory bulb straight to the posterior and anterior regions of the brain (Dhuria et al., 2010; Renner et al., 2012; Hanson et al., 2004; Kristensson and Olsson, 1971; Thorne and Frey, 2001). The additional drug absorption is also induced through the trigeminal nerve (Lochhead and Thorne, 2012) and rostral migratory stream (Scranton et al., 2011). In spite of the fact that only about 3% of the nasal cavity is lined by olfactory epithelium (Morrison and Costanzo, 1990), this is the most direct route to the olfactory bulb and is also a novel approach to drugs delivery (Ding and Dahl, 2003), when drug easily passing brain blood barrier (BBB). This route of drug delivery demonstrates extraordinary research achievements especially during the last ten years (Gozes and Divinski, 2007; Frey, 2002; Greco et al., 2009).

The specific carriers used up to now for the intranasal drugs delivery are: (a) mucoadhesive polymers, which are including carboxymethylcellulose, carbopol, hydroxylpropylmethyl cellulose, carbopol, liposomes (especially PEGylated immunoliposomes), (b) didanosine-loaded chitosan nanoparticles, Chitosan-N-acetyl-L-cysteine, chitosan/alginate nanoparticles (Charlton et al., 2007; Wang et al., 2009; Goycoolea et al., 2009), (c) phenylephrine hydrochloride, (d) erythropoietin (EPO) (especially Neuro-EPO, for intranasal drug delivery for neurodegenerative disorders) (Garcia-Rodriguez and Sosa-Teste, 2009), (e) DNA plasmids encoding therapeutic or antigenic genes (Han et al., 2007), (f) peptides and proteins, including insulin-like growth factor-1 and γ-interferon (Liu et al., 2004; Thorne et al., 2004; Thorne et al., 2008), (g) stem cells for a potential treatment of malignant Gliomas (Ahmed and Lesniak, 2011), and (h) some types penetration enhancers such as peppermint oil, which are significantly improving the drug delivery process (Vaka and Murthy, 2010). The Intranasal drug delivery rout used for treatment of different types of chronic disease, has been demonstrated to be effective and generally safe and is presently going through over 100 clinical trials. According to results of clinical meta-analysis of the intranasal drugs administration safety, the intranasal route of drug delivery has a few low risk side-effects in comparison with oral (ingestible) drugs and injectables. Such side-effects include but not limited to: increase in brain insulin level, systemic blood pressure, enhanced hypothalamo–pituitary–adrenal activity and a few others (Woods et al., 2003; Kern et al., 2005; Fruehwald-Schultes et al., 2001).

4.3.3   THE CELL-MEDIATED DRUG DELIVERY

The novel cell-mediated drug delivery, which basically takes advantage of cell properties (cellular signalling, metabolism, morphology, surface properties and circulation in blood flow time) provide an amazing opportunity to maximize therapeutic targeting and minimize drugs side effects. This frontier drug targeted delivery is obtained using specific cells ('trojan horse cells'), such as immunocytes, phagocytes, neutrophils (polymorphonuclear granulocytes (PMNs)), erythrocytes (RBC), lymphocytes (involved in various immune responses), monocytes (mononuclear leukocytes), platelets (fragments from megakaryocytes of the bone marrow), Sertoli cells, neural stem cells, bone marrow derived mesenchymalstem cells producing anti-tumor proteins (MSCs), bone marrow derived macrophages (BMM), albumens and others, which can safely pass tissue barriers and effectively deliver and release the drug, as well as are totally biodegradable and producing no toxic by-products (Thiele et al., 2003, Ikehara et al., 2006; Orlic et al., 2001; Orlic et al., 2001; Garcia et al., 2010; Akerud et al., 2001; Nakamizo et al., 2005; Garcia et al., 2010). The main obstacle for cell-mediated drug delivery is consistency of drug concentration in plasma and blood flow. The cell-mediated drug delivery system is used for treatment of cancer, HIV, lung injury, Alzheimer's and neurodegeneration, epilepsy, fungal and microbial manifestation, and others. Examples of cells-based drug delivery systems shown on Figure 4.5.

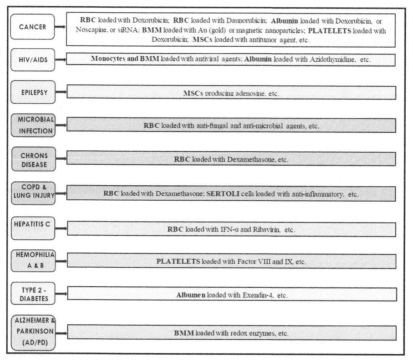

FIGURE 4.5 Examples of cells-based drug delivery systems (RBC – Red Blood Cells, BMM – bone marrow derived macrophages, MSCs – bone marrow derived mesenchymal stem cells producing anti-tumor proteins (Copyright @ Emerald, 2016).

4.3.4 THE TRANSDERMAL DRUG DELIVERY (TDD)

Transdermal delivery, which presently consist over 19 transdermal delivery systems (TDS) used for different drugs (such as peptides, DNA, small-interfering RNA (siRNA), virus-based and other vaccines, has variety of advantages in comparison to the oral route (Foldvari, et al., 2006). The TDS were introduced to North America market in 1970s (Prausnitz and Langer, 2008), and since that time, the transdermal delivery technology went through significant improvement and extraordinary progress. First scopolamine transdermal patch was introduced to US patients in 1979, and has been followed by clonidine, estradiol, testosterone, fentanyl, methylphenidate, rivastigmine, rotigotine, lidocaine and other patches, as well as aerosols sprays (Morgan et al., 1998) and gels (Prausnitz et al., 2004; Wilson et al., 1998). A variety of molecular enhancers can increase skin permeability and provide an added driving force for transport by increasing drug partitioning into the skin (Williams and Barry, 2004). Modern chemical transdermal delivery enhancers (cavitational ultrasound, thermal ablation, microdermabrasion, electroporation, as well as 500 different pairs of chemical enhancers combined with anti-irritants, biochemical enhancers, electroporation and ultrasound, can deliver through the skin barrier even macromolecules, such as proteins and vaccines.

The stratum corneum composed of lipid blocks and keratin represents natural barrier for microbes, drugs and other substances and there are over 300 chemical substances which can facilitate drug permeation through the skin. The group of chemical substances, which is used for transdermal drugs delivery permeation and which are able to increase the delivery of small molecules through the skin, could be divided to: alcohols, amines, fatty acids, amines, hydrocarbons (alkanes and squalene), hydrocarbons, surfactants, terpenes, phospholipids, propylene glycols and others (Karande et al., 2006; Karande et al., 2004). Many of them are used in combination with anti-irritants and soothing anti-inflammatory agents. In difference, physical transdermal drug delivery methods (such as hypodermic needles, jet injections under high pressure, microdermabrasion, thermal ablation, laser, microneedles, electroporation – high-voltage electrical pulses, iontophoresis, ultrasound) provide more effective for skin transdermal delivery (Burkoth et al., 1999; Denne et al., 1992; Gill et al., 2009; Park et al., 2008; Lee et al., 2001; Riviere and Heit, 1997; Wermeling et al., 2008; Prausnitz, 1999; Mitragotri, 2001; Mitragotri, 1996).

Extensively developed during last decade nanoparticles (lipids, sugars, metals, polymers, etc.) demonstrate a multiple advantage over chemical penetration enhancers, including continuous drug release and molecular protection of the drug from degradation caused by physical and chemical factors. The physicochemical characteristics of nanoparticles such as charge, size, shape, loading efficiency, application, etc., are crucial for permeability and successful transdermal delivery of encapsulated drug (de Leeuw et al., 2009). Liposomes (including ceramides, solid lipid nanoparticles (SLNs) and nanostructure lipid carriers (NLCs)), transferosomes, niosomes, ethosomes, bicelles, magnetic/metallic nanoparticles and quantum dots, titanium dioxide (TiO_2) and zinc oxide (ZnO)-based nanoparticles are those effective micovehicles, which can penetrate the skin, accumulate into hair follicles, where drug can release slowly into the circulatory and lymphatic system (Lademann et al., 2008). The liposomes penetrate the skin via intact penetration, osmotic gradient and hydration force, lateral diffusion, transappendageal pathway, and others (El Maghraby et al., 2008, El Maghraby et al., 2001). The magnetic nanoparticles, which are usually smaller than 10 nm can be easily transported through skin lipid matrix and hair follicles to the stratum granulosum, where its condensing between corneocytes (Baroli et al., 2007) and could be especially helpful for skin disease diagnostics. The nanoparticles containing vaccines, penetrates beyond hair follicles, and cannot flush out by sebum, initiating great immune response during prolong time (Jung et al., 2006).

Consisting less than 30 amino acids, cell-penetrating peptides (CPPs) or membrane transduction peptides (MTPs) (Rothbard et al., 2000) can cross the cell membrane and act as transacting activator of transcription (TAT) and pAntp (penetratin), performing an effective transdermal delivery of proteins and peptides (for example Polyarginine-7, R7-CyclosporineA (CsA) conjugates, antigenic peptide, biotin-transporter-9 conjugate, heat shock protein 20, polylysine-9 (K9), etc.) and following drug release in a dose-dependent manner in vitro and in vivo (Wender et al., 2002).

4.3.5 LIQUID CRYSTALS AS DRUG DELIVERY SYSTEMS

Lyotropic liquid crystals which are formed by polar lipids and when they are contacting water, spontaneously reorganize into three-dimensional structures crystalline phase, represent a revolutionary lipophilic drugs delivery system, which is a permeability enhancer for the transdermal drug delivery (Burrows et al., 1994; Guo et al., 2010; Yariv et al., 2010). The liquid crystalline phases of monoolein (a natural polar lipid that swells in water and forms various lyotropic liquid crystals), are depending on temperature, water content, drug or additive and other factors, and properties in the liquid crystalline phase of monoolein/water systems can be changed by addition of solvents and drugs (Alfons and Engström, 1998; Costa-Balogh et al., 2010; Phan et al., 2011; Sallam et al., 2002). The liquid crystals according to a structure, can exist in lamellar, cubic and hexagonal phase, with a large potential of continuous release of active pharmaceuticals of a different polarities and sizes (Boyd et al., 2006; Spicer et al., 2001), and protect peptides, nucleic acids, proteins and other molecules from any type of potential degradation (Clogston and Caffrey, 2005). The liquid crystals cubic phases improve drug bioavailability, decrease toxicity and improve the transdermal/topical delivery of small and larger molecules, for example: δ-aminolevulinic acid (Bender et al., 2005), calcein (Yamada et al.), acyclovir (Helledi and Schubert, 2001), paeonol (Luo et al., 2011), cyclosporin A (Chen et al., 2012). Using cubosomes as a drug delivery system significantly prolonged release of indomethacin anti-inflammatory activity when e□posedto 6 h UVB (ultraviolet B) irradiation (Esposito et al., 2005). The hexosomes has been found especially effective enhancers of the penetration of vitamins, complex drugs, such as sodium diclofenac and hormones, such as progesterone, as well as increase stability of some drugs, such as irinotecan (Lopes et al., 2007; Swarnakar et al., 2007; Boyd et al., 2006). During last decade scientists developed a variety of stimuli (pH, temperature, etc.) responsive cubic phase (fast release) and hexagonal phase (slow release) liquid crystals-based DDS for intestine- or colon-targeted delivery of glucose, hormones, and other bio actives (Fong et al., 2009; Negrini and Mezzenga, 2011).

4.4 DDS AND BIOSAFETY

DDS safety is still a concern due to: (a) potential toxicity of adjuvant materials, (b) drug leakage before reaching the lesion, (c) rapid elimination and instability in blood environment, (d) non-targeted drug delivery in vivo and others. Biosafety is a major issue in the design of novel drugs and researchers must give careful consideration to this when developing nanoscale materials for medical research. The physicochemical properties of nanoparticles, material they are produced from, their size, shape, charge, colloidal stability, and their interaction with environmental compounds, could be related to their potential toxicity. including geno- and cytotoxicity, in vivo bio-distribution, biodegradability, system clearance, immune-toxicity, side effects of treatment. It was unstudied enough and much more work need to be done to prove that nanomaterials are biocompatible and biosafe. From another side, biodegradable nanomaterials, such as proteins, nucleic acids, phospholipids, are approved by the Food and Drug Administration (FDA) (e.g., Doxil® and Abraxane®). However, generally bio safe and biodegradable nanomaterials and carriers may cause immune response and should be clinically tested before use. Inorganic nanomaterials, can also undergo biodegradation and have tissue accumulation and systemic clearance problem, can cause genomic instability, inflammatory response, and protein phosphorylation and could cause some concern before their use and a potential application. The bio safe nanomaterial and drug delivery systems must have low toxicity and immunogenicity, be safe and would not cause any damage on genetic, cellular and systemic levels in vivo.

4.5 SUMMARY

The emergence of nano- and biotechnology, and development of controlled drug delivery systems, have significant impact on pharmaceutical, medical and life science and contributes a lot to improvement of quality of life, therapeutics and modern medicine. In addition to development of nano-biotechnology, the rapid advances in materials science and novel drugs delivery and carrier-mediated systems represent a powerful technology for the treatment of various chronic conditions and health pathologies. Nanoparticles can be used for pharmaceuticals as well as diagnostics purpose, and can be used in drug delivery and for creation and development of personalized medicine. An exponential development of new Multiple Drug Delivery Systems (MDDS), new modes of applications and huge efforts to develop multidrug delivery systems with one formulation, when multiple agents can simultaneously deliver to a specific target, shows great effectiveness, lower cytotoxicity and enhanced internalization compared to commercially available liposomes and another DDS. Developed polymerization technique allows control the size, and polymers composition in relatively short time, and offers an opportunity to incorporate the polymers into functional networks which can be modified according to the particular drug delivery requirements. Nanoparticles are considered to be important in refining drug delivery, they can be pharmaceuticals as well as diagnostics. Refinements in drug delivery will facilitate the development of personalized medicine in which targeted drug delivery will play an important role. Novel DDS are also used in development of nanoscale diagnostic, drug discovery devices, microchips, sequencing of nanopores, etc. The targeting of drugs to an intracellular structure, cells, organs and tissues, could be done-based on control over release kinetics and protection of drugs involved. An extraordinary variety of nanoparticles, discovered over 30 years ago provides an opportunity to have often non-invasive localized drug release using molecular drug depots close to specific target, distribute the drug evenly and use smaller quantity of drug and low dosing to treat a variety of disease with increased effectiveness and success.

KEYWORDS

- chitosan
- CRS
- dispersions
- drugs delivery systems (DDS)
- liposomes
- magnetic/metallic nanoparticles
- MDDS
- micelles
- microcapsules
- microspheres
- nanoparticles (NP)
- nanostructure lipid carriers (NLCs)
- niosomes
- pharmacosomes
- quantum dots
- resealed erythrocytes
- solid lipid nanoparticles (SLNs)
- solid lipid nanoparticles (SLNs)
- titanium dioxide (TiO$_2$)

REFERENCES

Ahmed, A. U., & Lesniak, M. S., (2011). Glioblastoma multiforme: can neural stem cells deliver the therapeutic payload and fulfill the clinical promise? *Expert Review of Neurotherapeutics, 11*(6), 775–777.

Aimetti, A. A., Machen, A. J., & Anseth, K. S., (2009). Poly (ethylene glycol) hydrogels formed by thiolene photo polymerization for enzyme-responsive protein delivery. *Biomaterials, 30*, 6048–6054.

Akerud, P., Canals, J. M., Snyder, E. Y., et al. (2001). Neuroprotection through delivery of glial cell line-derived neurotrophic factor by neural stem cells in a mouse model of Parkinson's disease. *J. Neurosci., 21*(20), 8108–8118.

Alfons, K., & Engström, S., (1998). Drug compatibility with the sponge phases formed in monoolein, water and propyleneglycol or poly (ethylene glycol). *J. Pharm. Sci., 87*, 1527–1530.

Anders, R., & Merkle, H. P., (1989). Evaluation of laminated mucoadhesive patches for buccal drug delivery. *Int. J. Pharm., 49*, 231–240.

Bangham, A. D., & Horne, R. W., (1964). Negative staining of phospholipids and their structural modification by surface-active agents as observed in the electron microscope. *J. Mol. Biol., 8*, 660–668.

Banting, F. G., Best, C. H., Collip, J. B., Campbell, W. R., & Fletcher, A. A., (1922). Pancreatic extracts in the treatment of diabetes mellitus. *Canadian Medical Association Journal, 12*, 141–146.

Baroli, B., Ennas, M. G., Loffredo, F., Isola, M., Pinna, R., & López-Quintela, M. A., (2007). Penetration of metallic nanoparticles in human full-thickness skin. *J. Invest. Dermatol., 127*, 1701–1712.

Bender, J., Ericson, M. B., Merclin, N., et al. (2005). Lipid cubic phases for improved topical drug delivery in photodynamic therapy. *Journal of Controlled Release., 106*(3), 350–360.

Biswas, A., Joo, K. I., Liu, J., Zhao, M., Fan, G., Wang, P., et al. (2011). Endoprotease-mediated intracellular protein delivery using nanocapsules. *ACS Nano, 5*, 1385–1394.

Blake, J. C., Palmer, J. L., Minton, N. A., & Burroughs, A. K., (1993). The pharmacokinetics of intravenous ondansetron in patients with hepatic impairment. *Br. J. Clin. Pharmac., 35*, 441–443.

Bombardelli, E., & Spelta, M., (1991). Phospholipid-polyphenol complex a new concept in skin care ingredients. *Cosmet Toiletries, 106*, 69–76.

Bombardelli, E., Curri, S. B., Della, R. L., Del, N. P., Tubaro, A., & Gariboldi, P., (1989). Complexes between phospholipids and vegetal derivatives of biological interest. *Fitoterapia, 60*, 1–9.

Boyd, B. J., Whittaker, D. V., Khoo, S. M., & Davey, G., (2006). He␣osomes formed from glycerate surfactants-formulation as a colloidal carrier for irinotecan. *International Journal of Pharmaceutics, 318*(1–2), 154–162.

Boyd, B., Whittaker, D. V., Khoo, S. M., & Davey, G., (2006). Lyotropic liquid crystalline phases formed from glycerate surfactants as sustained release drug delivery systems. *International Journal of Pharmaceutics, 309*(1–2), 218–226.

Brunton, L., Chabner, B., Knollman, B., Goodman & Gilman, (2010). The Pharmacological Basis of Therapeutics. In: Buxton, I. L. O., Benet, L. Z., editors. Pharmacokinetics: The dynamics of drug absorption, distribution, metabolism, and elimination. McGraw-Hill, pp. 17–40.

Burkoth, T. L., Bellhouse, B. J., Hewson, G., Longridge, D. J., Muddle, A. G., et al. (1999). Transdermal and transmucosal powdered drug delivery. *Crit. Rev. Ther. Drug Carrier Syst.*, *16*(4), 331–384.

Burrows, R., Collett, J. H., & Attwood, D., (1994). The release of drugs from monoglyceride-water liquid crystalline phases. *Int. J. Pharm.*, *111*, 283–293.

Cai, L. L., Liu, P., Li, X., Huang, X., et al. (2011). RGD peptide-mediated chitosan-based polymeric micelles targeting delivery for integrin-overexpressing tumor cells. *Int. J. Nanomedicine*, *6*, 3499–3508.

Caruso, F., Hyeon, T., & Rotello, V. M. (2012). Nanomedicine. *Chem. Soc. Rev., 41*(7), 2537–2538.

Cevc, G, (1995). Material transport across permeability barriers by means of lipid vesicles, in: R. Lipowsky (Ed.), Handbook of Physics of Biological Systems, *I*, Elsevier Science., Chap. 9, pp. 441–466.

Challacombe, S. J., Rahman, D., & O'Hagan, D. T., (1997). Salivary, gut, vaginal and nasal antibody responses after oral immunization with biodegradable microparticles. *Vaccine*, *15*, 169–175.

Charcosset, C., El-Harati, A., & Fessi, H., (2006). Preparation of solid lipid nanoparticles using a membrane contactor. *J. Control Release*, *108*, 112–120.

Charlton, S. T., Davis, S. S., & Illum, L., (2007). Evaluation of bio adhesive polymers as delivery systems for nose to brain delivery: in vitro characterization studies. *J. Control Release*, *118*(2), 225–234.

Chaurasia, M., & Jain, N. K., (2006). Cross linked guar gum microspheres: A viable approach for improved delivery of anticancer drugs for the treatment of colorectal cancer. *AAPS Pharm. Sci. Tech.*, *7*, E143–E151.

Chebli, C., & Cartilier, L., (1999). Release and swelling kinetics of substituted amylose matrices. *J. Pharm. Belg.*, *54*(2), 51–54.

Chen, J., Dai, W. T., He, Z. M., et al. (2013). Fabrication and evaluation of curcumin-loaded nanoparticles-based on solid lipid as a new type of colloidal drug delivery system. *Indian Journal of Pharmaceutical Sciences*, *75*(2), 178–184.

Chen, Y., Lu, Y., Zhong, Y., Wang, Q., Wu, W., & Gao, S., (2012). Ocular delivery of cyclosporine A-based on glyceryl monooleate/poloxamer 407 liquid crystalline nanoparticles: preparation, characterization, in vitro corneal penetration and ocular irritation. *Journal of Drug Targeting*, *20*(10), 856–863.

Clogston, J., & Caffrey, M., (2005). Controlling release from the lipidic cubic phase. Amino acids, peptides, proteins and nucleic acids. *Journal of Controlled Release*, *107*(1), 97–111.

Coelho, J.F., Ferreira, P.C., et al. (2010). Drug delivery systems: Advanced technologies potentially
applicable in personalized treatments. *EPMA Journal.* *1*:164–209.

Costa-Balogh, F. O. C., Sparr, E., Souza, J. J. S., & Paes, A. C., (2010). Drug release from lipid liquid crystalline phases: relation with phase behavior. *Drug Dev. Ind. Pharm.*, *36*, 470–481.

Darshana Nagda, K. S., Rathore, et al. (2010). Bucky balls: A novel drug delivery system. *J. Chem. Pharm. Res.*, *2*(2), 240–248.

Dash, S., Murthy, P. N., Nath, L., & Chowdhury, P., (2010). Kinetic modeling on drug release from controlled drug delivery systems. *Acta Pol. Pharm.*, *67*(3), 217–223.

de Leeuw, J., de Vijlder, H. C., Bjerring, P., & Neumann, H. A., (2009). Liposomes in dermatology today. *J. Eur. Acad. Dermatol. Venereol.*, *23*, 505–516.

de Vries, M. E., Bodde, H. E., Verhoef, J. C., & Junginger, H. E., (1991). Developments in buccal drug delivery, *Crit. Rev. Ther. Drug Carr. Sys.*, *8*, 271–303.

Dechant, K. L., Clissold, S. P., & Sumatriptan, (1992). A review of its pharmacodynamic and pharmacokinetic properties, and therapeutic efficacy in the acute treatment of migraine and cluster headache. *Drugs*, *43*, 776–798.

Denne, J. R., Andrews, K. L., Lees, D. V., & Mook, W., (1992). A survey of patient preference for insulin jet injectors versus needle and syringe. *Diabetes Educ.*, *18*(3), 223–227.

Dharap, S. S., Wang, Y., et al. (2005). Tumor-specific targeting of an anticancer drug delivery system by LHRH peptide. *Proc. Natl. Acad. Sci. USA*, *102*(36), 12962–12967.

Dhuria, S. V., Hanson, L. R., Frey, II W. H., (2010). Intranasal delivery to the central nervous system: mechanisms and experimental considerations. *J. Pharm. Sci.*, *99*(4), 1654–1673.

Diehl, P., Fricke, A., et al. (2012). Microparticles: Major transport vehicles for distinct microRNAs in circulation. *Cardiovasc. Res.*, *93*(4), 633–644.

Ding, X., & Dahl, A. R., (2003). Olfactory mucosa: composition, enzymatic localization, and metabolism (chapter 3). In: Handbook of Olfaction and Gustation, 2nd edition. Doty, R. L. (ed.)., pp. 33–52. Marcel Dekker: New York, USA.

Drummond, C. J., & Fong, C., (1999). Surfactant self-assembly objects as novel drug delivery vehicles. *Curr. Opin. Colloid Interface Sci.*, *4*, 449–456.

Dugan, L. L., Turetsky, D. M., Du, C., et al. (1997). Carboxyfullerenes as neuroprotective agents. *Proc. Natl. Acad. Sci. USA*, *94*(17), 9434–9439.

Eeckman, F., Moës, A. J., & Amighi, K., (2002). Evaluation of a new controlled-drug delivery concept-based on the use of thermo responsive polymers. *Int. J. Pharm.*, *241*, 113–125.

Efentakis, M., Al-Hmoud, H., Buchton, G., et al. (1991). The influence of surfactant on drug release from hydrophobic matrix. *Int. J. Pharm.*, *70*, 153–158.

El Maghraby, G. M., Barry, B. W., & Williams, A. C., (2008). Liposomes and skin: From drug delivery to model membranes. *Eur. J. Pharm. Sci.*, *34*, 203–222.

El Maghraby, G. M., Williams, A. C., & Barry, B. W., (2001). Skin hydration and possible shunt route penetration in controlled estradiol delivery from ultra deformable and standard liposomes. *J. Pharm. Pharmacol.*, *53*, 1311–1322.

Eldridge, J. H., Staas, J. K., Chen, D., Marx, P. A., Tice, T. R., & Gilley, R. M., (1993). NEW advances in Vaccine Delivery Systems, Oct. 30, *4*(Suppl. 4), 16–24, Discussion 25.

Enderling, H., Rejniak, K. A., (2013). Simulating cancer: computational models in oncology. *Front. Oncol.*, *3*, 233.

Esparza, I., & Kissel, T., (1992). Parameters affecting the immunogenicity of microencapsulated tetanus toxoid. *Vaccine*, *10*, 714–720.

Esposito, E., Cortesi, R., Drechsler, M., et al. (2005). Cubosome dispersions as delivery systems for percutaneous administration of indomethacin. *Pharmaceutical Research.*, *22*(12), 2163–2173.

Fathima, A., Vedha Hari, B. N., & Ramya Devi, D., (2011). Micro particulate drug delivery system for anti- retroviral drugs: A review. *International Journal of Pharmaceutical and Clinical Research*, *4*, 11–16.

Feynman, R. P. (1960). Engineer and Science., *23*, 22–36.

Foldvari, M., Babiuk, S., & Badea, I., (2006). DNA delivery for vaccination and therapeutics through the skin. *Curr. Drug Deliv.*, *3*, 17–28.

Fong, W. K., Hanley, T., & Boyd, B. J., (2009). Stimuli responsive liquid crystals provide 'on-demand' drug delivery in vitro and in vivo. *Journal of Controlled Release*, *135*(3), 218–226.

Ford Versypt, A. N., Pack, D. W., & Braatz, (2013). Mathematical modeling of drug delivery from auto catalytically degradable plga microspheres: A Review. *J Control Release*, January 10, *165*(1), 29–37.

Frey, II W. H., (2002). Chiron Corporation (assignee), Method for administering a cytokine to the central nervous system and the lymphatic system, United States Patent 6991785. Mar. 20.

Friedman, A. J., Phan, J., Schairer, D. O., et al. (2013). Antimicrobial and anti-inflammatory activity of chitosan-alginate nanoparticles: A targeted therapy for cutaneous pathogens. *Journal of Investigative Dermatology*, *133*(5), 1231–1239.

Fruehwald-Schultes, B., Kern, W., Born, J., Fehm, H. L., & Peters, A., (2001). Hyper-insulinemia causes activation of the hypothalamus intranasal drug delivery 2483 pituitary-adrenal axis in humans. *Int. J. Obes. Relat. Metab. Disord.*, *25* Suppl 1, S38–40.

Gao, X. Y., He, C. L., Xiao, C. S., Zhuang, X. L., & Chen, X. S., (2013). Biodegradable pH-responsive polyacrylic acid derivative hydrogels with tunable swelling behavior for oral delivery of insulin. *Polymer*, *54*, 1786–1793.

Garcia, P., Youssef, I., Utvik, J. K., et al. (2010). Ciliary neurotrophic factor cell-based delivery prevents synaptic impairment and improves memory in mouse models of Alzheimer's disease. *J. Neurosci.*, *30*(22), 7516–7527.

Garcia-Rodriguez, J. C., & Sosa-Teste, I., (2009). The nasal route as a potential pathway for delivery of erythropoietin in the treatment of acute ischemic stroke in humans. *Scientific World Journal.*, *9*, 970–981.

Gasco, M.R., Trotta, M. (1986). Nanoparticles from microemulsions. *International Journal of Pharmaceutics, 29(2-3)*: 267.

Genç, L., Bilaç, H., & Güler, E., (1998). Preparation of controlled release dosage forms of diphenhydramine was prepared with different polymers. *International Journal of Pharmaceutics*, *169*(2), 232–235.

Gill, H. S., Andrews, S. N., Sakthivel, S. K., et al. (2009). Selective removal of stratum corneum by micro-dermabrasion to increase skin permeability. *Eur. J. Pharm. Sci.*, *38*(2), 95–103.

Giunchedi, P., Juliano, C., Gavini, E., Cossu, M., Sorrenti, M., (2002). Formulation and in vivo evaluation of chlorhexidine buccal tablets prepared using drug loaded chitosan microspheres. *Eur. J.Pharm. Biopharm.*, *53*, 233–239.

Giunchedi, P., Maggi, L., Manna, A. L., et al. (1999). Modification of the dissolution behavior of a water-insoluble drug, naftazine, for zero-order release matrix preparation. *J. Pharm. Pharmacol.*, *46*, 476–480.

Goycoolea, F. M., Lollo, G., Remuñán-López, C., Quaglia, F., & Alonso, M. J., (2009). Chitosan-alginate blended nanoparticles as carriers for the transmucosal delivery of macromolecules. *Biomacromolecules*, *10*(7), 1736–1743.

Gozes, I., & Divinski, I., (2007). NAP, a neuroprotective drug candidate in clinical trials, stimulates microtubule assembly in the living cell. *Curr. Alzheimer Res.*, 4, 507–509.

Greco, M. A. K., Frey, II, W. H., DeRose, J., Matthews, R. B., Hanson, L. R. B (inventors), and SRI International and Health Partners Research Foundation (assignees), (2009). Intranasal delivery of modafinil, United States Patent 20100204334. Feb. 6.

Gu, X., Wang, J., Wang, Y., Wang, Y., Gao, H., & Wu, G., (2013). Layer-by-layer assembled polyaspartamide nanocapsules for pH-responsive protein delivery. *Colloids and surfaces. Biointerfaces*, *108*, 205–211.

Guo, J., Skinner, G. W., & Harcum, W. W., (1998). Pharmaceutical application of naturally occurring water soluble polymers. *Pharmaceutical Science and Technology Today, 1,* 254–261.

Guo, J., Wang, F., Cao, R. J., Lee, & Zhai, G., (2010). Lyotropic liquid crystal systems in drug delivery. *Drug Discovery Today, 15*(23–24), 1032–1040.

Gurny, R., Boye, T., & Ibrahim, H., (1985). Ocular therapy with nanoparticulate systems for controlled drug delivery, *J. Control. Release, 2,* 353–361.

Han, I. K., Kim, M. Y., Byun, H. M., et al. (2007). Enhanced brain targeting efficiency of intranasally administered plasmid DNA: an alternative route for brain gene therapy. *J. Mol. Med., 85*(1), 75–83.

Handjani, R.M. (1989). Cosmetic and pharmaceutical compositions containing niosomes and a water-soluble polyamide, and a process for preparing these compositions. US patent No.-US4830857 A.

Hanson, L. R., Martinez, P. M., Taheri, S., Kamsheh, L., Mignot, E., et al. (2004). Intranasal administration of hypocretin 1 (orexin A) bypasses the blood–brain barrier and targets the brain: A new strategy for the treatment of narcolepsy. *Drug Del. Tech., 4,* 66–70.

Hanson, M. A., & Rouan, S. K. E., (1992). In: Stability of protein pharmaceuticals. Part B: In vivo pathways of degradation and strategies for protein stabilization. Ahern, T. J., Manning, M. C., (Eds.), Plenum: New York, pp. 209–233.

Harland, R. S., Gazzaniga, A., Sangalli, M. E., et al. (1988). Drug/polymer matrix swelling and dissolution. *Pharm. Res., 5*(8), 488–494.

Harris, D., Robinson, J. R., (1992). Drug delivery via the mucous membranes of the oral cavity. J. Pharm. Sci., *81,* 1–10.

Heasman, P. A., Offenbacher, S., Collins, J., Edwards, G. G., & Seymour, R. A. (1993). Flurbiprofen in the prevention and treatment of experimental gingivitis. *J. Clin. Periodontol. 20,* 732–738.

Heiati, H., Tawashi, R., & Phillips, N. C. (1998). Solid lipid nanoparticles as drug carriers II. Plasma stability and biodistribution of solid lipid nanoparticles containing the lipophilic prodrug 3'-azido-3'-deoxythymidine palmitate in mice. *Int. J. Pharm. 174,* 71–80.

Helledi, L. S., & Schubert, L. (2001). Release kinetics of acyclovir from a suspension of acyclovir incorporated in a cubic phase delivery system. *Drug Development and Industrial Pharmacy. 27*(10), 1073–1081.

Hermann, J., & Bodmeier, R. (1995). The effect of particle micro structure on the somatostatin release from poly (lactide)microspheres prepared by a w/o/ w solvent evaporation method. *J Control Release. 36,* 63–71.

Herper, M. (2012). The Truly Staggering Cost of Inventing New Drugs. Forbes/Pharma and HealthCare. February 10.

Hou, D. Z., Xie, C. S., Huang, K., et al. (2003). The production and characteristics of solid lipid nanoparticles (SLNs). *Biomaterials. 24,* 1781–1785.

Howard, C., Ansel Nicholos, G., & Popvich Lyold, V. (1995). Allen, pharmaceutical dosage forms and Drug Delivery system. 1st ed., p.78.

Ikehara, Y., Niwa, T., Biao, L., et al. (2006). A carbohydrate recognition-based drug delivery and controlled release system using intraperitoneal macrophages as a cellular vehicle. *Cancer Res. 66*(17), 8740–8748.

Irving, B. (2007). Nanoparticle drug delivery systems. *Inno. Pharm. Biotechnol., 24,* 58–62.

Janes, K. A., Fresneau, M. P., Marazuela, A., Fabra, A., & Alonso, M. A. J. (2001). Chitosan nanoparticles as delivery systems for doxorubicin. *J. Control Release., 73,* 255.

Jeyanthi, R., Mehta, R. C., Thanoo, B. C., & DeLuca, P. P. (1997). Effect of processing parameters on the properties of peptide-containing PLGA microspheres. *J. Microencapsul., 14*(2), 163–174.

Jung, S., Otberg, N., Thiede, G., Richter, H., Sterry, W., Panzner, S., et al. (2006). Innovative liposomes as a transfollicular drug delivery system: Penetration into porcine hair follicles. *J. Invest. Dermatol., 126*, 1728–1732.

Kalal, J., Drobnik, J., Kopecek, J., Exner, J. (1978). Water soluble polymers for medicine. *British Polymer Journal, 10*, pp. 111–114, 1978.

Kalantzi, L. E., Karavas, E., Koutris, E. X., et al. (2009). Recent advances in oral pulsatile drug delivery. *Recent Patents on Drug Delivery and Formulation, 3*, 49–63.

Kalita, B., Das, M., & Sharma, A. (2013). Novel Phytosome formulations in making herbal extracts more effective. *Res. J. Pharm. Technol., 6*(11), 1295–1301.

Kanwar, J. R., Sriramoju, B., & Kanwar, R. K. (2012). Neurological disorders and therapeutics targeted to surmount the blood brain barrier. *International Journal of Nanomedicine, 7*, 3259–3278.

Karande, P., Jain, A., & Mitragotri, S. (2004). Discovery of transdermal penetration enhancers by high throughput screening. *Nat. Biotechnol., 22*(2), 192–197.

Karande, P., Jain, A., & Mitragotri, S. (2006). Insights into synergistic interactions in binary mixtures of chemical permeation enhancers for transdermal drug delivery. *J. Control Release, 115*(1), 85–93.

Kean, T., Thanou, M. (2010). Biodegradation, biodistribution and toxicity of chitosan. *Adv. Drug Delivery Rev. 62*, 3.

Kern, W., Peters, A., Born, J., Fehm, H. L., & Schultes, B. (2005). Changes in blood pressure and plasma catecholamine levels during prolonged hyperinsulinemia. *Metabolism, 54*(3), 391–396.

Khandare, J. N, Madhavi, G., Tamhankar, B. M. (1994). Niosomes Novel Drug Delivery System. *The Eastern Pharmacist., 37*, 61–64.

Kianfar, F., Antonijevic, M. D., Chowdhry, B. Z., Boateng, J. S. (2011). Formulation development of a carrageenan-based delivery system for buccal drug delivery using ibuprofen as a model drug. *J Biomater. Nanobiotechnol., 2*, 582–595.

Kim, J. H., Kim, Y. S., Park, K., Lee, S., et al. (2008). Antitumor efficacy of cisplatin-loaded glycol chitosan nanoparticles in tumor-bearing mice. *J Control Release, 127*, 41.

Kim, S. Y., Doh, H. J., Jang, M. H., Ha, Y. J., Chung, S. I., & Park, H. J. (1999). Oral immunization with Helicobacter pylori-loaded poly (D, L-lactide-co-glycolide) nanoparticles. *Helicobacter, 4*, 33–39.

Koks, C., Meenhorst, P., Hillebrand, M., & Bult, A. (1996). Pharmacokinetics of fluconazole in saliva and plasma after administration of an oral suspension and capsules. *Antimicrob. Agents Chemother., 40*, 1935–1937.

Kreuter, J., & Speiser, P. (1976). In vitro studies of poly (methylmethacrylate) adjuvants, *J. Pharm. Sci., 65*, 1624–1627.

Krishna Sailaja, A., et al. (2016). Biomedical Applications of Microspheres. *J. Modern Drug Discovery And Drug Delivery Research*. Volume 4 / Issue 2, 1-5.

Kristensson, K., & Olsson, Y. (1971). Uptake of exogenous proteins in mouse olfactory cells. *Acta Neuropathol. (Berl)., 19*, 145–154.

Kroto, H. W., Heath, J. R., O'Brien, S. C., et al. (1985). C60, buckminsterfullerene. *Nature, 318*, 162–163.

Lademann, J., Knorr, F., Richter, H., et al. (2008). Hair follicles – an efficient storage and penetration pathway for topically applied substances. *Skin Pharmacol. Physiol., 21*, 150–155.

Lakshmi, P. K. (2010). Dissolution testing is widely used in the pharmaceutical industry for optimization of formulation and quality control of different dosage forms, *Pharma info.net*.

Landau, A. J., Eberhardt, R. T., & Frishman, W. H. (1994). Intranasal delivery of cardiovascular agents: an innovative approach to cardiovascular pharmacotherapy. *American Heart Journal, 127*(6), 1594–1599.

Langer, R. (1990). New methods of drug delivery. *Science, 249*(4976), 1527–1533.

Lawrence, M. J., & Rees, G. D. (2000). Microemulsion-based media as novel drug delivery systems. *Adv. Drug Deliv. Rev., 45*, 89–121.

Lee, E. S., Gao, Z., Kim, D., et al. (2008). Super pH-sensitive multifunctional polymeric micelle for tumor pH(e) specific TAT exposure and multidrug resistance. *J. Control Release, 129*, 228–236.

Lee, P. I., & Li, J. X. (2010). Evolution of oral controlled release dosage forms, in: Wen, H., Park, K. (Eds.), Oral Controlled Release Formulation Design and Drug Delivery, John Wiley and Sons, Inc.: Hoboken, NJ, USA, pp. 21–31.

Lee, W. R., Shen, S. C., Lai, H. H., Hu, C. H., & Fang, J. Y. (2001). A study of the structural disruption of skin subjected to high temperature. Transdermal drug delivery enhanced and controlled by erbium: YAG laser: a comparative study of lipophilic and hydrophilic drugs. *J. Control Release, 75*(1–2), 155–166.

Lemarchand, C., Couvreur, P., Besnard, M., Costantini, D., & Gref, R. (2003). Novel polyesterpolysaccharide nanoparticles. *Pharmaceutical Research, 20*, 1284–1292.

Liu, C., Krishnan, J., Stebbing, J., & Xu, X. Y. (2011). Use of mathematical models to understand anticancer drug delivery and its effect on solid tumors. *Pharmacogenomics, 12*(9), 1337–48, Epub 2011/09/1710.2217/pgs.11.71.

Liu, X. F., Fawcett, J. R., Hanson, L. R., Frey, II W. H. (2004). The window of opportunity for treatment of focal cerebral ischemic damage with noninvasive intranasal insulin-like growth factor-I in rats. *J Stroke Cerebrovasc. Dis., 13*, 16–23.

Lochhead, J., & Thorne, R. (2012). Intranasal delivery of biologics to the central nervous system. *Advanced Drug Delivery Reviews, 64*, 614–628.

Loew, S., Fahr, A., & May, S. (2011). Modeling the Release Kinetics of Poorly Water-soluble Drug Molecules from Liposomal Nanocarriers. *J. Drug Deliv., 2011*, 376548.

Lopes, L. B., Speretta, F. F. F., & Bentley, M. V. L. B. (2007). Enhancement of skin penetration of vitamin K using monoolein-based liquid crystalline systems. *European Journal of Pharmaceutical Sciences, 32*(3), 209–215.

Luo, M., Shen, Q., Chen, J. (2011). Transdermal delivery of paeonol using cubic gel and microemulsion gel. *International Journal of Nanomedicine, 6*, 1603–1610.

Ma, Z., Lim, T. M., & Lim, L. Y. (2005). Pharmacological activity of peroral chitosan–insulin nanoparticles in diabetic rats. *Int. J. Pharm., 293*, 271.

Maloy, K. J., Donachie, A. M., O'Hagan, D. T., & Mowat, A. M. (1994). Induction of mucosal and systemic immune responses by immunization with ovalbumin entrapped in poly (lactide-co-glycolide) microparticles. *Immunology, 81*, 661–667.

Martinez Gonzalez, I., & Villafuerte Robles, L. (2003). Influence of enteric citric acid on the release profile of 4-aminopyridine from HPMC matrix tablets. *Int. J. Pharm., 251*(1–2), 183–193.

Mashino, T., Shimotohno, K., Ikegami, N., et al. (2005). Human immunodeficiency virus-reverse transcriptase inhibition and hepatitis C virus RNA-dependent RNA polymerase inhibition activities of fullerene derivatives. *Bioorg. Med. Chem. Lett., 15*(4), 1107–1109.

Matsumura, Y., & Maeda, H. (1986). A new concept for macromolecular therapeutics in cancer chemotherapy: mechanism of tumoritropic accumulation of proteins and the antitumor agent smancs. *Cancer Research, 46*, 6387–6392.

Mitragotri, S., (2001). Effect of therapeutic ultrasound on partition and diffusion coefficients in human stratum corneum. *J. Control Release, 71*(1), 23–29.

Mitragotri, S., Blankschtein, D., & Langer, R. (1996). Transdermal drug delivery using low-frequency sonophoresis. *Pharm. Res., 13*(3), 411–420.

Mittal, G., Sahana, D. K., Bhardwaj, V., & Ravi Kumar, M. N. V. (2007). Estradiol loaded PLGA nanoparticles for oral administration: effect of polymer molecular weight and copolymer composition on release behavior in vitro and in *vivo. J Control Rel, 119*(1), 77–85.

Mizrahi, B., & Domb, A. J., (2007). Mucoadhesive tablet releasing iodine for treating oral infections. *J. Pharm. Sci., 96*, 3144–3150.

Mo, R., Sun, Q., Xue, J., Li, N., Li, W., Zhang, C., et al. (2012). Multistage pH-responsive liposomes for mitochondrial-targeted anticancer drug delivery. *Advanced materials, 24*, 3659–3665.

Morgan, T. M., Reed, B. L., & Finnin, B. C. (1998). Enhanced skin permeation of sex hormones with novel topical spray vehicles. *J. Pharm. Sci., 87*, 1213–1218.

Morrison, E. E., Costanzo, R. M. (1990). Morphology of the human olfactory epithelium. *J. Comp. Neurol., 297*(1), 1–13.

Moser, C., & Amacker, M. (2013). Influenza Virosomes as Antigen Delivery System. In: Novel Immune Potentiators and Delivery Technologies for Next Generation Vaccines, Springer: USA, 287–307.

Mueller-Goymann, C. C. (2002). Liquid crystals in drug delivery. In: Swarbrick, J., Boylan, J.C., (Eds.), Encyclopedia of Pharmaceutical Technology. Marcel Dekker: New York and Basel, pp. 834–853.

Müller, R. H., Mader, K., & Gohla, S. (2000). Solid lipid nanoparticles (SLN) for controlled drug delivery – a review of the state of the art, *Eur. J. Pharm. Biopharm., 50*(1), 161–177.

Muller, R., Lucks, J. (1996). Drug Carriers from Solid Lipid Particles, Solid Lipid Nanospheres (SLN). European Patent.

Mura, P., Corti, G., Cirri, M., Maestrelli, F., Mennini, N., & Bragagni, M. (2010). Development of mucoadhesive films for buccal administration of flufenamic acid: Effect of cyclodextrin complexation. *J. Pharm. Sci., 99*, 3019–3029.

Muzib, Y. I., & Kumari, K. S. (2011). Mucoadhesive buccal films of glibenclamide: Development and evaluation. *Int. J. Pharma. Investig., 1*, 42–47.

Nagai, T. (1988). Drug delivery systems by controlled release. *Yakugaku Zasshi, 108*(7), 613–624.

Nakamizo, A., Marini, F., Amano, T., et al. (2005). Human bone marrow-derived mesenchymal stem cells in the treatment of gliomas. *Cancer Res., 65*(8), 3307–3318.

Narisawa, S., Nagata, M., Danyoshi, C., et al. (1994). An organic acid-induced sigmoidal release system for oral controlled-release preparations. *Pharm. Res., 11*, 111–116.

Negrini, R., & Mezzenga, R. (2011). PH-responsive lyotropic liquid crystals for controlled drug delivery. *Langmuir, 27*(9), 5296–5303.

Ngoepe, M., Choonara, Y. E., et al. (2013). Integration of biosensors and drug delivery technologies for early detection and chronic management of illness. *Sensors, 13*, 7680–7713, doi: 10.3390/s130607680.

Nineth World Drug Delivery Summit, June 30–July 02, 2016. New Orleans, Louisiana, USA.

Nitta, S. K., & Numata, K. (2013). Biopolymer-based nanoparticles for drug/gene delivery and tissue engineering. *International Journal of Molecular Sciences, 14*(1), 1629–1654.

Okamoto, H., Taguchi, H., Iida, K., & Danjo, K. (2001). Development of polymer film dosage forms of lidocaine for buccal administration: I. Penetration rate and release rate. *J. Control Release, 77*, 253–260.

Orlic, D., Kajstura, J., Chimenti, S., et al. (2001). Bone marrow cells regenerate infarcted myocardium. *Nature, 410*(6829), 701–705.

Orlic, D., Kajstura, J., Chimenti, S., et al. (2001). Mobilized bone marrow cells repair the infarcted heart, improving function and survival. *Proc. Natl. Acad. Sci. USA, 98*(18), 10344–10349.

Park, J. H., Lee, J. W., Kim, Y. C., & Prausnitz, M. R. (2008). The effect of heat on skin permeability. *Int. J. Pharm. 359*(1–2), 94–103.

Park, J. H., Saravanakumar, G., Kim, K., & Kwon, I. C. (2010). Targeted delivery of low molecular drugs using chitosan and its derivatives. *Adv. Drug Delivery Rev., 62*, 28.

Patil, S. (2008). Cross linking of polysaccharides: Methods and applications. *Latest Reviews, 6*, 1–18.

Peppas, N. A., & Klier, J. (1991). Controlled release by using poly (methacrylic acid-g-ethylene glycol) hydrogels. *J Control Release, 16*, 203–214.

Peppas, N. A., & Sahlin, J. J. (1989). A simple equation for the description of solute release (III): coupling of diffusion and relaxation. *Int. J. Pharm., 57*(2), 169–172.

Phan, S., Fong, W. K., Kirby, N., Hanley, T., Boyd, B. J. (2011). Evaluating the link between self-assembled mesophase structure and drug release. *Int J Pharm., 421*, 176–182.

Pongjanyakul, T., & Suksri, H. (2009). Alginate-magnesium aluminum silicate films for buccal delivery of nicotine. *Colloids Surf B Biointerfaces, 74*, 103–113.

Prausnitz, M. R. (1999). A practical assessment of transdermal drug delivery by skin electroporation. *Adv. Drug Deliv. Rev., 35*(1), 61–76.

Prausnitz, M. R., & Langer, R. (2008). Transdermal drug delivery. *Nat. Biotech., 26*(11), 1261–1268.

Prausnitz, M. R., Mitragotri, S., & Langer, R. (2004). Current status and future potential of transdermal drug delivery. *Nat. Rev. Drug Discov., 3*(2), 115–124.

Puratchikody, A., Prasanth, V. V., Mathew, S. T., & Ashok, K. B. (2011). Buccal drug delivery: Past, present and future: A Review. *Int. J. Drug Deliv., 3*, 171–184.

Ramishetti, S., & Huang, L. (2012). Intelligent design of multifunctional lipid-coated nanoparticle platforms for cancer therapy. *Therapeutic Delivery, 3*(12), 1429–1445.

Rao, V. M., Haslam, J. L., & Stella, V. J. (2001). Controlled and complete release of a model poorly water-soluble drug, prednisolone, from hydroxypropyl methylcellulose matrix tablets using (SBE) (7m)-beta-cyclodextrin as a solubilizing agent. *J. Pharm. Sci., 90*(7), 807–816.

Rapoport, N. Y., Christensen, D. A., Fain, H. D., Barrows, L., & Gao, Z. (2004). Ultrasound-triggered drug targeting of tumor in vitro and in vivo. *Ultrasonics, 42*, 943–950.

Renner, D. B., Svitak, A. L., Gallus, N. G., Ericson, M. E., Frey, W. H. II, & Hanson, L. R. (2012). Intranasal delivery of insulin via the olfactory nerve pathway. *J. Pharm. Pharmacol.* doi:10.1111/j.2042–7158.2012. 01555.x.

Rigon, R. B., Oyafuso, M. H., et al. (2015). Nanotechnology-based drug delivery systems for melanoma antitumoral therapy: A Review. *Biomed Res Int., 2015*, 841817.

Riviere, J. E., & Heit, M. C. (1997). Electrically-assisted transdermal drug delivery. *Pharm. Res., 14*(6), 687–697.

Rothbard, J. B., Garlington, S., Lin, Q., et al. (2000), Conjugation of arginine oligomers to cyclosporin A facilitates topical delivery and inhibition of inflammation. *Nat. Med., 6*, 1253–1257.

Sainsbury, F., Zeng, B, Middelberg, A. P. J. (2014). Towards designer nanoemulsions for precision delivery of therapeutics. *Curr. Opin. Chem. Eng., 4*, 11–17.

Sallam, A., Khalil, E., Ibrahim, H., & Freij, I. (2002). Formulation of an oral dosage form using the properties of cubic liquid crystalline phases of glyceryl monooleate. *Eur. J. Pharm. Biopharm., 53*, 343–352.

Sambathkumar, R., Venkateswaramurthy, N., & Vijayabaskaran, M. (2011). Formulation of Clarithromycin loaded Mucoadhesive microspheres by emulsification-internal gelation technique for anti-helicobacter pylori therapy. *International Journal of Pharmacy and Pharmaceutical Sciences, 3*(2), 173–177.

Schatzlein, A., & Cevc, G. (1995). Skin penetration by phospholipids vesicles, Transfersomes as visualized by means of the Confocal Scanning Laser Microscopy, in characterization, metabolism, and novel biological applications, AOCS Press, 191–209.

Schroeder, A., Honen, R., Turjeman, K., Gabizon, A., Kost, J., & Barenholz, Y. (2009). Ultrasound triggered release of cisplatin from liposomes in murine tumors. *J. Control Release, 137*, 63–68.

Scranton, R. A., Fletcher, L., Sprague, S., Jimenez, D. F., & Digicaylioglu, M. (2011). The rostral migratory stream plays a key role in intranasal delivery of drugs into the CNS. *PLoS One, 6*(4), e18711.

Sekhar, K. C., Naidu, K. V., Vishnu, Y. V., Gannu, R., Kishan, V., & Rao, Y. M. (2008). Transbuccal delivery of chlorpheniramine maleate from mucoadhesive buccal patches. *Drug Deliv., 15*, 185–191.

Shargel, L., & Yu, A. B. C. (1999). Modified release drug products. In: Applied Biopharmaceutics and Pharmacokinetics. 4th ed. McGraw Hill, 169–171.

Shuo Yang, Xianzhen Yin, Caifen Wang, et al. (2014). Release Behavior of Single Pellets and Internal Fine 3D Structural Features Co-define the in Vitro Drug Release Profile. *AAPS J., 16*(4), 860–871.

Sjoblom, J., Lindberg, R., & Friberg, S.E. (1996). Microemulsions: Phase equilibria characterization, structures, applications and chemical reactions. *Advances in Colloid and Interface Science, 95*, 125-287.

Soppimath, K. S., Aminabhavi, T. M. et al. (2001). Biodegradable polymeric nanoparticles as drug delivery devices. *Journal of Controlled Release, 70*(1–2), 1–20.

Spicer, P. T., Hayden, K. L., Lynch, M. L., Ofori-Boateng, A., & Burns, J. L. (2001). Novel process for producing cubic liquid crystalline nanoparticles (cubosomes). *Langmuir, 17*(19), 5748–5756.

Stevenson, C.L, Bennett, D. B., Ballesteros, D. L. (2005). Pharmaceutical liquid crystals: the relevance of partially ordered systems. *J. Pharm. Sci., 94*, 1861–79.

Streubel, A., Siepmann, J., Dashevsky, A., et al. (2000). pH-independent release of a weakly basic drug from water-insoluble and -soluble matrix tablets. *J. Control Release, 67*(1), 101–110.

Stuart, M. A., Huck, W. T., Genzer, J., et al. (2010). Emerging applications of stimuli-responsive polymer materials. *Nature materials, 9*, 101–113.

Suhaimi, N. F., & Jalaludin, J. (2015). Biomarker as a research tool in linking exposure to air particles and respiratory health. *BioMed Research International, V.* Article ID 96285.

Swarnakar, N. K., Jain, V., Dubey, V., Mishra, D., & Jain, N. K. (2007). Enhanced oromucosal delivery of progesterone via hexosomes. *Pharmaceutical Research, 24*(12), 2223–2230.

Swierniak, A., Kimmel, M., & Smieja, J. (2009). Mathematical modeling as a tool for planning anticancer therapy. *Eur. J. Pharmacol., 625*(1–3), 108–121.

Takada, S., Uda, Y., Toguchi, H., & Ogawa, Y. (1995). Application of a spray drying technique in the production of TRH-containing injectable sustained-release microparticles of biodegradable polymers. *J. Pharm. Sci. Technol, 49*(4), 180–184.

Tan, M. L., Choong, P. F., & Dass, C. R. (2010). Recent developments in liposomes, microparticles and nanoparticles for protein and peptide drug delivery. *Peptides, 31*(1), 184–193.

Tanna, S., Sahota, T., Clark, J., & Taylor, M. J. (2002). Covalent coupling of concanavalin A to a Carbopol 934P and 941P carrier in glucose-sensitive gels for delivery of insulin. *The Journal of Pharmacy and Pharmacology, 54*, 1461–1469.

Tanner, P., Baumann, P., Enea, R., Onaca, O., Palivan, C., & Meier, W. (2011). Polymeric vesicles: from drug carriers to nanoreactors and artificial organelles. *Acc. Chem. Res., 44*, 1039–1049.

Thiele, L., Merkle, H. P., & Walter, E. (2003). Phagocytosis and phagosomal fate of surface-modified microparticles in dendritic cells and macrophages. *Pharm Res., 20*(2), 221–228.

Thorne, R. G., & Frey II W. H. (2001). Delivery of neurotrophic factors to the central nervous system: pharmacokinetic considerations. *Clin. Pharmacokinet, 40*(12), 907–946.

Thorne, R. G., Hanson, L. R., Ross, T. M., Tung, D., & Frey, W. H. II. (2008). Delivery of interferon-beta to the monkey nervous system following intranasal administration. *Neuroscience, 152*(3), 785–797.

Thorne, R. G., Pronk, G. J., Padmanabhan, V., & Frey II W. H. (2004). Delivery of insulin-like growth factor-I to the rat brain and spinal cord along olfactory and trigeminal pathways following intranasal administration. *Neuroscience, 127*, 481–496.

Thrash, T. P., Cagle, D. W., Alford, J. M., et al. (1999). Toward fullerene-based radiopharmaceuticals: high-yield neutron activation of endohedral 165Ho metallofullerenes. *Chem. Phys. Lett., 308*(3–4), 329–336.

Torchilin, V. P., & Papisov, M. I. (1994). Hypothesis: Why do PEG coated liposomes circulate so long?. *J Liposome Res., 4*, 725–739.

Tung, T., Kamm, W., Breitenbach, A., Hungerer, K. D., Hundt, E., & Kissel, T. (2001). Tetanus toxoid loaded nanoparticles from sulfobutylated poly (vinyl alcohol)-graft-poly (lactide-co-glycolide): Evaluation of antibody response after oral and nasal application in mice. *Pharm Res., 18*, 352–360.

Vaka, S. R., & Murthy, S. N. (2010). Enhancement of nose-brain delivery of therapeutic agents for treating neurodegenerative diseases using peppermint oil. *Pharmazie, 65*(9), 690–692.

Van der Pol, A., Boing, A. N., Harrison, P., et al. (2012). Classification, functions, and clinical relevance of extracellular vesicles. *Pharmacological Reviews, 64*(3), 676–705.

Victor, S. P., Paul, W., & Sharma, C. P. (2013). Chitosan self-aggregates and micelles in drug delivery. *J. Nanopharm. Drug Deliv., 1*, 193.

Vinod, K. R., Sandhya, S., Chandrashekar, J., Swetha, R., Rajeshwar, T., Banji, D., et al. (2010). A review on genesis and characterization of phytosomes. *Int. J. Pharm. Sci. Rev. Res., 4*(3), 69–75.

Vyas, S. P., & Khar, R. K. (2002). Controlled Drug delivery: Concepts and Advances. Concepts and Advances. 1st ed. Vallabh Prakashan. pp. 156–189.

Wang, H., & Park, J. (2011). A possible approach for the desire to innovate. Available at: www.ondrugdelivery.com, Ghosh T, Ghosh, A. Drug delivery through osmotic systems: An overview. *Journal of Applied Pharmaceutical Science, 1*, 38–49.

Wang, J., Lu, Z., et al. (2014). Tumor priming enhances siRNA delivery and transfection in intraperitoneal tumors. *Journal of Control Release, 178*, 79–85.

Wang, X., Zheng, C., Wu, Z., Teng, D., Zhang, X., Wang, Z., et al. (2009). Chitosan-NAC nanoparticles as a vehicle for nasal absorption enhancement of insulin. *J. Biomed. Mater. Res. B. Appl Biomater, 88*(1), 150–161.

Watanabe, K., Yakou, S., Takayama, K., Machida, Y., & Nagai, T. (1991). Drug release behaviors from hydrogel prepared with water soluble dietary fibers, *J. Pharm. Sci. Techn. Jpn, 51*, 29–35.

Welling, P. G., & Dobrinska, M. R. (1978). Dosing consideration and bioavailability assessment of controlled drug delivery system, Chapter 7, Controlled drug delivery, fundamentals and applications, 2nd edition, Robinson, J. R., & Lee, V. H. L. (Eds.), Marcel Dekker Inc.: New York, USA, *29*, pp. 254, 373.

Wender, P. A., Rothbard, I. B., et al. (2002). Oligocarbamate molecular transporters: Design, synthesis, and biological evaluation of a new class of transporters for drug delivery. *J. Am. Chem. Soc., 124*, 13382–13383.

Wermeling, D. P., Banks, S. L., Hudson, D. A. et al. (2008). Microneedles permit transdermal delivery of a skin impermeant medication to humans. *Proc. Natl. Acad. Sci. USA, 105*(6), 2058–2063.

Williams, A. C., & Barry, B. W. (2004). Penetration enhancers. *Adv. Drug Deliv. Rev., 56*, 603–618.

Wilson, D. E., Kaidbey, K., Boike, S. C., & Jorkasky, D. K. (1998). Use of topical corticosteroid pretreatment to reduce the incidence and severity of skin reactions associated with testosterone transdermal therapy. *Clin. Ther., 20*(2), 299–306.

Woods, S. C., Seeley, R. J., Baskin, D. G., & Schwartz, M. W. (2003). Insulin and the blood–brain barrier. *Curr. Pharm. Des., 9*(10), 795–800.

Woolfson, A. D., McCafferty, D. F., & Moss, G. P. (1998). Development and characterization of a moisture-activated bioadhesive drug delivery system for percutaneous local anesthesia. *Int. J. Pharm., 169*, 83–94.

Woolfson, A. D., McCafferty, D. F., McCallion, C. R., McAdams, E. T., & Anderson, J. (1995). Moisture-activated, electrically-conducting bioadhesive hydrogels as interfaces for bio-electrodes: Effect of film hydration on cutaneous adherence in wet environments. *J. Appl. Polym. Sci., 58*, 1291–1296.

Yaghmur, A., De Campo, L., Sagalowicz, L., Leser, M. E., & Glatter, O. (2005). Emulsified micro emulsions and oIL-containing liquid crystalline phases. *Langmuir, 21*(2), 569–577.

Yamada, K., Yamashita, J., Todo, H., et al. (2011). Preparation and evaluation of liquid-crystal formulations with skin-permeation enhancing abilities for entrapped drugs. *Journal of Oleo Science, 60*(1), 31–40.

Yamsani, V. V., Gannu, R., Kolli, C., Rao, M. E., & Yamsani, M. R. (2007). Development and in vitro evaluation of buccoadhesive carvedilol tablets. *Acta Pharm., 57*, 185–197.

Yariv, D., Efrat, R., Libster, D., Aserin, A., & Garti, N. (2010). *In vitro* permeation of diclofenac salts from lyotropic liquid crystalline systems. *Colloids Surf. B: Biointerfaces, 78*, 185–192.

Yehia, S. A., El-Gazayerly, O. N., & Basalious, E. B. (2008). Design and in vitro/in vivo evaluation of novel mucoadhesive buccal discs of an antifungal drug: relationship between swelling, erosion, and drug release. *AAPS Pharm. Sci. Tech., 9,* 1207–1217.

Yokoyama, M., Kwon, G. S., Okano, T., Sakurai, Y., Seto, T., & Kataoka, K. (1992). Preparation of micelle-forming polymer-drug Conjugates. *Bioconjugate Chemistry, 3,* no. 4, pp. 295–301

Yoshida, T., Tasaki, H., Maeda, A., et al. (2008). Mechanism of controlled drug release from a salting-out taste-masking system. *J. Control Release, 131,* 47–53.

Yoshida, T., Tasaki, H., Maeda, A., et al. (2009). Salting-out taste-masking system generates lag time with subsequent immediate release. *Int. J. Pharm., 365,* 81–88.

Zhang, W., Tichy, S. E., Pérez, L. M., Maria, G. C., Lindahl, P. A., & Simanek, E. E. (2003). Evaluation of multivalent dendrimers-based on melamine: kinetics of thiol–disulfide exchange depends on the structure of the dendrimer. *Journal of the American Chemical Society, 125,* 5086–5094.

Zhao, Y., Trewyn, B. G., & Slowing Lin, V. S. (2009). Mesoporous silica nanoparticle-based double drug delivery system for glucose-responsive controlled release of insulin and cyclic AMP. *Journal of the American Chemical Society, 131,* 8398–8400.

CHAPTER 5

APPLICATION OF NANOBIOFORMULATIONS FOR CONTROLLED RELEASE AND TARGETED BIODISTRIBUTION OF DRUGS

JOSEF JAMPÍLEK[1] and KATARÍNA KRÁĽOVÁ[2]

[1]Department of Pharmaceutical Chemistry, Faculty of Pharmacy, Comenius University, Odboj6rov 10, 83232, Bratislava, Slovakia, E-mail: josef.jampilek@gmail.com

[2]Institute of Chemistry, Faculty of Natural Sciences, Comenius University, Ilkovičova 6, 84215, Bratislava, Slovakia

CONTENTS

This chapter is sincerely dedicated to Professor Ferdinand Devínsky on the occasion of his 70th birthday.

5.1 INTRODUCTION

Nanotechnology can be regarded as one of the key technologies of the 21st century. U.S. National Nanotechnology Initiative defines nanoparticles (NPs) in the range 1–100 nm (National Nanotechnology Initiative, 2008). However, in pharmacy particles of 10–500 nm have been used, rarely up to 700 nm (Mody et al., 2010). From the aspect of passage through vessels, the inside diameter of which is in the range from 25 mm (aorta) to 5 μm (capillaries), the ideal size of NPs should be <300 nm to ensure efficient transport for targeted distribution of drugs (Sivasankar and Kumar, 2010; Couvreur, 2013; Suffredini and Levy, 2014). Nanoscale materials exhibit unusual physical, chemical, and biological properties, differing in important ways from the properties of bulk materials and single atoms or molecules (Medina et al., 2007; National Nanotechnology Initiative, 2008), therefore different properties of active pharmaceutical ingredients have been modified in such a way (Bamrungsap et al., 2012; Nekkanti et al., 2012; Dolez, 2015).

NPs in biomedical branches can be used as nanodiagnostics, as nanomaterials for tissue engineering, as drug carriers for specific delivery/targeted biodistribution or controlled release, and as agents/drugs for prevention/ treatment of diseases. Application of nanotechnology thus can be considered as an e□cellent tool for modification of parameters of drugs. Due to modification of properties using nanosystems/nanoformulations, enhanced bioavailability of active substances can be ensured, and route of administration can be modified. Due to these facts, smaller amount of substance can be used, for example, dose-dependent toxicity and various side effects decrease. In addition, many formulations also protect drugs from degradation (Allen and Cullis, 2004; Bamrungsap et al., 2012; Kamaly et al., 2012; Nekkanti et al., 2012; Parveen et al., 2012; Dolez, 2015).

Penetration through cell wall to cells and tissues, including brain, is probably the most appreciated changed property of drug NPs in pharmacy/ galenics. Especially NPs with particle size <100 nm are critical, because they can practically unlimitedly permeate through biomembranes. This property is connected not only with particle size but also with particle shape. Unfortunately this "free" permeation is related to the toxic effect of NPs and general risks of NPs for human health and environmental burden due to undesirable permeation of NPs to non-target organs and tissues (Buzea et al., 2007; Hagens et al., 2007; De Jong and Borm, 2008; Keck and Müller, 2013; Nehoff et al., 2014).

Nanoformulations can be considered as a complex of various materials in nanoscale. The biodegradability (ability of a compound to be degraded in organism/environment) of matrices/formulating stabilizers is connected not only with toxicity but also with formulation stability. Therefore, a system classifying NPs according to their particle size and biodegradability into four classes was suggested: (i) size >100 nm and biodegradable, (ii) size >100 nm and non-biodegradable, (iii) size <100 nm and biodegradable, (iv) size <100 nm and non-biodegradable (Keck and Müller, 2013). It is evident that the last class can be regarded as the most toxic. However, also degraded nanomaterial can damage cells by indirect toxicity, when fragments are accumulated in the cell, and their amount leads to intracellular changes such as disruption of organelle integrity, gene alterations, etc. (Medina et al., 2007; Lewinski et al., 2008; Fröhlich, 2013; Lungu et al., 2015).

5.2 NANOFORMULATIONS AND BIODEGRADABLE MATRICES

NPs can be generated by either top–down methods (dispergation, fluidization, homogenization processes or emulsifying technologies) or bottom–up methods (precipitation/condensations processes, evaporation techniques, various controlled sol–gel syntheses) (Nalwa, 2004–2011; Bhushan et al., 2014). Mechanical approaches are capable of producing NPs, typically, in the 100–1000 nm range, chemical and bottom–up methods tend to produce 10–100 nm particles (Acosta, 2009; Bhushan et al., 2014; Nalwa, 2004–2011). Innovative biotechnological approaches related to synthesis of NPs are summarized in Singh (2015).

NPs can be prepared from both inorganic and organic materials (Nalwa, 2004–2011; Bhushan et al., 2014; Singh, 2015). Nanonized substances must be stabilized (protected from aggregation) by some matrix/stabilizer, therefore these nanoformulations usually consist of two basic components, an active organic or inorganic ingredient and a matrix/stabilizer/nanocarrier that stabilizes the nanonized active ingredient (Perlatti et al., 2013; Nehoff et al., 2014). According to the stabilizer, NPs can be grouped into three major categories: (i) inorganic-based (mostly non-biodegradable), (ii) organic-based (frequently biodegradable), and (iii) hybrid NPs. As mentioned above, water-soluble biodegradable biocompatible polymers are more popular due to a decrease of toxic effects and now frequently used. A number of organic-based natural or synthetic/semisynthetic homopolymers, copolymers as well as combinations of various polymers and conjugates with other compounds can be found. The organic polymers/stabilizers most frequently generate the following types of nanosystems: liposomes, solid lipid particles, polymeric particles, micelles, capsules, spheres, and gels (De Jong and Borm, 2008; Mozafari, 2006; Zhang et al., 2013a). In general, biodegradable materials are materials of natural, synthetic or semisynthetic origin, degradable chemically (non-enzymatically) or enzymatically *in vivo* to biocompatible, toxicologically safe products that are eliminated from the human body by the standard metabolic pathways. A number of such materials have been used as excipients/matrices in controlled drug delivery. The basic categories of biomaterials used in drug delivery can be distinguished as follows: (i) synthetic biodegradable relatively hydrophobic polymers, for e□ample, α-hydro□y acids, polyanhydrides, and (ii) naturally occurring polymers (such as various carbohydrate-based materials, protein-based polymers, nucleic acid-based carriers) or inorganics (hydroxyapatite) (Makadia and Siegel, 2011; Nair and Laurencin, 2007; Uhrich et al., 1999; Zhang et al., 2013a).

The formulations based on biodegradable organic-based matrices prepared by encapsulation technology allow designing controlled-release nanocarriers. Controlled release formulations are based on interactions between an active ingredient and a carrier. The active ingredient can also be immobilized in the space of the cross-linked polymer. The release of the active ingredient is connected with the nature of the formulation and can proceed via diffusion and disassembling of polymeric nanomatrices after contact with water. In nanoformulations, in which polymer-active substance bonds occur, a hydrolysis reaction is needed for the release of the active ingredient, and the release is affected by the strength of chemical bonds, the chemical properties of both molecules, and by the size and the structure of the macromolecule. Thus encapsulation has been widely used in the pharmaceutical drug delivery field, because controlled release formulations prepared using the encapsulation technology protect the drug, maintain the concentration of the drug within an optimum range by adjusting the release of active components, and provide the sustainability of the effect (Kawabata et al., 2011; Margulis-Goshen and Magdassi, 2012; Zhao et al., 2012; Zhang et al., 2013b). Poly(lactic-co-glycolic) acid (PLGA) provides a good example of the above facts. This biocompatible and biodegradable copolymer has become one of the materials most frequently used as a matrix for drug delivery. PLGA degrades by hydrolysis (cleavage of its backbone ester bonds) in aqueous medium into oligomers and finally to monomers. During hydration, water penetrates into the amorphous region of PLGA and disrupts the van der Waals forces and hydrogen bonds, which is connected with the initiation of degradation – cleavage of covalent bonds and with a decrease in the molecular weight of the copolymer. A constant degradation phase after that follows, for example, free terminal carboxylic groups auto catalyze the degradation process, and mass loss begins by massive cleavage of the backbone covalent bonds resulting in loss of integrity. The last step of PLGA solubilization is subsequent fragmentation (further cleavage to molecules that are soluble in the aqueous environment) (Engineer et al., 2011; Gentile et al., 2014).

After administration, the prepared NPs can be distributed based on passive or active targeting. The passive accumulation of nanoscale particles can be accomplished, for example, due to the pathophysiological characteristics of tumor tissues or trapping of non-PEGylated NPs by reticuloendothelial tissues (Yu et al., 2012; Bazak et al., 2014). Besides, the surface of prepared NPs can be modified by various other molecules or specific compounds for active targeting of NPs, such as proteins (antibodies and their fragments), peptides, nucleic acids (aptamers), small molecules (vitamins, folic acid, carbohydrates, lectins, etc.). This approach enables enhanced bioavailability of encapsulated active ingredients, but also ensures targeted biodistribution and thus reduces the administered amount of drugs and decreases toxicity for non-target tissues and "pill burden" (Mozafari, 2006; De Jong and Borm, 2008; Silva et al., 2013; Passeleu-Le Bourdonnec et al., 2013; Zhang et al., 2013a; Xu et al., 2013; Yamashita and Hashida, 2013; Lu et al., 2014; Yu et al., 2012; Bazak et al., 2015; Yang et al., 2015).

Following sections are focused, except synthetic PLGA, on natural-based biocompatible and biodegradable materials that are used as matrices/carriers for controlled release or targeted biodistribution of nanoscale active pharmaceutical ingredients with low toxicity and favorable pharmacokinetics. All below discussed natural materials used as drug carriers have a long history.

5.3   α-HYDROXY ACID-BASED MATRICES

5.3.1   POLY(LACTIC-CO-GLYCOLIC) ACID

Poly(lactic-co-glycolic) acid (PLGA) is a linear copolymer that is composed of different ratios of lactic and glycolic acids. Depending on the ratio of lactic and glycolic acids, PLGA possesses different physicochemical properties and different crystallinity from fully amorphous to fully crystalline. PLGA degrades by hydrolysis of its ester bonds in aqueous environments (Engineer et al., 2011; Danhier et al., 2012). It belongs to the most effective biodegradable polymeric NPs. It has been approved by the U.S. Food and Drug Administration and the European Medicine Agency for use in drug delivery systems for parenteral administration due to controlled and sustained-release properties, low toxicity, and biocompatibility with tissue and cell (Makadia and Siegel, 2011; Danhier et al., 2012). *At present,* PLGA is widely used for the controlled delivery of therapeutic drugs, proteins, peptides, oligonucleotides, and genes, whereby by surface functionalization of PLGA NPs, a wide variety of combined properties and functions such as prolonged residence time in blood circulation, enhanced oral bioavailability, site-specific drug delivery, and tailored release characteristics could be achieved (Jain et al., 2011a). It is also known as a "smart polymer" due to its stimuli sensitive behavior (Kapoor et al., 2015).

The review paper of Astete and Sabliov (2006) quantitatively and comprehensively described the top–down synthesis techniques available for PLGA NPs formation. PLGA NPs are frequently used for controlled delivery of anticancer agents (e.g., Dinarvand et al., 2011; Acharya and Sahoo, 2011; Mirakabad et al., 2014) and antituberculotics (e.g., Malathi and Balasubramanian, 2011; Choonara et al., 2011; Horvati et al., 2015). Targeting dendritic cells with PLGA cancer vaccine nanoformulations was reported by Hamdy et al. (2011). *Findings related to the use of* PLGA NPs as vaccine delivery systems were summarized also by Akagi et al. (2012) and Silva et al. (2013). Siddiqui et al. (2012) in their review paper discussed the applications of nanotechnology for cancer detection, imaging, treatment, and prevention with particular emphasis on nanochemoprevention. Fredenberg et al. (2011) focused their attention on the mechanisms of drug release in PLGA-based drug delivery systems in a review paper and provided a survey and analysis of the processes determining the release rate. A critical review of biodegradation, biocompatibility, tissue/material interactions, selected examples of PLGA microsphere controlled release systems and the polymer and microsphere characteristics that modulate the degradation behavior and the foreign body reaction to the microspheres was presented by Anderson and Shive (1997, 2012).

PLGA NPs loaded with doxorubicin (DXR) exhibited sustained DXR release profiles over a 1-month period, whereas those containing unconjugated free DXR showed a rapid drug release within 5 days, the NPs also showed increased uptake within human hepatocellular carcinoma HepG2 cells, and a single injection of these NPs had comparable activity to that of free DXR administered by daily injection (Yoo et al., 2000). Orally administered spherically shaped DXR-loaded PLGA NPs (particle size 160 nm) significantly suppressed the growth of breast tumor in female Sprague Dawley rats and caused marked reduction in cardio toxicity when compared with intravenously (i.v.) injected bulk DXR (Jain et al., 2011b). The incorporation of paclitaxel (PTX) into PLGA NPs strongly enhanced the cytotoxic effect of the drug as compared to commercial formulation Taxol®, this effect being more relevant for prolonged incubation times. The release behavior of PTX from these NPs exhibited a biphasic pattern characterized by an initial fast release during the first 24 h, followed by a slower and continuous release (Fonseca et al., 2002). PLGA NPs conjugated with polyethylene glycol (PEG) and loaded with PTX showed greater tumor growth inhibition effect *in vivo* on transplantable liver tumor, compared with Taxol® (Danhier et al., 2009). Biodegradable PLGA nanofibrous membranes that facilitated a sustained release of carmustine, irinotecan, and cisplatin (CPT), released high concentrations of drugs for more than 8 weeks in the cerebral cavity of rats and caused progressive atrophy of the brain tissues without inflammatory reactions (Tseng et al., 2015). Temozolomide-loaded PLGA NPs exhibited a sustained release of the drug and a higher cellular uptake and prolonged the activity against C6 glioma cells, while retaining the anti-metastatic activity (Jain et al., 2014). Gemcitabine demonstrated sustained drug release from the PLGA NPs, *in vivo* ca. 20-fold bioavailability in rats in comparison with the bulk drug, and marked cytotoxicity on chronic myelogenous leukemia K-562 cells was estimated (Joshi et al., 2014).

Yallapu et al. (2014) reported that PLGA-curcumin NPs were efficiently internalized in prostate cancer cells and released biologically active drug in cytosolic compartment of cells for effective therapeutic activity. They more effectively inhibited proliferation and colony formation ability of prostate cancer cells than bulk drug and showed superior tumor regression compared to curcumin in xenograft mice. They also inhibited nuclear β-catenin and androgen receptor expression in cells and in tumor xenograft tissues. High potential of curcumin-loaded PLGA nanospheres as an adjuvant therapy for clinical application in prostate cancer was reported also by Mukerjee and Vishwanatha (2009). Ramalho et al. (2015) informed that the encapsulation of calcitriol in PLGA NPs enhanced its inhibitory effect on cell growth for the human pancreatic S2–013 cells (91%) and for adenocarcinoma human alveolar basal epithelial A549 cells (70%) compared to the free calcitriol, while the inhibitory effect on non-tumor human pancreatic duct hTERT-HPNE cells decreased to 65%.

Corrigan and Li (2009) investigated the mechanism of release of keto-profen, indomethacin (IND), coumarin-6 and macromolecules such as human serum albumin and ovalbumin from PLGA NPs (400-700 nm) having drug loadings less than 10% in phosphate buffer pH 7.4. They found that the release profiles e□hibited an initial burst release phase (reflecting diffusion-controlled dissolution of drug accessible to the solid/dissolution medium interface), a slower lag phase and a second increased release rate phase (reflecting release of drug entrapped in the polymer, the release of which is dependent on the degradation of the bulk polymer). The polymer erosion related parameters indicated that increased drug loading accelerated this phase of active pharmaceutical ingredient release, whereby small acidic hydrophobic drugs have a much greater effect on these parameters than larger hydrophilic more neutral proteins. The drug release rate from the fenbufen-loaded PLGA/chitosan (CS) nanofibrous scaffolds increased with the increase of CS content, because the addition of CS enhanced the hydrophilicity of the PLGA/CS composite scaffold, and nanofiber arrangement was found to influence the release behavior: the aligned PLGA/CS nanofibrous scaffold e□hibitedlower release rate than the randomly oriented one (Meng et al., 2011).

For targeted delivery of antibiotics to intracellular chlamydial infections, rifampin and azithromycin encapsulated in PLGA were tested and showed enhanced effectiveness in reducing microbial burden, and a combination of the two drugs was more effective than the individual drugs (Toti et al., 2011). In NPs prepared by incorporating mycobacterial cell wall mycolic acids as targeting ligands into isoniazid (INH)-encapsulating PLGA NPs, the inclusion of mycolic acids in the nanoformulations resulted in their expression on the outer surface and a significant increase in phagocytic uptake of the NPs. It was estimated that the NPs-containing phagosomes were rapidly processed into phagolysosomes, and NPs-containing phagolysosomes did not fuse with non-matured mycobacterium-containing phagosomes, but fusion events with mycobacterium-containing phagolysosomes were estimated (Lemmer et al., 2015).

Cell-penetrating peptide-assisted PLGA NPs for improving DXR delivery and overcoming cancer multidrug resistance were developed by Wang et al. (2014a). This delivery system could boost intracellular and intranuclear delivery, thereby circumventing drug efflu□ After treatment with this nano-formulation, enhanced uptake and penetration within the tumor and effective arrest of tumor growth in mice harboring drug-resistant breast tumors without detectable toxicity symptoms were estimated. It is also important to note that Tahara et al. (2011) investigated intracellular delivery of DXR using polysorbate 80-modified PLGA nanospheres to human glioblastoma A172 cells and observed that the drug release pattern was dependent on the pH of the medium and that DXR was released from NPs following accumulation in the cell nuclei. Moreover, such DXR-loaded NPs significantly inhibited both DXR efflu□from the cells and cell proliferation compared with a DXR solution. Guo et al. (2014) solved the co-delivery of CPT and rapamycin for enhanced anticancer therapy through synergistic effects and microenvironment modulation. The researchers coated CPT by dioleoylphosphatidic acid and co-encapsulated these particles together with rapamycin in PLGA NPs. These fi□ combinations were found to inhibit the growth of human melanoma A375 cells in a xenograft tumor model through modulation of the tumor vasculature and to allow enhanced penetration of NPs into the tumor.

L-Valine-conjugated PLGA NPs showed greater insulin uptake as compared to non-conjugated NPs and *in vivo* studies performed on streptozotocin-induced diabetic rabbits showed that an oral suspension of insulin-loaded PLGA NPs significantly reduced blood glucose and produced hypoglycemic effect within 4 h of administration, and the hypoglycemic effect prolonged till 12 h of oral administration (Jain and Jain, 2015). Pulmonary delivery of insulin with nebulized PLGA nanospheres to prolong hypoglycemic effect was also described (Kawashima et al., 1999).

Suh et al. (1998) formulated PTX into a biodegradable PEG-PLGA nanosphere as a sustained drug delivery system and studied its effects on vascular smooth muscle cells in culture. This copolymer is very suitable for the construction of micelles as carriers for insoluble drugs (Zhang et al., 2014a). The sustained drug release profile and cellular internalization results suggested that nanospheres loaded with PTX may be potentially used as an endocytizable, local sustained drug delivery system for the prevention of restenosis. The PTX-loaded PEG-PLGA block copolymer NPs with an average diameter of 70 nm were found to be able to diffuse 100-fold faster than similarly sized PTX-loaded PLGA NPs without PEG coating. PEGylated NPs significantly delayed tumor growth following local administration to established brain tumors as compared to drug-loaded PLGA NPs or bulk drug (Nance et al., 2014). Spherical methoxyPEG-PLGA NPs with size <100 nm showed biphasic drug release pattern with a fast release rate followed by a slow one, and it can be expected that such formulations with PTX could have long-circulating effects in circulation (Dong and Feng, 2004). CPT-cross linked carboxymethyl (CM) cellulose (CLS) core NPs made from methoxyPEG-PLGA copolymers exhibited controlled release of CPT in a sustained manner, and the dose-dependent treatment efficacy of CPT-loaded NPs against CPT-resistant ovarian cancer IGROV1-cP cells was confirmed (Cheng et al., 2011). PLGA and PLGA-PEG NPs with the size of ca. 180 nm and 260 nm that entrapped methotrexate (MTX) with an efficiency of >51% were found to be more effective than the bulk drug and increased the activity of caspase-3 in breast cancer MCF-7, A549 cell lines and human gastric adenocarcinoma AGS cells (Afshari et al., 2014). The release profile of docetaⰃel (DTX)-loaded PEG-PLGA NPs showed a burst release of approⰃ 20% followed by a sustained release profile of the loaded drug over 288 h, a greater cytotoxicity against ovarian carcinoma SKOV3 cells than the free drug as well as enhanced tumor-suppression effects (Koopaei et al., 2014).

Cholic acid-functionalized NPs of star-shaped PLGA-(D-α-tocopheryl-PEG succinate) copolymer used for DTX delivery to cervical cancer showed the enhanced cellular uptake and increased antitumor efficacy compared with non-functionalized NPs and PLGA NPs (Zeng et al., 2013). Moreover, DTX formulated in the cholic acid-functionalized PLGA NPs in combination with autophagy inhibitors, such as 3-methyladenine and chloroquine, showed great enhancement of the therapeutic effect, and both the volume and the weight of the shrunk tumor of the mice were found to be only about a half after 20 day treatment compared to the application of PLGA NPs formulation (Zhang et al., 2014b).

Bone is a favorable microenvironment for tumor growth and a frequent destination for metastatic cancer cells, therefore it is very important to target cancers within the bone marrow, securing drug availability, and to overcome microenvironment-induced resistance (Černíková and Jampílek, 2014). The NPs consisting of PEG-PLGA and alendronate as a targeting ligand were prepared by Swami et al. (2014). These long-circulating NPs showed enhanced bone homing due to bone mineral-targeting capabilities. The bone-targeted NPs with sustained release delivered bortezomib (a proteasome inhibitor that has shown marked antitumor effects in patients with multiple myeloma) to bone marrow microenvironment specifically and produced the antimyeloma effects similar to the free drug. Furthermore it was estimated that bortezomib as a pretreatment regimen modified the bone microenvironment and enhanced bone strength and volume.

Composite PLGA-magnetic (Fe_3O_4) NPs for co-delivery of hydrophobic and hydrophilic anticancer drugs and magnetic resonance imagining for cancer therapy were developed by Singh et al. (2011). The use of a ligand such as herceptin for targeted delivery of drugs resulted in enhanced cellular uptake of such a conjugated system and an augmented synergistic effect in an *in vitro* system when compared with native drugs. Moreover, the above-mentioned NPs showed a better contrast effect than commercial contrast agents due to higher T2 relaxation time with a blood circulation half-life similar to 47 min in the rat mode.

5.4 CARBOHYDRATE-BASED MATRICES

5.4.1 CYCLODEXTRINS

Cyclodextrins (CDs) are naturally occurring cyclic oligosaccharides consisting of α-(1→4)-linked glucopyranose units, possessing a basket-shaped topology with an "inner-outer" amphiphilic character, for example, with a hydrophilic outer surface and a somewhat lipophilic central cavity. They are able to form water-soluble inclusion complexes with many lipophilic water-insoluble drugs. In addition, the primary and secondary hydroxyl groups of the native α-, β- and γ-CDs are potential sites for chemical modification. They have been widely used as safe and effective carriers for both hydrophilic and lipophilic drugs (Frömming and Szejtli, 1994; Loftsson and Stefansson, 2007; Kurkov and Loftsson, 2013). The recent progress and future perspectives of CD-based supramolecular systems for drug delivery were summarized by Zhang and Ma (2013), Kurkov and Loftsson (2013), Yameogo et al. (2014), and Dash and Konkimalla (2015). Reviews focused on the recent advances in CD-functionalized polymers as drug carriers for cancer therapy were published by Wei and Yu (2015) and Gidwani and Vyas (2015).

Nanosponges can be referred to as solid porous particles having a capacity to load drugs into their nanocavity, they can be formulated as oral, parenteral, topical, or inhalation dosage forms, they offer high drug loading compared to other nanocarriers and are thus suitable for solving issues related to stability, solubility, and delayed release of actives (Tejashri et al.,

2013). CD-based nanosponges are innovative cross-linked CD polymers nanostructured within a three-dimensional (3D) network that are proposed as new nanoscale delivery systems. They form porous NPs with sizes <500 nm, spherical shape, and negative surface charge and show a good capacity for incorporating small molecules, macromolecules, ions, and gases. CD-based nanosponges can form complexes with different types of lipophilic or hydrophilic molecules and be used also for site-specific targeting by conjugating various ligands on their surface (Trotta et al., 2012, 2014). The insight into CD-based nanosponges, including their physical and chemical properties, methods of their preparation and characterization, biocompatibility, and their use as effective drug formulations, was presented by Chilajwar et al. (2014).

Supramolecular systems formed by the binding of several CDs to polymers or lipids via (non)-covalent bonds open a wide range of possibilities for the delivery of active substances. CDs can serve as multifunctionalizable cores, to which small molecules or macromolecules can be conjugated. They can be grafted with amphiphilic molecules, can polymerize with other CDs or can be used to functionalize pre-existing polymers to form polymers/networks, and thus CDs can be exploited also as transient cross-linkers to form poly(pseudo)rotaxane-based networks or zipper-like assemblies (Tan et al., 2014; Simoes et al., 2015). Ding et al. (2014) reviewed diverse non-covalent interactions that are applied to enhance polymeric micelles as drug nanocarriers and summarized future directions and perspectives of this field.

Smart stimuli-responsive drug carriers (STRDCs) are a hot topic in current chemical and pharmaceutical research, owing to their merits of controlled release of drugs relying on unique stimuli-responsive mechanisms. Many biocompatible materials are employed to fabricate STRDCs in combination with CDs to secure unique characteristics of self-assembly, molecular recognition, and dynamical reversibility, whereby drug controlled release from these CD-based STRDCs can be accomplished upon the regulation of their physicochemical properties in response to external stimuli. These can be endogenous (pH, redox agents, enzyme concentration, etc.) or exogenous (temperature, light, magnetic force, ultrasound, voltage stimulation, etc.) (Hu et al., 2014; Liao et al., 2015a).

The incorporation of hydrophilic CD moieties into the polymeric network of nanogels provides them with a drug loading and release mechanism that is based on the formation of inclusion complexes without decreasing the hydrophilicity of the network, because the covalent attachment of CDs to the chemically cross linked networks enables them to display fully their ability to form complexes, while simultaneously preventing drug release upon media dilution (Moya-Ortega et al., 2012). The conjugation of CDs with hydrogels may also allow the achievement of an optimal wound-dressing material, because the hydrogel component will maintain the moist environment required for the healing process, and the CD moiety has the ability to protect and modulate the release of bioactive molecules (Pinho et al., 2014).

Camptothecin (CTH)-loaded β-CD-based nanosponges of the size ranging from 450 nm to 600 nm exhibited slow and prolonged drug release over a period of 24 h, secured protection of the lactone ring of CTH and showed higher cytotoxicity towards human colon adenocarcinoma HT-29 cells after 24 h of incubation than plain drug (Swaminathan et al., 2010).

Spherical flurbiprofen-loaded hydro□ypropyl-β-CD (HP-β-CD) nano-liposomes prepared by co-evaporation significantly improved the relative bioavailability of flurbiprofen (Zhang et al., 2015a), indicating that drug-in-CD-in-liposomes could be a perspective way for delivery of insoluble drugs. Toomari et al. (2015) encapsulated MTX in a new dendrimer, having 14 β-CD residues attached to the core β-CD in secondary face and found that the release process was noticeably pH dependent.

The grafting of CD onto CS can result in the formation of a molecular carrier that possess the cumulative effects of inclusion, size specificity and transport properties of CDs as well as the controlled release ability of the polymeric matrix (Prabaharan and Mano, 2006). Moreover, nanocarriers prepared by incorporation of β-CD derivatives into CS were found to be more stable than CS NPs at pH 6.8 (Chen et al., 2015a). Complexation of triclosan and furosemide with the HP-β-CD facilitated entrapment of drugs into the CS NPs, increasing up to 4- and 10-fold (for triclosan and furosemide, respectively) the final drug loading of the NPs. These NPs e□hibited an initial fast release followed by a delayed release phase indicating that the mentioned nanosystem can be used for the transmucosal delivery of hydrophobic compounds (Maestrelli et al., 2006).

Sulfiso□azole/HPβ-CD inclusion complex incorporated in electrospun hydro□ypropylCLS nanofibers as a drug delivery system that resulted in a higher amount of the released drug due to its increased solubility by inclusion comple□ationas well as in a slower release of sulfiso□azole was described by Aytac et al. (2015). By bioconjugation of folic acid to β-CD using a PEG spacer, NPs were prepared that exhibited a greater extent of cellular uptake against cervical cancer HeLa cells than A549 cells (Zhang et al.,

2012). A novel platform consisting of drug-containing PLGA polymer NPs, stably fashioned with a shell composed of drug complexed with cationic CD exhibited sequential release *in vitro*, owing to CD shell displacement and a subsequent sustained release of core-loaded drug, kinetics preserved in breast cancer cells following internalization (Ruiz-Esparza et al., 2014). In an alginate/CS NP system, insulin was protected by forming complexes with cationic β-CD polymers, because insulin was mainly retained in the core of the NPs and well protected against degradation in simulated gastric fluid (Zhang et al., 2010a).

Namgung et al. (2014) reported about a novel nano-assembled drug-delivery system, formed by multivalent host-guest interactions between a polymer-CD conjugate and a polymer-PTX conjugate that can efficiently deliver PTX into the targeted cancer cells via both passive and active targeting mechanisms and in which the ester linkages between PTX and the polymer backbone enable efficient release of PTX within the cell by degradation. This fact was confirmed e□perimentally previously by Agueeros et al. (2009), who incorporated the complex of PTX with HP-β-CD into poly(anhydride) NPs with the size of ca. 300 nm and found that this nanoformulation showed 12-fold higher permeability of PTX compared to Taxol®, which indicated that this nanosystem can induce a positive effect over the intestinal permeability of PTX, being the bioadhesion a mandatory condition in this phenomena. Also DXR-loaded NPs of the amphiphilic copolymer composed of HP-β-CD, poly(L-lactide) (PLA), and 1,2-dipalmitoyl-SN-glycero-3-phosphoethanolamine showed a continuous release after a burst release and anticancer efficacy against HepG2 and A549 cells comparable with the bulk drug (Wang et al., 2011a).

Porphyrin-CD conjugates as a nanosystem for versatile drug delivery and multimodal cancer therapy were reported by Králová et al. (2010). NPs of CD-based star polymers prepared using β-CD as hydrophilic core and methyl methacrylate and *tert*-butyl acrylate as hydrophobic arms showed prolonged release profile of idarubicin. NPs were internalized with A549 cells related to Caco-2 cells in time-dependent manner with preferential accumulation in cytoplasm with superior uptake of positively (CS-modified NPs) over negatively charged particles (Nafee et al., 2015).

The supramolecular gels of poly-α-CD and PEG-based copolymers were able to sustain the release of vancomycin (VMC) for at least 5 days at 37°C more efficiently than dispersions of each polymer component separately (Simoes et al., 2014). A supramolecular hydrogel based on complexation of triblock poly(caprolactone)-PEG-poly(caprolactone) copolymer with γ-CDs was found to be a suitable system for providing a sustained release of therapeutic proteins such as insulin with desirable flow behavior (Khodaverdi et al., 2015). The suitability of gemcitabine/DTX/γ-CD inclusion complexes co-loaded PEGylated liposomes as highly effective combinational therapy of osteosarcoma was reported by Sun et al. (2015). It was estimated that PEGylated liposomes successfully delivered anticancer drugs in the osteosarcoma tumor interstitial spaces via enhanced permeability and retention effect.

5.4.2 CELLULOSE

In higher plants, cellulose (CLS) is organized in a hierarchical structure consisting of β-(1→4)-linked D-glucopyranose chains that are laterally bound by hydrogen bonds to form microfibrils with a diameter in the nanoscale range, which are further organized in microfibril bundles (Klemm et al., 1998). CLS esters are characterized by very low toxicity, endogenous and/or dietary decomposition products, stability, high water permeability, high glass transition temperature, film strength, compatibility with a wide range of actives, and ability to form micro- and nanoparticles and therefore represent an appropriate material for drug delivery systems (Edgar, 2007). Bacterial cellulose (BC) is different from plant CLS. It is naturally produced by aerobic bacteria of genera *Acetobacter*, *Agrobacterium* and *Sarcina ventriculi*. BC has different properties from plant CLS and is characterized by high purity, strength, moldability, and increased water holding ability (Esa et al., 2014).

The preparations of "smart" materials based on CLS by chemical modifications and physical incorporating/blending, including responsiveness of these "smart" materials in the form of copolymers, NPs, gels, and membranes to pH, temperature, light, electricity, magnetic fields, mechanical forces, etc., and their applications as drug delivery systems, hydrogels, electronic active papers, sensors, shape memory materials, smart membranes, etc. were reviewed by Qiu and Hu (2013). Yu et al. (2013) published an approach to the systematic design and fabrication of novel biomaterials with structural characteristics for providing complicated and programmed drug release profiles using coaxial electro spinning. They developed polyvinylpyrrolidone/ethyl CLS core/sheath nanofibers, in which nanofibers could provide biphasic drug release profiles consisting of an immediate and sustained release, where the amount of drug released in the first phase was tailored by adjusting the sheath flow rate, and the remaining drug released in the second phase was controlled by a typical diffusion mechanism. Chang and Zhang (2011) presented a review paper dealing with recent progress in design and fabrication of hydrogels based on native CLS and its derivatives, including methyl, hydroxypropyl, hydroxypropylmethyl, and carboxymethyl cellulose (CMC), obtained by physical as well as chemical cross-linking strategies, composite hydrogels prepared by using CLS in conjunction with other polymers through blending, formation of polyelectrolyte complexes, and interpenetrating polymer networks technology as well as CLS-inorganic hybrid hydrogels prepared by embedding inorganic NPs in CLS matrices. Review papers providing overviews of nanocrystalline CLS from the nanotechnology perspective, focusing on its surface modification, properties, and applications were written by Peng et al. (2011) and Lam et al. (2012). CLS nanocrystals were found to be non-toxic to the human brain micro vascular endothelial cells, and the uptake studies showed minimal uptake of untargeted CLS nanocrystals (Roman et al., 2009). In addition, CLS nanowhiskers, for example, grafted with γ-aminobutyric acid to achieve controlled and rapid delivery of the targeting moiety, are a promising material in the biomedical field (Dash and Ragauskas, 2012).

Kolakovic et al. (2013) evaluated the interactions between nanofibrillar CLS and the model drugs beclomethasone, IND, itraconazole, and nafarelin. The permeation studies revealed the size dependent diffusion rate of the model drugs through the nanofibrillar CLS films, studied drugs were found to bind to the nanofibrillar CLS material, the pH dependence of the binding and electrostatic forces as the main mechanism were estimated. Also Valo et al. (2013) integrated beclomethasone NPs coated with amphiphilic hydrophobin proteins into the nanofibrillar CLS aerogels. Since the release of the drug is controlled by the structure and interactions between the NPs and the CLS matrix, modulation of the matrix enables a control of the drug release rate. While nano-CLS aerogel scaffolds made from red pepper and microcrystalline CLS released beclomethasone immediately, BC aerogels showed a sustained release of tested drug. Tungprapa et al. (2007) encapsulated napro□en(NPN), IND, ibuprofen (IBU) and sulfindac (SUL) in ultrafine fiber mats of cellulose acetate (CA) and found that the release of the drugs from CA fiber mats was greater than that from the control CA films, and the maximum release of the drugs could be ranked as follows: NPN > IBU > IND > SUL.

Supramolecular hydrogels used as a matrix for slow release of DXR from *in situ* host-guest inclusion between chemically modified CLS nanocrystals and β-CD were prepared by Lin and Dufresne (2013). Curcumin-loaded hydroxypropyl methylcellulose (HPMC) NPs was found to exhibit high cellular uptake, caused maximum ultrastructural changes related to apoptosis (presence of vacuoles) in prostate cancer cells, and showed improved anticancer efficacy compared to the free drug (Yallapu et al., 2012). Enhancement in drug dissolution rate due to the large surface area and smaller drug particle size was confirmed by Sievens-Figueroa et al. (2012), who prepared HPMC films stabilizing NPs of NPN, fenofibrate, or griseofulvin. Butun et al. (2011) demonstrated that CMC particles that were made magnetic responsive by encasing independently prepared ferrite particles (Fe_3O_4) in CMC polymeric particles during the synthesis could be further modified by introducing new functional groups to CMC networks and can be applied as a targeted drug delivery system in the biomedical field.

Polysaccharide nanocrystals, such as rod-like CLS nanocrystals and chitin whiskers and platelet-like starch nanocrystals when incorporated into alginate-based nanocomposite microspheres enhance mechanical strength and increase the stability of the cross linked network structure, nanocomposite microspheres also e☐hibit prominent sustained release profiles of drugs (Lin et al., 2011). Dong et al. (2014) prepared folic acid-conjugated CLS nanocrystals for the targeted delivery of chemotherapeutic agents to folate receptor-positive cancer cells and found that cellular binding/uptake of the conjugate by human glioma DBTRG-05MG, H4 and C6 cells was 1452-, 975-, and 46-fold higher, respectively, than that of non-targeted CLS nanocrystals. While DBTRG-05MG and C6 cells internalized the conjugate primarily via caveolae-mediated endocytosis, internalization of the conjugate by H4 cells was accomplished via clathrin-mediated endocytosis. Nanocrystalline CLS with cetyltrimethylammonium bromide-modified surfaces bound significant quantities of the hydrophobic anticancer drugs DTX, PTX, and etoposide, releasing the drugs in a controlled manner over a 2-day period. This nanoformulation was found to bind to human bladder cancer KU-7 cells, and evidence of cellular uptake was observed (Jackson et al., 2011). Amphiphilic cationic CLS derivatives carrying long chain alkyl groups as hydrophobic moieties and quaternary ammonium groups as hydrophilic moieties which could be self-assembled into cationic micelles in distilled water with the average hydrodynamic radius of 320–430 nm were prepared by Song et al. (2011). Incorporation of prednisone acetate in these self-assembled micelles, showing low cytotoxicity, resulted in controlled release of the drug. A novel drug delivery system was developed by modifying the surface of oxidized CLS nanocrystal with CS. The fast (up to 1 h) *in vitro* release of procaine·HCl from this formulation suggests that it can be used as biocompatible and biodegradable drug carrier for transdermal delivery applications (Akhlaghi et al., 2013). Biodegradable graft copolymer comprised of hydrophobic PLA segments and hydrophilic CLS segment (CLS-g-PLA) was prepared by Dong et al. (2008). This copolymer can self-assemble into nanospheric micelles in water with the size range of 30-80 nm, in which the hydrophobic PLA segments are situated in the cores of micelles and the hydrophilic CLS segments form the outer shells, and the drug-loaded micelles show sustained drug release. A series of amphiphilic CLS-PLA grafted copolymers showing the self-assembly behaviors successfully encapsulated PTX (Guo et al., 2012).

Valo et al. (2011) employed an engineered hydrophobin fusion protein, where hydrophobin was coupled with 2 CLS binding domains in order to facilitate drug NP binding to nanofibrillar CLS. Hydrophobins are a group of small cysteine-rich proteins e□pressedonly by filamentous fungi. Enclosing the functionalized protein coated itraconazole NPs to the e□ternd nanofibrillar CLS matrix notably increased their shelf life stability, and as a consequence of the formation of immobilized nanodispersion, dissolution rate of itraconazole was increased significantly, which also enhanced the *in vivo* performance of the drug.

Shi et al. (2014) prepared hybrid hydrogels composed of BC nanofibers and Na^+ alginate as a dual-stimuli responsive release system. The increase of pH from 1.5 to 11.8 resulted in an increase of the swelling ratio from less than 8-fold compared with its dry weight at acidic conditions to more than 13-fold, and thus the release of IBU from these hydrogels was faster in alkaline conditions and slower in acidic conditions.

Biocompatible hollow poly(methacrylic acid-co-*N*-isopropylacrylamide-co-ethyleneglycol dimethacrylate)@CLS succinate spheres (with size ca. 460 nm) were loaded with daunorubicin (DNR) by three ways: associated with the polymer chain by dipole-charge interactions, physically entrapped in the polymer matrix, and absorbed on the surface of nanocontainers. This indicates that this nanocontainer drug delivery system has the potential for the sustained release of DNR *in vivo*. The considerable dependence on the releasing behavior of drug from the containers on pH values (in acidic conditions, the drug was released progressively) may originate from the difference of the strength of intramolecular interaction between the carboxylic spheres and amine of DNR molecules under different pH conditions (Metaxa et al., 2012). As the pH of both extracellular tumor (pH 6.8) and endosomes (pH 5.5) is acidic in contrast with the normal tissues (pH 7.4), the nanoformulation is suitable as a drug delivery system for cancer treatment.

As a promising carrier for various pharmaceutical applications, especially as a poorly water-soluble drug delivery system, also mucoadhesive hydrophobic cationic amino CLS was recommended (Songsurang et al., 2015). Excellently biodegradable and biocompatible aerogels from polyethylenimine-grafted CLS nanofibrils e□hibiteda high drug loading capability, sustained and controlled release behavior of the aerogels being highly dependent on pH and temperature (Zhao et al., 2015a). For BC functionalized with the antiseptic drug octenidine, rapid release in the first 8 h followed by a slower release rate up to 96 h was estimated. NPs of octenidine loaded BC showing high biocompatibility in human keratinocytes, antimicrobial activity against *Staphylococcus aureus*, and preservation of drug antibacterial activity over a period of 6 months represent a ready-to-use wound dressing for the treatment of infected wounds (Moritz et al., 2014). Antimicrobial BC membranes obtained by chemical grafting of aminoalkyl groups onto the surface of its nanofibrillar network were found to be simultaneously lethal against *S. aureus* and *Escherichia coli* and non-toxic to human adipose-derived mesenchymal stem cells, indicating that they may be useful for biomedical applications (Fernandes et al., 2013). As BC does not exhibit antimicrobial properties, this approach intends to mimic intrinsic antimicrobial properties of CS.

CLS nanofibrous mats coated by lysozyme/pectin bilayers (with lysozyme on the outmost layer) possessed stronger antibacterial activity against both *E. coli* and *S. aureus* than CLS mats (Zhang et al., 2015b). CMC/ZnO nanocomposite hydrogel with ZnO NPs size range 10–20 nm demonstrated antibacterial effects against *E. coli* and *S. aureus* bacteria (Yadollahi et al., 2015a). Antibacterial CMC/Ag nanocomposite hydrogels cross-linked with layered double hydroxides showed a pH sensitive swelling behavior, the antibacterial activity of the formulation increased considerably after formation of AgNPs and was stable for more than one month (Yadollahi et al., 2015b). Other CMC hydrogels loaded with AgNPs for medical applications were developed by Hebeish et al.

(2013). Magnetic Fe_3O_4/CS-CMC biocomposites were found to improve the activity of currently used antibiotics belonging to penicillin, macrolide, aminoglycoside, rifamycin, and quinolone classes and could be considered as potential macromolecular carriers for these antimicrobial substances to achieve extracellular and intracellular targets. As they are not cytotoⵏic and does not influence the human colon carcinoma HCT-8 cell cycle, the delivery of drugs is safe and effective (Grumezescu et al., 2012). The use of Fe_3O_4/CMC NPs for delivery of norfloⵏacin and cefotaⵏim was studied by Vlad et al. (2014). Efficient drug delivery systems based on CLS nanofiber-TiO_2 nanocomposites grafted with three different types of model drugs, such as Na^+ diclofenac, penicillamine D, and phosphomycin, was developed by Galkina et al. (2015). Protamine-Fe_3O_4/CMC nanocapsules for enhanced delivery of DXR into drug resistant HeLa cells were designed by Elumalai et al. (2015).

5.4.3   STARCH

Starch is a miⵏture of amylose, a linear polymer of α-($1\rightarrow4$)-linked D-glucopyranosyl units (20–30 wt% of standard starch), and amylopectin constituted of linear segments connected by α-($1\rightarrow6$)-branched D-glucopyranosyl units (5–6 wt% of standard starch) and non-branched α-($1\rightarrow4$)-linked D-glucopyranosyl units (70–80 wt% of standard starch). Due to its clustered organization of short branches associated into double helices amylopectin is believed to be mostly responsible for the overall crystallinity (Pérez and Bertoft, 2010; Lancuški et al., 2015). An overview of the most up-to-date information regarding the starch NPs including the preparation processes and physicochemical characterization as well as the prospects and outlooks for the industrial utilization of starch NPs was proposed in a review paper of Kim et al. (2015). Recent progress in chemical modification of starch and its applications were reported by Chen et al. (2015b). The naturally derived nanomaterials, nanostructured starches, can be chemically modified to allow for the introduction of functional groups, enhancing their potential for drug delivery and other biotechnology applications. It was found that starch nanospheres solution expands the circulating blood volume, improves the hemodynamics, increases the effective circulating blood volume, and improves the shock symptoms of effective hypovolemia (Li et al., 2013). Amphiphilic starch-based polymeric micelles used as drug carriers can improve the solubility of hydrophobic drug, extend drug circulation time *in vivo*, reduce systemic side effects, and increase the preferential accumulation at the targeted sites by the enhanced permeability and retention effect, and by introduction of some groups with stimuli-responsive function to starch backbone, smart polymeric micelles realizing rapid and targeted drug release could be prepared (Engelberth et al., 2015).

Starch NPs for the delivery of gemcitabine·HCl that could reduce its dose related side effects and may prolong its retention time (24 h) for the treatment of pancreatic cancer was developed by Khaira et al. (2014). Ultra-small cationic starch nanospheres with size of 50 nm possessing a good capacity in delivering negatively charged molecules, biocompatibility, and biodegradability were prepared by Huang et al. (2013). Spherical particles of amylose-IBU inclusion complexes with the sizes ranging from 30 to 80 nm were stable in the simulated gastric fluid, IBU was sustainably released from the complexes in the simulated medium of the small intestine, which is connected with the hydrolysis of the inclusion complexes by amylase. Thus, these complexes can be used as carriers releasing drug in an intestinally targeting and controlled manner (Yang et al., 2013). Situ et al. (2015) developed an oral colon-specific controlled release system suitable for delivery of polypeptides or proteins coated with a resistant starch-based film through aqueous dispersion coating process. A dramatic increase in the resistibility against enzymatic digestion resulting in the formation of resistant starch can be achieved by high-temperature-pressure modification of starch, enzymatic debranching, and retrogradation. A novel oral pH-responsive protein drug delivery vehicle made of starch NPs as backbone and poly(L-glutamic acid) as graft chains was synthesized by Zhang et al. (2013b). The loaded insulin released from the copolymers more slowly in artificial gastric juice (pH = 1.2) than in artificial intestinal liquid (pH = 6.8) due to the e⬜cellent stability in acidic condition.

The release behaviors of drug-loaded acetylated starch NPs depended on the size of NPs and the degree of acetylating and these starch-based NPs with uniform size can be used for the encapsulation of hydrophobic drug as sustained and controllable drug release carriers (Han et al., 2013). Highly bioadhesive PEGylated starch acetate NPs can be used as a carrier system for controlled delivery of insulin or other proteins for various therapeutic applications (Minimol et al., 2013). Hydroxyethyl starch (HES) possesses improved stability against enzymatic degradation and can be used to prepare nanocapsules that can be loaded with hydrophilic guests in its aqueous core, tuned in size, chemically functionalized in various pathways, and show high shelf life stability. Moreover, functionalization of the surface of the HES nanocapsules, for example, with PEG results in substantially enhanced blood half-life time (Kang et al., 2015a). Moreover, Kang et al. (2015b) developed PEGylated HES nanocarriers modified with mannose that showed high specific affinity for dendritic cells binding. HES as a parenteral drug delivery nanocarrier was also evaluated by Narayanan et al. (2015). HES with a pH-sensitive hydrazone linkage provided a pH-sensitive drug delivery system for a controlled and sustained release of DXR over a period of more than 3 days in an acidic environment mimicking the tumor microenvironment. These conjugates were characterized with good biocompatibility, long circulation, and lower cytoto□icity they could be efficiently transferred into HeLa and HepG2 cells where the release of the conjugated drug occurs (Zhu et al., 2015). Liu et al. (2013) developed a novel drug carrier through grafting polymerization of HES and D, L-lactic acid that self-assembled into micelles with uniform sizes ranging from 65 to 130 nm. The size of micelles as well as the release rates of DTX is modulated by changing the chain length of the lactic acid segments.

Starch-poly(vinyl alcohol) nanocomposites for the controlled delivery of anticancer drugs were developed by Athira and Jyothi (2015). At the same time, Li et al. (2015) designed starch-hyaluronic acid NPs loaded with DXR for targeted therapy of tumors. These nanocarriers e□hibited significantly higher anti-tumor efficiency in reducing tumor size compared with free drugs combination or single drug-loaded NPs individually. Yang et al. (2014) developed a novel amphiphilic polymer forming micelles by grafting hydrophobic deoxycholic acid into methoxyPEGylated starch loaded with DXR that in *in vitro* tests showed pH-induced drug release. The starch functionalized graphene nanosheets (starch-GNS) are biocompatible and non-toxic to colorectal adenocarcinoma SW-620 cells. After loading of hydroxycamptothecin via physisorption on starch-GNS, the composite showed high toxicity to SW-620 cells and *in vitro* sustained release mediated by pH sensitivity (Liu et al., 2015a). A controlled release system consisting of starch: CA-co-acrylate nanocomposite with starch: cellulose ratio 90:10 designed as a carrier for antitumor drugs was prepared by Helaly et al. (2013). The amount of released drugs depended on the pH of the media in the following order: basic media > acidic media > neutral.

Starch-coated Fe_3O_4 liposomes as an inhalable carrier for accumulation of fasudil in the pulmonary vasculature were prepared by Nahar et al. (2014).

3-Fold higher uptake of liposomes by pulmonary arterial smooth muscle cells was estimated under an applied magnetic field compared with that observed in the absence of the magnetic field. Different magnetic Fe_3O_4-loaded montmorillonite (MMT) NPs coated with CM-starch/CS were prepared for controlled release applications. This composite showed good stability, biocompatibility and mucoadhesivity of the Fe_3O_4 NPs, and MMT addition was found to enhance the swelling, INH loading, and release as well as the cytotoxicity and mucoadhesivity of the NPs (Saikia et al., 2014; Saikia et al., 2015).

5.4.4 PULLULAN

Pullulan is a water soluble exo-polysaccharide, produced aerobically from starch by fungus *Aureobasidium pullulans*. In principle, it is a regularly repeating copolymer with the chemical structure $\{(\rightarrow6)$-α-D-glucopyranosyl-$(1\rightarrow4)$-α-D-glucopyranosyl-$(1\rightarrow4)$-α-D-glucopyranosyl $(1\rightarrow)\}_n$ containing both α-$(1\rightarrow4)$- and α-$(1\rightarrow6)$-linkage in its structure (Singh et.al., 2008, 2015). Due to some beneficial properties such as high water solubility, multiple hydroxylic groups that can readily be chemically modified, no toxicity, lack of immunogenicity, and usefulness as a plasma expander, the pullulan can be used as a drug carrier (Yuen, 1974; Jeanes, 1977; Jung et al., 2004). An overview of the biosynthesis, biological activity, and chemical modification as well as applications of pullulan and its derivatives in formulations mainly for targeted drug or gene delivery was published by Singh et al. (2008, 2015).

pH-Sensitive pullulan-DXR core-shell NPs released >75% of DXR within 2 h at pH 5.0, while <15% of DXR was released after 12 h at pH 7.4. Thus, this formulation may contribute to quick diffusion of DXR from the acidic endosome/lysosome and the intracellular transfer into the nucleus. Pullulan-decorated surface of NPs enabled active targeting of hepatic cells via specific interaction with asialoglycoprotein receptors on the membrane of hepatic cells without the necessity to introduce any extra ligand (Li et al., 2014a). Bishwambhar and Suneetha (2012) prepared microspheres of a pH sensitive NPN-pullulan acetate delivery system, in which the released amount of the drug from the microspheres at pH 7.2 was 75-fold higher than that at pH 1.2. Pullulan acetate was also recognized as a useful carrier for delivery of anticancer drugs (Kumar et al., 2012).

Cho et al. (2009) developed self-organized nanogels from pullulan-PLA, in which PLA provides a hydrophobic moiety for the formation of self-organized structures. If the degree of grafted PLA per glucopyranose unit in pullulan was 0.68, the DXR loading efficiencies of nanogels were more than 52%, and high loading resulted in slower DXR release due to increasing of hydrophobic interaction. Moreover, the total amount of released DXR from the DXR-loaded thermo-sensitive PLA-pullulan nanogels increased with increasing temperature for 50 h (Seo et al., 2012). On the other hand, self-assembled spherical polymeric NPs of PEG grafted pullulan acetate of the size of 193 nm, in which clonazepam was entrapped, showed sustained drug release from the NPs for 4 days, and the release rate was increased by introduction of PEG. This is connected with the fact that the introduction of PEG into pullulan induces a more hydrophilic microenvironment in the NPs, and the drug can diffuse out more easily through hydrophilic aqueous channels. Moreover, the introduction of PEG reduced hydrophobic interaction between clonazepam and the polymer chain, and the easier drug release was induced by the increased solubility of the drug (Jung et al., 2004).

Scomparin et al. (2011) developed novel folated pullulan bioconjugates with DXR that released less than 20% of DXR in plasma within 3 days, while at pH 5.5 both conjugates underwent complete drug release in about 40 h. The advantage of the conjugate was shown also in an *in vivo* experiment, ca. 40% of the administered DXR was present in the bloodstream in 4 h after i.v. administration of the conjugates to Balb/c mice, while 80% of the unconjugated DXR was cleared within 30 min.

5.4.5 ALGINATE

Alginate (ALG) is biodegradable, biocompatible, bioactive, low-toxic, and low-cost polysaccharide. It is quite abundant in nature as a part of the structural component of bacteria and brown algae (sea weed *Phaeophyceae*) (Zia et al., 2015). ALG is a linear, anionic polysaccharide that is built from $(1 \rightarrow 4)$-linked β-D-mannuronic acid (M) and α-L-guluronic acid (G) arranged in blocks of repeating M residues (MM blocks), blocks of repeating G residues (GG blocks), and blocks of mixed M and G residues (MG blocks). ALG possesses a number of free hydroxyl and carboxyl moieties on backbone, therefore natural ALG provides an excellent material for subsequent modification and yielding of various excipients with different properties in comparison with the parent compound (Yang et al., 2011). At low pH, hydration of alginic acid results in the formation of a high-viscosity "acid gel", however, ALG could be also easily gelled in the presence of a divalent cation (e.g., Ca^{2+}). By re-swelling of dried sodium ALG beads, diffusion barrier is formed that decreases the migration of small molecules. The ability of ALG to form an acid gel as well as an ionotropic gel confers unique properties to this biopolymer in comparison with neutral macromolecules (Tonnesen and Karlsen, 2002). ALG hydrogels retain structural similarity to the extracellular matrices in tissues and can be manipulated to play several critical roles, therefore they are utilizable in wound healing, drug delivery, and tissue engineering applications (Lee and Mooney, 2012). An overview of the most relevant applications of ALG for drug administration by the oral route was presented by Sosnik (2014).

Ahmad et al. (2006a) prepared ALG-based nanoformulations as a delivery system for frontline antituberculosis drugs, such as rifampicin (RIF), INH, pyrazinamide (PZA), and ethambutol, that were applied as three oral doses to TB-infected mice spaced 15 days apart and resulted in complete bacterial clearance from the organs compared to 45 conventional doses of orally administered free bulk drugs. The relative bioavailabilities of the above drugs encapsulated in ALG NPs were found to be significantly higher than those of the free drugs, and drug levels were maintained at or above the minimum inhibitory concentration (MIC) until day 15 in organs after administration of encapsulated drugs (Ahmad et al., 2006b). Similarly, the bioavailability of aerosolized ALG NPs encapsulating INH, RIF and PZA with the mean particle size of 235 nm was significantly higher than that of oral bulk drugs, and concentration levels of the drugs detected in lungs, liver, and spleen exceeded the MIC until 15 days post nebulization (Zahoor et al., 2005).

A DXR-loaded ALG core and a shell of peptide amphiphile nanofibres functionalized for targeting the folate receptor have shown a 60-fold higher cytotoxicity against invasive breast adenocarcinoma MDA-MB-231 cells than non-targeting particles (Boekhoven et al., 2015). Testosterone-loaded ALG nanocapsules were found to exhibit better bioavailability in comparison with bulk testosterone and commercial testosterone injection (Jana et al., 2015).

The octanol-grafted ALG was employed to prepare NPs that were subsequently used for encapsulation of propofol (Najafabadi et al., 2015). The spherical NPs with the average particle size of 180 nm improved the stability and solubility of propofol for i.v. administration.

The use of tannic acid as a bridging cross-linking agent in Ca^{2+}-ALG NPs resulted in a more than 4-fold increase in encapsulation efficiency of diltiazem·HCl and reduced burst drug release from 44% to around 10% within the first 30 min (Abulateefeh and Taha, 2015). Hydrophilic DXR·HCl and hydrophobic PTX co-encapsulated in a multi-drug delivery system based on ALG/CaCO$_3$ hybrid NPs exhibited significantly enhanced inhibitory effect, especially for drug resistant tumor cells as well as significantly enhanced cell uptake and nuclear localization as compared with the single drug loaded NPs (Wu et al., 2014). In another study performed by Zhao et al. (2014) DXR was loaded to PEGylated ALG brushes onto the graphene oxide NPs. This drug delivery platform for tumor microenvironment-responsive triggered release of DXR demonstrated favorable efficiency of cell apoptosis.

ALG hydrogels coated with CS were found to have intrinsic antimicrobial activity with improved sustained release characteristics and thus, they can be applied for wound dressing (Straccia et al., 2015). A new drug delivery biocomposite system based on polyanionic matrix (Na^+-ALG), polycationic matrix (CS), and silica was shown to maintain or improve the efficacy of antibacterial chemotherapeutics as follows: piperacillin-tazobactam, cefepime, piperacillin, imipenem, gentamicin, ceftazidime against *Pseudomonas aeruginosa* ATCC 27853 and cefazolin, cefaclor, cefuroxime, ceftriaxone, cefoxitin, trimethoprim/sulfamethoxazole against *E. coli* ATCC 25922 (Balaure et al., 2013). Benzoyl peroxide, a commonly used antiacne drug, encapsulated in the ALG/CS NPs demonstrated superior antimicrobial activity against *Propionibacterium acnes* compared with the drug alone but less toxicity to eukaryotic cells, suggesting that in the treatment of dermatologic conditions with infectious and inflammatory components, topical delivery of ALG/CS NP-encapsulated drug therapy can be used (Friedman, et al., 2013).

CPT-ALG conjugate liposomes for targeted delivery to epidermal growth factor receptor-positive ovarian cancer cells prepared by Wang et al. (2014b) exhibited a sustained release of drug. *In vivo* experiments confirmed enhanced delivery of CPT into ovarian tumor tissues and improvement of the antitumor efficacy of CPT, while reducing nephrotoxicity and body weight loss in mice after administration was estimated. Tamoxifen-loaded folate-targeted NPs based on the disulfide bond reduced bovine serum albumin and bovine serum albumin/ALG-cysteine mixtures designed by Martinez et al. (2014) demonstrated controlled release of drug and promising anticancer activity, and the cytotoxicity of unloaded systems was estimated as low.

Folate mediated self-assembled phytosterol-ALG NPs were evaluated for targeted cellular DXR delivery by Wang et al. (2015a). Drug entrapped into folate-phytosterol-ALG NPs had lower IC_{50} value to folate receptor over-expressing cancer cells than the bulk DXR and DXR/phytosterol-ALG NPs because of folate receptor-mediated endocytosis process, and higher intracellular uptake of the drug for DXR/folate-phytosterol-ALG as compared with pure DXR was observed. L-Arginine (Arg) grafted ALG hydrogel beads were found to be a pH-sensitive system for specific protein delivery. Grafting of ALG has improved both thermal and morphological properties of Arg-g-ALG hydrogel beads and improved release profile behavior of bovine serum albumin particularly in acidic media (Eldin et al., 2015).

The release of DXR from sodium ALG/CS/hydroxyapatite nanocomposite was pH sensitive, the samples showed higher release at pH 5 (Abou Taleb et al., 2015), and such beads could prolong the release of Na^+ diclofenac by 8 h compared with the pristine sodium ALG hydrogel beads (Zhang et al., 2010b). pH-Responsive nanocarriers obtained by assembling of biocompatible multilayers of ALG/CS on the surface of mesoporous silica NPs were found to improve blood compatibility over the bare mesoporous silica NPs in terms of low hemolytic and cytotoxic activity against human red blood cells. The release of DXR loaded into these carriers was much faster at lower pH than at higher pH, and nanocarriers provided a sustained intracellular DXR release and a prolonged DXR accumulation in the nucleus of HeLa cells, resulting in a prolonged therapeutic efficacy (Feng et al., 2014). pH-Sensitive Laponite®/DXR/ALG nanohybrids capable of carrying a high load of the cationic DXR through the cell membrane prepared by Goncalves et al. (2014) were sensitive to pH and displayed sustained drug release behavior. In cell culture experiments it was shown that these NPs containing DXR could be effectively internalized by osteosarcoma CAL-72 cells and exhibit a remarkable higher cytotoxicity to cancer cells than bulk DXR.

ALG-coated Fe_3O_4 hollow microspheres exhibited super paramagnetic characteristics, and DNR loaded into the carrier exhibited controlled release behavior (Dong and Jin, 2015). ALG-coated Fe_3O_4 NPs with the mean diameter of ca. 50 nm were found to be stable for up to 9 days (Castello et al., 2015). Treatment with magnetic field of the core-shell carboxylic acid-functionalized nanostructure consisting of Fe_3O_4 NPs as the core, ALG as the shell and D-galactosamine as cell-targeting ligands on the outer surface of NPs exhibited excellent hyperthermic efficacy in HepG2 cells owing to enhanced cellular uptake (Liao et al., 2015b).

5.4.6 PECTIN

Pectin is a natural heteropolysaccharide composed of poly-α-(1→4)-D-galacturonic acid alternating with α-(1→2)-L-rhamnopyranosyl-α-(1→4)-D-galacturonosyl sections. The backbone can be branched by substitution by neutral carbohydrates, such as L-arabinose, D-xylose, D-galactopyranose, D-glucopyranose, D-mannopyranose, L-fucopyranosese, D-gluconic acid, etc. The carboxyl groups of D-galacturonic acid can be methylated, properties of the pectin then depend on the degree of esterification. In the presence of bivalent cations, mainly Ca^{2+}, low-methylated pectins form a gel. High-methylated pectins (with more than 45% of esterified carboxyl groups) can also form a gel. However, this property results from formation of hydrogen bonds and hydrophobic interactions at pH 3 or in the presence of carbohydrates. Another method of modifying the basic skeleton of pectin consists in replacement of the methyl groups of the D-galacturonic acid ester by other alkyl or arylalkyl groups, in addition, there is a possibility of replacing the ester group by an amide, mono- or dialkylamide group. Pectins have so far been used predominantly as food additives, nevertheless, the pectin scaffold system has proved to be suitable for an intended use towards biomedical applications (also as active substances), such as drug delivery (as excipients) and tissue engineering (Král et al., 2011a, 2011b; Nair et al., 2015). Moreover, Tummalapalli et al. (2015) used oxidized pectin as a reducing agent as well as a stabilizing agent for the synthesis of AgNPs.

Zhang et al. (2015c) found that the pectin NPs with galactose residues had higher cytotoxicity in HepG2 cells that highly expressed asialoglycoprotein receptors (ASGR) than free honokiol. This fact was not confirmed in ASGR-negative A549 cells. Due to a specific active targeting ability of the pectin NPs with galactose residues to ASGR-positive HepG2 cells, they can be considered as a potential drug carrier for treatment of liver-related tumors. Nair et al. (2015) investigated a pectin-fibrin composite scaffold in a mouse implantation model in order to prove the compatibility of the scaffold system *in vivo*. The biocompatibility tests after the 12th week revealed no pathological inflammatory responses, which indicates that this scaffold system could be a promising targeted drug delivery system for the slow release of drugs in a mouse disease model.

NPs of sodium caseinate and pectin were prepared by Luo et el. (2015), and it was found that pectin in nanocomplexes delayed the hydrolysis of sodium caseinate by pepsin in *gastric* milieu and enabled the controlled release of model drugs in simulated intestinal conditions indicating the potential of this system for oral delivery of nutrients and medicines. Cross linked gellan gum/pectin beads are characterized by high mucoadhesiveness. Their erosion was greater in acid media, while swelling was more pronounced at pH 7.4. Ketoprofen release was pH-dependent, and while at pH 7.4 sustained drug release up to 360 min was observed, at pH 1.2 the released drug amounts ranged from 20% to 34% (Prezotti et al., 2014).

Pastorino et al. (2015) carried out for the first time the layer-by-layer (LbL) assembly of CS and pectin for the fabrication of hollow capsules and reported about its potential use for drug delivery applications. The LbL self-assembly technique was also used to alternately deposit the positively charged lysozyme and negatively charged pectin on the surface of the CLS nanofibrous mats, and it was found that the average diameter of nanofibers increased with the increasing number of lysozyme/pectin bilayers. The strongest inhibitory effect against both *E. coli* and *S. aureus* was estimated for nanofibrous mats coated by 10.5 lysozyme/pectin bilayers (with lysozyme on the outmost layer) (Zhang et al., 2015b). pH-Sensitive lysozyme/pectin nanogel with spherical particles with diameters about 109 nm and negative surface charge for tumor MTX delivery was developed by Lin et al. (2015). MTX-loaded nanogels were pH-dependent, accelerated release of MTX at a decreasing pH from 7.4 to 5.3, the encapsulated MTX exhibited higher anticancer activity than bulk MTX and comparing to bulk these nanoformulations showed enhanced cancer-cell apoptosis, because they could be effectively endocytosed by HepG2 cells.

Eudragit S100 coated citrus pectin NPs were prepared for the colon targeting of 5-fluorouracil (5-FU), in which citrus pectin also acts as a ligand for galectin-3 receptors that are overexpressed on colorectal cancer cells. Selective drug release in the colonic region was observed and more than 70% of drug was released after 24 h from this nanoformualtion. Moreover, the NPs showed 1.5-fold greater cytotoxicity potential against HT-29 cells *in vitro* compared to 5-FU solution, and the results of an *in vivo* experiment showed that Eudragit S100 successfully guarded NPs to reach the colonic region wherein NPs were taken up and showed drug release for an extended period of time (Subudhi et al., 2015).

Green composites based on pectins and nano-hybrids composed of halloysite nanotubes loaded with rosemary essential oil were found to be suitable for application in the active packaging field where controlled release of active ingredient is desirable (Gorrasi, 2015).

5.4.7 CARRAGEENAN

Carrageenans (CRGs) are naturally occurring hydrophilic, polyanionic polysaccharides with wide application in pharmaceutical industries for controlled drug delivery and as food additives (Liu et al., 2015b). CRG is a sulfated linear biopolymer of D-galactopyranose and 3,6-anhydro-D-galactopyranose obtained by extraction of certain red seaweeds of the *Rhodophyceae* class. The units are joined by alternating α-$(1{\rightarrow}3)$ and β-$(1{\rightarrow}4)$ glycosidic bonds. Main three classes of CRGs can be distinguished according to the number and position of the ester sulfate groups on the repeating galactose units: (i) κ-CRG forms in the presence of K^+ ions strong, rigid gels, (ii) ι-CRG forms in the presence of Ca^{2+} ions soft gels, and (iii) λ-CRG does not form gel. In general, it can be stated that higher levels of ester sulfate decrease the solubility temperature of CRGs and decrease the strength of gels, or contribute to inhibition of gelatinization (Raman et al., 2015). Li et al. (2014b) summarized recent applications of CRGs in drug delivery systems and accentuated that based on the strong negative charge and gelling properties of CRG, it has been used as a gelling agent/viscosity enhancing agent for controlled drug release and prolonged retention and found application in tissue regeneration with therapeutic biomacromolecules and in cell delivery. CRG-based matrices are demonstrated to increase drug loading and drug solubility, enabling release of orally administrated drugs in zero-order or in a significantly prolonged period, however, they can also have adverse influence on the human body, for example, by inciting unwanted immune responses and inhibiting blood coagulation (Liu et al., 2015b). Nanodrugs prepared by electrostatic complexation of drug molecules with oppositely charged polysaccharides represent a promising bioavailability enhancement strategy for poorly soluble drugs. For example, the higher charge density and chain flexibility of dextran sulfate coupled with the greater hydrophobicity of CRG can be considered as an ideal supersaturated drug delivery system (Cheow et al., 2015).

A formulation of CRG-CS NPs with particle size range 350–650 nm and positive zeta potentials of 50-60 mV suitable for controlled release of macromolecules was designed by Grenha et al. (2010). These NPs demonstrated excellent capacity to provide a controlled release of model protein ovalbumin for up to 3 weeks and were found to be non-toxic against normal mouse fibroblast L929 cells, indicating appropriate biocompatibility of these carriers. The CS-CRG NPs with particle diameters between 200 and 1000 nm, zeta potentials of 40–55 mV, and polydispersity between 0.2 and 0.35 showed an increased recombinant human erythropoietin encapsulation efficiency of 50% and a more sustained *in vitro* release of approximately 50% over a 2-week period, when compared to previous nano/microparticle delivery systems. Increasing of surface charge and CS molecular weight resulted in improved encapsulation efficiency and reduced release rate (Bulmer et al., 2012). The encapsulation of BSA was used as a model system for the controlled drug delivery from ALG-CS-CRG composite NPs, and the release profile indicated an initial burst in the first few hours of the trial, followed by a slower steady release over time (Cheng et al., 2015).

By complexation of model poorly water-soluble compounds with hydrophilic λ-CRG, compounds became amorphous in the comple□. Use of additional CRG as a stabilizer enabled to prepare a nanosuspension of a compound/CRG complex with the median particle size of ca. 300 nm that showed increased aqueous solubility of the compound from <1 µg/mL to 39 µg/mL and faster dissolution rate than that of the bulk compound and the nanosuspension of the free compound (Dai et al., 2007).

κ-CRG closely mimics the glycosaminoglycan structure, one of the most important constituents of native tissues extracellular matrix. By introduction of photocrosslinkable methacrylate moieties on the κ-CRG backbone physically and chemically crosslinked hydrogels showing highly versatile physical and chemical properties (controllable compressive moduli, swelling ratios and pore size distributions), while maintaining the viability of encapsulated cells were prepared (Mihaila et al., 2013). The swelling and release properties of functional κ-CRG hydrogel nanocomposites were described by da Silva et al. (2012). INH-loaded polyelectrolyte κ-CRG and CS nanocapsules prepared by encapsulation of neem seed oil showed the release rates of INH higher at acidic media compared to basic media (Devi and Maji, 2010).

Super paramagnetic magnetite hydrogel κ-CRG nanospheres with particle sizes 50 and 75 nm were prepared, and subsequently their surface was carboxymethylated. These carboxylated nanospheres were shown to be thermo-sensitive in the temperature range 37–45°C, suggesting the potential of this formulation as a thermally controlled delivery system for drugs and/or Fe_3O_4 particles (da Silva et al., 2009a). *In vitro* release studies revealed that using metallic nanocomposites hydrogels, such as Ag and Fe_2O_3 nanofillers modified κ-CRG hydrogels, can improve the drug release in intestine and minimize it in the stomach, and cross-linking and nanofiller loading can significantly improve the targeted release and gastrointestinal tract controlled drug delivery (Hezaveh and Muhamad, 2012). Controlled drug release magnetic/pH-responsive beads based on κ-CRG and CMCS were designed by Mahdavinia et al. (2015), and the response of these beads to external stimulus makes them good candidates for novel drug delivery systems. The development of novel nanocomposite by coupling the synergistic effects of the sulfated polysaccharide (ι-CRG) and a magnetic NP (maghemite, Fe_3O_4), in which ι-CRGs electrostatically entrap maghemite NPs in their sulfate groups, may offer new interesting applications in drug delivery and cancer therapy (Raman et al., 2015). Due to magnetic properties these NPs can be used in magnet aided targeted drug delivery, especially in cancer therapy.

Beads made from encapsulation of fluorescein isothiocyanate-labeled dextran (FD-4) in the carboxymethylated κ-CRG displayed pH-dependent swelling and encapsulation efficiency of 74%. The release of dextran was low (23%) in simulated gastric fluid and high (90%) in simulated intestinal fluid in a 2 h dissolution study. An application of λ-CRG coating layer on the surface of the beads further reduced the FD-4 release in simulated gastric fluid (Leong et al., 2011).

5.4.8 POLYSACCHARIDE GUMS

5.4.8.1 Gellan Gum

Gellan gum (GG) is a linear water-soluble anionic deacetylated exopolysaccharide obtained by aerobic fermentation of *Sphingomonas elodea*. It possesses a tetrasaccharide repeating unit of one α-L-rhamnopyranose, one β-D-glucuronic acid and two β-D-glucopyranoses with one free carboxyl group. GG is widely used in food and confectionary industry as a thickening and gelling agent (Jansson et al., 1983; Morris et al., 2012; Prajapati et al., 2013; Goswami and Naik, 2014). Dhar et al. (2008) used it as a reducing and stabilizing agent for the synthesis of AuNPs instead of usually applied citrate and borohydride, and DXR·HCl loaded on these NPs showed enhanced cytotoxic effects on human glioma LN-18 and LN-229 cells. Sophorolipid-conjugated GG reduced/capped AuNPs showed greater efficacy in killing LN-229 cells and, pleasingly, glioblastoma HNGC-2 cells, and further enhancing of these effects was obtained after conjugation of DXR·HCl to these AuNPs (Dhar et al., 2011a). The specific uptake of such GG-reduced AuNPs into cancer cells was observed, because they were found to be situated in LN-229 cells, while no uptake was observed in normal fibroblast NIH-3T3 cells (Dhar et al., 2011b). D'Arrigo et al. (2012) prepared nanohydrogels with the particle size of ca. 300 nm based on sonicated GG chains and bounded prednisolone to the carboxylic groups of GG. In addition, a novel GG nanohydrogel system able to carry and deliver simultaneously anti-cancer and anti-inflammatory drugs was developed by D'Arrigo et al. (2014). This system can act as drug solubility enhancer, favoring the drug uptake in the cells. While in these formulations prednisolone was chemically linked to the carboxylic groups of GG and served as a hydrophobic moiety promoting nanohydrogel formation, PTX was then physically entrapped in it. It should be noted that nanohydrogel exhibited an increased cytotoxic effect *in vitro* on several types of cancer cells due to the synergistic effect of the combination of anti-inflammatory and anti-cancer drugs. Low-acyl GG was used to coat gold nanorods (AuNRs) surface forming hydrogel-like shell with 7 nm thickness around individual AuNRs. The formulation was no toxic, and AuNRs-GG combined with osteogenic media enhanced the mineralization capacity 2-fold as compared to cells exposed to osteogenic media alone, indicating that the system has interesting features for osteogenesis (Vieira et al., 2015).

5.4.8.2 Xanthan Gum

Xanthan gum (XG) can be considered as a natural CLS derivative. It is an anionic polysaccharide produced by the plant pathogenic bacterium *Xanthomonas campestris*. The backbone of XG is (1→4)-linked β-D-glucopyranose. Every 2nd glucopyranose unit is substituted in position 3 by a trisaccharide side chain consisting of β-D-mannopyranose-(1→4)-β-D-glucuronic acid-(1→2)-α-D-mannopyranose. The mannose residues in the side chain are substituted by acetyl and pyruvyl moieties, the degree of substitution depends on fermentation conditions. XG possesses excellent rheological properties and no adverse effect on human health. Moreover, XG has been found to provide nanowires and nanofibers with the desired pattern on the substrate, therefore this material is attractive also for applications in nanotechnology (Kobori and Nakao, 2010; Palaniraja and Jayaraman, 2011; Goswami and Naik, 2014). Also XG has been used as a surface modifier of NPs to improve stability, impart biocompatibility, and promote long systemic circulation (Muddineti et al., 2015). Conjugation of XG and lysozyme provided NPs with favorable size distribution (62–108 nm), which, due to alkali-coupled thermal treatment, lost their natural structures, explored more hydrophobic areas accompanied by molecule rearrangement and gelation, and then interpenetrated in forming the NPs (Xu et al., 2015). In addition, a physical blend of XG and methylcellulose in aqueous solution forms thermo-responsive injectable hydrogel material for long-term drug delivery with excellent biocompatibility and biodegradability (Liu and Yao, 2015). AuNPs, in which XG was used as both reducing and stabilizing agent, were investigated as a drug delivery carrier for DXR·HCl. It was found that DXR-loaded AuNPs were stable at pH range between pH 5–9, were biocompatible and non-toxic, and exhibited 3-fold more cytotoxicity in A549 cells than bulk DXR (Pooja et al., 2014).

5.4.8.3 Gum Arabic

Gum arabic (GA) is the complex exudate, especially from trees *Acacia senegal* and *Acacia seyal*, that is characterized by non-digestibility, low solution viscosity, emulsifying, stabilizing, binding, and shelf-life enhancing properties. The backbone of GA is composed of $(1 \rightarrow 3)$-linked β-D-galactopyranose bearing side chains from L-arabinose, L-rhamnopyranose, D-galactopyranose, and D-glucuronic acid in position 6. GA occurs neutral or as a weakly acidic salt. Sometimes protein residues are covalently bound to the carbohydrate backbone. Due to its composition, GA can be used in a wide spectrum of applications, including antimicrobial, anti-inflammatory, and anticoagulant applications (Belitz et al., 2009; Goswami and Naik, 2014). GA is also used as an emulsifying and encapsulating agent (Kulkarni et al., 2012; Patel and Goyal, 2015). Magnetic field-sensitive modified GA-based hydrogels are suitable to be effectively applied as biomaterial either on remote controlled release or tissue engineering (Paulino et al., 2010). Fe_2O_3 NPs coated with GA showed significant cellular uptake in glioma 9L cells, their accumulation in the tumor tissue after i.v. administration to rats with 9L cells followed by the application of an external magnetic field was 12-fold higher in excised tumors when compared to contralateral normal brain (Zhang et al., 2009). Functionalized radioactive AuNPs in tumor therapy utilizable in treating prostate and various solid tumors in human cancer patients were described by Kannan et al. (2012). Epirubicin-loaded GA-capped AuNPs functionalized with folic acid (FA) showed enhanced cytotoxic effects on A549 cells, indicating that functionalization with FA resulted in cancer cell targeting and increased uptake of epirubicin by cancer cells (Devi et al., 2015).

5.4.9 HYALURONIC ACID

Hyaluronic acid (HA) is a linear non-sulfated polysaccharide formed from disaccharide units containing 2-acetamido-2-deoxy-D-glucopyranose (N-acetyl-β-D-glucosamine) and β-D-glucuronic acid, linked via (1→3) and (1→4) glycosidic bonds. It has a high molecular mass, usually in the order of millions of Daltons, and interesting viscoelastic properties influenced by its polymeric and polyelectrolyte characteristics. HA is present in almost all biological fluids and tissues (Kogan et al., 2007). Synthesis, physicochemical characterization, and biological properties of different NP delivery systems that include HA in their structures were summarized by Ossipov (2010). However, to make HA suitable as an i.v. targeting carrier, it is necessary to focus on the reduction of HA clearance from the blood and suppression of the HA uptake by liver and spleen and provide tumor-triggered mechanisms of release of an active drug from the HA carrier (Ossipov, 2010). Medically, HA is used as a surgical aid in ophthalmology. It also possesses therapeutic potential in the treatment of arthritis and wound healing (Yadav et al., 2008). An overview of the occurrence and physiological properties of HA as well as of the recent advances in production biotechnology and preparation of the HA-based materials for medical application was published by Kogan et al. (2007).

The development of injectable and biocompatible vehicles for delivery, retention, growth, and differentiation of stem cells is of paramount importance for regenerative medicine. Prestwich (2011) summarized the design criteria for "living" HA derivatives and the many uses of such in situ cross-linkable HA-based synthetic extracellular matrix hydrogel for 3D culture of cells in vitro and translational use in vivo and noted that recent advances allow rapid expansion and recovery of cells in 3D space, and the bioprinting of engineered tissue constructs. Hyaluronic acid-based hydrogels as 3D matrices for in vitro evaluation of chemotherapeutic drugs (CTH, DTX, and rapamycin, alone and in combination) using poorly adherent prostate cancer cells were developed by Gurski et al. (2009), and it was found that this 3D system better reflected the bone metastatic microenvironment of the cancer cells than a conventional monolayer system.

To target letrozole to letrozole-resistant (LTLT-ca) cells, HA-bound drug NPs of the maximum size of 100 nm were prepared by nanoprecipitation. IC_{50} values of NPs related to MCF-7 cells overexpressing aromatase (MCF-7/Aro) and LTLT-ca cells were estimated as 2 μmol/L and 5 μmol/L, respectively. These NPs more effectively inhibited tumor growth and showed greater potency in reduction *in vitro* cellular and *in vivo* tumor aromatase enzyme activity than the corresponding bulk letrozole. Moreover, NPs restored and maintained a prolonged sensitivity and targeted delivery to letrozole-resistant xenograft tumors in mice (Nair et al., 2011).

The use of amino acids (valine, leucine, and phenylalanine) as spacers between PTX and a HA carrier resulted in formation of NPs with a narrow size distribution and a spherical shape. Facilitated PTX release from the conjugates was observed. Moreover, prodrugs synthesized as HA-amino acid-PTX conjugates exhibited enhanced cytotoxicity in breast cancer cell lines, and it was found that MCF-7 cells treated with conjugates were arrested in the G_2/M phase of the cell cycle (Xin et al., 2010). The HA conjugates prepared by chemical conjugation of hydrophobic 5β-cholan-24-ioc acid to the backbone of HA formed nano-sized spherical self-aggregates with the particle size of 350–400 nm via hydrophobic interactions among 5β-cholan-24-ioc acids, the NPs were effective against squamous cell carcinoma (SCC7) cells overexpressing CD44, the receptor for HA. After systemic administration, the concentration of these HA NPs in tumor was found to be 4-fold higher than that of pure HA polymer (Choi et al., 2009). Independently on particle size, significant amounts of hydrophobically modified HA NPs circulated for 2 days in the bloodstream and were selectively accumulated into the tumor site, smaller HA NPs reach the tumor site more effectively than larger HA-NPs, and accumulation of HA NPs into the tumor site was found to be performed by a combination of passive and active targeting mechanisms (Choi et al., 2010). Nanosystems for cellular and intracellular targeting delivery of hydrophobic anticancer drugs were prepared by encapsulating PTX into the core of micelles consisting of HA NPs grafted with a hydrophobic octadecyl tail ($HA-C_{18}$) and covered by folic acid ($FA-HA-C_{18}$). Formulations e☐hibitedsignificantly higher cytoto☐ic activity against MCF-7 (CD44-overexpressing cell line) and A549 cells compared to Taxol® even at a lower PTX concentration. The $HA-C_{18}$ and $FA-HA-C_{18}$ NPs were efficiently taken up via CD44 receptor-mediated endocytosis, and the folate receptor-mediated endocytosis further enhanced internalized amounts of FA-HA-C$_{18}$ micelles in MCF-7 cells as compared with HA-C$_{18}$ micelles, indicating that HA-C$_{18}$ and FA-HA-C$_{18}$ copolymers could be considered as biodegradable, biocompatible, and cell-specific targetable nanostructure carriers (Liu et al., 2011). HA-Ceramide (HA-CE)-based self-assembled NPs developed for i.v. DTX delivery enhanced the intracellular DTX uptake in MCF-7 cells, and DTX cellular uptake study showed that the NPs were taken up by the HA-CD44 interaction, whereby the multidrug resistance-overcoming effects of DTX-loaded HA-CE/Pluronic® 85-based NPs were observed in cytotoxicity tests in multidrug-resistant breast cancer MCF-7/ADR cells as well (Cho et al., 2011).

Although PEGylation of HA NPs reduced their cellular uptake *in vitro*, larger amounts of NPs were taken up by cancer cells overexpressing CD44 than by normal fibroblast cells, and the PEGylation effectively reduced the liver uptake of HA NPs and increased their circulation time in the blood. At systemic administration of NPs into tumor-bearing mice, high tumor targetability of NPs was observed, and PEG-HA NPs were more effectively (up to 1.6-fold) accumulated in the tumor tissue, than bare HA NPs (Choi et al., 2011a). PEG-HA NPs that were applied as carriers of DXR and CTH were internalized into SCC7 and MDA-MB-231 cells via receptor-mediated endocytosis, but were rarely taken up by normal NIH-3T3 cells. After incubation with cancer cells PEG-HA NPs rapidly released drugs, while CTH-loaded PEG-HA NPs showed dose-dependent cytotoxicity to cancer cells and significantly lower cytoto□icityagainst fibroblasts NIH-3T3 than bulk CTH. After systemic administration into tumor-bearing mice, these NPs selectively accumulated in the tumor, primarily due to prolonged circulation in the blood and binding to a receptor (CD44) that was overexpressed on the cancer cells, and no significant increases in tumor size for at least 35 days was observed (Choi et al., 2011b).

Nanocomplexes based on HA and PEGylated tumor necrosis factor-related apoptosis, inducing ligand (TRAIL) (PEG-TRAIL) formed by *N*-terminal specific PEGylation, were designed as potential antirheumatics. In *in vivo* tests these NPs secured sustained delivery and caused a significant reduction of serum inflammatory cytokines and collagen-specific antibodies that are responsible for the pathogenesis of rheumatoid arthritis (Kim et al., 2010).

DXR-loaded HA-PLA and HA-PLA-PEG micelles prepared by Pitarresi et al. (2010) showed selective cytotoxicity toward CD44 overexpressing colorectal carcinoma HCT-116 cells compared to receptor deficient human dermal fibroblasts. PEGylated micelles showed better stability and drug loading capacity and were able to escape from macrophage phagocytosis. DXR-loaded HA-PEG-PLGA NPs were found to sustain the release for up to 15 days. After i.v. injection these NPs in Ehrlich ascites tumor-bearing mice were able to deliver a higher amount of drug, showed a higher concentration of DXR in the tumor, and reduced tumor volume more significantly than monomethoxyPEG-PLGA NPs (Yadav et al., 2007). 5-FU-loaded HA-PEG-PLGA copolymer NPs administered i.v. to Ehrlich ascites tumor-bearing mice were found to be less hemolytic but more cytotoxic as compared to free 5-FU. This nanoformulation was less immunogenic compared to plain drug. However, these NPs were able to deliver a higher concentration of 5-FU in the tumor and to reduce tumor volume significantly in comparison with bulk 5-FU. Thus, whereas the conjugation of HA imparts targetability of the formulation, and enhanced permeation and retention effect rule out its access to the non-tumor tissues, favored selective entry in tumors reducing side effects both *in vitro* and *in vivo* can be observed (Yadav et al., 2010).

Encapsulation of PTX into HA-coated poly(butyl cyanoacrylate) (PBCA) NPs provided a sustained release in subsequent 188 h and showed significantly reduced cytotoxicity, and the uptake of HA-PBCA NPs by Sarcoma-180 cells was ca. 10-fold higher than that of PBCA NPs (He et al., 2009). Hydrotropic HA conjugates prepared by chemical conjugation of an amine-terminated hydrotropic oligomer to a HA backbone were also taken up by the cancer cell line (SCC7) overexpressing CD44, and the *in vitro* release rate of PTX was found to be lower for NPs that contained larger amounts of the drug, whereby drug-loaded NPs exhibited stronger cytotoxicityto SCC7 than to normal fibroblast monkey kidney CV-1 cells (Saravanakumar et al., 2010).

DXR-loaded HA NPs mineralized by $Ca_3(PO_4)_2$ released DXR in a sustained manner at pH 7.4, while a rapid release of DXR was observed in the acidic solution, indicating that these NPs have potential as robust NPs that can release DXR at specific sites under mildly acidic conditions, such as the extracellular matrix of tumor tissue and intracellular compartments (e. g., endosome and lysosome) of the cell (Han et al., 2011). Kamat et al. (2010) designed Fe_2O_3-based magnetic HA coated NPs that were biocompatible and stable in the presence of serum to target activated macrophages. While the HA immobilized on the NPs retained their specific biological recognition with HA receptor CD44 situated on activated macrophages, the magnetite cores of the HA-coated NPs were only transiently present inside the cells, thus reducing the potential concerns of nanotoxicity. Also HA-conjugated mesoporous silica NPs were found to be capable of selectively targeting specific cancer cells by overe☐pressingCD44, leading to rapid and concentration-dependent uptake by the cancer cells through the receptor-mediated endocytosis mechanism. On the other hand, selective targeting of these NPs was observed to CD44 low-expressing cells, and CTH encapsulated into the NPs showed enhanced toxicity against HeLa cells compared to bulk CTH (Ma et al., 2012).

5.4.10   CHITIN

Chitin (CT) is the most abundant nature resource next to CLS, and chitin and CS are the most widely accepted biodegradable and biocompatible materials subsequent to CLS. CT, composed of $(1\rightarrow4)$-linked units of N-acetyl-β-D-glucosamine, some of which are deacetylated, exists in the form of nano-structures in living organisms (Muzzarelli, 2011). CT and its deacetylated derivative, CS, are non-toxic, antibacterial biodegradable, and biocompatible biopolymers often used in many biomedical applications, such as tissue engineering scaffolds, drug delivery, wound dressings, separation membranes, antibacterial coatings, stent coatings, and sensors (Jayakumar et al., 2010a; Jayakumar et al., 2010b). The advantages of these biomaterials are such that they can be easily processed into different forms such as membranes, sponges, gels, scaffolds, microparticles, NPs, and nanofibers (Anitha et al., 2014a). Recent progress of CT-based materials was reviewed by Shi et al. (2011) and Barikani et al. (2014). The advantage of CT as a tissue engineering biomaterial lies in the fact that it can be easily processed into gel and scaffold forms for a variety of biomedical applications. An overview of the current status of tissue engineering/regenerative medicine research using chitin scaffolds for bone, cartilage, and wound healing applications was presented by Jayakumar et al. (2011a). Different derivatives of CT and CS have been prepared for treatment of wounds and burns in the form of hydrogels, fibers, membranes, scaffolds, and sponges, using the adhesive nature of CT and CS together with their antifungal and bactericidal character and their permeability to oxygen (Jayakumar et al., 2011b). For example, by introducing a CM group, water soluble derivatives carboxymethyl chitin (CMCT) and carboxymethyl chitosan (CMCS) can be prepared and versatile CMCT and CMCS nanomaterials are efficient vehicles for the delivery of DNA, proteins, and drugs (Narayanan et al., 2014).

While *in vivo*, CT occurs as a part of complex structures with other organic and inorganic compounds (e.g., in fungi it is covalently linked to glucans, in bacteria it is diversely combined depending on their type, and in arthropods CT is covalently linked to proteins and tanned by quinones), isolated and purified CT is a polysaccharide that, at the nano scale, presents itself as a highly associated structure and could be used as immunoadjuvant and non-allergenic drug carrier (Muzzarelli, 2010). Biodegradable CT nanogel of size <100 nm showing higher swelling and degradation in acidic pH was found to be retained inside cells and could be regarded as useful for the delivery of drugs, growth factors for drug delivery, and tissue engineering (Rejinold et al., 2012). Also locally injectable, biodegradable, hemocompatible, and pH sensitive CT-PLA composites exhibiting higher swelling and higher drug release at acidic pH compared to neutral pH, were found to be a promising anticancer drug delivery system for liver cancer therapy (Arunraj et al., 2014).

CT/Ag NPs exhibited bactericidal activity against *S. aureus* and *E. coli* and good blood clotting ability suggesting that they could be used for wound healing applications (Madhumathi et al., 2010). The NiNPs (120–150 nm) loaded CT nanogels were completely non-toxic against A549 and L929 cells and did not perturb their cellular constituents, while their antibacterial activity against *S. aureus* was significant (Kumar et al., 2013). RIF loaded amorphous CT NPs were found to be non-hemolytic and non-toxic against a variety of host cells. They released RIF to 60% in 24 h, followed by a sustained pattern till 72 h and showed 5–6-fold enhanced delivery of drug into the intracellular compartments of polymorphonuclear leukocytes that provide the primary host defense against invading pathogens by producing reactive oxygen species and microbicidal products and exhibited anti-microbial activity against *E. coli*, *S. aureus* and a variety of other bacteria (Smitha et al., 2015). Fluconazole loaded CT nanogels showing controlled release pattern that is ideal for the continuous availability of fluconazole over a longer period were found to be suitable for the treatment of corneal fungal infections. This nanoformulation is characterized by very good cell uptake in human dermal fibroblasts and penetration to the deeper sections of the porcine cornea with no signs of destruction or inflammation to corneal cells (Mohammed et al., 2013).

5-FU loaded chitin nanogels forming good, stable aqueous dispersion with spherical particles of 120-140 nm, pH responsive swelling, and drug release showed toxicity to melanoma cells (A375) in the concentration range of 0.4–2.0 mg/mL, but lower to☐icitytoward human dermal fibroblast cells as compared with bulk 5-FU. When applied at skin, loosening of the horny layer of epidermis by interaction of cationically charged chitin occurred, and these nanogels showed almost same steady state flu☐as control 5-FU, but the retention in the deeper layers of skin was found to be 4–5-fold higher for DXR (Sabitha et al., 2013). DXR loaded CT nanogels with the size of 130–160 nm exhibited higher toxicity to MCF-7, A549, HepG2 cells comparing to normal L929 cells and showed a significant uptake of DXR nanoformulation (Jayakumar et al., 2012). PTX loaded amorphous CT NPs of the size of ca. 150 nm showed a sustained release of the drug and increased cytotoxicity toward colon cancer cells (Smitha et al., 2013).

CMCT NPs prepared by cross-linking of CMCT solution with $CaCl_2$ and $FeCl_3$ were found to be non-toxic to L929 cells, and 5-FU loaded into these NPs with magnetic properties showed antibacterial activity and controlled and sustained drug release at pH 6.8 (Dev et al., 2010). Using lyophilization, a non-toxic nanocomposite scaffold was prepared by incorporation of nano-ZrO_2 onto the CT-CS scaffold. It demonstrated better swelling and controlled degradation in comparison with the control scaffold, and evaluation of the cell attachment of the composite scaffolds showed that cells were attached to the pore walls and spread uniformly throughout the scaffolds, indicating that it could be used for various tissue-engineering applications (Jayakumar et al., 2011c). The CT/PLGA NPs that were surface modified with Au, Fe_3O_4, CdTe/ZnTe-QDs, and umbelliferone, showed radio frequency heating. The chitin/PLGA composites could be useful for microbial monitoring and radio frequency application for cancer therapy, and multifaceted chitin/PLGA composite nanogels could be applied for hyperthermia for cancer treatment and microbial labeling and imaging (Rejinold et al., 2014).

Redox responsive cystamine conjugated HA-CT nanogels for the intracellular delivery of DXR within colon cancer cells were found to be to be safe for i.v. administration, because they were non-hemolytic and did not interfere with the coagulation cascade. In these 150–200 nm sized nanoformulations, chitin, having a slow degrading property, could make HA to slowly degrade. Consequently, DXR is protected from a sudden burst release, and HA, being a ligand for CD44 receptor that is overexpressed in HT-29 cells (Ashwinkumar et al., 2014). DXR loaded in a pH responsive CT-poly(caprolactone) composite nanogels showed higher swelling and degradation in acidic pH and also higher drug release at acidic pH compared to neutral pH. The cellular internalization of the nanogel systems and their blood compatibility were confirmed as well, and they showed dose dependent cytotoxicity toward A549 cells (Arunraj et al., 2013).

5.4.11   CHITOSAN

Biopolymer CS, a high-molecular-weight linear polycationic heteropolysaccharide comprising copolymers of β-(1–4)-linked 2-amino-2-deo☐y-D-glucopyranose (β-D-glucosamine) and N-acetyl-β-D-glucosamine, is produced by partial alkaline N-deacetylation of chitin that is commercially extracted from shrimp and crab shells and is also found in nature in the cell walls of fungi of the class *Zygomycetes* and in insect cuticles. CS has a number of applications in agriculture, biomedical area, and in pharmacy (Raafat et al., 2008; Jayakumar et al., 2010b; Wang et al., 2011b; Malam et al., 2011; Koo et al., 2011; Jee et al., 2012; Shukla et al., 2013; Gibot et al., 2015; Thakur and Thakur, 2015). For example, the recent advances in drug delivery applications of CS-based polyelectrolyte complexes with other various natural polysaccharides were reviewed by Luo and Wang (2014), and a comprehensive review focusing on the current use of injectable *in situ* CS hydrogels in cancer treatment was presented by Ta et al. (2008). CS-based polyelectrolyte complexes that are formed in solution by spontaneous association of positively charged CS at low pH values with negatively charged polyions preserve biocompatible characteristics of CS and can be used as excipients in drug delivery systems such as NPs and microparticles, beads, fibers, sponges, and matrix type tablets (Hamman, 2010).

CS e☐hibits significant antibacterial properties, and more negatively charged cell surfaces had a greater interaction with CS, resulting in its higher antibacterial activity, because more adsorbed CS would result in greater changes in the structure of the cell wall and in the permeability of the cell membrane (Chung et al., 2004). Raafat et al. (2008) reported that the antimicrobial activity of CS is connected with the fact that it causes a simultaneous permeabilization of the cell membrane to small cellular components coupled to a significant membrane depolarization, CS treatment induces multiple changes in the expression profiles of *S. aureus* SG511 genes involved in the regulation of stress and autolysis as well as genes associated with energy metabolism. According to the researchers, binding of CS to teichoic acids coupled with a potential extraction of membrane lipids (predominantly lipoteichoic acid) results in a sequence of events that eventually lead to bacterial death.

Glycyrrhizin-loaded CS NPs of the mean particle size of ca. 180 nm prepared by ionotropic gelation technique showed controlled release of the drug from the formulation over a period of 24 h after primarily burst release (Rani et al., 2015). The pH-sensitive behavior of MTX-loaded CS NPs resulted in the accelerated release of MTX in an acidic medium as well as membrane-lytic pH-dependent activity, which facilitated the cytosolic delivery of endocytosed materials. These NPs were more active against tumor HeLa and MCF-7 cell lines, causing greater cell cycle arrest and apoptotic effects than the free drug (Nogueira et al., 2013).

The release of RIF from the oleoyl-CMCS NPs of the mean size of 160 nm, in which an amide linkage between amino groups of CMCS and oleoyl chloride was formed, was found to be sensitive to the pH of the release media, and it could be retarded by addition of the crosslinker sodium tripolyphosphate (Li et al., 2011). Larger oleoylchitosan self-assembled NPs with mean diameter about 275 nm released RIF in solution with pH 6.0 and 6.8 faster than in solution with pH 3.8, and the sample with low concentration of RIF released the drug faster and entirely. Similarly to above mentioned oleoyl-CMCS NPs, the increase of sodium tripolyphosphate resulted in a slower release of RIF (Li et al., 2009). For complete eradication of *Helicobacter pylori* and prevention of the development of antibiotic resistance, triple therapy (amoxicillin, clarithromycin, and omeprazole) is used in clinical practice. An interesting nanoformulation was designed by Ramteke et al. (2009). The controlled release amoxicillin-, clarithromycin- and omeprazole-entrapped CS-glutamate nanosystem (particle sizes 550–900 nm) enabled selective targeting and delivery of drugs not only to bacterial cell proximity but also navigated it into the cellular interiors, where effective killing and eradicating of the pathogen occurs. The contact with the mucous layer was secured by mucoadhesive properties of the system, while interaction of *H. pylori* with the specific ligand present in the system resulted in eradication of the bacterium from gut. Chang et al. (2010) developed CS/poly-γ-glutamic acid NPs incorporated into pH-sensitive hydrogels as an efficient carrier for amoxicillin delivery. While the amount of amoxicillin released from NPs incorporated in hydrogels at pH 1.2 was significantly lower than from only NPs (14% and 50%, respectively), by the incorporation of antibiotic-loaded NPs in a hydrogel, amoxicillin was protected from the actions of the gastric juice, and its interaction with the site of *H. pylori* infection was facilitated. Spherical positively charged nanoemulsion CS-heparin particles loaded with amoxicillin exhibited controlled release of amoxicillin in the gastrointestinal dissolution medium, and amoxicillin-loaded nanoemulsion particles had a significantly greater *H. pylori* clearance effect in the gastric infection mouse model than the amoxicillin solution alone (Lin et al., 2012).

Negatively charged mucoadhesive CS-coated ALG (CS-ALG) NPs with a size range of 380–420 nm were used as a delivery system for daptomycin permeation across ocular epithelia. The drug that is suitable for the treatment of endophthalmitis caused by methicillin-resistant *S. aureus (MRSA)* can be effectively encapsulated into CS-ALG NPs, and encapsulation did not negatively affect the antibacterial activity of daptomycin (Costa et al., 2015). Gellan ☐anthangels along with CS NPs, basic fibroblast growth factor, and bone morphogenetic protein 7 employed in a dual growth factor delivery system showed significantly improved cell proliferation and differentiation of human fetal osteoblasts due to the sustained release of growth factors. Moreover, application of these formulations resulted in higher alkaline phosphatase and calcium deposition than the use of single growth factor loaded gels, and antibacterial effects of gellan xanthan gels against the common pathogens in implant failure, namely *P. aeruginosa*, *S. aureus*, and *S. epidermidis*, were estimated (Dyondi et al., 2013).

Nanoformulation of arginylglycylaspartic acid (RGD)-tripeptide encapsulated into NPs of glycol CS hydrophobically modified with 5β-cholan-24-ioc acid significantly decreased tumor growth and micro vessel density compared to native RGD-peptide injected either intravenously or intratumorally, because this formulation strongly enhanced the antiangiogenic and antitumoral efficacy of RGD-peptide by affording prolonged and sustained RGD-peptide delivery locally and regionally in solid tumors (Kim et al., 2008a).

DNA fragmentation that is characteristic of apoptosis and elevated caspase-3-like activity in CS-treated bladder cancer cells were observed by Hasegawa et al. (2001). Eudragit S 100-coated CS NPs encapsulating a caspase 3 activator of the mean size of 260 nm with irregular surface due to coating showed better tumor regression ability than the uncoated NPs and demonstrated the continuous release profile of caspase 3 activator and comprehensive residence time, indicating that they may be applied as a potential delivery system for the targeting and treatment of colon cancer (Jain et al., 2015).

Su et al. (2015) synthesized a self-assembled biodegradable nanoscaled DXR delivery micellar system that consists of a stearic acid-grafted CS via disulfide linkers. This conjugate showed higher cytoto☐icity against

DXR-resistant cells than free DXR with reversal ability up to 35-fold, secured selective drug accumulation in tumor, and reduced its non-specific accumulation in hearts. NPs of a polyelectrolyte complex composed of CS/CMCS as a pH responsive carrier for oral delivery of DXR·HCl were investigated by Feng et al. (2013), who confirmed that this nanoformulation can effectively deliver DXR into blood, giving an absolute bioavailability of 42%. An experiment with rats showed low level of DXR in heart and kidney, and obviously decreased cardiac and renal toxicities of DXR. Nam et al. (2013) described targeting efficiency and site-specific targeted delivery of target ligand-modified lauric acid-O-CMCS-transferrin micelles of spherical shape with the particle size of approx. 140–649 nm, in which PTX was encapsulated. CS was also used in combination with PLGA, HA, and Pluronic® F127 for co-delivery nanoformulation of multiple anticancer drugs to target the drug resistance mechanisms of cancer stem-like cells (Wang et al., 2015b). These NPs encapsulated with DXR and irinotecan inhibited the activity of topoisomerases II and I, respectively, and can fight against the cancer stem-like cell drug resistance associated with their enhanced DNA repair and anti-apoptosis. In addition, mechanisms of CS-based nanocarriers prepared by grafting cancer-specific ligands onto the CS NPs, which could also respond to external or internal physical and chemical stimulus in targeted tumors related to tumor-targeting drug delivery, was described by Ghaz-Jahanian et al. (2015) and Wu et al. (2015). Hierarchical targeted hepatocyte mitochondrial multifunctional CS NPs for anticancer drug delivery were reported by Chen et al. (2015c). These intelligent CS NPs could exhibit targeted drug release after the progressively shedding of functional groups, thus realizing the efficient intracellular delivery and mitochondrial localization, inhibit the growth of tumor, elevate the antitumor efficacy, and reduce the toxicity of anticancer drugs.

Anitha et al. (2014b) evaluated the combinatorial anticancer effects of curcumin/5-FU loaded thiolated CS NPs on HT-29 cells and observed a 2.5- to 3-fold increase of anticancer effects as well as improved plasma concentrations of both drugs up to 72 h, unlike bare curcumin and 5-FU. A sustained release profile over a period of 4 days was observed, the release being higher in acidic pH.

PTX loaded PEGylated CS-PLGA NPs showed dramatic prolongation in blood circulation, as well as reduced macrophage uptake, with only a small amount of the NPs sequestered in the liver, when compared to PLGA-CS and PLGA NPs (Parveen and Sahoo, 2011). The PTX-incorporated photo cross linkable CS hydrogel inhibited the growth of subcutaneously induced tumors with Lewis lung cancer cells for at least 11 days and markedly reduced the number of CD34-positive vessels in subcutaneous tumors, indicating a strong inhibition of angiogenesis (Obara et al., 2005). PTX-loaded glycol CS NPs hydrophobically modified by linking of 5β-cholan-24-ioc acid showed a sustained release of the incorporated drug, were less cytotoxic to musculus skin melanoma B16F10 cells than bulk PTX formulated in Cremophor EL, and prevented increases in tumor volume for 8 days after injection into the tail vein of tumor-bearing mice (Kim et al., 2006). Similar results were obtained also for DTX-loaded hydrophobically modified glycol CS NPs that showed higher antitumor efficacy, including reduced tumor volume and increased survival rate in lung cancer A549 cells-bearing mice, and strongly reduced the anticancer drug toxicity compared to that of free drug in tumor-bearing mice (Hwang et al., 2008). Amphiphilic glycidol-CS-deoxycholic acid NPs as a drug carrier for DXR were proposed by Zhou et al. (2010). PTX-incorporated *N*-acetyl histidine-conjugated glycol CS self-assembled NPs tested as a promising system for intracytoplasmic delivery of drugs released PTX into cytosol and were effective in inducing arrest of cell growth (Park et al., 2006). *N*-octyl-*O*-sulfate CS micelles that were developed as a delivery system for PTX were found to be safe for intravenous injection and e□hibitedsimilar antitumor efficacy as Ta□ol®, but significantly reduced the toxicity and improved the bioavailability of PTX (Zhang et al., 2008). Based on the experiments in tumor-bearing mice, glycol CS NPs were considered as a promising carrier for CPT (Kim et al., 2008b).

Gaspar et al. (2015) synthesized multifunctional NPs decorated with FA-PEG and dual amino acid-modified CS comple□ed with DNA. These NPs showed hemocompatibility and were internalized by target cells, achieving a ca. 4-fold increase in gene expression. In *in vivo*-mimicking 2D co-cultures they e□hibiteda real affinity towards cancer cells and a negligible uptake in normal cells and penetrated into 3D organotypic tumor spheroids to a higher extent than non-targeted nanocarriers.

Cross linked CMCS-montmorillonite NPs that were tested for controlled delivery of INH showed increased swelling and release of the drug from the NPs with the decrease in the montmorillonite content, the percentage swelling degree and cumulative release were higher at pH 1.2 than at pH 7.4, and the NPs containing montmorillonite showed lower cytotoxicity than montmorillonite free NPs (Banik et al., 2014). The release profiles of streptomycin-CS-magnetic NP-based antibacterial chemotherapeutics showed an initial fast release that became slower as time progressed, and practically total drug amount was released after 350 minutes. This nanocomposite e□hibited significantly enhanced antibacterial activity against *MRSA* (Hussein-Al-Ali et al., 2014). CS functionalized LaF_3:Yb, Er upconverting nanotransducers with controlled size and shape that show bright up conversion fluorescence upon e□citation with 974 nm near infrared (NIR) region can be used as luminescent probes for bioimaging and deep tissue cancer therapeutic applications (Gayathri et al., 2015).

Conjugate of graphene o□idemodified with CMCS followed by conjugation of HA and fluorescein isothiocyanate was found to e□hibitsignificantly higher release rate of encapsulated DXR under tumor cell microenvironment of pH 5.8 than under physiological conditions of pH 7.4 and specifically targeted cancer cells overexpressing CD44 receptors and effectively inhibited their growth (Yang et al., 2016). DXR loaded thermo sensitive nanogel incorporating CS-modified chemically reduced graphene o□idereleased DXR faster at 42°C than at 37°C, whereby its incubation with cancer cells at 37°C resulted in DXR expression in the cytoplasm, while incubation at 42°C led to e□pression in nucleus. Moreover, significantly greater cytoto□icity was estimated after irradiation with NIR owing to a NIR-triggered increase in temperature leading to nuclear DXR release (Wang et al., 2013).

5.4.12   CHONDROITIN SULFATE

Chondroitin is an unbranched polysaccharide belonging to a glycosaminoglycan family. It is composed of (1→3)-linked 2-acetamido-2-deo□y-D-galactopyranose (*N*-acetyl-D-galactosamine) and (1→4)-linked β-D-glucuronic acid. Chondroitin is sulfated in various quantities and positions of both sugars. Chondroitin sulfate (CsA) can be also bonded with amino acids or proteins (Zhao et al., 2015b). Its properties predestine it to be widely applied as a carrier. Naturally, it is a major constituent of articular cartilage, and Bishnoi et al. (2014) reported that CsA conjugated solid lipid NPs could be potentially an effective vector for the treatment or management of osteoarthritis. DXR loaded CsA nanogel interacted with HeLa cells and was internalized together with the entrapped drug within the cytoplasm, probably via an endocytic mechanism exploited by sugar receptors (Park et al., 2010). CsA-capped AuNPs of the mean size of ca. 50 nm after loading with insulin showed the mean size of 123 nm, and oral administration of these drug-loaded NPs provided an efficient regulation of glucose level, compared to insulin solution-treated group in the streptozotocin-induced diabetic rat model. (Cho et al., 2014). The release of terbinafine from CsA microspheres with embedded hydroxyapatite nanowhiskers loaded with the drug in simulated gastric fluid and simulated intestinal fluid was found to be sustained (Cellet et al., 2015). Chondroitin nanocapsules enhancing DXR induced apoptosis against leishmaniasis via Th1 immune response were designed by Chaurasia et al. (2015). The drug-loaded nanocapsules improved cell cycle arrest at G1-S phase (ca. 2-fold) and apoptosis against promastigotes (ca. 6-fold), and *in vivo* antileishmanial activity in hamsters significantly increased parasitic inhibition by these formulations (ca. 1.4-fold).

5.5   PROTEIN-BASED MATRICES

5.5.1   COLLAGEN

Collagen constitutes the most frequent structural protein of the extracellular matrix of various connective tissues in animals. *In vitro*, natural collagen can be formed into highly organized, 3D scaffolds that are intrinsically biocompatible, biodegradable, non-toxic upon exogenous application, and endowed with high tensile strength (Chattopadhyay and Raines, 2014; Mateescu et al., 2015). As collagen gels are flowable, they can be considered as an easily injectable, biocompatible drug delivery matrix. However, the effective pore size of several tens of nanometers in fibrillar collagen gels is too large to control release by hindered diffusion, therefore, it is necessary to bind the drug to collagen, either by covalent or non-covalent bonds, or by sequestering in a secondary matrix. On the other hand, non-fibrillar collagen is characterized by a lower effective pore size of 4–6 nm, but it dissolves rapidly *in vivo* (Wallace and Rosenblatt, 2003). Collagen has broad utility as gene activated matrices, capable of delivering large quantities of DNA in a direct, localized manner (Dang and Leong, 2006; Mateescu et al., 2015). The structure and molecular interactions of collagen *in vivo*, the recent use of natural collagen in sponges, injectables, films, membranes, dressings, and skin grafts, and the on-going development of synthetic collagen were reported by Chattopadhyay and Raines (2014) and Mateescu et al. (2015).

Swatschek et al. (2002) prepared spherical collagen NPs of 120–300 nm using marine sponge collagen loaded with all-*trans* retinol, and it was found that the dermal penetration of retinol nanosystem into hair-less mice skin increased significantly (appro□ 2-fold as compared with bulk retinol). Nicklas et al. (2009) developed NPs of *Chondrosia reniformis* sponge collagen as penetration enhancers for the transdermal drug delivery of 17β-estradiol hemihydrate in hormone replacement therapy. The hydrogel with estradiol-loaded collagen NPs provided a prolonged estradiol release compared to a commercial gel as well as a considerably enhanced estradiol absorption. Duraipandy et al. (2015) developed a novel wound dressing material based on cross-linking of collagen with plumbagin caged AgNPs, which resulted in uniform alignment of collagen fibrils to form orderly aligned porous structured scaffolds with potent antibacterial activity that in turn enhanced its ability to promote cell proliferation and wound healing.

Chan et al. (2008) described nano-fibrous collagen spheres for protein delivery and demonstrated the possibility to use photochemical cross-linking as the secondary retention mechanism for proteins. Castaneda et al. (2008) cross-linked tiopronin-protected AuNPs to collagen, and this material can be used for the delivery of small molecule drugs as well as AuNPs for photo thermal therapies, imaging, and cell targeting. DXR-conjugated collagen hybrid gels for metastasis-associated drug delivery systems were described by Kojima et al. (2013). A collagen peptide-modified dendrimer that attached DXR via a pH-degradable linkage was synthesized as a polymer prodrug. Compared with bulk DXR, the diffusion of the dendrimer prodrug from the collagen gel was suppressed. Highly invasive MDA-MB-231 cells were more sensitive to the prodrug collagen gel than poorly invasive MCF-7 cells. The dendrimer prodrug/collagen hybrid gel not only suppressed tumor growth but also attenuated metastatic activity *in vivo*.

Sustainable release of VMC, gentamicin and lidocaine was estimated from novel electrospun sandwich-structured PLGA/collagen nanofibrous membranes. The biodegradable nanofibrous membranes released high concentrations of VMC and gentamicin (above MIC) for 4 and 3 weeks, respectively, and lidocaine for 2 weeks, and the nanofibrous membranes were functionally active in responses in human fibroblasts (Chen et al., 2012). Kong et al. (2013) incorporated CS nanospheres into collagen coatings. The nanosystem facilitated the sustained release of VMC and had a clear antibacterial effect. *In vitro* release of pilocarpine has demonstrated the possibility to use CS-collagen hydrogels as an ophthalmic drug delivery matrix (Flocea et al., 2010).

A biomimetic and injectable hydrogel scaffold based on nano-hydroxy-apatite, collagen and CS showing some features of natural bone in the main composition and microstructure was synthesized by Huang et al. (2011). Sotome et al. (2004) investigated *in vivo* hydroxyapatite-collagen-ALG as a bone filler and a drug delivery carrier of bone morphogenetic protein. Active bone formations around the material and tissue invasion into the material were observed throughout the experiment in bone, compared to those observed for simple ALG and porous hydroxyapatite. A Fe_2O_3-enriched collagen/hydroxyapatite composite material was also developed by Andronescu et al. (2010) for bone cancer therapy by induction of tumor cell apoptosis via magnetite-induced hyperthermia by electromagnetic field application. Highly metastatic MDA-MB-231 cells were more sensitive to PTX-loaded hydroxyapatite/collagen drug delivery nanosystems developed by Watanabe et al. (2013) than poorly metastatic MCF-7 cells. The cytotoxic, anti-proliferative, and anti-invasive activities of collagen-hydroxyapatite/CPT drug delivery systems on G292 osteosarcoma cells depended on the CPT concentration released in culture medium, and these systems could be considered as a feasible approach for locoregional chemotherapy of bone cancer (Andronescu et al., 2013).

The collagen coated cholesterol-free liposome NP matrix, in which collagen stabilizes the original liposomal structure, was found to be more stable *in vitro* than control liposome, and mineralized collagen composite matrix could be useful for bone and dental implants (Krishnamoorthy et al., 2011).

da Silva et al. (2009b) described a bioresorbable and biodegradable collagen/carbon nanotube (CCNT) biocomposite that incorporates the advantages of both collagen and CNTs and meets most of the criteria of an ideal biomaterial (mechanical rigidity, 3D nanostructured surface, functionalizable surface) in order to be used as an osteoinductive agent. Tan et al. (2010) developed CCNT composites with two types of functionalization, carboxylated and covalently functionalized nanotubes (CFNTs). It was found that the surface modification and the loading concentration of NTs determined interactions between NTs and collagen fibrils, thus altering the structure and the properties of NT-collagen composites, and the incorporation of CFNTs in collagen-based constructs was an effective means of restructuring collagen fibrils, because CFNTs strongly bound to collagen molecules inducing the formation of larger fibril bundles. Biocompatibility was highly dependent on nanotube loading concentration, and at a low loading level, CFNTs showed higher endothelial coverage than the other tested constructs or materials.

Hybrid collagen-based hydrogel nanocomposites with organo-modified clays Dellite® 67G and Cloisite® 93A had superior collagenase resistance compared to hydrogels without the inorganic NPs (Nistor et al., 2013). Silicification significantly improved the mechanical and thermal stability of the collagen network within the hybrid systems. These nanocomposites favor the metabolic activity of immobilized human dermal fibroblasts, while decreasing the hydrogel contraction. At *in vivo* implantation of bulk hydrogels in subcutaneous sites of rats over the vascular inflammatory period, these materials were colonized and vascularized without inducing strong inflammatory response (Desimone et al., 2011). According to Lee et al. (2013), the delivery system of nerve growth factor using enlarged-pored mesoporous silica NPs in combination with collagen may be potentially useful as therapeutic neural engineering matrices, and the concept can also be extended to other growth factor delivering systems. Silicate-collagen nanocomposites loaded with tetracycline were found to inhibit *S. aureus* cell growth and exhibited no cytotoxic effects on osteoblast-like cells indicating that these composites act as suitable bioactive carriers of some antibacterial drugs (Rivadeneira et al., 2014). Also Alvarez et al. (2014) investigated silica-collagen nanocomposite hydrogels loaded with gentamycin and rifamycin for prolonged antimicrobial activity of wound infection prevention. They found that high silica dose gentamycin-loaded silica NPs (500 nm) immobilized in concentrated collagen hydrogels without modifying their fibrillar structure or impacting their rheological behavior increase their proteolytic stability and sustained drug release from the nanocomposites over 7 days and ensure excellent prolonged antibacterial activity. On the other hand, rifamycin-loaded NPs (100 nm) significantly altered the collagen hydrogel structure at high silica doses, and these nanocomposites did not show antibacterial activity due to strong adsorption of rifamycin on collagen fibers.

5.5.2   GELATIN

Gelatin is a soluble protein obtained by partial hydrolysis of collagen, cartilages and skins, therefore, the source, the age of the animal, and the type of collagen are all intrinsic factors influencing the properties of gelatins (Johnston-Banks, 1990). Gómez-Guillén et al. (2011) reviewed functional and bioactive properties of collagen and gelatin from alternative sources. Cell-adhesive structure, low cost, off-the-shelf availability, high biocompatibility, biodegradability, and low immunogenicity belong to main advantages of gelatin that can be formulated in NPs, employed as a size-controllable porogen, adopted as a surface coating agent, and mixed with synthetic or natural biopolymers forming composite scaffolds, thus it represents an excellent biomaterial for tissue engineering and drug delivery (Su and Wang, 2015). Gelification properties of gelatin as well as the strong dependence of gelatin ionization on pH makes this compound an interesting candidate to be used for effective intracellular delivery of active biomacromolecules (Moran et al., 2015). While gelatin microparticles can serve as vehicles for cell amplification and for delivery of large bioactive molecules, gelatin NPs are better suitable for i.v. delivery or for drug delivery to the brain. Gelatin fibers have a high surface area-to-volume ratio, while gelatin hydrogels can trap molecules between polymer crosslink gaps, allowing these molecules to diffuse into the blood stream (Foox and Zilberman, 2015). Daear et al. (2015) studied interactions of gelatin NPs with the major lipids of model lung surfactant and found that gelatin NPs interact stronger with negatively charged phosphatidyl-glycerols compared to zwitterionic phosphatidyl-cholines and that the addition of the NPs results in concomitantly significant effects on the lateral organization of the monolayers.

High-molecular-weight antibiotics colistin and VMC were released in a sustained manner from oppositely charged gelatin carriers for more than 14 days, contrary to low-molecular-weight antibacterial chemotherapeutics gentamicin and moᐨifloᐨacinthat were released in a burst-like manner. Song et al. (2015) found that electrostatic and hydrophobic interactions between these drugs and gelatin matrix determined the release manner.

Alam and Shubhra (2015) designed unique thin film composed of a silk-gelatin blend that was capable of delivering drug to heal a wound in a rat model. PEGylation of film reduced a degradation rate of the film and cipro-floᐨacin-loadedPEG-film was able to heal the wound within 1 week, while unmodified film was without any curative effect.

The nano-sized graphene oxide (200 nm and 300 nm) did not enable carboplatin to kill the cancer cells efficiently, but gelatin modified nanographene oxide NPs with particle size 100 nm loaded with carboplatin showed low toxicity against non-target cells, and a synergistic activity of all components resulted in increasing the local concentration of carboplatin inside the cancer cells (Makharza et al., 2015). Folate-grafted non-aggregated gelatin NPs improved CPT delivery in HeLa cells, and it was found that they preferably accumulated in the cytoplasm of HeLa cells nearby nucleus by following receptor-mediated endocytosis pathway (Dixit et al., 2015). It is important to note that cross linked gelatin-montmorillonite NPs showed less cytotoxicity than montmorillonite NPs alone, they were very sensitive to the pH environment and were found to be suitable matrices for controlled drug delivery applications (Sarmah et al., 2015).

5.6   CONCLUSION

Nanoscale science and nanotechnology have unambiguously demonstrated to have a great potential in providing novel and improved solutions. Rather biodegradable biocompatible natural-based matrices have been used for pharmaceutical formulations due to a decrease of the risk of toxic effect on non-target organisms, organs, and tissues. As discussed above, various native or modified/functionalized matrices have been applied. Due to space limitation in this chapter, other promising native materials, such as hydroxyapatite, guar gum, karaya gum, tragacanth, and lignin, are not discussed. In relation to the rapid development of nanotechnology and applications of functionalized biodegradable biocompatible nanomaterials, controlled release and targeted delivery nanoformulations not only enhance bioavailability, decrease toxicity for non-target tissues and "pill burden," but also protect active ingredients against decomposition and deactivation. Additionally, in many cases, nanoencapsulation of active ingredients has restored their biological effect against resistant tissues/receptors or cells/strains. Thus, the efficacy of many chemotherapeutics or natural bioactive agents can be intensified by preparation of nanoformulations. Moreover, the applicability of drugs with different physicochemical properties in combination therapy can contribute to overcoming current difficulties in the therapy of many diseases and even to the development of new special delivery systems based on multifunctional smart nanoformulations.

KEYWORDS

- carbohydrates
- controlled release
- drug carriers
- nanoformulations

- **nanoparticles**
- **proteins**
- **targeted delivery**
- **α-hydroxy acids**

ACKNOWLEDGMENTS

This study was supported by Sanofi-Aventis Pharma, Slovakia, s.r.o.

REFERENCES

Abou Taleb, M. F., Alkahtani, A., & Mohamed, S. K., (2015). Radiation synthesis and characterization of sodium alginate/chitosan/hydroxyapatite nanocomposite hydrogels: A drug delivery system for liver cancer. *Polym. Bull., 72*, 725–742.

Abulateefeh, S. R., & Taha, M. O., (2015). Enhanced drug encapsulation and extended release profiles of calcium-alginate nanoparticles by using tannic acid as a bridging cross-linking agent. *J. Microencapsul., 32*, 96–105.

Acharya, S., & Sahoo, S. K., (2011). PLGA nanoparticles containing various anticancer agents and tumour delivery by EPR effect. *Adv. Drug Deliv. Rev., 63,* 170–183.

Acosta, E., (2009). Bioavailability of nanoparticles in nutrient and nutraceutical delivery. *Curr. Opin. Colloid Interface Sci., 14*, 3–15.

Afshari, M., Derakhshandeh, K., & Hosseinzadeh, L., (2014). Characterisation, cytotoxicity and apoptosis studies of methotrexate-loaded PLGA and PLGA-PEG nanoparticles. *J. Microencapsul., 31*, 239–245.

Agueeros, M., Ruiz-Gaton, L., Vauthier, C., Bouchemal, K., Espuelas, S., Ponchel, G., et al., (2009). Combined hydroxypropyl-beta-cyclodextrin and poly(anhydride) nanoparticles improve the oral permeability of paclitaxel. *Eur. J. Pharm. Sci., 38*, 405–413.

Ahmad, Z., Pandey, R., Sharma, S., & Khuller, G. K., (2006a). Alginate nanoparticles as antituberculosis drug carriers: formulation development, pharmacokinetics and therapeutic potential. *Ind. J. Chest Dis. Allied Sci., 48*, 171–176.

Ahmad, Z., Pandey, R., Sharma, S., & Khuller, G. K., (2006b). Pharmacokinetic and pharmacodynamic behaviour of antitubercular drugs encapsulated in alginate nanoparticles at two doses. *Int. J. Antimicrob. Agents, 27*, 409–416.

Akagi, T., Baba, M., & Akashi, M., (2012). Biodegradable nanoparticles as vaccine adjuvants and delivery systems: regulation of immune responses by nanoparticle-based vaccine. In *Polymers in Nanomedicine*, Kunugi, S., Yamaoka, T., Eds.; *Advances in Polymer Science, 247*, Springer-Verlag: Berlin-Heidelberg, 31–64.

Akhlaghi, S. P., Berry, R. C., & Tam, K. C., (2013). Surface modification of cellulose nanocrystal with chitosan oligosaccharide for drug delivery applications. *Cellulose, 20*, 1747–1764.

Alam, A. K. M. M., & Shubhra, Q. T. H., (2015). Surface modified thin film from silk and gelatin for sustained drug release to heal wound. *J. Mater. Chem. B, 3*, 6473–6479.

Allen, T. M., & Cullis, P. R., (2004). Drug delivery systems: Entering the mainstream. *Science, 303*, 1818–1822.

Alvarez, G.S, Helary, C., Mebert, A. M., Wang, X. L., Coradin, T., & Desimone, M. F., (2014). Antibiotic-loaded silica nanoparticle-collagen composite hydrogels with prolonged antimicrobial activity for wound infection prevention. *J. Mater. Chem. B, 2*, 4660–4670.

Anderson, J. M., & Shive, M. S., (1997). Biodegradation and biocompatibility of PLA and PLGA microspheres. *Adv. Drug Deliv. Rev., 28*, 5–24.

Anderson, J. M., & Shive, M. S., (2012). Biodegradation and biocompatibility of PLA and PLGA microspheres. *Adv. Drug Deliv. Rev., 64*, 72–82.

Andronescu, E., Ficai, M., Voicu, G., Ficai, D., Maganu, M., & Ficai, A., (2010). Synthesis and characterization of collagen/hydroxyapatite: Magnetite composite material for bone cancer treatment. *J. Mater. Sci. Mater. Med., 21*, 2237–2242.

Andronescu, E., Ficai, A., Georgiana, M., Mitran, V., Sonmez, M., Ficai, D., et al., (2013). Collagen-hydroxyapatite/cisplatin drug delivery systems for locoregional treatment of bone cancer. *Technol. Cancer Res. Treat., 12*, 275–284.

Anitha, A., Sowmya, S., Kumar, P. T. S., Deepthi, S., Chennazhi, K. P., Ehrlich, H., et al., (2014a). Chitin and chitosan in selected biomedical applications. *Progr. Polym. Sci., 39*, 1644–1667.

Anitha, A., Deepa, N., Chennazhi, K. P., Lakshmanan, V. K., & Jayakumar, R., (2014b). Combinatorial anticancer effects of curcumin and 5-fluorouracil loaded thiolated chitosan nanoparticles towards colon cancer treatment. *BBA General Subjects, 1840*, 2730–2743.

Arunraj, T. R., Rejinold, N. S., Kumar, N. A., & Jayakumar, R., (2013). Doxorubicin-chitin-poly(caprolactone) composite nanogel for drug delivery. *Int. J. Biol. Macromol., 62*, 35–43.

Arunraj, T. R., Rejinold, N. S., Kumar, N. A., & Jayakumar, R., (2014). Bio-responsive chitin-poly(L-lactic acid) composite nanogels for liver cancer. *Colloids Surf. B Biointerfaces, 113*, 394–402.

Ashwinkumar, N., Maya, S., & Jayakumar, R., (2014). Redox-responsive cystamine conjugated chitin-hyaluronic acid composite nanogels. *RSC Adv., 4*, 49547–49555.

Astete, C. E., & Sabliov, C. M., (2006). Synthesis and characterization of PLGA nanoparticles. *J. Biomater. Sci. Polym. (Ed.), 17*, 247–289.

Athira, G. K., & Jyothi, A. N., (2015). Cassava starch-poly(vinyl alcohol) nanocomposites for the controlled delivery of curcumin in cancer prevention and treatment. *Starch–Stärke, 67*, 549–558.

Aytac, Z., Sen, H. S., Durgun, E., & Uyar, T., (2015). Sulfisoxazole/cyclodextrin inclusion complex incorporated in electrospun hydroxypropyl cellulose nanofibers as drug delivery system. *Colloids Surf. B Biointerfaces, 128*, 331–338.

Balaure, P. C., Andronescu, E., Grumezescu, A. M., Ficai, A., Huang, K. S., Yang, C. H., et al., (2013). Fabrication, characterization and *in vitro* profile-based interaction with eukaryotic and prokaryotic cells of alginate-chitosan-silica biocomposite. *Int. J. Pharm., 441*, 555–561.

Bamrungsap, S., Zhao, Z., Chen, T., Wang, L., Li, C., Fu, T., et al., (2012). Nanotechnology in therapeutics: a focus on nanoparticles as a drug delivery system. *Nanomedicine, 7*, 1253–1271.

Banik, N., Ramteke, A., & Maji, T. K., (2014). Carboxymethyl chitosan-montmorillonite nanoparticles for controlled delivery of isoniazid: evaluation of the effect of the glutaraldehyde and montmorillonite. *Polym. Adv. Technol.*, *25*, 1580–1589.

Barikani, M., Oliaei, E., Seddiqi, H., & Honarkar, H., (2014). Preparation and application of chitin and its derivatives: a review. *Iranian Polym. J.*, *23*, 307–326.

Bazak, R., Houri, M., El Achy, S., Hussein, W., & Refaat, T., (2014). Passive targeting of nanoparticles to cancer: A comprehensive review of the literature. *Mol. Clin. Oncol.*, *2*, 904–908.

Bazak, R., Houri, M., El Achy, S., Kamel, S., & Refaat, T., (2015). Cancer active targeting by nanoparticles: a comprehensive review of literature. *J. Cancer Res. Clin. Oncol.*, *141*, 769–784.

Belitz, H. D., Grosch, W., & Schieberle, P., (2009). *Food Chemistry, 4th revised and extended ed.*; Springer-Verlag: Berlin-Heidelberg.

Bhushan, B., Luo, D., Schricker, S. R., Sigmund, W., & Zauscher, S., (2014). Handbook of Nanomaterials Properties; Springer-Verlag: Berlin-Heidelberg.

Bishnoi, M., Jain, A., Hurkat, P., & Jain, S. K., (2014). Aceclofenac-loaded chondroitin sulfate conjugated SLNs for effective management of osteoarthritis. *J. Drug Target*, *22*, 805–812.

Bishwambhar, M., & Suneetha, V., (2012). Release study of naproxen, a modern drug from pH sensitive pullulan acetate microsphere. *Int. J. Drug Dev. Res.*, *4*, 184–191.

Boekhoven, J., Zha, R. H., Tantakitti, F., Zhuang, E., Zandi, R., Newcomb, C. J., et al., (2015). Alginate-peptide amphiphile core-shell microparticles as a targeted drug delivery system. *RSC Adv.*, *5*, 8753–8756.

Bulmer, C., Margaritis, A., & Xenocostas, A., (2012). Encapsulation and controlled release of recombinant human erythropoietin from chitosan-carrageenan nanoparticles. *Curr. Drug Deliv.*, *9*, 527–537.

Butun, S., Ince, F. G., Erdugan, H., & Sahiner, N., (2011). One-step fabrication of biocompatible carboxymethyl cellulose polymeric particles for drug delivery systems. *Carbohydr. Polym.*, *86*, 636–643.

Buzea, C., Pacheco, I., & Robbie, K., (2007). Nanomaterials and nanoparticles: Sources and toxicity. *Biointerphases.* *2*, MR17–71.

Castaneda, L., Valle, J., Yang, N., Pluskat, S., & Slowinska, K., (2008). Collagen cross-linking with Au nanoparticles. *Biomacromolecules*, *9*, 3383–3388.

Castello, J., Gallardo, M., Busquets, M. A., & Estelrich, J., (2015). Chitosan (or alginate)-coated iron oxide nanoparticles: A comparative study. *Colloids Surf. A Physicochem. Eng. Aspects*, *468*, 151–158.

Cellet, T. S. P., Pereira, G. M., Muniz, E. C., Silva, R., & Rubira, A. F., (2015). Hydroxyapatite nanowhiskers embedded in chondroitin sulfate microspheres as colon targeted drug delivery systems. *J. Mater. Chem. B*, *3*, 6837–6846.

Černíková, A., & Jampílek, J., (2014). Structure modification of drugs influencing their bioavailability and therapeutic effect. *Chem. Listy.*, *108*, 7–16.

Chan, O. C. M., So, K. F., & Chan, B. P., (2008). Fabrication of nano-fibrous collagen microspheres for protein delivery and effects of photochemical cross linking on release kinetics. *J. Control Release, 129*, 135–143.

Chang, C. H., Lin, Y. H., Yeh, C. L., Chen, Y. C., Chiou, S. F., Hsu, Y. M., et al., (2010) Nanoparticles incorporated in pH-sensitive hydrogels as amoxicillin delivery for eradication of *Helicobacter pylori*. *Biomacromolecules, 11*, 133–142.

Chang, C. Y., & Zhang, L. N., (2011). Cellulose-based hydrogels: Present status and application prospects. *Carbohydr. Polym., 84*, 40–53.

Chattopadhyay, S., & Raines, R. T., (2014). Collagen-based biomaterials for wound healing. *Biopolymers, 101*, 821–833.

Chaurasia, M., Pawar, V. K., Jaiswal, A. K., Dube, A., Paliwal, S. K., & Chourasia, M. K., (2015). Chondroitin nanocapsules enhanced doxorubicin induced apoptosis against leishmaniasis via Th1 immune response. *Int. J. Biol. Macromol., 79*, 27–36.

Chen, D. W., Hsu, Y. H., Liao, J. Y., Liu, S. J., Chen, J. K., & Ueng, S. W. N., (2012). Sustainable release of vancomycin, gentamicin and lidocaine from novel electrospun sandwich-structured PLGA/collagen nanofibrous membranes. *Int. J. Pharm., 430*, 335–341.

Chen, Y. Z., Huang, Y. K., Chen, Y., Ye, Y. J., Lou, K. Y., & Gao, F., (2015a). Novel nanoparticles composed of chitosan and beta-cyclodextrin derivatives as potential insoluble drug carrier. *Chin. Chem. Lett., 26*, 909–913.

Chen, Q., Yu, H. J., Wang, L., ul Abdin, Z., Chen, Y. S., Wang, J. H., et al., (2015b). Recent progress in chemical modification of starch and its applications. *RSC Adv., 5*, 67459–67474.

Chen, Z. P., Zhang, L. J., Song, Y., He, J. Y., Wu, L., Zhao, C., et al., (2015c). Hierarchical targeted hepatocyte mitochondrial multifunctional chitosan nanoparticles for anticancer drug delivery. *Biomaterials, 52*, 240–250.

Cheng, L. H., Jin, C. M., Lv, W., Ding, Q. P., & Han, X., (2011). Developing a highly stable PLGA-mPEG nanoparticle loaded with cisplatin for chemotherapy of ovarian cancer. *PLOS One, 6*, Article ID e25433.

Cheng, L. D., Bulmer, C., & Margaritis, A., (2015). Characterization of novel composite alginate chitosan-carrageenan nanoparticles for encapsulation of BSA as a model drug delivery system. *Curr. Drug Deliv., 12*, 351–357.

Cheow, W. S., Kiew, T. Y., & Hadinoto, K., (2015). Amorphous nanodrugs prepared by complexation with polysaccharides: Carrageenan versus dextran sulfate. *Carbohydr. Polym., 117*, 549–558.

Chilajwar, S. V., Pednekar, P. P., Jadhav, K. R., Gupta, G. J. C., & Kadam, V. J., (2014). Cyclodextrin-based nanosponges: A propitious platform for enhancing drug delivery. *Expert Opin. Drug Deliv., 11*, 111–120.

Cho, J. K., Park, W., & Na, K., (2009). Self-organized nanogels from pullulan-g-poly(L-lactide) synthesized by one-pot method: Physicochemical characterization and *in vitro* doxorubicin release. *J. Appl. Polym. Sci., 113*, 2209–2216.

Cho, H. J., Yoon, H. Y., Koo, H., Ko, S. H., Shim, J. S., Lee, J. H., et al., (2011). Self-assembled nanoparticles based on hyaluronic acid-ceramide (HA-CE) and Pluronic® for tumor-targeted delivery of docetaxel. *Biomaterials, 32*, 7181–7190.

Cho, H. J., Oh, J., Choo, M. K., Ha, J. I., Park, Y., & Maeng, H. J., (2014). Chondroitin sulfate-capped gold nanoparticles for the oral delivery of insulin. *Int. J. Biol. Macromol, 63*, 15–20.

Choi, K. Y., Min, K. H., Na, J. H., Choi, K., Kim, K., Park, J. H., et al., (2009). Self-assembled hyaluronic acid nanoparticles as a potential drug carrier for cancer therapy: synthesis, characterization, and *in vivo* biodistribution. *J. Mater. Chem., 19*, 4102–4107.

Choi, K. Y., Chung, H., Min, K. H., Yoon, H. Y., Kim, K., Park, J. H., et al., (2010). Self-assembled hyaluronic acid nanoparticles for active tumor targeting. *Biomaterials, 31*, 106–114.

Choi, K. Y., Min, K. H., Yoon, H. Y., Kim, K., Park, J. H., Kwon, I. C., et al., (2011a). PEGylation of hyaluronic acid nanoparticles improves tumor targetability *in vivo*. *Biomaterials, 32*, 1880–1889.

Choi, K. Y., Yoon, H. Y., Kim, J. H., Bae, S. M., Park, R. W., & Kang, Y. M., (2011b). Smart nanocarrier based on PEGylated hyaluronic acid for cancer therapy. *ACS Nano, 5*, 8591–8599.

Choonara, Y. E., Pillay, V., Ndesendo, V. M. K., du Toit, L. C., Kumar, P., Khan, R. A., et al., (2011). Polymeric emulsion and crosslink-mediated synthesis of super-stable nanoparticles as sustained-release anti-tuberculosis drug carriers. *Colloids Surf. B Biointerfaces, 87*, 243–254.

Chung, Y. C., Su, Y. P., Chen, C. C., Jia, G., Wang, H. I., Wu, J. C. G., et al., (2004). Relationship between antibacterial activity of chitosan and surface characteristics of cell wall. *Acta Pharmacol. Sin., 25*, 932–936.

Corrigan, O. I., & Li, X., (2009). Quantifying drug release from PLGA nanoparticulates. *Eur. J. Pharm. Sci., 37*, 477–485.

Costa, M. R., Silva, N. C., Sarmento, B., & Pintado, M., (2015). Potential chitosan-coated alginate nanoparticles for ocular delivery of daptomycin. *Eur. J. Clin. Microbiol. Infect. Dis., 34*, 1255–1262.

Couvreur, P., (2013). Nanoparticles in drug delivery: past, present and future. *Adv. Drug Deliv. Rev., 65*, 21–23.

D'Arrigo, G., Di Meo, C., Gaucci, E., Chichiarelli, S., Coviello, T., Capitani, D., et al., (2012). Self-assembled gellan-based nanohydrogels as a tool for prednisolone delivery. *Soft Matter, 8*, 11557–11564.

D'Arrigo, G., Navarro, G., Di Meo, C., Matricardi, P., & Torchilin, V., (2014). Gellan gum nanohydrogel containing anti-inflammatory and anti-cancer drugs: a multi-drug delivery system for a combination therapy in cancer treatment. *Eur. J. Pharm. Biopharm., 87*, 208–216.

da Silva, A. L. D., Fateixa, S., Guiomar, A. J., Costa, B. F. O., Silva, N. J. O., Trindade, T., et al., (2009a). Biofunctionalized magnetic hydrogel nanospheres of magnetite and kappa-carrageenan. *Nanotechnology, 20*, Article ID 355602.

da Silva, E. E., Della Colleta, H. H. M., Ferlauto, A. S., Moreira, R. L., Resende, R. R., Oliveira, S., et al., (2009b). Nanostructured 3-D collagen/nanotube biocomposites for future bone regeneration scaffolds. *Nano Res, 2*, 462–473.

da Silva, A. L. D., Salgueiro, A. M., Fateixa, S., Moreira, J., Estrada, A. C., Gil, A. M., et al., (2012). Swelling and Release Properties of Functional kappa-carrageenan Hydrogel Nanocomposites. In *Multifunctional Polymer-based Materials*; Lendlein, A., Behl, M., Feng, Y., Guan, Z., Xie, T., Eds.; Materials Research Society Symposium Proceedings (1403). Cambridge University Press: New York, pp. 213–219.

Daear, W., Lai, P., Anikovskiy, M., & Prenner, E., (2015). Differential interactions of gelatin nanoparticles with the major lipids of model lung surfactant: Changes in the lateral membrane organization. *J. Phys. Chem. B, 119*, 5356–5366.

Dai, W. G., Dong, L. C., & Song, Y. Q., (2007). Nanosizing of a drug/carrageenan complex to increase solubility and dissolution rate. *Int. J. Pharm., 342*, 201–207.

Dang, J. M., & Leong, K. W., (2006). Natural polymers for gene delivery and tissue engineering. *Adv. Drug Deliv. Rev., 58*, 487–499.

Danhier, F., Lecouturier, N., Vroman, B., Jerome, C., Marchand-Brynaert, J., Feron, O., et al., (2009). Paclitaxel-loaded PEGylated PLGA-based nanoparticles: *In vitro* and *in vivo* evaluation. *J. Control Release, 133*, 11–17.

Danhier, F., Ansorena, E., Silva, J. M., Coco, R., Le Breton, A., & Preat, V., (2012). PLGA-based nanoparticles: An overview of biomedical applications. *J. Control Release, 161*, 505–522.

Dash, R., & Ragauskas, A. J. (2012). Synthesis of a novel cellulose nanowhisker-based drug delivery system. *RSC Adv.*, *2*, 3403–3409.

Dash, T. K., & Konkimalla, V. B. (2015). Modification of Cyclode☐trin for Improvement of Complexation and Formulation Properties. In *Handbook of Polymers for Pharmaceutical Technologies – Biodegradable Polymers,* Thakur, V. K., Thakur, M. K., Eds.; Scrivener Publishing and J. Wiley and Sons: Hoboken, 205–224.

De Jong, W. H., & Borm, P. J. ,(2008). Drug delivery and nanoparticles: Applications and hazards. *Int. J. Nanomedicine, 3,* 133–149.

Desimone, M. F., Helary, C., Quignard, S., Rietveld, I. B., Bataille, I., Copello, G. J., et al., (2011). *In vitro* studies and preliminary *in vivo* evaluation of silicified concentrated collagen hydrogels. *ACS Appl. Mater. Interfaces, 3,* 3831–3838.

Dev, A., Mohan, J. C., Sreeja, V., Tamura, H., Patzke, G. R., Hussain, F., et al., (2010). Novel carboxymethyl chitin nanoparticles for cancer drug delivery applications. *Carbohydr. Polym., 79,* 1073–1079.

Devi, N., & Maji, T. K., (2010). Genipin cross linked chitosan-κ-carrageenan polyelectrolyte nanocapsules for the controlled delivery of isoniazid. *Int. J. Polym. Mater., 59,* 828–841.

Devi, P. R., Kumar, C. S., Selvamani, P., Subramanian, N., & Ruckmani, K. (2015). Synthesis and characterization of Arabic gum capped gold nanoparticles for tumor-targeted drug delivery. *Mater. Lett., 139,* 241–244.

Dhar, S., Reddy, E. M., Shiras, A., Pokharkar, V., & Prasad, B. L. V., (2008). Natural gum reduced/stabilized gold nanoparticles for drug delivery formulations. *Chem. Eur. J., 14,* 10244–10250.

Dhar, S., Reddy, E. M., Prabhune, A., Pokharkar, V., Shiras, A., & Prasad, B. L. V., (2011a). Cytotoxicity of sophorolipid-gellan gum-gold nanoparticle conjugates and their doxorubicin loaded derivatives towards human glioma and human glioma stem cell lines. *Nanoscale, 3,* 575–580.

Dhar, S., Mali, V., Bodhankar, S., Shiras, A., Prasad, B. L. V., & Pokharkar, V., (2011b). Biocompatible gellan gum-reduced gold nanoparticles: cellular uptake and subacute oral toxicity studies. *J. Appl. Toxicol., 31,* 411–420.

Dinarvand, R., Sepehri, N., Manoochehri, S., Rouhani, H., & Atyabi, F., (2011). Polylactide-co-glycolide nanoparticles for controlled delivery of anticancer agents. *Int. J. Nanomed., 6,* 877–895.

Ding, J. X., Chen, L. H., Xiao, C. S., Chen, L., Zhuang, X. L., & Chen, X. S., (2014). Non-covalent interaction-assisted polymeric micelles for controlled drug delivery. *Chem. Commun., 50,* 11274–11290.

Di☐it,N., Vaibhav, K., Pandey, R. S., Jain, U. K., Katare, O. P., Katyal, A., et al., (2015). Improved cisplatin delivery in cervical cancer cells by using folate-grafted non-aggregated gelatin nanoparticles. *Biomed. Pharmacother., 69,* 1–10.

Dolez, P. I., (2015). Nanoengineering: Global Approaches to Health and Safety Issues, Elsevier: Amsterdam.

Dong, Y. C., & Feng, S. S., (2004). Methoxy poly(ethylene glycol)-poly(lactide) (MPEG-PLA) nanoparticles for controlled delivery of anticancer drugs. *Biomaterials, 25,* 2843–2849.

Dong, H. Q., Xu, Q., Li, Y. Y., Mo, S. B., Cai, S. J., & Liu, L. J., (2008). The synthesis of biodegradable graft copolymer cellulose-graft-poly(L-lactide) and the study of its controlled drug release. *Colloids Surf. B Biointerfaces, 66,* 26–33.

Dong, S. P., Cho, H. J., Lee, Y. W., & Roman, M., (2014). Synthesis and cellular uptake of folic acid-conjugated cellulose nanocrystals for cancer targeting. *Biomacromolecules, 15,* 1560–1567.

Dong, L. J., & Jin, G., (2015). Alginate-coated Fe_3O_4 hollow microspheres for drug delivery. *Chinese, J. Chem. Phys., 28*, 193–196.

Duraipandy, N., Lakra, R., Srivatsan, K. V., Ramamoorthy, U., Korrapati, P. S., & Kiran, M. S., (2015). Plumbagin caged silver nanoparticle stabilized collagen scaffold for wound dressing. *J. Mater. Chem. B, 3*, 1415–1425.

Dyondi, D., Webster, T. J., & Banerjee, R., (2013). A nanoparticulate injectable hydrogel as a tissue engineering scaffold for multiple growth factor delivery for bone regeneration. *Int. J. Nanomed., 8*, 47–59.

Edgar, K. J., (2007). Cellulose esters in drug delivery. *Cellulose, 14*, 49–64.

Eldin, M. S. M., Kamoun, E. A., Sofan, M. A., & Elbayomi, S. M., (2015). L-Arginine grafted alginate hydrogel beads: A novel pH-sensitive system for specific protein delivery. *Arabian J. Chem., 8*, 355–365.

Elumalai, R., Patil, S., Maliyakkal, N., Rangarajan, A., Kondaiah, P., & Raichur, A. M., (2015). Protamine-carboxymethyl cellulose magnetic nanocapsules for enhanced delivery of anticancer drugs against drug resistant cancers. *Nanomed. Nanotechnol. Biol. Med., 11*, 969–981.

Engelberth, S. A., Hempel, N., & Bergkvist, M., (2015). Chemically modified dendritic starch: A novel nanomaterial for siRNA delivery. *Bioconjug. Chem., 26*, 1766–1774.

Engineer, C., Parikh, J., & Raval, A., (2011). Review on hydrolytic degradation behavior of biodegradable polymers from controlled drug delivery system. *Trends Biomater. Artif. Organs, 25*, 79–85.

Esa, F., Tasirin, S. M., & Rahman, N. A., (2014). Overview of bacterial cellulose production and application. *Agric. Agric. Sci. Proc., 2*, 113–119.

Feng, C., Wang, Z. G., Jiang, C. Q., Kong, M., Zhou, X., Li, Y., et al., (2013). Chitosan/o-carboxymethyl chitosan nanoparticles for efficient and safe oral anticancer drug delivery: *In vitro* and *in vivo* evaluation. *Int. J. Pharm., 457*, 158–167.

Feng, W., Nie, W., He, C. L., Zhou, X. J., Chen, L., Qiu, K. X., et al., (2014). Effect of pH-responsive alginate/chitosan multilayers coating on delivery efficiency, cellular uptake and biodistribution of mesoporous silica nanoparticles-based nanocarriers. *ACS Appl. Mater. Interfaces, 6*, 8447–8460.

Fernandes, S. C. M., Sadocco, P., Aonso-Varona, A., Palomares, T., Eceiza, A., Silvestre, A. J. D., et al., (2013). Bioinspired antimicrobial and biocompatible bacterial cellulose membranes obtained by surface functionalization with aminoalkyl groups. *ACS Appl. Mater. Interfaces, 5*, 3290–3297.

Flocea, P., Verestiuc, L., Popa, M., Sunel, V., & Lungu, A., (2010). Cross linked networks based on polysaccharides and collagen for pilocarpine sustained release. *J. Macromol. Sci. A Pure Appl. Chem., 47*, 616–625.

Fonseca, C., Simoes, S., & Gaspar, R., (2002). Paclitaxel-loaded PLGA nanoparticles: preparation, physicochemical characterization and *in vitro* anti-tumoral activity. *J. Control Release, 83*, 273–286.

Foox, M., & Zilberman, M., (2015). Drug delivery from gelatin-based systems. *Expert Opin. Drug Deliv., 12*, 1547–1563.

Fredenberg, S., Wahlgren, M., Reslow, M., & Axelsson, A., (2011). The mechanisms of drug release in poly(lactic-co-glycolic acid)-based drug delivery systems–A review. *Int. J. Pharm., 415*, 34–52.

Friedman, A. J., Phan, J., Schairer, D. O., Champer, J., Qin, M., Pirouz, A., et al., (2013). Antimicrobial and anti-inflammatory activity of chitosan alginate nanoparticles: A targeted therapy for cutaneous pathogens. *J. Invest. Dermatol., 133*, 1231–1239.

Fröhlich, E., (2013). Cellular targets and mechanisms in the cytotoxic action of non-biodegradable engineered nanoparticles. *Curr. Drug Metab., 14,* 976–988.

Frömming, K. H., & Szejtli, J., (1994). Cyclodextrins in Pharmacy. Kluwer Academic Publishers: Dordrecht.

Galkina, O. L., Ivanov, V. K., Agafonov, A. V., Seisenbaeva, G. A., & Kessler, V. G., (2015). Cellulose nanofiber-titania nanocomposites as potential drug delivery systems for dermal applications. *J. Mater. Chem. B, 3,* 1688–1698.

Gaspar, V. M., Costa, E. C., Queiroz, J. A., Pichon, C., Sousa, F., & Correia, I. J., (2015). Folate-targeted multifunctional amino acid-chitosan nanoparticles for improved cancer therapy. *Pharm. Res., 32,* 562–577.

Gayathri, S., Ghosh, O. S. N., Sudhakara, P., & Viswanath, A. K., (2015). Chitosan conjugation: A facile approach to enhance the cell viability of LaF$_3$, Yb, Er upconverting nanotransducers in human breast cancer cells. *Carbohydr. Polym., 121,* 302–308.

Gentile, P., Chiono, V., Carmagnola, I., & Hatton, P. V., (2014). An overview of poly(lactic-co-glycolic) acid (PLGA)-based biomaterials for bone tissue engineering. *Int. J. Mol. Sci., 15,* 3640–3659.

Ghaz-Jahanian, M. A., Abbaspour-Aghdam, F., Anarjan, N., Berenjian, A., & Jafarizadeh-Malmiri, H., (2015). Application of chitosan-based nanocarriers in tumor-targeted drug delivery. *Mol. Biotechnol., 57,* 201–218.

Gibot, L., Chabaud, S., Bouhout, S., Bolduc, S., Auger, F. A., & Moulin, V. J., (2015). Anticancer properties of chitosan on human melanoma are cell line dependent. *Int. J. Biol. Macromol., 72,* 370–379.

Gidwani, B., & Vyas, A., (2015). A comprehensive review on cyclodextrin-based carriers for delivery of chemotherapeutic cytotoxic anticancer drugs. *BioMed Res. Int., 2015,* 15, Article ID 198268.

Gómez-Guillén, M. C., Giménez, B., López-Caballero, M. E., & Montero, M. P., (2011). Functional and bioactive properties of collagen and gelatin from alternative sources: A review. *Food Hydrocolloids, 25,* 1813–1827.

Goncalves, M., Figueira, P., Maciel, D., Rodrigues, J., Qu, X., Liu, C. S., et al., (2014). pH-sensitive Laponite®/doxorubicin/alginate nanohybrids with improved anticancer efficacy. *Acta Biomater., 10,* 300–307.

Gorrasi, G., (2015). Dispersion of halloysite loaded with natural antimicrobials into pectins: Characterization and controlled release analysis. *Carbohydr. Polym., 127,* 47–53.

Goswami, S., & Naik, S., (2014). Natural gums and its pharmaceutical application. *J. Sci. Innov. Res., 3,* 112–121.

Grenha, A., Gomes, M. E., Rodrigues, M., Santo, V. E., Mano, J. F., Neves, N. M., et al., (2010). Development of new chitosan/carrageenan nanoparticles for drug delivery applications. *J. Biomed. Mater. Res. A,. 92A,* 1265–1272.

Grumezescu, A. M., Andronescu, E., Ficai, A., Bleotu, C., Mihaiescu, D. E., & Chifiriuc, M. C., (2012). Synthesis, characterization and *in vitro* assessment of the magnetic chitosan-carboxymethylcellulose biocomposite interactions with the prokaryotic and eukaryotic cells. *Int. J. Pharm., 436,* 771–777.

Guo, Y. Z., Wang, X. H., Shu, X. C., Shen, Z. G., & Sun, R. G., (2012). Self-assembly and paclitaxel loading capacity of cellulose-graft-poly(lactide) nanomicelles. *J. Agric. Food Chem., 60,* 3900–3908.

Guo, S. T., Lin, C. M., Xu, Z. H., Miao, L., Wang, Y. H., & Huang, L., (2014). Co-delivery of cisplatin and rapamycin for enhanced anticancer therapy through synergistic effects and microenvironment modulation. *ACS Nano, 8,* 4996–5009.

Gurski, L. A., Jha, A. K., Zhang, C., Jia, X. Q., & Farach-Carson, M. C., (2009). Hyaluronic acid-based hydrogels as 3D matrices for *in vitro* evaluation of chemotherapeutic drugs using poorly adherent prostate cancer cells. *Biomaterials, 30,* 6076–6085.

Hagens, W. I., Oomen, A. G., de Jong, W. H., Cassee, F. R., & Sips, A. J. A. M., (2007). What do we (need to) know about the kinetic properties of nanoparticles in the body? *Regul. Toxicol. Pharmacol., 49,* 217–229.

Hamdy, S., Haddadi, A., Hung, R. W., & Lavasanifar, A., (2011). Targeting dendritic cells with nano-particulate PLGA cancer vaccine formulations. *Adv. Drug Deliv. Rev., 63,* 943–955.

Hamman, J. H. (2010). Chitosan-based polyelectrolyte complexes as potential carrier materials in drug delivery systems. *Mar. Drugs, 8,* 1305–1322.

Han, S. Y., Han, H. S., Lee, S. C., Kang, Y. M., Kim, I. S., & Park, J. H., (2011). Mineralized hyaluronic acid nanoparticles as a robust drug carrier. *J. Mater. Chem., 21,* 7996–8001.

Han, F., Gao, C. M., & Liu, M. Z., (2013). Fabrication and characterization of size-controlled starch-based nanoparticles as hydrophobic drug carriers. *J. Nanosci. Nanotechnol., 13,* 6996–7007.

Hasegawa, M., Yagi, K., Iwakawa, S., & Hirai, M., (2001). Chitosan induces apoptosis via caspase-3 activation in bladder tumor cells. *Jpn. J. Cancer Res., 92,* 459–466.

He, M., Zhao, Z. M., Yin, L. C., Tang, C., & Yin, C. H., (2009). Hyaluronic acid coated poly(butyl cyanoacrylate) nanoparticles as anticancer drug carriers. *Int. J. Pharm., 373,* 165–173.

Hebeish, A., Hashem, M., Abd El-Hady, M. M., & Sharaf, S., (2013). Development of CMC hydrogels loaded with silver nano-particles for medical applications. *Carbohydr. Polym., 92,* 407–413.

Helaly, F. M., Khalaf, A. I., & El Nashar, D., (2013). Starch cellulose acetate co-acrylate (SCAA) polymer as a drug carrier. *Res. Chem. Intermed, 39,* 3209–3220.

Hezaveh, H., & Muhamad, I. I., (2012). The effect of nanoparticles on gastrointestinal release from modified kappa-carrageenan nanocomposite hydrogels. *Carbohydr. Polym., 89,* 138–145.

Horvati, K., Bacsa, B., Szabo, N., Fodor, K., Balka, G., Rusvai, M., et al., (2015). *Mycobacterium tuberculosis* in a series of *in vitro* and *in vivo* models. *Tuberculosis, 95,* S207–S211.

Hu, Q. D., Tang, G. P., & Chu, P. K., (2014). Cyclodextrin-based host-guest supramolecular nanoparticles for delivery: From design to applications. *Acc. Chem. Res., 47,* 2017–2025.

Huang, Z., Feng, Q. L., Yu, B., & Li, S. J., (2011). Biomimetic properties of an injectable chitosan/nano-hydroxyapatite/collagen composite. *Mater. Sci. Eng. C Mater. Biol. Appl., 31,* 683–687.

Huang, Y. J., Liu, M. Z., Gao, C. M., Yang, J. L., Zhang, X. Y., Zhang, X. J., et al., (2013). Ultra-small and innocuous cationic starch nanospheres: Preparation, characterization and drug delivery study. *Int. J. Biol. Macromol., 58,* 231–239.

Hussein-Al-Ali, S. H., El Zowalaty, M. E., Hussein, M. Z., Ismail, M., & Webster, T. J., (2014). Synthesis, characterization, controlled release, and antibacterial studies of a novel streptomycin chitosan magnetic nanoantibiotic. *Int. J. Nanomed., 9,* 549–557.

Hwang, H. Y., Kim, I. S., Kwon, I. C., & Kim, Y. H., (2008). Tumor targetability and antitumor effect of docetaxel-loaded hydrophobically modified glycol chitosan nanoparticles. *J. Control Release, 128,* 23–31.

Jackson, J. K., Letchford, K., Wasserman, B. Z., Ye, L., Hamad, W. Y., & Burt, H. M., (2011). The use of nanocrystalline cellulose for the binding and controlled release of drugs. *Int. J. Nanomed.*, *6*, 321–330.

Jain, A. K., Das, M., Swarnakar, N. K., & Jain, S., (2011a). Engineered PLGA nanoparticles: An emerging delivery tool in cancer therapeutics. *Crit. Rev. Ther. Drug Carrier Syst.*, *28*, 1–45.

Jain, A. K., Swarnakar, N. K., Das, M., Godugu, C., Singh, R. P., Rao, P. R., et al., (2011b). Augmented anticancer efficacy of doxorubicin-loaded polymeric nanoparticles after oral administration in a breast cancer induced animal model. *Mol. Pharm.*, *8*, 1140–1151.

Jain, D. S., Athawale, R. B., Bajaj, A. N., Shrikhande, S. S., Goel, P. N., Nikam, Y., et al., (2014). Unraveling the cytotoxic potential of temozolomide loaded into PLGA nanoparticles. *DARU-J. Pharm. Sci.*, *22*, 9, Article ID 18.

Jain, A., & Jain, S. K., (2015). L-Valine appended PLGA nanoparticles for oral insulin delivery. *Acta Diabet.*, *52*, 663–676.

Jain, A., Jain, S., Jain, R., & Kohli, D. V., (2015). Coated chitosan nanoparticles encapsulating caspase 3 activator for effective treatment of colorectral cancer. *Drug Deliv. Transl. Res.*, *5*, 596–561.

Jana, S., Gangopadhaya, A., Bhowmik, B. B., Nayak, A. K., & Mukherjee, A., (2015). Pharmacokinetic evaluation of testosterone-loaded nanocapsules in rats. *Int. J. Biol. Macromol.*, *72*, 28–30.

Jansson, P. E., Lindberg, B., & Sandford, P. A., (1983). Structural studies of gellan gum, an extracellular polysaccharide elaborated by *Pseudomonas elodea*. *Carbohydr. Res.*, *124*, 135–139.

Jayakumar, R., Prabaharan, M., Nair, S. V., & Tamura, H., (2010a). Novel chitin and chitosan nanofibers in biomedical applications. *Biotechnol. Adv.*, *28*, 142–150.

Jayakumar, R., Menon, D., Manzoor, K., Nair, S. V., & Tamura, H., (2010b). Biomedical applications of chitin and chitosan-based nanomaterials-A short review. *Carbohydr. Polym.*, *82*, 227–232.

Jayakumar, R., Chennazhi, K. P., Srinivasan, S., Nair, S. V., Furuike, T., & Tamura, H., (2011a). Chitin scaffolds in tissue engineering. *Int. J. Mol. Sci.*, *12*, 1876–1887.

Jayakumar, R., Prabaharan, M., Kumar, P. T. S., Nair, S. V., & Tamura, H., (2011b). Biomaterials based on chitin and chitosan in wound dressing applications. *Biotechnol. Adv.*, *29*, 322–337.

Jayakumar, R., Ramachandran, R., Kumar, P. T. S., Divyarani, V. V., Srinivasan, S., Chennazhi, K. P., et al., (2011c). Fabrication of chitin-chitosan/nano ZrO_2 composite scaffolds for tissue engineering applications. *Int. J. Biol. Macromol.*, *49*, 274–280.

Jayakumar, R., Nair, A., Rejinold, N. S., Maya, S., & Nair, S. V., (2012). Doꞔorubicin-loaded pH-responsive chitin nanogels for drug delivery to cancer cells. *Carbohydr. Polym.*, *87*, 2352–2356.

Jeanes, A., (1977). Dextrans and pullulans: Industrially significant. *ACS Symp. Ser.*, *45*, 284–298.

Jee, J. P., Na, J. H., Lee, S., Kim, S. H., Choi, K., Yeo, Y., et al., (2012). Cancer targeting strategies in nanomedicine: Design and application of chitosan nanoparticles. *Curr. Opin. Solid State Mater. Sci.*, *16*, 333–342.

Johnston-Banks, F. A., (1990). *Gelatin*. In *Food Gels*; Harris, P., (Ed.), Elsevier Applied Science Publishers: London, 233–289.

Joshi, G., Kumar, A., & Sawant, K., (2014). Enhanced bioavailability and intestinal uptake of Gemcitabine HCl loaded PLGA nanoparticles after oral delivery. *Eur. J. Pharm. Sci.*, *60*, 80–89.

Jung, S. W., Jeong, Y. I., Kim, Y. H., & Kim, S. H., (2004). Self-assembled polymeric nanoparticles of poly(ethylene glycol) grafted pullulan acetate as a novel drug carrier. *Arch. Pharm. Res.*, *27*, 562–569.

Kamaly, N., Xiao, Z., Valencia, P. M., Radovic-Moreno, A. F., & Farokhzad, O. C., (2012). Targeted polymeric therapeutic nanoparticles: design, development and clinical translation. *Chem. Soc. Rev.*, *41*, 2971–3010.

Kamat, M., El-Boubbou, K., Zhu, D. C., Lansdell, T., Lu, X. W., Li, W., et al., (2010). Hyaluronic acid immobilized magnetic nanoparticles for active targeting and imaging of macrophages. *Bioconjug. Chem.*, *21*, 2128–2135.

Kang, B., Okwieka, P., Schoettler, S., Seifert, O., Kontermann, R. E., Pfizenmaier, K., et al., (2015a). Tailoring the stealth properties of biocompatible polysaccharide nanocontainers. *Biomaterials*, *49*, 125–134.

Kang, B., Okwieka, P., Schoettler, S., Winzen, S., Langhanki, J., Mohr, K., et al., (2015b). Carbohydrate-based nanocarriers exhibiting specific cell targeting with minimum influence from the protein corona. *Angew. Chem. Int. (Ed.)*, *54*, 7436–7440.

Kannan, R., Zambre, A., Chanda, N., Kulkarni, R., Shukla, R., Katti, K., et al., (2012). Functionalized radioactive gold nanoparticles in tumor therapy. *Wiley Interdisc. Rev. Nanomed. Nanobiotechnol.*, *4*, 42–51.

Kapoor, D. N., Bhatia, A., Kaur, R., Sharma, R., Kaur, G., & Dhawan, S., (2015). PLGA: a unique polymer for drug delivery. *Ther. Deliv.*, *6*, 41–58.

Kawabata, Y., Wada, K., Nakatani, M., Yamada, S., & Onoue, S., (2011). Formulation design for poorly water-soluble drugs based on biopharmaceutics classification system: Basic approaches and practical applications. *Int. J. Pharm.*, *420*, 1–10.

Kawashima, Y., Yamamoto, H., Takeuchi, H., Fujioka, S., & Hino, T., (1999). Pulmonary delivery of insulin with nebulized DL-lactide/glycolide copolymer (PLGA) nanospheres to prolong hypoglycemic effect. *J. Control Release*, *62*, 279–287.

Keck, C. M., & Müller, R. H., (2013). Nanotoxicological classification system (NCS)–a guide for the risk-benefit assessment of nanoparticulate drug delivery systems. *Eur. J. Pharm. Biopharm.*, *84*, 445–448.

Khaira, R., Sharma, J., & Saini, V., (2014). Development and characterization of nanoparticles for the delivery of gemcitabine hydrochloride. *Sci. World J.*, *2014*, 6, Article ID 560962.

Khodaverdi, E., Heidari, Z., Tabassi, S. A. S., Tafaghodi, M., Alibolandi, M., Tekie, F. S. M., et al., (2015). Injectable supramolecular hydrogel from insulin-loaded triblock PCL-PEG-PCL copolymer and gamma-cyclodextrin with sustained-release property. *AAPS Pharm. SciTech.*, *16*, 140–149.

Kim, J. H., Kim, Y. S., Kim, S., Park, J. H., Kim, K., Choi, K., et al., (2006). Hydrophobically modified glycol chitosan nanoparticles as carriers for paclitaxel. *J. Control Release*, *111*, 228–234.

Kim, J. H., Kim, Y. S., Park, K., Kang, E., Lee, S., Nam, H. Y., et al., (2008a). Self-assembled glycol chitosan nanoparticles for the sustained and prolonged delivery of antiangiogenic small peptide drugs in cancer therapy. *Biomaterials*, *29*, 1920–1930.

Kim, J. H., Kim, Y. S., Park, K., Lee, S., Nam, H. Y., Min, K. H., et al., (2008b). Antitumor efficacy of cisplatin-loaded glycol chitosan nanoparticles in tumor-bearing mice. *J. Control Release*, *127*, 41–49.

Kim, Y. J., Chae, S. Y., Jin, C. H., Sivasubramanian, M., Son, S., Choi, K. Y., et al., (2010). Ionic complex systems based on hyaluronic acid and PEGylated TNF-related apoptosis-inducing ligand for treatment of rheumatoid arthritis. *Biomaterials, 31,* 9057–9064.

Kim, H. Y., Park, S. S., & Lim, S. T., (2015). Preparation, characterization and utilization of starch nanoparticles. *Colloids Surf. B Biointerfaces, 126,* 607–620.

Klemm, D., Heinze, T., Wagenknecht, W., Phillip, B., & Heinze, U., (1998). Comprehensive Cellulose Chemistry, Wiley-VCH: Weinheim.

Kobori, T., & Nakao, H., (2010). Xanthan gum–Basic properties, applications, and future perspective in nanotechnology. In *Polysaccharides: Development, Properties and Applications,* Tiwari, A., (Ed.), Polymer Science and Technology Series, Nova Science Publishers, pp. 379–393.

Kogan, G., Soltes, L., Stern, R., & Gemeiner, P., (2007). Hyaluronic acid: A natural biopolymer with a broad range of biomedical and industrial applications. *Biotechnol. Lett., 29,* 17–25.

Kojima, C., Suehiro, T., Watanabe, K., Ogawa, M., Fukuhara, A., Nishisaka, E., et al., (2013). Doxorubicin-conjugated dendrimer/collagen hybrid gels for metastasis-associated drug delivery systems. *Acta Biomater., 9,* 5673–5680.

Kolakovic, R., Peltonen, L., Laukkanen, A., Hellman, M., Laaksonen, P., Linder, M. B., et al., (2013). Evaluation of drug interactions with nanofibrillar cellulose. *Eur. J. Pharm. Biopharm., 85,* 1238–1244.

Kong, Z. Q., Yu, M. F., Cheng, K., Weng, W. J., Wang, H. M., Lin, J., et al., (2013). Incorporation of chitosan nanospheres into thin mineralized collagen coatings for improving the antibacterial effect. *Colloids Surf. B Biointerfaces, 111,* 536–541.

Koo, H., Huh, M. S., Sun, I. C., Yuk, S. H., Choi, K., Kim, K., et al., (2011). *In vivo* targeted delivery of nanoparticles for theranosis. *Acc. Chem. Res., 44,* 1018–1028.

Koopaei, M. N., Khoshayand, M. R., Mostafavi, S. H., Amini, M., Khorramizadeh, M. R., Tehrani, M. J., et al., (2014). Docetaxel loaded PEG-PLGA nanoparticles: Optimized drug loading, *in-vitro* cytotoxicity and *in-vivo* antitumor effect. *Iranian J. Pharm. Res., 13,* 819–833.

Král, V., Oktábec, Z., Jampílek, J., Pekárek, T., Proksa, B., Dohnal, J., et al. (2011a). Pectin Complexes of Steroids and Pharmaceutical Compositions based Thereon. WO/2011/063774 A2.

Král, V., Oktábec, Z., Jampílek, J., Pekárek, T., Proksa, B., Dohnal, J., et al., (2011b). Pectin Complexes of Sartans and Pharmaceutical Compositions based Thereon. WO/2011/063775 A2.

Králová, J., Kejík, Z., Bříza, T., Poučková, P., Král, A., Martásek, P., et al., (2010). Porphyrin-cyclodextrin conjugates as a nanosystem for versatile drug delivery and multimodal cancer therapy. *J. Med. Chem., 53,* 128–138.

Krishnamoorthy, G., Krithica, N., Sehgal, P. K., Mandal, A. B., & Sadulla, S. (2011). Preparation and characterization of collagen coated cholesterol-free liposome nanoparticles matrix for drug carriers and tissue engineering application. *Transact. Ind. Inst. Metals, 64,* 199–204.

Kulkarni, V. S., Butte, K. D., & Rathod, S. S. (2012). Natural polymers: A comprehensive review. *Int. J. Res. Pharm. Biomed. Sci., 3,* 1597–1613.

Kumar, B. S., Kumar, M. G., Suguna, L, Sastry, T. P., & Mandal, A. B., (2012). Pullulan acetate nanoparticles-based delivery system for hydrophobic drug. *Int. J. Pharma Bio Sci., 3,* 24–32.

Kumar, N. A., Rejinold, N. S., Anjali, P., Balakrishnan, A., Biswas, R., & Jayakumar, R., (2013). Preparation of chitin nanogels containing nickel nanoparticles. *Carbohydr. Polym.*, *97*, 469–474.

Kurkov, S. V., & Loftsson, T., (2013). Cyclode☐trins. *Int. J. Pharm.*, *453*, 167–180.

Lam, E., Male, K. B., Chong, J. H., Leung, A. C. W., & Luong, J. H. T., (2012). Applications of functionalized and nanoparticle-modified nanocrystalline cellulose. *Trends Biotechnol.*, *30*, 283–290.

Lancuški, A., Vasilyev, G., Putau☐ J. L., & Zussman, E., (2015). Rheological properties and electrospinnability of high-amylose starch in formic acid. *Biomacromolecules, 16*, 2529–2536.

Lee, K. Y., & Mooney, D. J., (2012). Alginate: Properties and biomedical applications. *Progr. Polym. Sci., 37*, 106–126.

Lee, J. H., Park, J. H., Eltohamy, M., Perez, R., Lee, E. J., & Kim, H. W., (2013). Collagen gel combined with mesoporous nanoparticles loading nerve growth factor as a feasible therapeutic three-dimensional depot for neural tissue engineering. *RSC Adv.*, *3*, 24202–24214.

Lemmer, Y., Kalombo, L., Pietersen, R. D., Jones, A. T., Semete-Makokotlela, B., Van Wyngaardt, S., et al., (2015). Mycolic acids, a promising mycobacterial ligand for targeting of nanoencapsulated drugs in tuberculosis. *J. Control Release, 211*, 94–104.

Leong, K. H., Chung, L. Y., Noordin, M. I., Mohamad, K., Nishikawa, M., Onuki, Y., et al., (2011). Carboxymethylation of kappa-carrageenan for intestinal-targeted delivery of bioactive macromolecules. *Carbohydr. Polym.*, *83*, 1507–1515.

Lewinski, N., Colvin, V., & Drezek, R., (2008). Cytoto☐icity of nanoparticles. *Small, 4*, 26–49.

Li, Y. Y., Chen, X. G., Zhang, J., Liu, C. S., Xue, Y. P., Sun, G. Z., et al., (2009). *In vitro* release of rifampicin and biocompatibility of oleoylchitosan nanoparticles. *J. Appl. Polym. Sci., 111*, 2269–2274.

Li, Y. Y., Zhang, S. S., Meng, X. J., Chen, X. G., & Ren, G. D., (2011). The preparation and characterization of a novel amphiphilic oleoyl-carboxymethyl chitosan self-assembled nanoparticles. *Carbohydr. Polym.*, *83*, 130–136.

Li, Y., Li, X. M., & Xu, L. X., (2013). Effects of starch nanospheres solution on hemodynamic values in rats with hemorrhagic shock: A preliminary study in hemorrhagic shock resuscitation. *J. Surg. Res.*, *181*, 142–145.

Li, H. N., Bian, S. Q., Huang, Y. H., Liang, J., Fan, Y. J., & Zhang, X. D., (2014a). High drug loading pH-sensitive pullulan-DOX conjugate nanoparticles for hepatic targeting. *J. Biomed. Mater. Res. A, 102*, 150–159.

Li, L., Ni, R., Shao, Y., & Mao, S. R., (2014b). Carrageenan and its applications in drug delivery. *Carbohydr. Polym.*, *103*, 1–11.

Li, K., Liu, H., Gao, W., Chen, M., Zeng, Y., Liu, J. J., et al., (2015). Mulberry-like dual-drug complicated nanocarriers assembled with apogossypolone amphiphilic starch micelles and doxorubicin hyaluronic acid nanoparticles for tumor combination and targeted therapy. *Biomaterials, 39*, 131–144.

Liao, R. Q., Liu, M. S., Liao, X. L., & Yang, B., (2015a). Cyclodextrin-based smart stimuli-responsive drug carriers. *Progr. Chem., 27*, 79–90.

Liao, S. H., Liu, C. H., Bastakoti, B. P., Suzuki, N., Chang, Y., Yamauchi, Y., et al., (2015b). Functionalized magnetic iron oxide/alginate core-shell nanoparticles for targeting hyperthermia. *Int. J. Nanomed.*, *10*, 3315–3328.

Lin, N., Huang, J., Chang, P. R., Feng, L. D., & Yu, J. H., (2011). Effect of polysaccharide nanocrystals on structure, properties, and drug release kinetics of alginate-based microspheres. *Colloids Surf. B Biointerfaces, 85*, 270–279.

Lin, H. Y., Chiou, S. F., Lai, C. H., Tsai, S. C., Chou, C. W., Peng, S. F., et al., (2012). Formulation and evaluation of water-in-oil amoxicillin-loaded nanoemulsions using for *Helicobacter pylori* eradication. *Proc. Biochem., 47*, 1469–1478.

Lin, N., & Dufresne, A., (2013). Supramolecular hydrogels from *in situ* host-guest inclusion between chemically modified cellulose nanocrystals and cyclodextrin. *Biomacromolecules, 14*, 871–880.

Lin, L. F., Xu, W., Liang, H. S., He, L., Liu, S. L., Li, Y., et al., (2015). Construction of pH-sensitive lysozyme/pectin nanogel for tumor methotrexate delivery. *Colloids Surf. B Biointerfaces, 126*, 459–466.

Liu, Y. H., Sun, J., Cao, W., Yang, J. H., Lian, H., Li, X., et al., (2011). Dual targeting folate-conjugated hyaluronic acid polymeric micelles for paclitaxel delivery. *Int. J. Pharm., 421*, 160–169.

Liu, Q. Y., Yang, X. L., Xu, H. B., Pan, K. J., & Yang, Y. J., (2013). Novel nanomicelles originating from hydroxyethyl starch-g-polylactide and their release behavior of docetaxel modulated by the PLA chain length. *Eur. Polym. J., 49*, 3522–3529.

Liu, Z. J., & Yao, P., (2015). Injectable thermo-responsive hydrogel composed of xanthan gum and methylcellulose double networks with shear-thinning property. *Carbohydr. Polym., 132*, 490–498.

Liu, K. P., Wang, Y. M., Li, H. M., & Duan, Y. X., (2015a). A facile one-pot synthesis of starch functionalized graphene as nano-carrier for pH sensitive and starch-mediated drug delivery. *Colloids Surf. B Biointerfaces, 128*, 86–93.

Liu, J. J., Zhan, X. D., Wan, J. B., Wang, Y. T., & Wang, C. M., (2015b). Review for carrageenan-based pharmaceutical biomaterials: Favourable physical features versus adverse biological effects. *Carbohydr. Polym., 121*, 27–36.

Loftsson, T., & Stefansson, E., (2007). Cyclodextrins in ocular drug delivery: theoretical basis with dexamethasone as a sample drug. *J. Drug Deliv. Sci. Technol., 17*, 3–9.

Lu, C. T., Zhao, Y. Z., Wong, H. L., Cai, J., Peng, L., & Tian, X. Q., (2014). Current approaches to enhance CNS delivery of drugs across the brain barriers. *Int. J. Nanomedicine, 9*, 2241–2257.

Lungu, M., Neculae, A., Bunoiu, M., & Biris, C., (2015). *Nanoparticles' Promises and Risks: Characterization, Manipulation, and Potential Hazards to Humanity and the Environment*, Springer: Heidelberg.

Luo, Y., & Wang, Q., (2014). Recent development of chitosan-based polyelectrolyte complexes with natural polysaccharides for drug delivery. *Int. J. Biol. Macromol., 64*, 353–367.

Luo, Y. C., Pan, K., & Zhong, Q. X., (2015). Casein/pectin nanocomplexes as potential oral delivery vehicles. *Int. J. Pharm., 486*, 59–68.

Ma, M., Chen, H. R., Chen, Y., Zhang, K., Wang, X., Cui, X. Z., & Shi, J. L., (2012). Hyaluronic acid-conjugated mesoporous silica nanoparticles: excellent colloidal dispersity in physiological fluids and targeting efficacy. *J. Mater. Chem., 22*, 5615–5621.

Madhumathi, K., Kumar, P. T. S., Abhilash, S., Sreeja, V., Tamura, H., Manzoor, K., et al., (2010). Development of novel chitin/nanosilver composite scaffolds for wound dressing applications. *J. Mater. Sci. Mater. Med., 21*, 807–813.

Maestrelli, F., Garcia-Fuentes, M., Mura, P., & Alonso, M. J., (2006). A new drug nanocarrier consisting of chitosan and hydoxypropylcyclodextrin. *Eur. J. Pharm. Biopharm., 63*, 79–86.

Mahdavinia, G. R., Etemadi, H., & Soleymani, F., (2015). Magnetic/pH-responsive beads based on caboxymethyl chitosan and kappa-carrageenan and controlled drug release. *Carbohydr. Polym., 128*, 112–121.

Makadia, H. K., & Siegel, S. J., (2011). Polylactic-co-glycolic acid (PLGA) as biodegradable controlled drug delivery carrier. *Polymers, 3*, 1377–1397.

Makharza, S., Vittorio, O., Cirillo, G., Oswald, S., Hinde, E., Kavallaris, M., et al., (2015). Graphene oxide–gelatin nanohybrids as functional tools for enhanced carboplatin activity in neuroblastoma cells. *Pharm. Res., 32*, 2132–2143.

Malam, Y., Lim, E. J., & Seifalian, A. M., (2011). Current trends in the application of nanoparticles in drug delivery. *Curr. Med. Chem., 18*, 1067–1078.

Malathi, S., & Balasubramanian, S., (2011). Synthesis of biodegradable polymeric nanoparticles and their controlled drug delivery for tuberculosis. *J. Biomed. Nanotechnol., 7*, 150–151.

Margulis-Goshen, K., & Magdassi, S., (2012). Nanotechnology: An Advanced Approach to the Development of Potent Insecticides. In *Advanced Technologies for Managing Insect Pests;* Ishaaya, I., Horowitz, A. R., Palli, S. R., Eds.; Springer: Dordrecht, pp. 295–314.

Martinez, A., Olmo, R., Iglesias, I., Teijon, J. M., & Blanco, M. D., (2014). Folate-targeted nanoparticles based on albumin and albumin/alginate mixtures as controlled release systems of tamoxifen: Synthesis and *in vitro* characterization. *Pharm. Res., 31*, 182–193.

Mateescu, M. A., Ispas-Szabo, P., & Assaad, E., (2015). Controlled Drug Delivery: The Role of Self-Assembling Multi-Task Excipients. Woodhead Publishing and Elsevier, Walham.

Medina, C., Santos-Martinez, M. J., Radomski, A., Corrigan, O., & Radomski, M. W., (2007). Nanoparticles: pharmacological and toxicological significance. *Br. J. Pharmacol., 150*, 552–558.

Meng, Z. X., Zheng, W., Li, L., & Zheng, Y. F., (2011). Fabrication, characterization and *in vitro* drug release behavior of electrospun PLGA/chitosan nanofibrous scaffold. *Mater. Chem. Phys., 125*, 606–611.

Metaxa, A. F., Efthimiadou, E. K., Boukos, N., & Kordas, G., (2012). Polysaccharides as a source of advanced materials: Cellulose hollow microspheres for drug delivery in cancer therapy. *J. Colloid Interface Sci., 384*, 198–206.

Mihaila, S. M., Gaharwar, A. K., Reis, R. L., Marques, A. P., Gomes, M. E., & Khademhosseini, A., (2013). Photocrosslinkable kappa-carrageenan hydrogels for tissue engineering applications. *Adv. Healthcare Mater., 2*, 895–907.

Minimol, P. F., Paul, W., & Sharma, C. P., (2013). PEGylated starch acetate nanoparticles and its potential use for oral insulin delivery. *Carbohydr. Polym., 95*, 1–8.

Mirakabad, F. S. T., Nejati-Koshki, K., Akbarzadeh, A., Yamchi, M. R., Milani, M., Zarghami, N., et al., (2014). PLGA-based nanoparticles as cancer drug delivery systems. *Asian Pac. J. Cancer Prev., 15*, 517–535.

Mody, V. V., Siwale, R., Singh, A., & Mody, H. R., (2010). Introduction to metallic nanoparticles. *J. Pharm. Bioallied Sci., 2*, 282–289.

Mohammed, N., Rejinold, N. S., Mangalathillam, S., Biswas, R., Nair, S. V., & Jayakumar, R., (2013). Fluconazole loaded chitin nanogels as a topical ocular drug delivery agent for corneal fungal infections. *J. Biomed. Nanotechnol., 9*, 1521–1531.

Moran, M. C., Rosell, N., Ruano, G., Busquets, M. A., & Vinardell, M. P., (2015). Gelatin-based nanoparticles as DNA delivery systems: Synthesis, physicochemical and biocompatible characterization. *Colloids Surf. B Biointerfaces, 134*, 156–168.

Moritz, S., Wiegand, C., Wesarg, F., Hessler, N., Mueler, F. A., Kralisch, D., et al., (2014). Active wound dressings based on bacterial nanocellulose as drug delivery system for octenidine. *Int. J. Pharm., 471*, 45–55.

Morris, E. R., Nishinari, K., & Rinaudo, M., (2012). Gelation of gellan: A review. *Food Hydrocolloids.*, *28*, 373–411.

Moya-Ortega, M. D., Alvarez-Lorenzo, C., Concheiro, A., & Loftsson, T., (2012). Cyclodextrin-based nanogels for pharmaceutical and biomedical applications. *Int. J. Pharm.*, *428*, 152–163.

Mozafari, M. R., (2006). Nanocarrier Technologies: Frontiers of Nanotherapy, Springer: Dordrecht.

Muddineti, O. S., Ghosh, B., & Biswas, S., (2015). Current trends in using polymer coated gold nanoparticles for cancer therapy. *Int. J. Pharm.*, *484*, 252–267.

Mukerjee, A., & Vishwanatha, J. K., (2009). Formulation, characterization and evaluation of curcumin-loaded PLGA nanospheres for cancer therapy. *Anticancer Res.*, *29*, 3867–3875.

Muzzarelli, R. A. A., (2010). Chitins and chitosans as immunoadjuvants and non-allergenic drug carriers. *Mar. Drugs.*, *8*, 292–312.

Muzzarelli, R. A. A., (2011). Chitin nanostructures in living organisms. In *Chitin. Formation and Diagenesis*; Gupta, S. N., Ed.; Springer: New York, 1–34.

Nafee, N., Hirosue, M., Loretz, B., Wenz, G., & Lehr, C. M., (2015). Cyclodextrin-based star polymers as a versatile platform for nanochemotherapeutics: Enhanced entrapment and uptake of idarubicin. *Colloids Surf. B Biointerfaces, 129*, 30–38.

Nahar, K., Absar, S., Patel, B., & Ahsan, F., (2014). Starch-coated magnetic liposomes as an inhalable carrier for accumulation of fasudil in the pulmonary vasculature. *Int. J. Pharm.*, *464*, 185–195.

Nair, L. S., & Laurencin, C. T., (2007). Biodegradable polymers as biomaterials. *Prog. Polym. Sci.*, *32*, 762–798.

Nair, H. B., Huffman, S., Veerapaneni, P., Kirma, N. B., Binkley, P., Perla, R. P., et al., (2011). Hyaluronic acid-bound letrozole nanoparticles restore sensitivity to letrozole-resistant xenograft tumors in mice. *J. Nanosci. Nanotechnol.*, *11*, 3789–3799.

Nair, R. S., Reshmi, T., Reshmi, P., Sarika, C., Snima, K. S., Akk, U., et al., (2015). Biocompatibility studies of pectin-fibrin nanocomposite bearing BALB/c mice. *Cell. Chem. Technol.*, *49*, 55–60.

Najafabadi, A. H., Azodi-Deilami, S., Abdouss, M., Payravand, H., & Farzaneh, S., (2015). Synthesis and evaluation of hydroponically alginate nanoparticles as novel carrier for intravenous delivery of propofol. *J. Mater. Sci. Mater. Med.*, *26*, 11, Article ID 145.

Nalwa, H. S., (2004–2011). *Encyclopedia of Nanoscience and Nanotechnology*, American Scientific Publisher: Valencia.

Nam, J. P., Park, S. C., Kim, T. H., Jang, J. Y., Choi, C., Jang, M. K., et al., (2013). Encapsulation of paclitaxel into lauric acid-*O*-carboxymethyl chitosan-transferrin micelles for hydrophobic drug delivery and site-specific targeted delivery. *Int. J. Pharm.*, *457*, 124–135.

Namgung, R., Lee, Y. M., Kim, J., Jang, Y., Lee, B. H., Kim, I. S., et al., (2014). Poly-cyclodextrin and poly-paclitaxel nano-assembly for anticancer therapy. *Nature Commun. 5*, 12, Article ID 3702.

Nance, E., Zhang, C., Shih, T. Y., Xu, Q. G., Schuster, B. S., & Hanes, J., (2014). Brain-penetrating nanoparticles improve paclitaxel efficacy in malignant glioma following local administration. *ACS Nano*, *8*, 10655–10664.

Narayanan, D., Jayakumar, R., & Chennazhi, K. P., (2014). Versatile carboxymethyl chitin and chitosan nanomaterials: a review. *Wiley Interdiscip. Rev. Nanomed. Nanobiotechnol.*, *6*, 574–598.

Narayanan, D., Nair, S., & Menon, D., (2015). A systematic evaluation of hydroxyethyl starch as a potential nanocarrier for parenteral drug delivery. *Int. J. Biol. Macromol.*, *74*, 575–584.

National Nanotechnology Initiative. (2008). Big Things from a Tiny World. Arlington, VA.

Nehoff, H., Parayath, N. N., Domanovitch, L., Taurin, S., & Greish, K., (2014). Nanomedicine for drug targeting: Strategies beyond the enhanced permeability and retention effect. *Int. J. Nanomedicine*, *9*, 2539–2555.

Nekkanti, V., Vabalaboina, V., & Pillai, R., (2012). Drug nanoparticles: An overview. In *The Delivery of Nanoparticles*, Hashim, A. A., (Ed.), InTech: Rieka, pp. 111–132.

Nicklas, M., Schatton, W., Heinemann, S., Hanke, T., & Kreuter, J. (2009). Preparation and characterization of marine sponge collagen nanoparticles and employment for the transdermal delivery of 17 beta-estradiol-hemihydrate. *Drug Dev. Ind. Pharm.*, *35*, 1035–1042.

Nistor, M. T., Vasile, C., Chiriac, A. P., & Tartau, L., (2013). Biocompatibility, biodegradability, and drug carrier ability of hybrid collagen-based hydrogel nanocomposites. *J. Bioact. Compat. Polym.*, *28*, 540–556.

Nogueira, D. R., Tavano, L., Mitjans, M., Perez, L., Infante, M. R., & Vinardell, M. P., (2013). *In vitro* antitumor activity of methotrexate via pH-sensitive chitosan nanoparticles. *Biomaterials*, *34*, 2758–2772.

Obara, K., Ishihara, M., Ozeki, Y., Ishizuka, T., Hayashi, T., Nakamura, S., et al., (2005). Controlled release of paclitaxel from photocrosslinked chitosan hydrogels and its subsequent effect on subcutaneous tumor growth in mice. *J. Control Release*, *110*, 79–89.

Ossipov, D. A., (2010). Nanostructured hyaluronic acid-based materials for active delivery to cancer. *Expert Opin. Drug Deliv.*,*7*, 681–703.

Palaniraja, A., & Jayaraman, V., (2011). Production, recovery and applications of □anthan gum by *Xanthomonas campestris*. *J. Food Eng.*, *106*, 1–12.

Park, J. S., Han, T. H., Lee, K. Y., Han, S. S., Hwang, J. J., Moon, D. H., et al., (2006). *N*-acetyl histidine-conjugated glycol chitosan self-assembled nanoparticles for intracytoplasmic delivery of drugs: Endocytosis, exocytosis and drug release. *J. Control Release, 115,* 37–45.

Park, W., Park, S. J., & Na, K., (2010). Potential of self-organizing nanogel with acetylated chondroitin sulfate as an anti-cancer drug carrier. *Colloids Surf. B Biointerfaces*, *79*, 501–508.

Parveen, S., & Sahoo, S. K., (2011). Long circulating chitosan/PEG blended PLGA nanoparticle for tumor drug delivery. *Eur. J. Pharmacol.*, *670*, 372–383.

Parveen, S., Misra, R., & Sahoo, S. K., (2012). Nanoparticles: A boon to drug delivery, therapeutics, diagnostics and imaging. *Nanomedicine*, *8*, 147–166.

Passeleu-Le Bourdonnec, C., Carrupt, P. A., Scherrmann, J. M., & Martel, S., (2013). Methodologies to assess drug permeation through the blood–brain barrier for pharmaceutical research. *Pharm. Res.*, *30*, 2729–2756.

Pastorino, L., Erokhina, S., Ruggiero, C., Erokhin, V., & Petrini, P., (2015). Fabrication and characterization of chitosan and pectin nanostructured multilayers. *Macromol. Chem. Phys., 216,* 1067–1075.

Patel, S., & Goyal, A., (2015). Applications of natural polymer gum arabic: A review. *Int. J. Food Prop.*, *18*, 986–998.

Paulino, A. T., Guilherme, M. R., Mattoso, L. H. C., & Tambourgi, E. B., (2010). Smart hydrogels based on modified gum arabic as a potential device for magnetic biomaterial. *Macromol. Chem. Phys.*, *211*, 1196–1205.

Peng, B. L., Dhar, N., Liu, H. L., & Tam, K. C., (2011). Chemistry and applications of nano-crystalline cellulose and its derivatives: A nanotechnology perspective. *Can. J. Chem. Eng.*, *89*, 1191–1206.

Pérez, S., & Bertoft, E., (2010). The molecular structures of starch components and their contribution to the architecture of starch granules: A comprehensive review. *Starch–Starke*, *62*, 389–420.

Perlatti, B., de Souza Bergo, P. L., das Graças Fernandes da Silva, M. F., Fernandes, J. B., & Forim, M. R., (2013). Polymeric nanoparticle-based insecticides: A controlled release purpose for agrochemicals. In *Insecticides–Development of Safer and More Effective Technologies*; Trdan, S., (Ed.), Intech: Rijeka, 523–550.

Pinho, E., Grootveld, M., Soares, G., & Henriques, M., (2014). Cyclodextrin-based hydrogels toward improved wound dressings. *Crit. Rev. Biotechnol*, *34*, 328–337.

Pitarresi, G., Palumbo, F. S., Albanese, A., Fiorica, C., Picone, P., & Giammona, G., (2010). Self-assembled amphiphilic hyaluronic acid graft copolymers for targeted release of antitumoral drug. *J. Drug Target.*, *18*, 264–276.

Pooja, D., Panyaram, S., Kulhari, H., Rachamalla, S. S., & Sistla, R., (2014). Xanthan gum stabilized gold nanoparticles: Characterization, biocompatibility, stability and cytotoxicity. *Carbohydr. Polym.*, *110*, 1–9.

Prabaharan, M., & Mano, J. F., (2006). Chitosan derivatives bearing cyclodextrin cavities as novel adsorbent matrices. *Carbohydr. Polym.*, *63*, 153–166.

Prajapati, V. D., Jani, G. K., Zala, B. S., & Khutliwala, T. A., (2013). An insight into the emerging exopolysaccharide gellan gum as a novel polymer. *Carbohydr. Polym.*, *93*, 670–678.

Prestwich, G. D., (2011). Hyaluronic acid-based clinical biomaterials derived for cell and molecule delivery in regenerative medicine. *J. Control Release*, *155*, 193–199.

Prezotti, F. G., Cury, B. S. F., & Evangelista, R. C., (2014). Mucoadhesive beads of gellan gum/pectin intended to controlled delivery of drugs. *Carbohydr. Polym.*, *113*, 286–295.

Qiu, X. Y., & Hu, S. W., (2013). "Smart" materials based on cellulose: A review of the preparations, properties, and applications. *Materials*, *6*, 738–781.

Raafat, D., von Bargen, K., Haas, A., & Sahl, H. G., (2008). Insights into the mode of action of chitosan as an antibacterial compound. *Appl. Environ. Microbiol.*, *74*, 3764–3773.

Ramalho, M. J., Loureiro, J. A., Gomes, B., Frasco, M. F., Coelho, M. A. N., & Pereira, M. C., (2015). PLGA nanoparticles as a platform for vitamin D-based cancer therapy. *Beilstein J. Nanotechnol.*, *6*, 1306–1318.

Raman, M., Devi, V., & Doble, M., (2015). Biocompatible L-carrageenan-gamma-maghemite nanocomposite for biomedical applications – synthesis, characterization and *in vitro* anticancer efficacy. *J. Nanobiotechnol.*, *13*, 13, Article ID 18.

Ramteke, S., Ganesk, N., Battacharya, S., & Jain, N. K., (2009). Amoxicillin, clarithromycin, and omeprazole-based targeted nanoparticles for the treatment of *H. pylori*. *J. Drug Target.*, *17*, 225–234.

Rani, R., Dilbaghi, N., Dhingra, D., & Kumar, S., (2015). Optimization and evaluation of bioactive drug-loaded polymeric nanoparticles for drug delivery. *Int. J. Biol. Macromol.*, *78*, 173–179.

Rejinold, N. S., Nair, A., Sabitha, M., Chennazhi, K. P., Tamura, H., Nair, S. V., et al., (2012). Synthesis, characterization and *in vitro* cytocompatibility studies of chitin nanogels for biomedical applications. *Carbohydr. Polym.*, *87*, 943–949.

Rejinold, N. S., Biswas, R., Chellan, G., & Jayakumar, R., (2014). Multifaceted chitin/poly(lactic-co-glycolic) acid composite nanogels. *Int. J. Biol. Macromol.*, *67*, 279–288.

Rivadeneira, J., Di Virgilio, A. L., Audisio, M. C., Boccaccini, A. R., & Gorustovich, A. A., (2014). Evaluation of antibacterial and cytotoxic effects of nano-sized bioactive glass/collagen composites releasing tetracycline hydrochloride. *J. Appl. Microbiol.*, *116*, 1438–1446.

Roman, M., Dong, S. P., Hirani, A., & Lee, Y. W., (2009). Cellulose nanocrystals for drug delivery. In *Polysaccharide Materials: Performance by Design*; Edgar, K. J., Heinze, T., Buchanan, C. M., (Eds.), ACS Symposium Series (1017). American Chemical Society: Washington, DC, pp. 81–91.

Ruiz-Esparza, G. U., Wu, S. H., Segura-Ibarra, V., Cara, F. E., Evans, K. W., Milosevic, M., et al., (2014). Polymer nanoparticles encased in a cyclodextrin complex shell for potential site-and sequence-specific drug release. *Adv. Funct. Mater.*, *24*, 4753–4761.

Sabitha, M., Rejinold, N. S., Nair, A., Lakshmanan, V. K., Nair, S. V., & Jayakumar, R., (2013). Development and evaluation of 5-fluorouracil loaded chitin nanogels for treatment of skin cancer. *Carbohydr. Polym., 91*, 48–57.

Saikia, C., Hussain, A., Ramteke, A., Sharma, H. K., & Maji, T. K., (2014). Crosslinked thiolated starch coated Fe_3O_4 magnetic nanoparticles: Effect of montmorillonite and cross linking density on drug delivery properties. *Starch – Starke, 66*, 760–771.

Saikia, C., Hussain, A., Ramteke, A., Sharma, H. K., & Maji, T. K., (2015). Carboxymethyl starch-chitosan-coated iron oxide magnetic nanoparticles for controlled delivery of isoniazid. *J. Microencapsul., 32*, 29–39.

Saravanakumar, G., Choi, K. Y., Yoon, H. Y., Kim, K., Park, J. H., Kwon, I. C., et al., (2010). Hydrotropic hyaluronic acid conjugates: Synthesis, characterization, and implications as a carrier of paclitaxel. *Int. J. Pharm.*, *394*, 154–161.

Sarmah, M., Banik, N., Hussain, A., Ramteke, A., Sharma, H. K., & Maji, T. K., (2015). Study on crosslinked gelatin-montmorillonite nanoparticles for controlled drug delivery applications. *J. Mater. Sci.*, *50*, 7303–7313.

Scomparin, A., Salmaso, S., Bersani, S., Satchi-Fainaro, R., & Caliceti, P., (2011). Novel folated and non-folated pullulan bioconjugates for anticancer drug delivery. *Eur. J. Pharm. Sci., 42*, 547–558.

Seo, S., Lee, C. S., Jung, Y. S., & Na, K., (2012). Thermo-sensitivity and triggered drug release of polysaccharide nanogels derived from pullulan-g-poly(L-lactide) copolymers. *Carbohydr. Polym.*, *87*, 1105–1111.

Shi, X. W., Li, X. X., & Du, Y. M., (2011). Recent progress of chitin-based materials. *Acta Polym. Sin., 1*, 1–11.

Shi, X. N., Zheng, Y. D., Wang, G. J., Lin, Q. H., & Fan, J. S., (2014). pH-and electro-response characteristics of bacterial cellulose nanofiber/sodium alginate hybrid hydrogels for dual controlled drug delivery. *RSC Adv., 4*, 47056–47065.

Shukla, S. K., Mishra, A. K., Arotiba, O. A., & Mamba, B. B., (2013). Chitosan-based nanomaterials: A state-of-the-art review. *Int. J. Biol. Macromol., 59*, 46–58.

Siddiqui, I. A., Adhami, V. M., Christopher, J., Chamcheu, & Mukhtar, H., (2012). Impact of nanotechnology in cancer: emphasis on nanochemoprevention. *Int. J. Nanomed., 7*, 591–605.

Sievens-Figueroa, L., Bhakay, A., Jerez-Rozo, J. I., Pandya, N., Romanach, R. J., Michniak-Kohn, B., et al., (2012). Preparation and characterization of hydroxypropyl methyl cellulose films containing stable BCS Class II drug nanoparticles for pharmaceutical applications. *Int. J. Pharm.*, *423*, 496–508.

Silva, J. M., Videira, M., Gaspar, R., Preat, V., & Florindo, H. F.,(2013). Immune system targeting by biodegradable nanoparticles for cancer vaccines. *J. Control Release, 168*, 179–199.

Simoes, S. M. N., Veiga, F., Ribeiro, A. C. F., Figueiras, A. R., Taboada, P., Concheiro, A., et al., (2014). Supramolecular gels of poly-alpha-cyclodextrin and PEO-based copolymers for controlled drug release. *Eur. J. Pharm. Biopharm.*, *87*, 579–588.

Simoes, S. M. N., Rey-Rico, A., Concheiro, A., & Alvarez-Lorenzo, C., (2015). Supramolecular cyclodextrin-based drug nanocarriers. *Chem. Commun.*, *51*, 6275–6289.

Singh, R. S., Saini, G. K., & Kennedy, J. F.,(2008). Pullulan: Microbial sources, production and applications. *Carbohydr. Polym.*, *73*, 515–531.

Singh, A., Dinawaz, F., Mewar, S., Sharma, U., Jagannathan, N. R., & Sahoo, S. K.,(2011). Composite polymeric magnetic nanoparticles for co-delivery of hydrophobic and hydrophilic anticancer drugs and MRI imaging for cancer therapy. *Appl. Mater. Interfaces*, *3*, 842–856.

Singh, O. V., (2015). Bio-Nanoparticles: Biosynthesis and Sustainable Biotechnological Implications, Wiley-Blackwell: Hoboken.

Singh, R. S., Kaur, N., & Kennedy, J. F.(2015). Pullulan and pullulan derivatives as promising biomolecules for drug and gene targeting. *Carbohydr. Polym.*, *123*, 190–207.

Situ, W. B., Li, X. X., Liu, J., & Chen, L., (2015). Preparation and characterization of glycoprotein-resistant starch complex as a coating material for oral bioadhesive microparticles for colon-targeted polypeptide delivery. *J. Agric. Food Chem.*, *63*, 4138–4147.

Sivasankar, M., & Kumar, B. P., (2010). Role of nanoparticles in drug delivery system. *Int. J. Res. Pharm. Biol. Sci.*, *1*, 41–66.

Smitha, K. T., Anitha, A., Furuike, T., Tamura, H., Nair, S. V., & Jayakumar, R., (2013). *In vitro* evaluation of paclitaxel loaded amorphous chitin nanoparticles for colon cancer drug delivery. *Colloids Surf. B Biointerfaces*, *104*, 245–253.

Smitha, K. T., Nisha, N., Maya, S., Biswas, R., & Jayakumar, R., (2015). Delivery of rifampicin-chitin nanoparticles into the intracellular compartment of polymorphonuclear leukocytes. *Int. J. Biol. Macromol.*, *74*, 36–43.

Song, Y. B., Zhang, L. Z., Gan, W. P., Zhou, J. P., & Zhang, L. N.,(2011). Self-assembled micelles based on hydrophobically modified quaternized cellulose for drug delivery. *Colloids Surf. B Biointerfaces*, *83*, 313–320.

Song, J. K., Odekerken, J. C. E., Loewik, D. W. P. M., Lopez-Perez, P. M., Welting, T. J. M., Yang, F., et al., (2015). Influence of the molecular weight and charge of antibiotics on their release kinetics from gelatin nanospheres. *Macromol. Biosci.*, *15*, 901–911.

Songsurang, K., Siraleartmukul, K., & Muangsin, N., (2015). Mucoadhesive drug carrier based on functional-modified cellulose as poorly water-soluble drug delivery system. *J. Microencapsul.*, *32*, 450–459.

Sosnik, A., (2014). Alginate particles as platform for drug delivery by the oral route: State-of-the-art. *ISRN Pharm.*, 2014, 17, Article ID 926157.

Sotome, S., Uemura, T., Kikuchi, M., Chen, J., Itoh, S., Tanaka, J., et al., (2004). Synthesis and *in vivo* evaluation of a novel hydroxyapatite/collagen-alginate as a bone filler and a drug delivery carrier of bone morphogenetic protein. *Mater. Sci. Eng. C Biomim. Supramol. Syst.*, *24*, 341–347.

Straccia, M. C., d'Ayala, G. G., Romano, I., Oliva, A., Laurienzo, P., (2015). Alginate hydrogels coated with chitosan for wound dressing. *Mar. Drugs,*. 13, 2890–2908.

Su, K., & Wang, C. M., (2015). Recent advances in the use of gelatin in biomedical research. *Biotechnol. Lett.*, *37*, 2139–2145.

Su, Y. G., Hu, Y. W., Du, Y. Z., Huang, X., He, J. B., You, J., et al., (2015). Redox-responsive polymer drug conjugates based on doxorubicin and chitosan oligosaccharide-g-stearic acid for cancer therapy. *Mol. Pharm.*, *12*, 1193–1202.

Subudhi, M. B., Jain, A., Hurkat, P., Shilpi, S., Gulbake, A., & Jain, S. K., (2015). Eudragit S100 coated citrus pectin nanoparticles for colon targeting of 5-fluorouracil. *Materials, 8,* 832–849.

Suffredini, G., & Levy, L. M., (2014). Nanopolymers and nanoconjugates for central nervous system diagnostics and therapies. In *The Textbook of Nanoneuroscience and Nanoneurosurgery*; Kateb, B., Heiss, J. D., Eds.; CRC Press, Taylor and Francis Group: Boca Raton, pp. 39–50.

Suh, H. R., Jeong, B. M., Rathi, R., & Kim, S. W., (1998). Regulation of smooth muscle cell proliferation using paclitaxel-loaded poly(ethylene oxide)-poly(lactide/glycolide) nanospheres. *J. Biomed. Mater. Res., 42,* 331–338.

Sun, L., Zhou, D. S., Zhang, P., Li, Q. H., & Liu, P., (2015). Gemcitabine and gamma-cyclodextrin/docetaxel inclusion complex-loaded liposome for highly effective combinational therapy of osteosarcoma. *Int. J. Pharm., 478,* 308–317.

Swami, A., Reagan, M. R., Basto, P., Mishima, Y., Kamaly, N., Glavey, S., et al., (2014). Engineered nanomedicine for myeloma and bone microenvironment targeting. *Proc. Natl. Acad. Sci. USA. 111,* 10287–10292.

Swaminathan, S., Pastero, L., Serpe, L., Trotta, F., Vavia, P., & Aquilano, D., (2010). Cyclodextrin-based nanosponges encapsulating camptothecin: Physicochemical characterization, stability and cytotoxicity. *Eur. J. Pharm., 74,* 193–201.

Swatschek, D., Schatton, W., Muller, E. E. G., & Kreuter, J., (2002). Microparticles derived from marine sponge collagen (SCMPs): preparation, characterization and suitability for dermal delivery of all-*trans* retinol. *Eur. J. Pharm. Biopharm., 54,* 125–133.

Ta, H. T., Dass, C. R., & Dunstan, D. E., (2008). Injectable chitosan hydrogels for localised cancer therapy. *J. Control Release, 126,* 205–216.

Tahara, K., Kato, Y., Yamamoto, H., Kreuter, J., & Kawashima, Y., (2011). Intracellular drug delivery using polysorbate 80-modified poly(D,L-lactide-co-glycolide) nanospheres to glioblastoma cells. *J. Microcaps., 28,* 29–36.

Tan, W., Twomey, J., Guo, D., Madhavan, K., & Li, M., (2010). Evaluation of nanostructural, mechanical, and biological properties of collagen-nanotube composites. *IEEE Transact. Nanobiosci., 9,* 111–120.

Tan, S., Ladewig, K., Fu, Q., Blencowe, A., & Qiao, G. G., (2014). Cyclodextrin-based supramolecular assemblies and hydrogels: Recent advances and future perspectives. *Macromol. Rapid Commun., 35,* 1166–1184.

Tejashri, G., Amrita, B., & Darshana, J., (2013). Cyclodextrin-based nanosponges for pharmaceutical use: A review. *Acta Pharm., 63,* 335–358.

Thakur, V. K., & Thakur, M. K., (2015). Handbook of Polymers for Pharmaceutical Technologies – Biodegradable Polymers. Scrivener Publishing and J. Wiley & Sons: Hoboken, pp. *33*–60, 105–126, 275–298.

Tonnesen, H. H., & Karlsen, J., (2002). Alginate in drug delivery systems. *Drug Dev. Ind. Pharm., 28,* 621–630.

Toomari, Y., Namazi, H., & Entezami, A. A., (2015). Fabrication of biodendrimeric beta-cyclodextrin via click reaction with potency of anticancer drug delivery agent. *Int. J. Biol. Macromol., 79,* 883–893.

Toti, U. S., Guru, B. R., Hali, M., McPharlin, C. M., Wykes, S. M., Panyam, J., et al., (2011). Targeted delivery of antibiotics to intracellular chlamydial infections using PLGA nanoparticles. *Biomaterials, 32,* 6606–6613.

Trotta, F., Zanetti, M., & Cavalli, R., (2012). Cyclodextrin-based nanosponges as drug carriers. *Beilstein, J. Org. Chem., 8,* 2091–2099.

Trotta, F., Dianzani, C., Caldera, F., Mognetti, B., & Cavalli, R., (2014). The application of nanosponges to cancer drug delivery. *Expert Opin. Drug Deliv., 11,* 931–941.

Tseng, Y. Y., Wang, Y. C., Su, C. H., Yang, T. C., Chang, T. M., Kau, Y. C., et al., (2015). Concurrent delivery of carmustine, irinotecan, and cisplatin to the cerebral cavity using biodegradable nanofibers: *In vitro* and *in vivo* studies. *Colloids Surf. B Biointerfaces, 134,* 254–261.

Tummalapalli, M., Deopura, B. L., Alam, M. S., & Gupta, B., (2015). Facile and green synthesis of silver nanoparticles using oxidized pectin. *Mater. Sci. Eng. C Mater. Biol. Appl., 50,* 31–36.

Tungprapa, S., Jangchud, I., & Supaphol, P., (2007). Release characteristics of four model drugs from drug-loaded electrospun cellulose acetate fiber mats. *Polymer., 48,* 5030–5041.

Uhrich, K. E., Cannizzaro, S. M., Langer, R. S., & Shakesheff, K. M., (1999). Polymeric systems for controlled drug release. *Chem Rev., 99,* 3181–3198.

Valo, H., Kovalainen, M., Laaksonen, P., Hakkinen, M., Auriola, S., Peltonen, L., et al., (2011). Immobilization of protein-coated drug nanoparticles in nanofibrillar cellulose matrices-Enhanced stability and release. *J. Control Release, 156,* 390–397.

Valo, H., Arola, S., Laaksonen, P., Torkkeli, M., Peltonen, L., Linder, M. B., et al., (2013). Drug release from nanoparticles embedded in four different nanofibrillar cellulose aerogels. *Eur. J. Pharm. Sci., 50,* 69–77.

Vieira, S., Vial, S., Maia, F. R., Carvalho, M., Reis, R. L., Granja, P. L., et al., (2015). Gellan gum-coated gold nanorods: an intracellular nanosystem for bone tissue engineering. *RSC Adv., 5,* 77996–78005.

Vlad, M., Andronescu, E., Grumezescu, A. M., Ficai, A., Voicu, G., Bleotu, C., et al., (2014). Carboxymethyl-cellulose/Fe_3O_4 nanostructures for antimicrobial substances delivery. *Biomed. Mater. Eng., 24,* 1639–1646.

Wallace, D. G., & Rosenblatt, J., (2003). Collagen gel systems for sustained delivery and tissue engineering. *Adv. Drug Deliv., 55,* 1631–1649.

Wang, T. W., Zhang, C. L., Liang, X. J., Liang, W., & Wu, Y., (2011a). Hydroxypropyl-beta-cyclodextrin copolymers and their nanoparticles as doxorubicin delivery system. *J. Pharm. Sci., 100,* 1067–1079.

Wang, J. J., Zeng, Z. W., Xiao, R. Z., Xie, T. A., Zhou, G. L., & Zhan, X. R., (2011b). Recent advances of chitosan nanoparticles as drug carriers. *Int. J. Nanomed., 6,* 765–774.

Wang, C. Y., Mallela, J., Garapati, U. S., Ravi, S., Chinnasamy, V., Girard, Y., et al., (2013). A chitosan-modified graphene nanogel for noninvasive controlled drug release. *Nanomed. Nanotechnol. Biol. Med., 9,* 903–911.

Wang, H. X., Zhao, Y. Z., Wang, H. Y., Gong, J. B., He, H. N., Shin, M. C., et al., (2014a). Low-molecular-weight protamine-modified PLGA nanoparticles for overcoming drug-resistant breast cancer. *J. Control Release, 192,* 47–56.

Wang, Y. F., Zhou, J. H., Qiu, L. H., Wang, X. R., Chen, L. L., Liu, T., & Di, W., (2014b). Cisplatin-alginate conjugate liposomes for targeted delivery to EGFR-positive ovarian cancer cells. *Biomaterials, 35,* 4297–4309.

Wang, J. T., Wang, M., Zheng, M. M., Guo, Q., Wang, Y. F., Wang, H. Q., et al., (2015a). Folate mediated self-assembled phytosterol-alginate nanoparticles for targeted intracellular anticancer drug delivery. *Colloids Surf. B Biointerfaces, 129,* 63–70.

Wang, H., Agarwal, P., Zhao, S. T., Xu, R. X., Yu, J. H., Lu, X. B., et al., (2015b). Hyaluronic acid-decorated dual responsive nanoparticles of Pluronic® F127, PLGA, and chitosan for targeted co-delivery of doxorubicin and irinotecan to eliminate cancer stem-like cells. *Biomaterials, 72,* 74–89.

Watanabe, K., Nishio, Y., Makiura, R., Nakahira, A., & Kojima, C., (2013). Paclitaxel-loaded hydroxyapatite/collagen hybrid gels as drug delivery systems for metastatic cancer cells. *Int. J. Pharm., 446*, 81–86.

Wei, H., & Yu, C. Y., (2015). Cyclodextrin-functionalized polymers as drug carriers for cancer therapy. *Biomater. Sci., 3*, 1050–1060.

Wu, J. L., Wang, C. Q., Zhuo, R. X., & Cheng, S. X., (2014). Multi-drug delivery system based on alginate/calcium carbonate hybrid nanoparticles for combination chemotherapy. *Colloids Surf. B Biointerfaces, 123*, 498–505.

Wu, J. W., Song, X. F., Sun, H. W., Zhang, Y. C., Gu, X. L., Dong, P. X. et al., (2015). Chitosan nano carriers loading anti-tumor drugs. *J. Nano Res., 32*, 113–127.

Xin, D. C., Wang, Y., & Xiang, J. N., (2010). The use of amino acid linkers in the conjugation of paclitaxel with hyaluronic acid as drug delivery system: Synthesis, self-assembled property, drug release, and *in vitro* efficiency. *Pharm. Res., 27*, 380–389.

Xu, S., Olenyuk, B. Z., Okamoto, C. T., & Hamm-Alvarez, S. F., (2013). Targeting receptor-mediated endocytotic pathways with nanoparticles: Rationale and advances. *Adv. Drug Deliv. Rev., 65*, 121–138.

Xu, W., Jin, W. P., Li, Z. S., Liang, H. S., Wang, Y. T., Shah, B. R., et al., (2015). Synthesis and characterization of nanoparticles based on negatively charged xanthan gum and lysozyme. *Food Res. Int., 71*, 83–90.

Yadav, A. K., Mishra, P., Mishra, A. K., Mishra, P., Jain, S., & Agrawal, G. P., (2007). Development and characterization of hyaluronic acid-anchored PLGA nanoparticulate carriers of doxorubicin. *Nanomed. Nanotechnol. Biol. Med., 3*, 246–257.

Yadav, A. K., Mishra, P., & Agrawal, G. P., (2008). An insight on hyaluronic acid in drug targeting and drug delivery. *J. Drug Target., 16*, 91–107.

Yadav, A. K., Agarwal, A., Rai, G., Mishra, P., Jain, S., Mishra, A. K., Agrawal, H., et al., (2010). Development and characterization of hyaluronic acid decorated PLGA nanoparticles for delivery of 5-fluorouracil. *Drug Deliv., 17*, 561–572.

Yadollahi, M., Gholamali, I., Namazi, H., & Aghazadeh, M., (2015a). Synthesis and characterization of antibacterial carboxymethyl cellulose/ZnO nanocomposite hydrogels. *Int. J. Biol. Macromol., 74*, 136–141.

Yadollahi, M., Namazi, H., & Aghazadeh, M., (2015b). Antibacterial carboxymethyl cellulose/Ag nanocomposite hydrogels cross-linked with layered double hydroxides. *Int. J. Biol. Macromol., 79*, 269–277.

Yallapu, M. M., Dobberpuhl, M. R., Maher, D. M., Jaggi, M., & Chauhan, S. C., (2012). Design of curcumin loaded cellulose nanoparticles for prostate cancer. *Curr. Drug Metab., 13*, 120–128.

Yallapu, M. M., Khan, S., Maher, D. M., Ebeling, M. C., Sundram, V., Chauhan, N., et al., (2014). Anti-cancer activity of curcumin loaded nanoparticles in prostate cancer. *Biomaterials, 35*, 8635–8648.

Yamashita, F., & Hashida, M., (2013). Pharmacokinetic considerations for targeted drug delivery. Adv. *Drug Deliv. Rev., 65*, 139–147.

Yameogo, J. B. G., Geze, A., Choisnard, L., Putaux, J. L., Semde, R., & Wouessidjewe, D., (2014). Progress in developing amphiphilic cyclodextrin-based nanodevices for drug delivery. *Curr. Topics Med. Chem., 14*, 526–541.

Yang, J. S., Xie, Y. J., & He, W., (2011). Research progress on chemical modification of alginate: A review. *Carbohydr. Polym., 84*, 33–39.

Yang, L. Q., Zhang, B. F., Yi, J. Z., Liang, J. B., Liu, Y. L., & Zhang, L. M., (2013). Preparation, characterization, and properties of amylose-ibuprofen inclusion complexes. *Starch – Stärke. 65*, 593–602.

Yang, J. L., Gao, C. M., Lu, S. Y., Wang, X. G., Chen, M. J., & Liu, M. Z., (2014). Novel self-assembled amphiphilic mPEGylated starch-deoxycholic acid polymeric micelles with pH-response for anticancer drug delivery. *RSC Adv. 4*, 55139–55149.

Yang, Q., Parker, C. L., McCallen, J. D., & Lai, S. K., (2015). Addressing challenges of heterogeneous tumor treatment through bispecific protein-mediated pretargeted drug delivery. *J. Control Release, 220*, 715–726.

Yang, H. H., Bremner, D. H., Tao, L., Li, H. Y., Hu, J. A., & Zhu, L. M., (2016). Carboxy-methyl chitosan-mediated synthesis of hyaluronic acid-targeted graphene oxide for cancer drug delivery. *Carbohydr. Polym., 135*, 72–78.

Yoo, H. S., Lee, K. H., Oh, J. E., & Park, T. G., (2000). *In vitro* and *in vivo* anti-tumor activities of nanoparticles based on doxorubicin-PLGA conjugates. *J. Control Release, 68*, 419–431.

Yu, M. K., Park, J., & Jon, S., (2012). Targeting strategies for multifunctional nanoparticles in cancer imaging and therapy. *Theranostics., 2*, 3–44.

Yu, D. G., Wang, X., Li, X. Y., Chian, W., Li, Y., & Liao, Y. Z., (2013). Electrospun biphasic drug release polyvinylpyrrolidone/ethyl cellulose core/sheath nanofibers. *Acta Biomater., 9*, 5665–5672.

Yuen, S., (1974). Pullulan and its applications. *Process Biochem., 9*, 7–9.

Zahoor, A., Sharma, S., & Khuller, G. K., (2005). Inhalable alginate nanoparticles as anti-tubercular drug carriers against experimental tuberculosis. *Int. J. Antimicrob. Agents., 26*, 298–303.

Zeng, X. W., Tao, W., Mei, L., Huang, L. G., Tan, C. Y., & Feng, S. S., (2013). Cholic acid-functionalized nanoparticles of star-shaped PLGA-vitamin E TPGS copolymer for docetaxel delivery to cervical cancer. *Biomaterials, 34*, 6058–6067.

Zhang, C., Qu, G., Sun, Y., Wu, X., Yao, Z., Guo, Q., et al., (2008). Pharmacokinetics, biodistribution, efficacy and safety of *N*-octyl-*O*-sulfate chitosan micelles loaded with paclitaxel. *Biomaterials, 29*, 1233–1241.

Zhang, L., Yu, F. Q., Cole, A. J., Chertok, B., David, A. E., Wang, J., et al., (2009). Gum Arabic-coated magnetic nanoparticles for potential application in simultaneous magnetic targeting and tumor imaging. *AAPS J., 11*, 693–699.

Zhang, N., Li, J. H., Jiang, W. F., Ren, C. H., Li, J. S., Xin, J. Y., & Li, K., (2010a). Effective protection and controlled release of insulin by cationic beta-cyclodextrin polymers from alginate/chitosan nanoparticles. *Int. J. Pharm., 393*, 212–218.

Zhang, J., Wang, Q., & Wang, A., (2010b). In situ generation of sodium alginate/hydroxy-apatite nanocomposite beads as drug-controlled release matrices. *Acta Biomater., 6*, 445–454.

Zhang, H. H., Cai, Z. S., Sun, Y., Yu, F., Chen, Y. Q., & Sun, B. W., (2012). Folate-conjugated beta-cyclodextrin from click chemistry strategy and for tumor-targeted drug delivery. *J. Biomed. Mater. Res., 100A*, 2441–2449.

Zhang, J. X., & Ma, P. X., (2013). Cyclodextrin-based supramolecular systems for drug delivery: Recent progress and future perspective. *Adv. Drug Deliv. Rev., 65*, 1215–1233.

Zhang, Y., Chan, H. F., & Leong, K. W., (2013a). Advanced materials and processing for drug delivery: The past and the future. *Adv. Drug Deliv. Rev., 65*, 104–120.

Zhang, Z., Shan, H. L., Chen, L., He, C. L., Zhuang, X. L., & Chen, X. S., (2013b). Synthesis of pH-responsive starch nanoparticles grafted poly(L-glutamic acid) for insulin controlled release. *Eur. Polym. J., 49*, 2082–2091.

Zhang, X. D., Dong, Y. C., Zeng, X. W., Liang, X., Li, X. M., Tao, W., et al., (2014a). The effect of autophagy inhibitors on drug delivery using biodegradable polymer nanoparticles in cancer treatment. *Biomaterials, 35*, 1932–1943.

Zhang, K. R., Tang, X., Zhang, J., Lu, W., Lin, X., Zhang, Y., et al., (2014b). PEG-PLGA copolymers: Their structure and structure-influenced drug delivery applications. *J. Control Release, 183*, 77–86.

Zhang, L. N., Zhang, Q., Wang, X., Zhang, W. J., Lin, C. C., Chen, F., et al., (2015a). Drug-in-cyclodextrin-in-liposomes: A novel drug delivery system for flurbiprofen. *Int. J. Pharm., 492*, 40–45.

Zhang, T. T., Zhou, P. H., Zhan, Y. F., Shi, X. W., Lin, J. Y., Du, Y. M., et al., (2015b). Pectin/lysozyme bilayers layer-by-layer deposited cellulose nanofibrous mats for antibacterial application. *Carbohydr. Polym., 117*, 687–693.

Zhang, Y. X., Chen, T., Yuan, P., Tian, R., Hu, W. J., Tang, Y. L., et al., (2015c). Encapsulation of honokiol into self-assembled pectin nanoparticles for drug delivery to HepG2 cells. *Carbohydr. Polym. 133*, 31–38.

Zhao, M., Liu, L., Her, R., Kalantar, T., Schmidt, D., Mathieson, T., et al., (2012). Nano-sized delivery for agricultural chemicals. In *NanoFormulation*; Tiddy, G., Tan, R., Eds.; Royal Society of Chemistry: Cambridge, pp. 256–265.

Zhao, X. B., Liu, L., Li, X. R., Zeng, J., Jia, X., & Liu, P., (2014). Biocompatible graphene oxide nanoparticle-based drug delivery platform for tumor microenvironment-responsive triggered release of doxorubicin. *Langmuir, 30*, 10419–10429.

Zhao, J. Q., Lu, C. H., He, X., Zhang, X. F., Zhang, W., & Zhang, X. M., (2015a). Polyethylenimine-grafted cellulose nanofibril aerogels as versatile vehicles for drug delivery. *ACS Appl. Mater. Interfaces., 7*, 2607–2615.

Zhao, L. L., Liu, M. R., Wang, J., & Zhai, G. X., (2015b). Chondroitin sulfate-based nano-carriers for drug/gene delivery. *Carbohydr. Polym., 133*, 391–399.

Zhou, H. F., Yu, W. T., Guo, X., Liu, X. D., Li, N., Zhang, Y., et al., (2010). Synthesis and characterization of amphiphilic glycidol-chitosan-deoxycholic acid nanoparticles as a drug carrier for doxorubicin. *Biomacromolecules, 11*, 3480–3486.

Zhu, Y., Yao, X. M., Chen, X. F., & Chen, L., (2015). pH-sensitive hydroxyethyl starch-doxorubicin conjugates as antitumor prodrugs with enhanced anticancer efficacy. *J. Appl. Polym. Sci., 132*, 427–478.

Zia, K. M., Zia, F., Zuber, M., Rehman, S., & Ahmad, M. N., (2015). Alginate-based polyurethanes: A review of recent advances and perspective. *Int. J. Biol. Macromol., 79*, 377–387.

CHAPTER 6

TARGETED DELIVERY SYSTEMS BASED ON POLYMERIC NANOPARTICLES FOR BIOMEDICAL APPLICATIONS

SA YANG,[1,2] CAN HUANG,[1] ZHI-PING LI,[1] QIAN NING,[1] WEN HUANG,[3] WEN-QIN WANG,[4] and CUI-YUN YU[1,3]

[1]*Hunan Province Cooperative Innovation Center for Molecular Target New Drug Study, University of South China, Hengyang, 421001, China, E-mail: yucuiyunusc@hotmail.com*

[2]*Changsha Medical University Affiliated People's Hospital of Xiangxiang City, Hunan Province, 411400, China*

[3]*Learning Key Laboratory for Pharmacoproteomics of Hunan Province, Institute of Pharmacy and Pharmacology, University of South China, Hengyang, 421001, China*

[4]*Hunan Province Cooperative Innovation Center for Molecular Target New Drug Study University of South China, Hengyang, 421001, China*

CONTENTS

ABSTRACT

Cancer is becoming one of the most devastating malignant diseases threatening human health and the most terrible cause of death worldwide. For cancer, either conventional chemotherapy therapies or gene therapies are associated with poor therapy effect such as low specificity and non-selective biodistribution, not mention to the combination of both. The polymeric nanobiomaterial with the unique biological properties targeted delivery antineoplastic drugs or genes can overcome these problems. Co-delivery antineoplastic drugs and genes have been widely explored for enhancement of their therapeutic efficacy in treating cancers. Besides, stimuli-responsive polymeric nanobiomaterial can enhance the bioavailability of drugs and genes at the disease site thus further improve the effect of treatment. In this paper, we reviewed the current status of polymeric nanobiomaterial to delivery different antineoplastic drugs or genes for cancer via passive targeting strategies and active targeting strategies. These current advances show a bright future to the development of cancer therapy against combination of chemotherapy therapies and gene therapies.

6.1   INTRODUCTION

Malignant cancers are the worldwide concerns since they might reduce life quality and even lead to death, and the incidence of cancer increase with age (Cho et al., 2008; Jemal et al., 2011). According to the statistical analysis of American Cancer Society in National Cancer Institute of the US, it accounting for about 580,350 deaths, almost 1,600 people per day in 2013 (Masis et al., 2013). At present, surgical intervention is the most effective method to treat patients with early-stage cancer (Bupathi et al., 2015; Maluccio et al., 2012). However, diagnosed patients with cancer usually at advanced stages and do not benefit from surgical resection, chemotherapy or gene therapy become the main treatment (Cheng et al., 2012; Lacaze et al., 2015;

Magalhães et al., 2014). Conventional chemotherapy are associated with poor survival rates owing to the lacks selectivity for cancerous cells and causes damage to rapidly proliferating normal cells, leads to high intrinsic toxicity and low survival profile of patients (Bansal et al., 2014). Low transfection efficiency and nonselective delivery is the obstacle to gene therapy. Therefore, an effective therapeutic strategy that are effective, specific and safe should be developed to increase the chemotherapeutic efficacy in cancers (Ding et al., 2012; Liu et al., 2014).

In line with most solid tumors, tumor possess significantly different pathophysiological feature compared to the normal tissue. That is to say, high permeability of the tumor vasculature allows nanoparticles or macromolecules to enter the tumor interstitial space, while the compromised lymphatic filtration allows them to stay there. With time the tumor concentration of nanoparticles or macromolecules will build up reaching several folds higher than that of the plasma, produce enhanced permeation and retention (EPR) effects (Figure 6.1) (Maeda et al., 2000, 2001; Torchilin et al., 2011). Therefore, macromolecular drugs and drug carriers such as nanoparticles, micelles, liposomes, etc., can penetrate tumor cells into tumor tissue, and accumulated in tumor tissue, which give nanoparticles passive targeting properties (Maeda, 2015; Greish et al., 2007). Recent years, nanoparticles-based delivery systems have received extensive attention in cancer therapy (Kojima et al., 2015; Yu et al., 2014, 2015). Benefit from the unique biological properties of nanoparticles, such as small size and high surface area, nanoparticles could provide tiny vehicles to efficiently deliver various

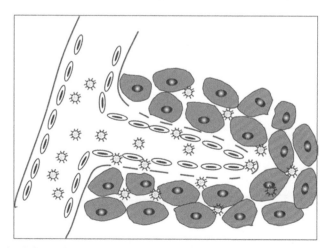

FIGURE 6.1 Schematic representation of the mechanism of passive targeting.

compounds including small molecule drugs, DNA, RNA, etc. (Yan et al., 2015). In addition, nanoparticles possess other excellent properties such as offer controlled drug release, mask physicochemical properties, reduce drug toxicity, improve bioavailability and enhance therapeutic efficacy and bio-distribution (Leyva-Gómez et al., 2015; Zhao et al., 2013;).

Besides the passive effect of nanoparticles-based on EPR effect, there have another important targeted approach that is active target approach. By modified on their surfaces with targeting ligands which have high affinity with receptors over expressed on tumor cells, such as antibodies or other molecules in order to improve the pharmacokinetic profile of a compound and selectivity for cancer tissues (Bergs et al., 2015; Branco et al., 2009). Through receptor-mediated endocytosis by malignant cells, nanoparticles could enter tumor tissue. As shown in Figure 6.2, with targeting ligands modified of surface, nanoparticles could be internalized by malignant cells through receptor-mediated endocytosis, leading to an intracellular drug release and a greater cytotoxic activity. And then the nanoparticles can be targeted to the tumor tissue passively due to the EPR effective.

Nanotechnology has been attending much attention since 1980s and has been adapted into many engineering fields such as electronics, mechanical, biomedical and space engineering. In particular, nanotechnology has

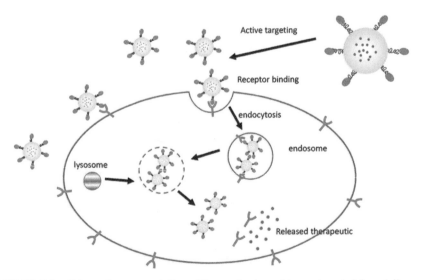

FIGURE 6.2   Schematic representation of the mechanism of tumor targeted drug delivery for treating solid tumor.

led to the significant progress in a biomedical field such as drug/gene delivery. Among nanotechnology strategies, polymeric nanoparticles have attracted most of the attention due to their good biocompatibility, nontoxic, biodegradability and ability to prolong drug release (Prabhu et al., 2015; Makino et al., 2014). Polymeric nanoparticles are frequently used as carriers to increase the therapeutic value of many drugs that are insoluble in water and bioactive molecules by improving their bioavailability, solubility, stability and control release (Seabra et al., 2015; Fonte et al., 2015).

In this chapter, we briefly introduced some polymeric nanoparticles targeted to tumor for delivery antineoplastic drug, gene, and co-delivery drugs and genes.

6.2 RECEPTOR-MEDIATED TARGETED TREATMENT OF TUMOR BY NANOSCALE-SIZED PLATFORMS

The therapeutic potential of most antineoplastic drugs are restricted because of rapid drug catabolism, non-specific, low bioavailability, poor solubility, multi drug resistance and serious side effects (Misra et al., 2014; Zhang et al., 2012; Ashwanikumar et al., 2014; Araki et al., 2012; Marelli et al., 2013). Targeted delivery systems of nanobiomaterials are currently in the process of developing nanoscale-sized platforms that surface modificated with active targeting ligands to achieve an optimum localization of drug into the tumor and to decrease systemic exposure (Jain et al., 2010). These modified nanoparticles could be internalized by malignant cells through receptor-mediated endocytosis, leading to an intracellular drug release and a greater cytotoxic activity.

6.2.1 SOMATOSTATIN RECEPTOR

Somatostatin receptors (SSTRs) are members of the superfamily of G-protein-coupled receptors (GPCR), which are widely distributed in a variety of tumors and cancer cell lines, including small cell lung cancer, neuroendocrine tumors, prostate cancer, breast cancer, colorectal carcinoma, gastric cancer and hepatocellular carcinoma. SSTRs have high affinity with somatostatin and somatostatin analogues (Schonbrunn et al., 1978; Stengel et al., 2013). This has led to two clinical applications. First, somatostatin analogues, such as octreotide or lanreotide, can functionally inhibit the excess hormone release and the further growth of these tumors. Second, radiolabeled somatostatin analogues can be used for diagnostic imaging of the tumors and targeted radiotherapy (Oberg et al., 2010; Waser et al., 2012). To date five SSTR subtypes have been cloned (SSTR1-SSTR5). The receptor genes are intron-less except for a SSTR2 intron that produces 2 spliced variants, SSTR2a and SSTR2b. Many tumor cells most frequently express SSTR2 at a high concentration, some also express SSTR1, 3, and 5, but fewer express SSTR4 (Sun et al., 2011; Froidevaux et al., 2002). In some reports, SSTR2 is upregulated in proliferative endothelial cells (ECs) lining the internal wall of tumor blood vessels, but undetectable in quiescent ECs (Reubi et al., 1996; Watson et al., 2001). Due to the aberrant expression in tumors and tumoral blood vessels, these SSTRs, specifically SSTR2, have been considered for ligand-based targeted therapy.

6.2.2   ASIALOGLYCOPROTEIN RECEPTOR

Asialoglycoprotein receptor (ASGPR) known as "The Ashwell–Morell Receptor" was the first cellular mammalian lectin discovered (Grewal et al., 2010). ASGPR is a C-type lectin, primary expressed on the sinusoidal surface of the hepatocyte (Roggenbuck et al., 2012). By regulating endocytosis and lysosomal degradation of glycoproteins, ASGPR plays critical roles in maintaining serum glycoprotein homeostasis (Shi et al., 2013). Since ASGPR is exclusively expressed by parenchymal hepatocytes and minimally expressed on extrahepatic, it could provide attractive advantage for hepatocyte-mediated delivery. ASGPR, also known as galactose receptor or hepatic lectin, exhibits high affinity for D-galactose (Gal) or Nacetylgalactosamine (GalNAc) (D'Souza et al., 2015). For example, pectin is polymer with galactose residues, which as the natural targeting ligands specific recognized ASGPR. Our lab developed to fabricate natural pectin-based nanoparticles as drug delivery systems to delivery 5-FU, in vitro cytotoxicity and cellular uptake studies showed higher cytotoxicity than free 5-FU on HepG2, pharmacokinetics evaluation in SD rats and biodistribution in healthy mice confirmed that 5-FU-NPs possessed a much longer half-life than free 5-FU in the circulation system (Yu et al., 2014).

6.2.3   VASCULAR ENDOTHELIAL GROWTH FACTOR RECEPTOR

In the 1960s, Folkman proposed that growth of tumor was angiogenesis dependent, and confirmed the close relationship between angiogenesis and tumor metastasis so as to death through experiment (Folkman et al., 1990). Vascular endothelial growth factor (VEGF) is one of the strongest and the highest specificity currently known vascular growth factors that regulate angiogenesis, is a potent angiogenic polypeptide and implicated as an essential for growth of vascular endothelial cells (Itakura et al., Zhang et al., 2015; Hoeben et al.). VEGF is a 40–45 kDadimeric glycoprotein containing a cysteine knot motif (Lohela et al., 2009). Current research exhibiting, VEGFR family has five members: VEGF-A, VEGF-B, VEGF-c, VEGF-D, and placental growth factor (Holmes et al., 2007). There are three receptors for VEGF: VEGFR-1, VEGFR-2 and VEGFR-3, belonging to the tyrosine kinase receptor superfamily (RTKs), mainly distributed on the surface of vascular endothelial cells (Grünwald et al., 2013; Man et al., 2006). Many types of nanoparticles modified by VEGFRs have been developed to target the tumor.

6.2.4   FOLATE RECEPTORS

Folate receptor (FR) is a secreted protein that either anchors to membrane via a glycosyl-phosphatidylinositol linkage or exists as a soluble form (Yan et al., 2015). The human FR e□istsin three isoforms, α and β equipped with GPI anchors and γ not possessing a GPI anchor (Mehra et al., 2013). It has been reported that FRs is over expressed on the surface of several cancers, including breast, kidney, lung, brain, and ovary cancers (Bahrami et al., 2015). FRs can bind folate and facilitate the transfer of folate-targeted NPs through receptor mediated endocytosis (Jin et al., 2014). Folate is an appropriate choice for targeting tumor cells in NP-based cancer therapies due to its small size, non-toxic, non-immunogenic nature, price cheapness, and easy conjugation to nanocarriers (Mansoori et al., 2010). Some research supported it. Hao et al., have prepared FA-conjugated DOX-loaded BSA-dextran NPs for cancer drug delivery. It should be noted that the dextran shell makes the NPs more dispersible in solution. They showed that these NPs allow to the administration of the higher doses of DOX and exert remarkable anti-tumor activity in murine ascites hepatoma H22 tumor-bearing mice (Shen et al., 2011).

6.2.5 EPIDERMAL GROWTH FACTOR RECEPTOR

Epidermal growth factor receptor (EGFR) has also captured much attention in tumor targeting. EGFR is a member of the protein kinase superfamily and a transmembrane glycoprotein receptor for epidermal growth factor (EGF). These are expressed on various normal cells at 1×104 per cell as well as over expressed on variety of tumor cells at 10–1000 fold more than normal cells (Heppner et al., 2015; LeMaistre et al., 1994). Ligands that can bind EGFR include epidermal growth factor (EGF), transforming growth factor, amphiregulin, epiregulin, betacellulin, heparin-binding EGF and epigen.

6.2.6 TRANSFERRIN RECEPTOR

Transferrin receptor (TfR) is ubiquitously expressed at low levels in most normal human tissues serving as the main route of entry for iron into cells, which plays an important role in iron regulation and cell growth (Voth et al., 2015; Kolhatkar et al., 2011). There are two TfRs in mammals, TfR1 and TfR2, both TfRs are homodimeric transmembrane glycoproteins that are specific for iron-loaded transferrin (Reyes-López et al., 2015). The combination of the iron-bound transferrin with the TfR initiates a receptor mediated endocytosis by which the iron-bound transferrin-receptor complex gets internalized (Kolhatkar et al., 2011). Transferrin is over expressed in various cancer types including liver cancer and has therefore been a target for alternative treatments (Tseng et al., 2009). TfR is another well-studied ligand catched much attention for the live cancer targeting, several studies reported that TfR ligand-conjugated nanoparticles had showed promising prospects for directly delivery antineoplastic drugs to tumor.

6.3 POLYMERIC NANOPARTICLES FOR TARGETED DELIVERY ANTICANCER THERAPEUTICS

Polymers are innovative nanobiomaterials to be engineered for the delivery system of drugs and genes (Kamaly et al., 2012). And a number of biodegradable polymers have made their impact on drug delivery, such as the polysaccharides, protein, polypeptide and aliphatic polyesters.

6.3.1 POLYSACCHARIDE NANOPARTICLES

Polysaccharides are long carbohydrate molecules of repeated monosaccharide units joined together by glycosidic bonds. Owing to the presence of diverse functional groups, polysaccharides have been readily modified for biomedical applications such as drug delivery systems. Most natural polysaccharides have hydrophilic groups such as hydroxyl, carboxyl and amino groups, which also affect the polymer charges and could form noncovalent bioadhesion or react with functional molecules. Thus therapeutic can be covalently conjugated at the backbone of polysaccharides or physically encapsulated into the polysaccharide-based nanoparticles (Han et al., 2015). Polysaccharides are highly stable, biocompatible and biodegradable (Aider et al., 2010).

6.3.1.1 Chitosan

The most commonly used polysaccharide for nanoparticle fabrication is chitosan. Chitosan is the only naturally occurring cationic polysaccharide. It is derived from the deacetylation (60–100%) of chitin, which is the second most abundant biopolymer after cellulose. As a linear polysaccharide, it consisting of repeating D-glucosamine and N -acetyl-D-glucosamine units that are linked via (1–4) glycosidic bonds (Hamidi et al., 2008). Chitosan is most likely degraded in vertebrates by lysozyme and by bacterial enzymes in the colon, and several studies have observed that chitosan can be degraded into smaller excretable fragments (Kean et al., Tammam et al., 2015). Besides, according to its chemical structure, chitosan is a potential pH-responsive polymer. The protonation of the amino groups makes it soluble and behave as a cationic polyelectrolyte at pH $<pKa_{chitosan}$ (6.5), while it is insoluble at pH over its pKa (Kumar et al., 2000; Rinaudo et al., 2006). In addition to pH, the degree of deacetylation (DDA), the average molecular weight (Mw) and the distribution of acetyl and amino groups along the polymer backbone also can significantly affect the physicochemical properties (Kiang et al., 2004; Lavertu et al., 2006; Liu et al., 2014). Owing to its good biodegradability, high biocompatibility, low toxicity, and effective antibacterial

activity, chitosan is employed in a wide range of applications in drug and gene delivery.

Chitosan is a useful material that upon appropriate nanoparticles modification with ligand can serve as tumor targeted therapeutic agents. Modificated chitosan-based nanoparticles holds great potential for liver-targeted intracellular delivery of 5-FU. Galactosylated chitosan (GC) was combined with 5-FU to synthesized GC/5-FU nanoparticles, the study in the mouse orthotropic liver cancer mouse model indicate that sustained release of GC/5-FU nanoparticles are more effective at targeting hepatic cancer cells than 5-FU monotherapy (Cheng et al., 2012). Otherwise, chitosan was modified by hepatoma cell specific binding molecule glycyrrhetinic acid (GA) to delivery 5-FU by Cheng et al., In vitro data found that the effective drug exposure time against hepatic cancer cells was increased in comparison with free 5-FU. Further in vivo studies on an orthotropic liver cancer mouse model demonstrated that GA-chitosan/5-FU significantly inhibited cancer cell proliferation, resulting in increased survival time (Cheng et al., 2014; Cheng et al., 2013).

Gene therapy has received much attention in the fields of medicine, and nonviral vectors have primarily been used to deliver genes. Chitosan is an attractive polymer for gene delivery, showing excellent biocompatibility, admirable biodegradability, ecological safety, low toxicity, antimicrobial activity and low immunogenicity (Rinaudo et al., 2008; Mourya et al., 2008). In spite of the high potential of chitosan as a gene carrier, its poor solubility and low transfection efficiency have greatly impeded its practical application (Jiang et al., 2014). Recently, a great deal of methods to enhance the transfection efficiency of chitosan has been studied and a more general understanding of parameters that may impact on the efficiency of transfection has been attained. Zhang et al., developed galactosylatedtrimethyl chitosan-cysteine (GTC) nanoparticles as an oral delivery siRNA carrier to the activated macrophages for treatment of ulcerative colitis. The cellular uptake of GTC/TPP NPs in activated macrophages was significantly enhanced compared to trimethyl chitosan-cysteine (TC)/TPP NPs owing to galactose receptor-mediated endocytosis. Moreover, compared to TC/TPP NPs, GTC/TPP NPs more efficiently promoted the distribution of siRNA in ulcerative colon after oral administration (Zhang et al., 2013). Otherwise, folated poly (ethylene glycol)-chitosan-graft-polyethylenimine (FPCP) was synthesized. The FPCP showed low cytotoxicity in various cell lines, and FPCP/DNA compleⵏes showed good cancer cell specificity as well as good transfection efficiency in the presence of serum. Further, the FPCP/Pdcd4 complexes reduced tumor numbers and progression more effectively than PEI 25kDa/Pdcd4 ones in H-ras12V liver cancer model mice after intravenous administration.

6.3.1.2 Alginate

Similar to chitosan, alginate is another polysaccharide that has drawn substantial attention in the biomedical field over the past decades. Chemically, alginate is a linear anionic polysaccharide composed of alternating blocks of 1,4-linked β-D-mannuronic acid (M) and α-l-guluronic acid (G) residues (Liew et al., 2006). Alginate has some advantages in its high mucoadhesiveness, aqueous solubility, and a tendency for gelation in proper condition, biocompatibility and non-toxicity (Weber et al., 2000; Sun et al., 2013). Besides, polyvalent cations such as calcium and barium can initiate cross-linking of alginate, forming insoluble alginate with the anionic polymer (Liew et al., 2006; Grant et al., 1973). The number of studies involving alginate-based nanoparticles is increasing in delivery of therapeutic agents (George et al., 2006; Kimberly et al., 2006).

6.3.2   PROTEIN-BASED NANOPARTICLES

With the advancement of novel drug-delivery systems, protein nanoparticles have drawn very much attention of formulation development scientists. In recent times nanoparticles have been developed for the delivery of the therapeutics using various proteins including albumin, gelatin, casein, silk proteins, elastin and lectins.

6.3.2.1   Albumin-Based Nanoparticles

Albumin is the most abundant plasma protein (35–50 g/l human serum) with a molecular weight of 66.5 kDa and molecular dimensions of 30×30×80 Å (33×3×8 nm) (Malhotra et al., 2014). Like most of the plasma proteins, it is synthesized in the liver and is important for various physiological processes including delivery of nutrients to cells, solubilization of long chain fatty acids, balancing plasma pH, providing colloidal osmotic pressure in the blood, and for binding to bilirubin and therapeutic agents. There are three types of albumin: ovalbumin, bovine serum albumin, human serum albumin (Elzoghby et al., 2012). In addition, albumin nanoparticles can be easily prepared under soft conditions by coacervation, controlled desolvation or emulsion formation, can remain stable under the typical processing conditions encountered during nanoparticle preparation. Because unlike other endogenous proteins, it is stable over a wide pH range (4–9) and is stable for upto 10 h at 60°C, it is unaltered by denaturing agents and solvents at moderate concentrations (Kim et al., 2011; Neumann et al., 2010). More importantly, the properties of albumin that preferential uptake in tumor, biodegradability, low toxicity, immunogenicity and suitable blood circulation with a half-time of 19 days make albumins an ideal material as a drug delivery carrier (Kratz et al., 2008).

6.3.2.2   Gelatin

Gelatin is an attractive material for nanoparticles as it is obtained from collagen by acid and alkaline hydrolysis consisting of glycine, proline and 4-hydroxyproline residues with typical structure of -Ala-Gly-Pro-Arg-Gly-Glu-4HyP-Gly-Pro. Gelatin has attracted a great interest for its biocompatibility and biodegradability and offers the advantages of being cheap and readily available (Elzoghby et al., 2012; Elzoghby et al., 2013). Besides, gelatin does not produce harmful byproducts upon enzymatic degradation, as it is derived from collagen, which is the most abundant protein in animals. Moreover, gelatin has relatively low antigenicity because of being denatured in contrast to collagen which is known to have antigenicity due to its animal origin (Elzoghby et al., 2012; Elzoghby et al., 2013). Moreover, their high number of functional groups on polymer backbone can be used for chemical modification such as cross-linking and addition of ligands (Lai et al., 2015). In a word, all these advantages make gelatin-based nanoparticles a promising carrier system for drug delivery.

Recent studies have shown the availability of the loading of therapeutic molecules into gelatin-based nanoparticles. Gelatin was conjugated with DOX after amine group-blockage using acetaldehyde. Gelatin–DOX nanoparticles exhibited much lower cytotoxicity and remarkably inhibited tumor growth and suppressed pulmonary metastasis compared to free DOX (Lee et al., 2006). Besides, gelatin nanoparticles loaded with paclitaxel were also used in intravesical bladder cancer therapy. Both the hydrophilicity of gelatin and the nanoencapsulation of paclitaxel in amorphous state within gelatin nanoparticles may be responsible for the enhanced solubility and rapid release of drug (87% in 2 h without enzymes) which is highly desirable in intravesical bladder therapy in which the drug formulation is typically maintained in the bladder for only a short duration (i.e., 2 h) (Lu et al., 2004).

6.3.2.3 Silk Fibroin

Silk fibroin (SF), a naturally occurring protein polymer with several unique properties that make it is a suitable material for incorporation into a variety of drug delivery vehicles capability of delivering a range of therapeutic agents (Mottaghitalab et al., 2015). SF has been produced by spiders and insects such as silkworms, and they form fibrous materials. The unique properties of silk fibroin (SF) such as slow biodegradation, superior mechanical properties, favorable processability in combination with biocompatibility, have fueled wide interest in this material for a variety of applications, ranging from textiles to biomedical use (Brooks et al., 2015; Elzoghby et al., 2012).

Thus unique properties of SF make it suitable for preparing drug delivery vehicles. For example, it was reported that by using a capillary-microdot technique, SF nanoparticles could be produced with an average size (b100 nm) appropriate for delivering curcumin as a therapeutic agent for cancer. The greatest entrapment, intracellular uptake, and controlled release of curcumin were observed by encapsulating them in SF nanoparticles and administering them to breast cancer cells (Gupta et al., 2009). Chen et al. (Zhang et al., 2006) prepared paclitaxel (PTX)-loaded SF nanoparticles ranging from 270 to 520 nm by addition of PTX-ethanol solution into regenerated SF solution under gentle stirring. The release time of PTX-SF nanoparticles can be as long as two weeks when the drug loading is about 3.0%. By a similar method, Wu et al., also prepared the PTX-SF nanoparticles with a diameter of 130 nm. PTX kept its pharmacological activity when incorporating into PTX-SF nanoparticles. The in vivo antitumor studies of PTX-SF nanoparticles on gastric cancer nude mice exnograft model indicated that locoregional delivery of PTX-SF nanoparticles demonstrated superior antitumor efficacy by delaying tumor growth and reducing tumor weights compared with systemic administration (Chen et al., 2012).

SF-based gene delivery systems have recently been reported to provide biodegradability, biocompatibility, high transfection efficiency, and DNase resistance. Recombinant spider silk-based nanoparticles containing DNA-binding domains poly (L-lysine) and tumor-homing peptides

(THPs) such as F3 (KDEPQRRSARLSAKPAPPKPEPKPKKAPAKK), Lyp1 (CGNKRTRGC), and CGKRK have been developed to deliver target-specific plasmid DNA (pDNA) to the tumor cells (MDA-MB-435 and MDA-MB-231) with low cytoto☐icity significant enhancement of target specificity to tumor cells and high efficiency (Numata et al., 2011; Numata et al., 2012).

6.3.3 POLYPEPTIDE NANOPARTICLES

Polypeptide has been extensively investigated to find out the possibilities to use them as drug or drug delivery or drug targeting systems over the past few decades. The most often fabricated polypeptides are elastin-like polypeptides (ELPs) that consist of alternating hydrophobic blocks and cross-linking domains (Sun et al., 2011; Callahan et al., 2012). ELPs are a class of temperature sensitive biopolymers-based on the structural motif found in mammalian tropoelastin. ELPs composed of short repeating peptide motifs, of which the most common is the pentapeptide VPGXG where X is a guest residue that is any amino acid except proline (Tatham et al., 2000; Urry et al., 1997). ELPs exhibit stimulus-responsive lower critical solution temperature phase transition behavior. They are soluble at temperatures below a characteristic cloud point temperature (Tt) (also known as the inverse transition temperature) and aggregate into micron scale coacervates above the Tt. ELPs are useful materials in a variety of applications since their stimulus-response is highly tunable (Rodríguez-cabello et al., 2015). Besides, the biological composition of ELPs also ensures their biodegradation, permitting their safe break down into peptides and amino acids that can be easily cleared from the body. Due to above merits, ELPs was suitable for used for drug delivery carrier. For example, Chilkoti et al. have reviewed the results on the development of macromolecular carriers for thermal targeting of therapeutics to solid tumors. The two thermally responsive polymers namely, poly (N-isopropylacrylamide-co-acrylamide) [poly (NIPAAm)] and an artificial ELP designed by them exhibited ITT behavior at a temperature slightly > 37°C. In vivo fluorescent video microscopy and radiolabel distribution studies of ELP delivery to human tumors implanted in nude mice demonstrated that hyperthermic targeting of the thermally responsive ELP for 1 h provides a 2-fold increase in tumor localization compared to the same polypeptide without hyperthermia. Similar results were also obtained for poly (NIPAAm) though the extent of accumulation was somewhat lesser than observed for ELP I (Chilkoti et al., 2002).

Except for ELPs, there are another polypeptides that most used as drug carrier called silk-elastin like protein polymers (SELPs). SELPs are a family of genetically engineered protein block copolymers, consisting of tandemly repeated units of silk-like (GAGAGS) and elastin-like (GXGVP) peptide blocks, where X in the elastin block can be any amino acid, except proline that affects the coacervation process of elastin (Lucas et al., 1957; Sandberg et al., 1981). SELPs are usually formed by two steps, (1) formation of micellar-like particles with the silk blocks as the core structure by hydrogen bonding, (2) the hydrophobic interaction between elastin blocks above a specific transition temperature, which leads to the ordered association of SELPs molecules (Oyarzun-Ampuero et al., 2011). Compared with elastin-based materials, the silk blocks in SELPs are able to crystallize into β-sheets via hydrogen bonding enabling robust materials formation. In addition, through tuning the ratio of silk and elastin blocks, material structures, strength and biodegradability can be controlled (Gustafson et al., 2010; Hu et al., 2010). SELPs form materials with unique mechanical properties, avoid a need for chemical cross-linking, can be processed in aqueous conditions, and have been explored for various biomedical applications.

Besides, somatostatin (SST) also widely used in drug delivery. SST is a neuropeptide that demonstrate powerful inhibitory effect against several endocrine systems. SST was originally isolated as an endocrine inhibitor of pituitary growth hormone secretion and is now recognized as a hormone capable of regulating fundamental processes, such as secretion, cell division, proliferation and apoptosis (Narayanan et al., 2016; Mariniello et al., 2011). The cellular actions of SST are mediated by five somatostatin receptor subtypes (termed SSTR 1–5). SST's therapeutic applications are limited due to its short half-life and various synthetic Somatostatin analogs (SSTA) such as octreotide, lanreotide, vapreotide, etc. have been developed successfully by shortening the sequence and introducing D-amino acids to replace natural L-amino acids, improved metabolic stability and increased affinity to SSTRs (Mulvey et al., 2016; Liapakis et al., 1996; Karashima et al., 1987; Lesche et al., 2009). Octreotide was the first SSA developed for clinical application. It was approved in 1988 for palliation of carcinoid syndrome as well as other hormonal syndromes caused by metastatic gastroenteropancreatic (GEP)-NETs including vasoactive intestinal peptide-producing tumors.

Apart from OCT, other long-acting SSTAs have been synthesized and investigated in vitro and in vivo. Lanreotide was licensed in Europe in 1998 for the treatment of symptoms associated with advanced neuroendocrine tumors (Raderer et al., 1999; Kvols et al., 1986; Narayanan et al., 2015; O'Toole et al., 2000). There are other SSTAs such as Vapreotide, pasireotide, SST dicarba-analog 11 and 14, etc. These SSTAs have been widely used to treat various tumors, particularly endocrine tumors.

6.3.4   ALIPHATIC POLYESTERS

Aliphatic polyesters derived from cyclic lactones [ε-caprolactone (ε-CL), lactide and glycolide] represent the major family of polymers currently used in medical applications. Aliphatic polyesters are thermoplastics that present numerous advantages like tailorable degradation kinetics and mechanical properties, ease of shaping and manufacture, and whose broad spectrum of properties can easily be tuned for instance by the use of lactic acid-based stereocopolymers, or copolymerization with the biocompatible hydrophilic poly (ethylene glycol). Biodegradable aliphatic polyesters such as poly (lactide) (PLA), poly (lactide-co-glycolide) (PLGA) and polycaprolactone (PCL) have attracted much interest for applications of drug carrier.

6.3.4.1   Poly (lactide-co-glycolide) (PLGA)

Poly (Lactide Co-Glycolic Acid), namely PLGA, was approved by US FDA and European Medicine Agency (EMA). PLGA is a widely used polymer due to biocompatibility, longstanding track record in biomedical functions and well documented utility for continued drug release compared to the conventional devices up to days, weeks or months, and ease of parenteral administration via injection (Biondi et al., 2008; Faraji et al., 2009). Besides, PLGA is one of the most effectively used biodegradable polymers for the development of nanomedicines because it undergoes hydrolysis in the body to produce the biodegradable metabolite monomers, lactic acid and glycolic acid (McCall et al., 2013). In the recent years, PLGA-based nanoparticles are successfully applied for drug delivery of the biomedical approaches for the development of translational medicine.

6.3.4.2   Poly (lactide) (PLA)

Poly (l-lactide) (PLA) is a aliphatic polyester that is produced from lactic acid, either by the direct polycondensation of lactic acid or via the ring-opening polymerization of lactide, approved by the US Food and Drug Administration and is widely studied to encapsulate hydrophobic anticancer drugs (Sharma et al., 2015; Yuan et al., 2015). PLA own properties of high biocompatibility, non-toxicity. Moreover, another merits of PLA used as nanoparticles is biodegradability, which is important to maintain clearance of the nanoconstruct after injection and delivery of the cargo and to minimize the risk of toxic buildup of the polymer carrier in tissue (Tang et al., 2013). PLA hydrolyzes into nontoxic hydroxyl-carboxylic acid through ester bond cleavage and enters into citric acid cycle, finally metabolizing into water and carbon dioxide (Garofal et al., 2014). Owing to above advantages, PCL become a commonly used material of nano-sized delivery carrier. Thu and colleagues successfully prepared paclitaxel loaded nanoparticles composed by PLA–TPGS copolymer through a simple modified modification/solvent evaporation method. The results showed that PTX loaded nanoparticles exhibit great advantages compared to free PTX and the folate decoration significantly improve the targeted delivery of drug to cancer cell in both in vitro and in vivo (Thu et al., 2015).

6.3.4.3   Polycaprolactone (PCL)

Polycaprolactone (PCL) is non-toxic, biodegradable aliphatic polyester, approved by Food and Drug Administration (FDA). This polyester is a biodegradable thermoplastic polymer that is relatively inexpensive and allows for the manufacture of drug delivery systems (Helland et al., 2007; Woodruff et al., 2010). The PCL is a hydrophobic poly (ester) due to the presence of a sequence of four hydrocarbons that ends up defining a more hydrophobic behavior to the molecule. Besides, PCL degrades through hydrolysis due to ester bonds with slow degradation rate, thus makes it suitable for long-term delivery extending over a period of more than 1 year (Pereira et al., 2013).

6.4   POLYMERIC NANOPARTICLES FOR CO-DELIVERY OF DRUGS AND GENES

The curative effects of conventional chemotherapy hitting single target in tumor cells are therefore severely limited. The combination of chemotherapy and gene therapy provides a promising modality to improve the therapeutic index through simultaneous modulation of multiple signaling pathways in tumor cells (Skandrani et al., 2014; Zhao et al., 2015). Rational drug combinations aim to exploit either additive or synergistic effects arising from the action of several species with the final goal to maximize therapeutic efficacy. It has been shown that an appropriate combination of chemotherapeutic drugs and gene agents can improve the therapeutic outcome and patient compliance due to reduced dose and decreased development of drug resistance (Gandhi et al., 2014). The most critical factor in co-delivery of drugs and genes is to ensure high drug loading efficiency and high level of transgene expression, give full play to therapy effect of drugs and genes. So find applicable carriers is the biggest challenge in co-delivery drugs and gene agents, since gene agents have higher molecular weight and negatively charged surface, while most frequently used anti-cancer drugs are hydrophobic small molecules (Dai et al., 2015).

Recently, there has been a remarkable progress in a drug and gene co-delivery system. Biodegradable nano-carriers was synthesized and assembled by diblock copolymers of poly (e-caprolectone) and linear poly (ethylene imine), for co-delivery of BCL-2 siRNA and DOX into hepatic cancer cells. The nanoparticles e□hibited high transfection efficiency and ideally controlled release of drug at certain ratios. Furthermore, the folate-targeted delivery of BCL-2 siRNA resulted in more significant gene suppression, inducing cancer cell apoptosis and improving the therapeutic efficacy of the co-administered DOX, compared to non-specific delivery (Cao et al., 2011). Besides, a folate modified multifunctional nanoassembly (FNA) loading both docetaxel and iSur-pDNA was constructed and evaluated as a therapeutic approach for HCC. Cytotoxicity of FNAs against hepatocellular carcinoma cell line BEL 7402 was much higher than either docetaxel or non-docetaxel FNAs (nFNAs) loading only iSurpDNA, and was also superior to the combined treatment with free docetaxel and nFNAs. Furthermore, in mouse HCC □enograft model, it observed a better antitumor efficacy of FNAs with low systemic toxicity (Xu et al., 2010).

6.5 STIMULI-RESPONSIVE POLYMERIC NANOPARTICLES AS DELIVERY SYSTEM

Although passive and active targeted-drug delivery has addressed a number of important issues, it is still essential to enhance the bioavailability of drugs and genes at the disease site, and especially upon cellular internalization by polymeric nanoparticles. To this end, various "intelligent" polymeric nanoparticles that release drugs in response to an internal or external stimulus such as pH, redox, or temperature would be amenable to address some of the systemic and intracellular delivery barriers, called stimuli-responsive nanoparticles. The central operating principle of stimuli-responsive nanoparticles lies in the fact that a specific cellular/extracellular stimulus of chemical, biochemical, or physical origin can modify the structural composition/conformation of the nanocarriers, thereby promoting release of the active species to specific biological environment.

6.5.1 *pH-SENSITIVE POLYMERIC NANOPARTICLES*

The lowered extracellular/interstitial pH is a hallmark of tumor malignancy, especially in solid tumors, which is caused by excessive metabolite, mainly lactic acid and CO_2 as well as the increased expression and activity of vacuolar-type (V-type) H(+)-ATPases (proton pumps) (Gerweck et al., 1996). The tumor's extracellular pH may drop to 6.5or less compared to about pH 7.4 in normal blood or most normal tissues. Therefore, the low pH of endosomal and lysosomal compartments has been exploited to trigger changes in material properties that induce lysis of intracellular vesicles (Vander et al., 2011; Crucho et al., 2015).

Usually, the rational design of pH-sensitive nanoparticles has combined pH-sensitive linkages or polymers that undergo a pH-dependent conformational change. Thus, by decomposition or destabilization of the nanocarriers, their payload is released at compartments with lower pH. Ulbrich et al., synthesized a HPMA (N-(2-hydroxypropyl) methacrylamide) polymer conjugate using hydrazone groups as pH responsive linkers for the attachment of the anticancer drug doxorubicin. These conjugates were stable at pH 7.4 and released the drug at pH 5. Furthermore, they attached an antibody to the polymer backbone to achieve an effective targeting to T cell lymphoma EL 4 cells (Ulbrich et al., 2004). Another possibility is to use acid degradable linker units to attach the drug to the hydrophobic block of the amphiphilic polymer. Shenoy et al., prepared pH-responsive polymeric nanoparticles from poly (b-amino ester) (PbAE), a biodegradable and pH-sensitive polymer. The polymeric nanoparticles were able to release the encapsulated material in the endosomal pH range through a dissolution mechanism due to protonation of amines in the polymer matrix (Shenoy et al., 2005).

6.5.2   TEMPERATURE-SENSITIVE POLYMERIC NANOPARTICLES

Temperature has been used as effective external stimuli for the design of responsive nanocarriers owing to its physiological significance. Temperature-responsive polymeric nanoparticles have drawn attention as smart materials for biomedical applications due to their phase-transition behavior in response to changes in temperature (Zhu et al., 2013). Polymeric materials containing thermo-responsive units are advantageous for constructing molecular architectures which reversibly respond to changes in temperature within a very sharp and narrow range. If a polymer is soluble in water below a specific temperature but becomes insoluble and causes phase separation above this temperature, it is said to have a "lower critical solution temperature (LCST)." A polymer with the opposite relationship is referred to as having an "upper critical solution temperature (UCST)." And most applications have relied on abrupt changes in aqueous solubility at either LCST or UCST (Abulateefeh et al., 2011; Roy et al., 2013).

To date, most of the temperature-sensitive polymers used in drug delivery systems are the LCST type. Poly (N-isopropylacrylamide) (PNIPAM), a thermo sensitive polymer which changes its conformation from a hydrophilic"coil" below its LCST (32°C) to a hydrophobic "globule" above its LCST in the aqueous environment, has been the most extensively studied (Kojima et al., 2010). In 1999, Chung and colleagues prepared temperature-sensitive nanomicells using poly (butyl methacrylate) pNIPAM as a hydrophoic core while pNIPAM was used as the thermo sensitive shell. It was suggested that pNIPAM-b-PBMA-based nanomicells loaded with doxorubicin released 15% of the drug after 15h at 30°C, compared to 90% drug release at 37°C. The cytotoxicity experiments indicate less than 5% cell death at 29°C, but 65% cell death at 37°C, owing to the temperature-dependent drug release (Chung et al., 1999).

6.5.3   REDOX-SENSITIVE POLYMERIC NANOPARTICLES

Recently, the synthesis and self-assembly of nanoparticles with reduction-responsive links has been recognized as another powerful and efficient strategy for tumor-targeted drug delivery. Considering the difference of redox potential (~100–1000 fold) existing between the extracellular space and the intracellular space, it has been well-established that extracellular space is oxidative while the intracellular is reductive, which is strongly related to the intra- and extracellular glutathione concentration (Meng et al., 2009; Cheng et al., 2011). The difference between two provide an opportunity to design redox-sensitive delivery systems for triggered intracellular release of drugs or genes.

The disulfide bond has been widely used as the cleavable/reversible linker in nanocarriers to render redox potential sensitivity to nanopreparations due to the disulfide-to-thiol reduction reaction (Kurtoglu et al., 2009). Navath and co-workers (2008) reported N-acetylcysteinedendrimer conjugates with disulfide linkages. Nacetylcysteine is prescribed for neuroinflammation, however, it reacts with cysteine residues of natural proteins resulting in e☐tremelylow bioavailability. The disulfide-linked dendrimer drug conjugates had much better antioxidant effects than the drug alone (Kurtoglu et al., 2009). In another approach, A new redox-responsive polymer-based on poly (ethylene glycol)-b-poly (lactic acid) (MPEG-SS-PLA) diblock copolymers was developed by Song et al., The synthesized copolymers were self-assembled into rice-grain-shaped PNPs. In vitro drug release studies showed that the paclitaxel-loaded PNPs accomplished rapid drug release under reducing conditions compared with the redox-insensitive MPEG-PLA PNPs. Furthermore, the in vitro activity of the PNPs against A549, MCF-7, and HeLa carcinoma cells demonstrate that these PNPs have low cytotoxicity, high cytocompatibility, and an ability to assist the endocytosis of antitumor drugs (Song et al., 2011).

6.6 CONCLUSIONS AND PROSPECTS

Cancer is worldwide anxious problem and is eager to solve. Clinic used chemotherapy agents frequently encounter important problems such as low specificity and non-selective biodistribution for advance solid tumor. In recent years, nanoparticles contain antineoplastic drugs or genes are becoming attractive in modern drug development due to various advantages compared with free drugs or genes. The most critical factor in drug or genes delivery is to ensure high level of aim drug or gene with greater specificity reducing the off-target effects on the normal cells. Drug delivery systems-based on polymeric nanoparticles equipped with ligand provide novel opportunities to overcome the above limitations. Co-delivery of gene and chemotherapy drugs with nanocarries has a good synergistic effect compared with the traditional chemotherapy which can increase the amount of the drug distribution in target organ in order to reduce the toxic side effects thereby enhancing efficacy. Besides, stimuli-responsive nanoparticles can enhance the bioavailability of drugs and genes at the disease site thus further improve the effect of treatment. MutIL-stimuli responsive polymeric nanoparticles equipped with multiple ligand for co-delivery antineoplastic drugs could be an attractive development trends in future and should be pay more attention to. However, it is still far away to be applied in clinic, and preclinical studies such as animal experiment, clinical research trials should be conducted to get more precise and valid results.

ACKNOWLEDGEMENTS

Financial support from National Natural Science Foundation of China (81471777, 81102409), Program for Outstanding Young Scientists was given to Cui-Yun Yu by the Natural Science Foundation of Hunan province (2017JJ1024), Support of the Innovation Team of Antitumor Drugs (NHCXTD05) was given to Cui-Yun Yu by University of South China, the Young Talent Program was given to Cui-Yun Yu by University of South China, and 225 Talent Project was given to Cui-Yun Yu by the Health Department of Hunan Province, and all above support is gratefully acknowledged.

KEYWORDS

- antineoplastic drugs
- biomedical applications
- genes
- polymeric nanobiomaterial
- target delivery system

REFERENCES

Abulateefeh, S. R., Spain, S. G., Aylott, J. W., Chan, W. C., Garnett, M. C., & Alexander, C. (2011). Thermoresponsive polymer colloids for drug delivery and cancer therapy. *Macromol. Biosci. 11*, 1722–1734.

Aider, M. (2010). Chitosan application for active bio-based films production and potential in the food industry: Review. *LWT. Food. Sci. Technol. 43*, 837–842.

Araki, T., Kono, Y., Ogawara, K., Watanabe, T., Ono, T., Kimura, T., & Higaki, K. (2012). Formulation and evaluation of paclitaxel-loaded polymeric nanoparticles composed of polyethylene glycol and polylactic acid block copolymer. *Biol. Pharm. Bull. 35*, 1306–1313.

Ashwanikumar, N., Kumar, N. A., Nair, S. A., & Kumar, G. S. (2014). 5-Fluorouracil-lipid conjugate: potential candidate for drug delivery through encapsulation in hydrophobic polyester-based nanoparticles. *Acta. Biomater. 10*, 4685–4694.

Bahrami, B., Mohammadnia-Afrouzi, M., Bakhshaei, P., Yazdani, Y., Ghalamfarsa, G., Yousefi, M., Sadreddini, S., et al., (2015). Folate-conjugated nanoparticles as a potent therapeutic approach in targeted cancer therapy. *Tumour. Biol. 36*, 5727–5742.

Bansal, D., Yadav, K., Pandey, V., Ganeshpurkar, A., Agnihotri, A., Dubey, N. (2014). Lactobionic acid coupled liposomes: an innovative strategy for targeting hepatocellular carcinoma. *Drug. Deliv. 23*, 140–146.

Bergs, J. W., Wacker, M. G., Hehlgans, S., Piiper, A., Multhoff, G., Rödel, C., et al., (2015). The role of recent nanotechnology in enhancing the efficacy of radiation therapy. *Biochim. Biophys. Acta. 1856*, 130–143.

Biondi, M., Ungaro, F., Quaglia, F., & Netti, P. A. (2008). Controlled drug delivery in tissue engineering. *Adv. Drug. Deliv. Rev. 14*, 60, 229–242.

Branco, M. C., & Schneider, J. P. (2009). Self-assembling materials for therapeutic delivery. *Acta. Biomater. 5*, 817–831.

Brooks, A. E. (2015). The Potential of Silk and Silk-Like Proteins as Natural Mucoadhesive Biopolymers for Controlled Drug Delivery. *Front. Chem. 3*, 65.

Bupathi, M., Kaseb, A., Meric-Bernstam, F., & Naing, A. (2015). Hepatocellular carcinoma: Where there is unmet need. *Mol. Oncol. 9*, 1501–1509.

Callahan, D. J., Liu, W., Li, X., Dreher, M. R., Hassouneh, W., Kim, M., et al., (2012). Triple stimulus-responsive polypeptide nanoparticles that enhance intratumoral spatial distribution. *Nano. Lett. 12*, 2165–2170.

Cao, N., Cheng, D., Zou, S., Ai, H., Gao, J., & Shuai, X. (2011). The synergistic effect of hierarchical assemblies of siRNA and chemotherapeutic drugs co-delivered into hepatic cancer cells. *Biomaterials. 32*, 2222–2232.

Chen, M., Shao, Z., & Chen, X. (2012). Paclitaxel-loaded silk fibroin nanospheres. *J. Biomed. Mater. Res. A. 100*, 203–210.

Cheng, M., Chen, H., Wang, Y., Xu, H., He, B., Han, J., & Zhang, Z. (2014). Optimized synthesis of glycyrrhetinic acid-modified chitosan 5-fluorouracil nanoparticles and their characteristics. *Int. J. Nanomedicine. 9*, 695–710.

Cheng, M., He, B., Wan, T., Zhu, W., Han, J., Zha, B., Chen, H., et al., (2012). 5-Fluorouracil Nanoparticles Inhibit Hepatocellular Carcinoma via Activation of the p53 Pathway in the Orthotopic Transplant Mouse Model. *PLoS. One. 7*, e47115.

Cheng, M., He, B., Wan, T., Zhu, W., Han, J., Zha, B., et al., (2012). 5-fluorouracil Nanoparticles Inhibit Hepatocellular Carcinoma via Activation of the p53 Pathway in the Orthotopic Transplant Mouse Model. *PLoS. One. 7*, e47115.

Cheng, M., Xu, H., Wang, Y., Chen, H., He, B., Gao, X., et al., (2013). Glycyrrhetinic acid-modified chitosan nanoparticles enhanced the effect of 5-fluorouracil in murine liver cancer model via regulatory T-cells. *Drug. Des. Devel. Ther. 7*, 1287–1299.

Cheng, R., Feng, F., Meng, F., Deng, C., Feijen, J., & Zhong, Z. (2011). Glutathione-responsive nano-vehicles as a promising platform for targeted intracellular drug and gene delivery. *J. Control Release. 152*, 2–12.

Chilkoti, A., Dreher, M. R., Meyer, D. E., & Raucher, D. (2002). Targeted drug delivery by thermally responsive polymers. *Adv. Drug. Deliv. Rev. 54*, 613–630.

Cho, K., Wang, X., Nie, S., Chen, Z. G., & Shin, D. M. (2008). Therapeutic nanoparticles for drug delivery in cancer. *Clin. Cancer. Res. 14*, 1310–1316.

Chung. J. E., Yokoyama, M., & Yamato, M. (1999). Thermo-responsive drug delivery from polymeric micells constructed using block copolymers of poly(butylmethacrylate). *J. Control Releas. 62*, 115–127.

Crucho, C. I. (2015). Stimuli-Responsive polymeric nanoparticles for nanomedicine. *Chem. Med. Chem.* 10, 24–38.

D'Souza, A. A., & Devarajan, P. V. (2015). Asialoglycoprotein receptor mediated hepatocyte targeting - Strategies and applications. *J. Control Release. 203*, 126–139.

Dai, X., & Tan, C. (2015). Combination of microRNA therapeutics with small molecule anti-cancer drugs: mechanism of action and co-delivery nanocarriers. *Adv. Drug. Deliv. Rev. 81*, 184–197.

Ding, B., Li, T., Zhang, J., Zhao, L., & Zhai, G. (2012). Advances in Liver-Directed Gene Therapy for Hepatocellular Carcinoma by Non-Viral Delivery Systems. *Curr. Gene. Ther. 12*, 92–102.

Elzoghby, A. O. (2013). Gelatin-based nanoparticles as drug and gene delivery systems: Reviewing three decades of research. *J. Control Release. 172*, 1075–1091.

Elzoghby, A. O., Samy, W. M., & Elgindy, N. A. (2012). Albumin-based nanoparticles as potential controlled release drug delivery systems. *J. Control Release. 157*, 168–182.

Elzoghby, A. O., Samy, W. M., & Elgindy, N. A. (2012). Protein-based nanocarriers as promising drug and gene delivery systems. *J. Control Release. 161*, 38–49.

Elzoghby, A. O., Samy, W. M., & Elgindy, N. A. (2012). Protein-based nanocarriers as promising drug and gene delivery systems. *J. Control Release. 161*, 38–49.

Faraji, A. H., & Wipf, P. (2009). Nanoparticles in cellular drug delivery. *Bioorg. Med. Chem. 15*, 17, 2950–2962.

Folkman, J. (1990). What is the evidence that tumors are angiogenesis dependent? *J. Nati. Cancer. Inst. 82*, 4–6.

Fonte, P., Araújo, F., Silva, C., Pereira, C., Reis, S., Santos, H. A., et al., (2015). Polymer-based nanoparticles for oral insulin delivery: Revisited approaches. *Biotechnol. Adv. 33*, 1342–1354.

Froidevaux, S., & Eberle, A. N. (2002). Somatostatin analogs and radiopeptides in cancer therapy. *Biopolymers. 66*, 161–183.

Gandhi, N. S., Tekade, R. K., & Chougule, M. B. (2014). Nanocarrier mediated delivery of siRNA/miRNA in combination with chemotherapeutic agents for cancer therapy: current progress and advances. *J. Control Release. 194*, 238–256.

Garofalo, C., Capuano, G., Sottile, R., Tallerico, R., Adami, R., Reverchon, E., Carbone, E., et al., (2014). Different insight into amphiphilic PEG-PLA copolymers: influence of macromolecular architecture on the micelle formation and cellular uptake. *Biomacromolecules. 15*, 403–415.

George, M., & Abraham, T. E. (2006). Polyionic hydrocolloids for the intestinal delivery of protein drugs: Alginate and chitosan - a review. *J. Control Release. 114*, 1–14.

Gerweck, L. E., & Seetharaman, K. (1996). Cellular pH gradient in tumor versus normal tissue: potential exploitation for the treatment of cancer. *Cancer. Res. 56*, 1194–1198.

Grant, G. T., Morris, E. R., Rees, D. A., Smith, P. J. C., & Thom, D. (1973). Biological Interactions between Polysaccharides and Divalent Cations-Egg-Box Model. *FEBS. Lett. 32*, 195–198.

Greish, K. (2007). Enhanced permeability and retention of macromolecular drugs in solidtumors: A royal gate for targeted anticancer nanomedicines. *J. Drug. Target. 15*, 457–464.

Grewal, P. K., (2010). Chapter 13—The Ashwell-Morell receptor, in: F. Minoru (Ed.), Methods in Enzymology. Academic Press. 223–241.

Grünwald, V., & Merseburger, A. S. (2013). The progression free survival-plateau with vascular endothelial growth factor receptor inhibitorseis there more to come?. *Eur. J. Cancer. 49*, 2504–2511.

Gupta, V., Aseh, A., Ríos, C. N., Aggarwal, B. B., & Mathur, A. B. (2009). Fabrication and characterization of silk fibroin-derived curcumin nanoparticles for cancer therapy. *Int. J. Nanomedicine. 4*, 115–122.

Gustafson, J. A., & Ghandehari, H. (2010). Silk-elastin like protein polymers for matrix-mediated cancer gene therapy. *Adv. Drug Delivery Rev. 62*, 1509–1523.

Hamidi, M., Azadi, A., & Rafiei, P. (2008). Hydrogel nanoparticles in drug delivery. *Adv. Drug. Deliv. Rev. 60*, 1638.

Han, H. S., Thambi, T., Choi, K. Y., Son, S., Ko, H., Lee, M. C., et al., (2015). Bioreducible shell-cross-linked hyaluronic acid nanoparticles for tumor-targeted drug delivery, *Biomacromolecules. 16*, 447–456.

Helland, A., Wick, P., Koehler, A., Schmid, K., & Som, C. (2007). Reviewing the environmental and human health knowledge base of carbon nanotubes. *Environ. Health. Perspect. 115*, 1125–1131.

Heppner, D. E., & van der Vliet, A. (2015). Redo☐-dependent regulation of epidermal growth factor receptor signaling. *Redox. Biol. 8*, 24–27.

Hoeben, A., Landuyt, B., Highley, M. S., Wildiers, H., Van Oosterom, A. T., & De Bruijn, E. A. (2004) Vascular endothelial growth factor and angiogenesis. *Pharmacol. Rev. 56*, 549–580.

Holmes, K., Roberts, O. L., Thomas, A. M., & Cross, M. J. (2007). Vascular endothelial growth factor receptor-2: structure, function, intracellular signaling and therapeutic inhibition. *Cell. Signal. 19*, 2003–2012.

Hu, X., Lu, Q., Sun, L., Cebe, P., Wang, X., Zhang, X., (2010). *Biomacromolecules. 11*, 3178–3188.

Itakura, J., Ishiwata, T., Friess, H., Fujii, H., Matsumoto, Y., Buchler, M. W., et al., (1997). Enhanced expression of vascular endothelial growth factor in human pancreatic cancer correlates with local disease progression. *Clin. Cancer. Res. 3*, 1309–1316.

Jain, R. K., & Stylianopoulos, T. (2010). Delivering nanomedicine to solid tumors. *Nat. Rev. Clin. Oncol., 7*, 653–664.

Jemal, A., Bray, F., Center, M. M., Ferlay, J., Ward, E., Forman, D. (2011). Global cancer statistics. *CA. Cancer. J. Clin. 61*, 69–90.

Jiang, H. L., Cui, P. F., Xie, R. L., & Cho, C. S. (2014). Chemical modification of chitosan for efficient gene therapy. *Adv. Food. Nutr. Res. 73*, 83–101.

Jin, S. E., Jin, H. E., & Hong, S. S. (2014). Targeted delivery system of nanobiomaterials in anticancer therapy: from cells to clinics. *Biomed. Res. Int. 2014*, 814208.

Kamaly, N., Xiao, Z., Valencia, P. M., Radovic-Moreno, A. F., &Farokhzad, O. C. (2012). Targeted polymeric therapeutic nanoparticles: design, development and clinical translation. *Chemical Society. Reviews. 41*, 2971–3010.

Karashima, T., & Schally, A. V. (1987). Inhibitory effects of somatostatin analogs on prolactin secretion in rats pretreated with estrogen or haloperidol. *Proc. Soc. Exp. Biol. Med. 185*, 69–75.

Kean, T., & Thanou, M. (2010). Biodegradation, biodistribution and toxicity of chitosan. *Adv. Drug Deliv. Rev. 62*, 3.

Kiang, T., Wen, J., Lim, H. W., & Leong, K. W. (2004). The effect of the degree of chitosan deacetylation on the efficiency of gene transfection. *Biomaterials. 25*, 5293–5301.

Kim, T. H., Jiang, H. H., Youn, Y. S., Park, C. W., Lim, S. M., & Jin, C. H. (2011). Preparation and characterization of water-soluble albumin-bound curcumin nanoparticles with improved antitumor activity. *Int. J. Pharmaceut. 403*, 285–291.

Kim, Y. K., Minai-Tehrani, A., Lee, J. H., Cho, C. S., Cho, M. H., & Jiang, H. L. (2013). Therapeutic efficiency of folated poly (ethylene glycol)-chitosan-graftpolyethylenimine-Pdcd4 comple☐es in H-ras12Vmice with liver cancer. *Int. J. Nanomed. 8*, 1489–1498.

Kimberly, L., & Douglas, Ciriaco. A. (2006). Effects of alginate inclusion on the vector properties of chitosan-based nanoparticles. *J. Control Release.* 115, 354–361.

Kojima, C. (2010). Design of stimuli-responsive dendrimers. *Expert. Opin. Drug. Deliv. 7*, 307–319.

Kojima, R., Aubel, D., & Fussenegger, M. (2015). Novel theranostic agents for next-generation personalized medicine: small molecules, nanoparticles, and engineered mammalian cells. *Curr. Opin. Chem. Biol. 28*, 29–38.

Kolhatkar, R., Lote, A., & Khambati, H. (2011). Active tumor targeting of nanomaterials using folic acid, transferrin and integrin receptors. *Curr. Drug. Discov. Technol. 8*, 197–206.

Kratz, F. (2008). Albumin as a drug carrier: Design of prodrugs, drug conjugates and nanoparticles. *J. Control Release.* 132, 171–183.

Kumar, M. (2000). A review of chitin and chitosan applications. *React. Funct. Polym. 46*, 1–27.

Kurtoglu, Y. E., Navath, R. S., & Wang, B. (2009). Poly(amidoamine) dendrimer–drug conjugates with disulfide linkages for intracellular drug delivery. *Biomaterials. 30*, 2112–2121.

Kurtoglu, Y. E., Navath, R. S., Wang, B., Kannan, S., Romero, R., & Kannan, R. M. (2009). Poly(amidoamine) dendrimer–drug conjugates with disulfide linkages for intracellular drug delivery. *Biomaterials. 30*, 2112–2121.

Kvols, L. K., Moertel, C. G., O'Connell, M. J., Schutt, A. J., Rubin, J., & Hahn, R. G. (1986). Treatment of the malignant carcinoid syndrome. Evaluation of a long-acting somatostatin analogue. *N. Engl. J. Med. 315*, 663–666.

Lacaze, L., & Scotté, M. (2015). Surgical treatment of intra hepatic recurrence of hepatocellular carcinoma. *World. J. Hepatol. 7*, 1755–1760.

Lai, P., Daear, W., Löbenberg, R., & Prenner, E. J. (2015). Overview of the preparation of organic polymeric nanoparticles for drug delivery based on gelatine, chitosan, poly(d, l-lactide-co-glycolic acid) and polyalkylcyanoacrylate. *Colloids. Surf. B. Biointerfaces. 118*, 154–163.

Lavertu, M., Methot, S., Tran-Khanh, N., & Buschmann, M. D. (2006). High efficiency gene transfer using chitosan/DNA nanoparticles with specific combinations of molecular weight and degree of deacetylation. *Biomaterials. 27*, 4815–4824.

Lee, G. Y., Park, K., Nam, J. H., Kim, S. Y., & Byun, Y. (2006). Anti-tumor and anti-metastatic effects of gelatin–doxorubicin and PEGylated gelatin–doxorubicin nanoparticles in SCC7 bearing mice. *J. Drug. Target. 14*, 707–716.

LeMaistre, C. F., Meneghetti, C., Howes, L., & Osborne, C. K. (1994). Targeting the EGF receptor in breast cancer treatment. *Breast. Cancer. Res. Treat. 32*, 97–103.

Lesche, S., Lehmann, D., Nagel, F., Schmid, H. A., & Schulz, S. (2009). Differential effects of octreotide and pasireotide on somatostatin receptor internalization and trafficking in vitro. *J. Clin. Endocrinol. Metab. 94*, 654–661.

Leyva-Gómez, G., Cortés, H., Magaña, J. J., Leyva-García, N., Quintanar-Guerrero, D., & Florán, B. (2015). Nanoparticle technology for treatment of Parkinson's disease: the role of surface phenomena in reaching the brain. *Drug. Discov. Today. 20*, 824–837.

Liapakis, G., Hoeger, C., Rivier, J., &Reisine, T. (1996). Development of a selective agonist at the somatostatin receptor subtype sstr1. *J. Pharmacol. Exp. Ther. 276*, 1089–1094.

Liew, C. V., Chan, L. W., Ching, A. L., & Heng, P. W. S. (2006). Evaluation of sodium alginate as drug release modifier in matrix tablets. *Int. J. Pharm. 309*, 25–37.

Liu, L., Dong, X., Zhu, D., Song, L., Zhang, H., & Leng, X. G. (2014). TAT-LHRH conjugated low molecular weight chitosan as a gene carrier specific for hepatocellular carcinoma cells. *Int. J. Nanomedicine. 9*, 2879–2889.

Liu, W., Sun, S., Cao, Z., Zhang, X., Yao, K., Lu, W. W., et al., (2005). An investigation on the physicochemical properties of chitosan/DNA polyelectrolyte complexes. *Biomaterials. 26*, 2705–2711.

Lohela, M, Bry, M., Tammela, T., & Alitalo, K. (2009). VEGFs and receptors involved in angiogenesis versus lymphangiogenesis. *Curr. Opin. Cell. Biol. 21*, 154–165.

Lu, Z., Yeh, T. K., Tsai, M., Au, J. L., &Wientjes, M. G. (2004). Paclitaxel-loaded gelatin nanoparticles for intravesical bladder cancer therapy. *Clin. Cancer. Res. 10*, 7677–7684.

Lucas, F., Shaw, J. T., & Smith, S. G. (1957). The amino acid sequence in a fraction of the fibroin of Bombyxmori. *Biochem. J. 66*, 468–479.

Maeda, H. (2001). The enhanced permeability and retention (EPR) effect in tumor vasculature: the key role of tumor-selective macromolecular drug targeting. *Adv. Enzyme. Regul. 41*, 189–207.

Maeda, H. (2010). Tumor-Selective Delivery of Macromolecular Drugs via the EPR Effect: Background and Future Prospects. *Bioconjug. Chem. 21*, 797–802.

Maeda, H., Wu, J., Sawa, T., Matsumura, Y., & Hori, K. Tumor (2000). vascular permeability and the EPR effect in macromolecular therapeutics: a review. *J. Contr. Rel. 65*, 271–284.

Magalhães, M., Farinha, D., PedrosodeLima, M. C.,& Faneca, H. (2014). Increased gene delivery efficiency and specificity of a lipid-based nanosystem incorporating a glycolipid. *Int. J. Nanomedicine. 9*, 4979–4989.

Makino, A., & Kimura, S. (2014). Solid Tumor-Targeting Theranostic Polymer Nanoparticle in Nuclear Medicinal Fields. *Scientific. World. Journal. 2014*, 424513.

Malhotra, A., & Mittal, B. R. (2014). SiRNA gene therapy using albumin as a carrier. *Pharmacogenet. Genomics. 24*, 582–587.

Maluccio, M., & Covey, A. (2012). Recent progress in understanding, diagnosing, and treating hepatocellular carcinoma. *CA. Cancer. J. Clin. 62*, 394–399.

Man, X. Y., Yang, X. H., Cai, S. Q., Yao, Y. G., & Zheng, M. (2006). Immunolocalization and e□pressionof vascular endothelial growth factor receptors (VEGFRs) and neuropilins (NRPs) on keratinocytes in human epidermis. *Mol. Med. 12*, 127–136.

Mansoori, G. A., Brandenburg, K. S., & Shakeri-Zadeh, A. (2010). A comparative study of two folate-conjugated gold nanoparticles for cancer nanotechnology applications. *Cancers. 2*, 1911–1928.

Marelli, U. K., Rechenmacher, F., Sobahi, T. R., Mas-Moruno, C., & Kessler, H. (2013). Tumor targeting via integrin ligands. *Front. Oncol. 3*, 222.

Mariniello, B., Finco, I., Sartorato, P., Patalano, A., Iacobone, M., Guzzardo, V., Fassina, A., & Mantero, F. (2011). Somatostatin receptor expression in adrenocortical tumors and effect of a new somatostatin analog SOM230 on hormone secretion in vitro and in ex vivo adrenal cells. *J. Endocrinol. Invest. 34*, e131–138.

Masis, N., Reed, D., McCool, B., Cooper, J., & Lyford, C. (2013). Assessment of cancer risk in two rural West Texas communities using anthropometrics, diet, and physical activity. *Open. J. Prev. Med. 3*, 285–292.

McCall, R. L., & Sirianni, R. W. (2013). PLGA nanoparticles formed by single- or double-emulsion with vitamin E-TPGS. *J. Vis. Exp. 27*, 51015.

Mehra, N. K., Mishra, V., & Jain, N. K. (2013). Receptor-based targeting of therapeutics. *Ther. Deliv. 4*, 369–394.

Meng, F., Hennink, W. E., & Zhong, Z. (2009). Reduction-sensitive polymers and bioconjugates for biomedical applications. *Biomaterials. 30*, 2180–2198.

Misra, R., Das, M., Sahoo, B. S., & Sahoo, S. K. (2014). Reversal of multidrug resistance in vitro by co-delivery of MDR1 targeting siRNA and DOX using a novel cationic poly(lactide-co-glycolide) nanoformulation. *Int. J. Pharm. 475*, 372–384.

Mottaghitalab, F., Farokhi, M., Shokrgozar, M. A., Atyabi, F., & Hosseinkhani, H. (2015). Silk fibroin nanoparticle as a novel drug delivery system. *J. Control Release. 206*, 161–176.

Mourya, V. K., & Inamdar, N. N. (2008). Chitosan-modifications and applications: opportunities galore. *React. Funct. Polym. 6*, 1013–1051.

Mulvey, C. K., & Bergsland, E. K. (2016). Systemic Therapies for Advanced Gastrointestinal Carcinoid Tumors. *Hematol. Oncol. Clin. North. Am. 30*, 63–82.

Narayanan, S., & Kunz, P. L. (2015). Role of somatostatin analogues in the treatment of neuroendocrine tumors. *J. Natl. Compr. Canc. Netw. 13*, 109–117.

Narayanan, S., & Kunz, P. L. (2016). Role of Somatostatin Analogues in the Treatment of Neuroendocrine Tumors. *Hematol. Oncol. Clin. North . Am. 30*, 163–177.

Navath, R. S., Kurtoglu, Y. E., Wang, B., Kannan, S., Romero, R., & Kannan, R. M. (2008). Dendrimer-drug conjugates for tailored intracellular drug release based on glutathione levels. *Bioconjugate. Chem. 19*, 2446–2455.

Neumann, E., Frei, E., Funk, D., Becker, M. D., & VSchrenk, H. H. (2010). Native albumin for targeted drug delivery. *Exp. Opin. Drug. Deliv. 7*, 915–925.

Numata, K., Mieszawska-Czajkowska, A. J., Kvenvold, L. A., & Kaplan, D. L. (2012). Silk-based nanocomplexes with tumor-homing peptides for tumor-specific gene delivery. *Macromol. Biosci. 12*, 75–82.

Numata, K., Reagan, M. R., Goldstein, R. H., Rosenblatt, M., & Kaplan, D. L. (2011). Spider silk-based gene carriers for tumor cell-specific delivery. *Bioconjug. Chem. 22*, 1605–1610.

O'Toole, D., Ducreux, M., Bommelaer, G., Wemeau, J. L., Bouché, O., Catus, F., Blumberg, J., & Ruszniewski, P. (2000). Treatment of carcinoid syndrome: a prospective crossover evaluation of lanreotide versus octreotide in terms of efficacy, patient acceptability, and tolerance. *Cancer. 15*, 88, 770–776.

Oberg, K. (2010). Cancer: antitumor effects of octreotide LAR, a somatostatin analog. *Nat. Rev. Endocrinol. 6*, 188–189.

Oyarzun-Ampuero, F. A., Goycoolea, F. M., Torres, D., & Alonso, M. J. (2011). A new drug nanocarrier consisting of polyarginine and hyaluronic acid. *Eur. J. Pharm. Biopharm. 79*, 54–57.

Pereira, Ade, F., Pereira, L. G., Barbosa, L. A., Fialho, S. L., & Pereira, B. G. (2013). Efficacy of methotrexate-loaded poly(e-caprolactone) implants in Ehrlich solid tumor-bearing mice. *Drug. Deliv. 20*, 168–179.

Prabhu, R. H., Patravale, V. B., & Joshi, M. D. (2015). Polymeric nanoparticles for targeted treatment in oncology: current insights. *Int. J. Nanomedicine. 10*, 1001–1018.

Raderer, M., Hamilton, G., Kurtaran, A., Valencak, J., & Haberl, I. (1999). Treatment of advanced pancreatic cancer with the long-acting somatostatin analogue lanreotide: in vitro and in vivo results. *Br. J. Cancer. 79*, 535–537.

Reubi, J. C., Mazzucchelli, L., Hennig, I., & Laissue, J. A. (1996). Local upregulation of neuropeptide receptors in host blood vessels around human colorectal cancers. *Gastroenterology. 110*, 1719–1726.

Reyes-López, M., Piña-Vázquez, C., & Serrano-Luna, J. (2015). Transferrin: Endocytosis and Cell Signaling in Parasitic Protozoa. *Biomed. Res. Int. 2015*, 641392.

Rinaudo, M. (2006). Chitin and chitosan: Properties and applications. *Prog. Polym. Sci. 31*, 603–632.

Rinaudo, M. (2008). Main properties and current applications of some polysaccharides as biomaterials. *Polym. Int. 3*, 397–430.

Rodríguez-Cabello, J. C., Arias, F. J., Rodrigo, M. A., & Girotti, A. (2016). Elastin-like polypeptides in drug delivery. *Adv. Drug. Deliv. Rev. 97*, 85–100.

Roggenbuck, D., Mytilinaiou, M. G., Lapin, S. V., Reinhold, D., & Conrad, K. (2012). Asialoglycoprotein receptor (ASGPR), a peculiar target of liver-specific autoimmunity. *Auto. Immun. Highlights. 3*, 119–125.

Roy, D., Brooks, W. L., & Sumerlin, B. S. (2013). New directions in thermoresponsive polymers. *Chem. Soc. Rev. 42*, 7214–7243.

Sandberg, L. B., Soskel, N. T., & Leslie, J. G. (1981). Elastin structure, biosynthesis, and relation to disease states. *N. Engl. J. Med. 304*, 566–579.

Schonbrunn, A. H., & Tashijan, A. H. (1978). Characterization of functional receptors for somatostatin in rat pituitary cells in culture. *J. Biol. Chem. 253*, 6473–6483.

Seabra, A. B., Justo, G. Z., & Haddad, P. S. (2015). State of the art, challenges and perspectives in the design of nitric oxide-releasing polymeric nanomaterials for biomedical applications. *Biotechnol. Adv. 33*, 1370–1379.

Sharma, G., Park, J., Sharma, A. R., Jung, J. S., Kim, H., Chakraborty, C., et al., (2015). Methoxy poly(ethylene glycol)-poly(lactide) nanoparticles encapsulating quercetin act as an effective anticancer agent by inducing apoptosis in breast cancer. *Pharm. Res. 32*, 723–735.

Shen, Z., Li, Y., Kohama, K., Oneill, B., & Bi, J. (2011). Improved drug targeting of cancer cells by utilizing actively targetable folic acidconjugated albumin nanospheres. *Pharmacol. Res. 63*, 51–58.

Shenoy, D., Little, S., Langer, R., & Amiji, M. (2005). Poly(ethylene oxide)-modified poly(beta-amino ester) nanoparticles as a pH-sensitive system for tumor-targeted delivery of hydrophobic drugs. 1. In vitro evaluations. *Mol. Pharm. 2*, 357–366.

Shi, B., & Abrams, M. (2013). Sepp-Lorenzino L: Expression of asialoglycoprotein receptor 1 in human hepatocellular carcinoma. *J. Histochem. Cytochem. 61*, 901–909.

Skandrani, N., Barras, A., Legrand, D., Gharbi, T., Boulahdour, H., & Boukherroub, R. (2014). Lipid nanocapsules functionalized with polyethyleneimine for plasmid DNA and drug co-delivery and cell imaging. *Nanoscale. 6*, 7379–7390.

Song, N., Liu, W., Tu, Q., Liu, R., Zhang, Y., & Wang, J. (2011). Preparation and in vitro properties of redox-responsive polymeric nanoparticles for paclitaxel delivery. *Colloids. Surf. B. 87*, 454–463.

Stengel, A., Rivier, J., & Taché, Y. (2013). Central actions of somatostatin-28 and oligosomatostatin agonists to prevent components of the endocrine, autonomic and visceral responses to stress through interaction with different somatostatin receptor subtypes. *Curr. Pharm. Des. 19*, 98–105.

Sun, G., Hsueh, P. Y., Janib, S. M., Hamm-Alvarez, S., & Andrew MacKay, J. (2011). Design and cellular internalization of genetically engineered polypeptide nanoparticles displaying adenovirus knob domain. *J. Control Release. 155*, 218–226.

Sun, J., & Tan, H. (2013). Alginate-based biomaterials for regenerative medicine applications. *Materials. 6*, 1285–1309.

Sun, L. C., & Coy, D. H. (2011). Somatostatin Receptor-Targeted Anti-Cancer Therapy. *Curr. Drug. Deliv. 8*, 2–10.

Tammam, S. N., Azzazy, H. M., & Lamprecht, A. (2015). Biodegradable Particulate Carrier Formulation and Tuning for Targeted Drug Delivery. *J. Biomed. Nanotechnol. 11*, 555–577.

Tang, X., Cai, S., Zhang, R., Liu, P., Chen, H., Zheng, Y., & Sun, L. (2013). Paclitaxel-loaded nanoparticles of star-shaped cholic acid-core PLA-TPGS copolymer for breast cancer treatment. *Nanoscale. Res. Lett. 8*, 420.

Tatham, A. S., & Shewry, P. R. (2000). Elastomeric proteins: Biological roles, structures and mechanisms. *Trends. Biochem. Sci. 25*, 567–571.

Thu, H. P., Nam, N. H., Quang, B. T., Son, H. A., Toan, N. L., & Quang, D. T. (2015). In vitro and in vivo targeting effect of folate decorated paclitaxel loaded PLA-TPGS nanoparticles. *Saudi. Pharm. J. 23*, 683–688.

Torchilin, V. (2011). Tumor delivery of macromolecular drugs based on the EPR effect. *Adv. Drug. Deliv. Rev. 63*, 131–135.

Tseng, H. H., Chang, J. G., Hwang, Y. H., Yeh, K. T., Chen, Y. L., & Yu, H. S. (2009). Expression of hepcidin and other iron-regulatory genes in human hepatocellular carcinoma and its clinical implications. *J. Cancer. Res. Clin. Oncol. 135*, 1413–1420.

Ulbrich, K., Etrych, T., Chytil, P., Jelínková, M., & Ríhová, B. (2004). Antibody-targeted polymer-doxorubicin conjugates with pH-controlled activation. *J. Drug. Target. 12*, 477–489.

Urry, D. W. (1997). Physical chemistry of biological free energy transduction as demonstrated by elastic protein-based polymers. *J. Phys. Chem. B. 101*, 11007–11028.

Vander, Heiden, M. G. (2011). Targeting cancer metabolism: a therapeutic window opens. *Nat. Rev. Drug. Discov. 10*, 671–684.

Voth, B., Nagasawa, D. T., Pelargos, P. E., Chung, L. K., Ung, N., Gopen, Q., Tenn, S., et al., (2015). Transferrin receptors and glioblastomamultiforme: Current findings and potential for treatment. *J. Clin. Neurosci. 22*, 1071–1076.

Waser, B., Cescato, R., Liu, Q., Kao, Y. J., & Körner, M. (2012). Phosphorylation of sst2 Receptors in Neuroendocrine Tumors after Octreotide Treatment of Patients. *Am. J. Pathol. 180*, 1942–1949.

Watson, J. C., Balster, D. A., Gebhardt, B. M., O'Dorisio, T. M., O'Dorisio, M. S., Espenan, G. D., et al., (2001). Growing vascular endothelial cells express somatostatin subtype 2 receptors. *Br. J. Cancer. 85*, 266–272.

Weber, C., Reiss, S., & Langer, K. (2000). Preparation of surface modified protein nanoparticles by introduction of sulfhydryl groups. *Int. J. Pharm. 211*, 67–78.

Woodruff, M. A., & Hutmacher, D. W. (2010). The return of a forgotten polymer-polycaprolactone in the 21st century. *Prog. Polym. Sci. 35*, 10233–10237.

Xu, Z., Zhang, Z., Chen, Y., & Chen, L. (2010). The characteristics and performance of a multifunctional nanoassembly system for the co-delivery of docetaxel and iSur-pDNA in a mouse hepatocellular carcinoma model. *Biomaterials. 31*, 916–922.

Yan, J. J, Liao, J. Z., Lin, J. S., & He, X. X. (2015). Active radar guides missile to its target: receptor-based targeted treatment of hepatocellular carcinoma by nanoparticulate systems. *Tumour. Biol. 36*, 55–67.

Yan, J. J., Liao, J. Z., Lin, J. S., & He, X. X. (2015). Active radar guides missile to its target: receptor-based targeted treatment of hepatocellular carcinoma by nanoparticulate systems. *Tumor. Biol. 36*, 55–67.

Yu, C. Y., Li, N. M., Yang, S., Ning, Q., Huang, C., Huang, W., et al., (2015). Fabrication and characterization of GC-FUA based nanoparticle drug delivery systems and their slow release profiles. *J. Appl. Polym. Sci. 132*, 11854.

Yu, C. Y., Wang, Y. M., Li, N. M., Liu, G. S., Yang, S., Tang, G. T., et al., (2014). In vitro and in vivo evaluation of pectin-based nanoparticles for hepatocellular carcinoma drug chemotherapy. *Mol. Pharm. 11*, 638–644.

Yuan, Z., Qu, X., Wang, Y., Zhang, D. Y., Luo, J. C., Jia, N., & Zhang, Z. T. (2015). Enhanced antitumor efficacy of 5-fluorouracil loaded methoxy poly(ethylene glycol)-poly(lactide) nanoparticles for efficient therapy against breast cancer. *Colloids. Surf. B. Biointerfaces. 128*, 489–497.

Zhang, C. Y., Yang, Y. Q., Huang, T. X., Zhao, B., Guo, X. D., Wang, J. F., et al., (2012). Self-assembled pH-responsive MPEG-b-(PLA-co-PAE) block copolymer micelles for anticancer drug delivery. *Biomaterials. 33*, 6273–6283.

Zhang, J., Jiang, X., Jiang, Y., Guo, M., Zhang, S., Li, J., et al., (2015). Recent advances in the development of dual VEGFR and c-Met small molecule inhibitors as anticancer drugs. *Eur. J. Med. Chem. 108*, 495–504.

Zhang, J., Lan, C. Q., Post, M., Simard, B., Deslandes, Y., & Hsieh, T. H. (2006). Design of nanoparticles as drug carriers for cancer therapy. *Cancer. Genom. Proteom. 3*, 147–157.

Zhang, J., Tang, C., & Yin, C. (2013). Galactosylatedtrimethyl chitosan-cysteine nanoparticles loaded with Map4k4 siRNA for targeting activated macrophages. *Biomaterials. 34*, 3667–3677.

Zhao, J., & Feng, S. S. (2015). Nanocarriers for delivery of siRNA and co-delivery of siRNA and other therapeutic agents. *Nanomedicine. 10*, 2199–2228.

Zhao, Z., Meng, H., Wang, N., Donovan, M. J., Fu, T., You, M., et al., (2013). A controlled-release nanocarrier with extracellular pH value driven tumor targeting and translocation for drug delivery. Angew. *Chem. Int. Ed. Engl. 52*, 7487–7491.

Zhu, L., & Torchilin, V. P. (2013). Stimulus-responsive nanopreparations for tumor targeting. *Integr. Biol. 5*, 96–107.

CHAPTER 7

SMART DELIVERY OF NANOBIOMATERIALS IN DRUG DELIVERY

MIRZA SARWAR BAIG,[1] RAJ K. KESERVANI,[2]
MOHD FASIH AHMAD,[3] and MIRZA EHTESHAM BAIG[3]

[1]Department of Biosciences, Jamia Millia Islamia
(A Central University), New Delhi, 110025, India

[2]Faculty of Pharmaceutics, Sagar Institute of Research and
Technology-Pharmacy, Bhopal-462041, India,
E-mail: rajksops@gmail.com

[3]Barrocyte Pvt. Ltd., 105, First Floor, Block No-CHBS, Sukhdev Vihar
CSC, Near Fortis-Escorts Hospital, New Delhi, 110025, India

CONTENTS

7.1 INTRODUCTION

One of the most critical problems in the treatment of many diseases is due to the disproportionate delivery of drugs and therapeutic agents to the target sites. The conventional approach of medication is characterized by insufficient cellular uptake, limited effectiveness, poor bio-distribution, low solubility, lack of selectivity and rapid drug elimination (Nevozhay et al., 2007). These limitations can be overcome by targeted nano-scale drug delivery systems (DDS), where the drug and therapeutic agents is transported to the exact site of action in effective concentration for right period of time, thus influences only vital tissues and reduces the undue side effects. Nano-scale DDS can be optimized for designed parameters to tune release kinetics, regulate delivery location, size of nanobiomaterial (NM), surface properties of NM, and minimize toxic side effects, thereby enhancing the therapeutic index of a given drug (Kayser et al., 2005).

Drug-encapsulated nanobiomaterials (NMs) are typically 20–500 nm hydrophobic spherical particles. The transport properties of these NMs are comparable with those of macromolecules like LDL (MW $\approx$ 2500 kDa), which are 25–50 nm and are likewise hydrophobic in nature. The specific NMs have the potential to revolutionize therapies for cardiovascular diseases such as diffuse lesions, multi vessel disease and vulnerable plaque. It can also be used for other drug delivery therapies apart from cardiovascular domain, for example, cancer, diabetic nephropathy, congestive heart failure, as well as oral biologics delivery. However, researchers have shed light on the possible behavior of these drug-encapsulated NMs under diseased conditions regarding the general transport properties in healthy conditions. This next generation medication is especially important when there is a discrepancy between the size and shape, dose or concentration of drug-encapsulated NMs and its therapeutic results or toxic effects (Gratton et al., 2008). There should be an import regulatory checkpoints to measure drug delivery efficiency, and thus effectiveness of various drug delivery systems.

The aim of this chapter is firstly to give readers a holistic preview of NMs or nanoparticles (NPs) and its application in life sciences and medicine, secondly an insight into the drug delivery system (DDS) plus most recent developments in this field, and finally to discuss the limitations and future of NMs.

7.2   THE BEGINNING OF NANOBIOMATERIALS (NMS) AND NANOPARTICLES (NPS)

Firstly, understand what is a biomaterial? Biomaterials are generally substances of varying shapes and sizes other than food or drugs contained in therapeutic or diagnostic systems that are in contact with tissue or biological fluids (Langer and Peppas, 2003). However, the prefix "nano" deals with sizes of 100 nm or smaller in size in at least one dimension. Now, we are able to define NMs, which are highly specific structures often spherical, cylindrical and plate-like, composed of inorganic or polymeric NPs. NMs can be used as a convenient surface for molecular assembly, and may be used as vehicle for the drug molecules or therapeutic compounds (Salata, 2004). It can also be in the form of nano-vesicle surrounded by a membrane or a layer. The shape, size and size distribution of NMs are important in the context of positive effect and side effects for example, the effect of size and shape on the toxicity of gold NPs has been investigated by Zhang and his research group.

According to National Nanotechnology Initiative (NNI), NPs are defined as the structures ranging from 1 to 100 nm in size in at least one dimension (1D). Therefore, NPs have been defined as submicron-sized polymeric colloidal system with a therapeutic agent of interest encapsulated within their polymeric matrix or adsorbed or conjugated onto the surface (Panyam and Labhasetwar, 2003). The NPs may be biodegradable or not. The first example of lipid vesicles which later became known as liposomes were described in 1965 (Bangham et al., 1965), the first controlled release polymer system of macromolecules was described in 1976 (Langer and Folkman, 1976; Birrenbach and Speiser, 1976), the first long circulating stealth polymeric nanoparticle was described in 1994 (Gref et al., 1994), the first quantum dot bioconjugate was described in 1998 (Bruchez et al., 1998; Chan et al., 1998), and the first nanowire nanosensor dates back to 2001 (Cui et al., 2001).

Generally, NPs have been categorized in polymers, ceramics, metals, and biological materials with various forms. Each structure offers unique characteristics that make it a suitable drug delivery candidate for a particular therapy. NPs have even been shown to cross the blood–brain barrier (Roney et al., 2005). The drug is dissolved, entrapped, encapsulated, or attached to a NP matrix. Depending upon the process used for the preparation of NP, nanospheres or nanocapsules, nanotubes, nanogels, silver and gold NPs can be obtained (Kumar, 2000; Lambert et al., 2001; Soppimath et al., 2001). While spherical NPs are the simplest to create, other shapes and constructions offer advantages for certain applications. These supramolecular NPs include nanocapsules, nanotubes, nanogels and dendrimers. Nanocapsules are vesicular systems in which the drug is confined to a cavity (an oil or aqueous core) surrounded by a unique polymeric membrane. Nanospheres are matrix systems in which the drug is physically and uniformly dispersed throughout the particles.

Polymeric NPs can be fabricated in a wide range of sizes and varieties (Raghuvanshi et al., 2001; Kreuter et al., 2003). In addition to the steady drug release for weeks, biodegradable polymeric NPs do not accumulate in the body (Soppimath et al., 2001; Panyam and Labhasetwar, 2003). Polymeric micelles are another form of non-biodegradable polymeric NPs. Because of their minute size (<100 nm) polymeric micelles can be an efficiently and effectively engineer to deliver drugs to target solid tumors and destroy only the cancer cells (Panyam and Labhasetwar, 2003). These nanostructures are physiologically stable in biological environment, and can be engineer their hydrophobic segments thus can deliver water-insoluble drugs securely to the targeted location.

Liposomes (80–300 nm in size) are nano-spherical biological materials that are synthesized from steroids and non-toxic phospholipids bilayer (Sunderland et al., 2006). Because they are natural materials, liposomes are considered as harmless drug delivery carriers that can circulate in the blood stream for a long time. They increase the solubility of drugs and improve their pharmacokinetic properties, such as the therapeutic index of chemotherapeutic agents, rapid metabolism, and reduction of side effects and increase of in vitro and in vivo anticancer activity (Giuberti et al., 2011). There are a lot of drug examples in liposomal formulations, such as anticancer drugs (Hofheinz et al., 2005; Giuberti et al., 2011), neurotransmitters (serotonin) (Afergan et al., 2008), antibiotics (Turkova et al., 2011; Yukihara et al., 2011), anti-inflammatory (Paavola et al., 2000) and antirheumatic drugs (van den Hoven et al., 2011).

Ceramic NPs are unique porous inorganic molecules which are having compatibility with the biological system. Therefore, these biocompatible ceramic NPs such as silica, titania, alumina are used in cancer therapy. A recent study suggests that, because of their physiological stability, hybridizing ceramic NPs with DNA offers a potential therapy to target liver cancer cells (Dey, 2005). Current investigation on entrapping protein in ceramic NPs shows very promising results for orally administered drugs. For example, the release profile of insulin entrapped in hydro yapatite NPs shows very good results for orally administered insulin instead of repeatable injections (Paul and Sharma, 2001).

Metal like gold or silver NPs can be synthesized in extremely small sizes (<50 nm), and have ability to carry a relatively high drug dose. Recent research reveals that gold NPs can be functionalized into a composite system to carry both an anti-angiogenic and anticancer agent in one gold NP (Priyabrata et al., 2005). By doing so, the tumor's cells can be destroyed at the same time the survival of other possible tumors can be prevented. Metal nanoshells are also being explored to encapsulate drugs inside their hollow core (Sun et al., 2002). These metal NPs release their drug into the targeted location using an external exciting source like an infrared light or a magnetic field.

Other NPs such as dendrimers (<5 nm) are emerging as a new class of polymeric nanosystems with applications in drug delivery. These systems are built from a series of branches around an inner core. They can be synthesized from almost any core molecule and the branches similarly constructed from any bi-functional molecule. The star-shaped distinctive architecture of dendrimers has capabilities to load drug molecules in either the interior or attached to the surface groups. In a recent in vivo study, dendrimic polymers have been loaded with an anticancer drug to selectively target tumor cells (Latallo et al., 2005). Carbon nanotubes are extremely small tubes with either a single or multi wall carbon structure. The exclusive structures of these tiny tubes make them an appealing candidate to encapsulate drugs inside their cavities. At the present, these nanotubes are still under widespread investigation in laboratories. However, the toxicity of these carbon-based tubes is of concern, particularly if they circulate in the blood stream or accumulate in the targeted tissue.

By the advent of advancement in medicinal chemistry, computationl chemistry, bioinformatics and nanoinformatics varieties of materials were used to synthesize NPs (Hawker and Wooley, 2005). Recent novel examples of nonpolymer-based NPs, include calcium carbonate, calcium-deficient hydro yapatite, chitosan, oligo-3-hydro ybutyrates and porous hollow silica (Chen et al., 2004; Piddubnyak et al., 2004;

Ueno et al., 2005; Liu et al., 2005; Bodnar et al., 2005). Polymer micelles have emerged as a very important system for targeted drug delivery. Self-assembly of amphiphilic block in aqueous media leads to NPs with hydrophobic cores for drug encapsulation and hydrophilic shells for stabilization and specific targeting (Nasongkla et al., 2004). The potential utility of polymer micelle-aptamer bioconjugates that encapsulate docetaxel for therapeutic applications, were applied on prostate cancer as a model system (Farokhzad et al., 2006). Intelligent core-shell NPs have also been produced: thermo-responsive, pH-responsive and biodegradable NPs were made from poly (D, Llactide)-graft-poly (N-isopropyl acrylamide-co-methacrylic acid) (PLA-g-P(NIPAm-co-MAA)) to yield a hydrophilic outer shell and a hydrophobic inner core that exhibited a phase transition temperature above 37°C, rendering them apposite for biomedical applications (Lo et al., 2005).

In recent years, biodegradable polymeric NMs are extensively used as potential nano-scale drug delivery system (DDS). Their capability of controlled drug release and to target particular organs or tissue, they are attractively applied as carriers of oligonucleotides in antisense therapy, DNA in gene therapy. Similarly, these biodegradable polymeric NPs easily deliver proteins, peptides and genes through oral administration (Langer, 2000). These applications of nanoparticulate biomaterial systems will be discussed below in more detail.

7.3 NANOBIOMATERIALS AS CARRIERS OF THERAPEUTIC MOLECULES

The "magic bullet" concept was first theorized by Paul Ehrlich in 1891 which represents the first early description of the drug-targeting paradigm (Gensini et al., 2006). The aim of nano-scale drug targeting is to deliver drugs and therapeutic molecules to the right place, at the right concentration, for the right period of time. Nanobiomaterials (NMs) or nanoparticles (NPs) can be functionalized with therapeutic molecules through various techniques, including physical adsorption, electrostatic binding, complementary recognition and covalent coupling (Katz et al., 2004). Drugs can be adsorbed onto the surface of NMs, entrapped inside the polymer NPs, or dissolved within the particle matrix. Interestingly, NMs can carrier more than one drug within the particles.

Most of the time, nanocarriers used for medical applications have to be biocompatible without eliciting immune response or any negative effects and nontoxic. There are several types of nanocarriers for encapsulating or immobilizing or covalently attaching or adsorbing therapeutic molecules. For example, dendrimer nanocarriers, polymeric nanocarriers, silica-based nanocarriers, carbon nanocarriers, magnetic nanocarriers, etc. Undesirable effects of NPs strongly depend on their hydrodynamic size, shape, amount, surface chemistry, the route of administration, reaction of the immune system (especially by macrophages and granulocytes) and residence time in the bloodstream. We are discussing these nanocarriers in details stating their characteristic properties, applications and their limitations.

Dendrimers are mono-dispersed symmetric NM built around a small molecule with an internal cavity surrounded by a large number of reactive end groups. Biocompatibility and physicochemical properties of dendrimers are determined by surface reactive or functional groups (Caminade et al., 2005). Some of the examples of nanometric molecules possessing dendritic structure include: glycogen, amylopectin, and proteoglycans (Svenson and Tomalia, 2005). Cytotoxicity of dendrimers and their so-called polyvalence (number of active groups on a dendrimers surface) is particularly relevant for biomedical purposes. For example, changing the surface amine groups into hydroxyl ones may result in lower levels of cytotoxicity. The drug may be encapsulated in the internal structure of dendrimers or it can be chemically attached or physically adsorbed on dendrimers surface (Emanuele and Attwood, 2005; Menjoge et al., 2010). The choice of the immobilization method depends on the properties of drugs. Encapsulation is used when drugs are labile, toxic, or poorly soluble. The surface of dendrimers provides an e□cellentplatform for an attachment of specific ligands, which may include folic acid (Singh et al., 2008), antibodies (Wängler et al., 2006), cyclic targeting peptides – arginine-glycine-aspartic acid (RGD) (Waite et al., 2009), selective A3 adenosine receptor (Tosh et al., 2010), silver salts complexes antimicrobial agents (Balogh et al., 2001), or poly ethylene glycol (PEG) (Lope et al., 2009). Poly amido amide (PAMAM) is a dendrimer which is frequently used in biomedical applications. An example of a drug immobilization in PAMAM dendrimer is cisplatin. This complex compared with free cisplatin exhibits several advantages, such as slower rate of drug release, higher accumulation of the drug in solid tumors, and lower toxicity in all organs (Malik et al., 1999; Emanuele and Attwood, 2005). Some other examples of PAMAM dendrimer as nanocarriers are anticancer drugs, including methotrexate (Natali et al., 2010), doxorubicin (Han et al., 2010), 5-FU (Singh et al., 2008), and anti-inflammatory drugs, for e☐ample,ibuprofen (Tanis et al., 2009), piroxicam (Prajapati et al., 2009), or indomethacin (Chauhan et al., 2009). Dendrimers can modulate cytokine and chemokine release. The PAMAM generation of 3,5-glucosamine dendrimers induces a synthesis of pro-inflammatory chemokines, such as MIP-1a, MIP-1h, and cytokines TNF-a, IL-1h, IL-6, IL-8 in human dendritic cells and macrophages, which exhibits an immunomodulatory effect.

The polymeric nanoparticle (PNP)-based carriers are obtained from synthetic polymers, such as poly-ε-caprolactone, polyacrylamide and polyacrylate, or natural polymers, for example, albumin, DNA, chitosan, gelatin.-based on in vivo behavior, PNPs may be classified as biodegradable, for e☐ample, poly (L-lactide) (PLA), polyglycolide (PGA), and non-biodegradable, for example, polyurethane (Park et al., 2009; Fritzen-Garcia et al., 2009; Mainardes et al., 2010). Drugs can be immobilized on PNPs surface after a polymerization reaction or can be encapsulated on PNP structure during a polymerization step (Luo et al., 2010; Mora-Huertas et al., 2010). PNPs are usually coated with nonionic surfactants in order to reduce immunological interactions (e.g., opsonization or presentation PNPs to CD8 T-lymphocytes) as well as intermolecular interactions between the surface chemical groups of PNPs (e.g., van der Waals forces, hydrophobic interaction or hydrogen bonding). Moreover, drugs may be released by desorption, diffusion, or nanoparticle erosion in target tissue (Torchilin, 2008). The application of biodegradable nanosystems for the development of nanomedicines is one of the most successful ideas. Nanocarriers composed of biodegradable polymers undergo hydrolysis in the body, producing biodegradable metabolite monomers, such as lactic acid and glycolic acid. The examples of drug-polymeric nanocarrier conjugates used as nano-drug delivery systems (DDS) are shown in Table 7.1.

Carbon nanocarriers used in DDS are differentiated into nanotubes (CNTs) and nanohorns (CNH). CNTs are characterized by unique architecture formed by rolling of single (SWNCTs – single walled carbon nanotubes) or multi (MWCNTs – multi walled carbon nanotubes) layers of graphite with an enormous surface area and an excellent electronic and thermal conductivity (Beg et al., 2011). There are three ways of drug immobilization in carbon nanocarriers, which are: encapsulation of a drug in the carbon nanotube (Tripisciano et al., 2010; Arsawang et al., 2011), chemical adsorption on the surface or in the spaces between the nanotubes (by electrostatic, hydrophobic, P-p interactions and hydrogen bonds) (Chen et al., 2011; Zhang et al.,

TABLE 7.1 Polymer Nanocarriers as Drug Delivery Systems (DDS)

Drug	Therapeutic activity	Nanocarrier	Reference
Analogue of b-lactam	Antibiotic, infection G(+)bacteria	Polyacrylate	Turos et al., 2007
Capecitabine	Pro-drug of fluorouracil, metastatic colorectal and breast cancer	CS-poly (ethylene oxide-g-acrylamide)	Agnihotri et al., 2006
Carboplatin	Antineoplastic drug, ovarian, head, neck and lung cancer	Sodium alginate	Nanjwade et al., 2010
Clotrimazole	Antifungal drug	PLA-co-PLG	Pandey et al., 2005
Cyclosporine A	Cyclic polypeptide, immunosuppressant	GMO/poloxamer 407 cubic nanoparticles	Lai et al., 2010
Doxorubicin	Antineoplastic agent, wide spectrum of tumors	PEGylatedPLGA	Park et al., 2009
Lamivudine	Anti-HIV drug	PLA/CS	Dev et al., 2010
Mitomycin C	Chemotherapeutic agent, bladder tumors	CS, CS-PCL, PLL-PCL	Bilensoy et al., 2009
Retinyl acetate EC	Photoaging, severe acne and skin inflammation	EC	Arayachukeat et al., 2011
Rifampicin	Antitubercular drugs, latent *M. tuberculosis* infection in adults	Gelatin	Saraog et al., 2010
Tacrine	Anti-Alzheimer drug	CS	Wilson et al., 2010
5-Fluorouracil (5-FU)	Anticancer drug, colon cancer	CS-g-poly (N-vinyl caprolactam)	Rejinold et al., 2011

CS – chitosan, PEGylated PLGA – PEGylated poly (lactic-co-glycolic acid), PCL – polycaprolactone, PLL – poly (L-lysine), GMO – glyceryl monooleate, EC – ethyl cellulose, PLA – polylactide, PLG – polyglycolide.

2011), and attachment of active agents to functionalized carbon nanotubes (f-cNTs). Drug release from carbon nanotubes can be electrically or chemically controlled. The toxicity of carbon nanomaterials also depends on their unique well-defined geometric structure (Jia et al., 2005). In addition, some impurities, such as residual metal and amorphous carbon, contribute to the level increase of reactive oxygen species (ROS), thus, inducing the oxidative stress in cells (Dobrovolskaia and McNeil, 2007). The examples of drugs that were attached to CNTs are listed in Table 7.2.

Magnetic nanoparticles (MNPs) exhibit a wide variety of attributes, which make them highly promising nanocarriers for DDS. MNPs, for

TABLE 7.2   Carbon Nanotubes as Drug Delivery Systems (DDS)

Type of nanotubes	Drug	Method of immobilization	Reference
f-cNTs	Amphotericin B	Conjugated to carbon nanotubes	Prajapati et al., 2011
f-cNTs	Sulfamethoxazole	Adsorption	Zhang et al., 2011
MWCNTs	Cisplatin	Encapsulation *via* capillary forces	Tripisciano et al., 2010
MWCNTs	Dexamethasone	Encapsulation	Luo et al., 2011
MWCNTs @ poly (ethylene glycol-*b*-propylene sulfide)	Doxorubicin	Adsorption	Di Crescenzo et al., 2011
MWNTs	Epirubicin	Hydrochloride Adsorption	Cho et al., 2009
SWCNTs	Gemcitabine	Encapsulation	Arsawang et al., 2011
SWNTs-PL-PEG- NH2	Pt (IV) prodrug-FA	Covalent amide linkages	Dhar et al., 2008
SWNTs	Cisplatin – EGF	Attachment to carbon nanotubes *via* amide linkages	Bhirde et al., 2009

f-cNTs – functionalized carbon nanotubes, MWNTs – multi walled nanotubes, SWNTs – single-walled carbon nanotubes, PL-PEG-NH2 amine-functionalized single-walled carbon nanotubes.

instance, tend to aggregate into larger clusters loosing the specific properties connected with their small dimensions and making physical handling difficult. In turn, magnetic force may not be strong enough to overcome the force of blood flow and to accumulate magnetic drugs only at target site (Neuberger et al., 2005). Depending on magnetic properties, MNPs can be divided into pure metals such as cobalt (Co), nickel (Ni), manganese (MN), and iron (Fe), their alloys and oxides (Meng et al., 2011; Sayed et al., 2011; Smolensky et al., 2011; Kale et al., 2012). Iron oxide nanoparticles, due to the favourable features they exhibit, are the only type of MNPs approved for clinical use by Food and Drug Administration (FDA). These attributes are: facile single step synthesis by alkaline co-precipitation of Fe^{2+} and Fe^{3+} (Figuerola et al., 2010), chemical stability in physiological conditions (Asmatulu et al., 2005) and possibility of chemical modification by coating the iron o☐ide cores with various shells, for example, golden, polymeric, silane, or dendrimeric. Connecting a drug with MNP may be achieved by covalent binding, electrostatic interactions, adsorption, or encapsulation process (Figuerola et al., 2010; Xiong et al., 2011; Yallapu et al., 2011). Therapeutic activity of diverse drugs incorporated into iron oxide nanocarriers have been tested and reported in Table 7.3.

TABLE 7.3 Magnetic Nanoparticles as Drug Delivery Systems (DDS)

Drug	Therapeutic activity	Nanocarrier (core@shell)	Reference
Anti-*b*-HCG monoclonal, antibody	Choriocarcinoma-specific gene vector	Fe_3O_4@dextran	Jingting et al., 2011
Cisplatin	Chemotherapeutic drug	Fe_3O_4@poly *e*-caprolactone	Yang et al., 2006
Ciprofloxacin	Anti-infective agents (antibiotic)	Fe_3O_4@poly (vinyl alcohol)-*g*-poly (methyl methacrylate)	Bajpai and Gupta, 2011
Dopamine	Catecholamine neurotransmitter, Parkinson's disease	Fe_3O_4@silica (diatom)	Losic et al., 2010
Doxorubicin	Antineoplastic agent	Fe_3O_4@gelatin	Gaihre et al., 2009
Daunorubicin	Chemotherapeutic leukemia drug	Fe_3O_4	Wu et al., 2010
Gemcitabine	Antimetabolites, cancer chemotherapy	Fe_3O_4@poly (ethylene glycol)	Tong et al., 2011
Paclitaxel	Mitotic inhibitor used in cancer, chemotherapy	Fe_3O_4@poly[aniline-co-sodium *N*-(1-butyric acid)	Hua et al., 2010
t-PA	Tissue plasminogen activator, thrombolytic therapy	Fe_3O_4@tetraethyl orthosilicate	Kempe and Kempe 2010
1, 3-Bis (2-chloroethyl)-1-nitrosourea (BCNU)	Anti-cancer chemotherapy drug	Fe_3O_4@poly[aniline-co-*N*-(1-butyric acid) aniline]	Hua et al., 2011
5-Fluorouracil	Antimetabolites, anticancer drug	Fe_3O_4@ ethylcellulose	Arias et al., 2010

@ – functionalization.

Novel nanobiomaterials (NMs) formulations are having enhanced capabilities of site-specific targeting and controlled release of traditional therapeutic molecules, recombinant proteins, vaccines and nucleic acids. Therefore, there is urgent need to develop advanced nano-scale drug-delivery systems (DDS) that can be devised to tune release kinetics, to regulate bio-distribution and to minimize toxic side effects, thereby enhancing the therapeutic index of a given drug (Kayser et al., 2005). These are the reasons why many drugs, for example, antibiotics, antiviral and antiparasitic drugs, cytostatics, vitamins, protein and peptides, including enzymes and hormones have been associated with NMs or NPs. For example, incorporating aminoglycoside antibiotics into NPs have reduced drug-induced nephrotoxicity (Schiffelers et al., 2001; Pinto-Alphandary, 2000). There is another formulation Amikacin-liposome which is undergone in clinical trial, to maximize the therapeutic index and minimize the toxicity of this antibiotic. It was found that overall therapeutic index is improved compared to conventional amikacin. Amikacin encapsulated in unilamellar liposomes (MiKasome®) has prolonged circulation time and sustained efficacy in animals infected with *Mycobacterium* spp., *Klebsiella* spp., & *Pseudomonas* spp. (Fielding et al., 1998). The reason behind the effectiveness of these NM formulations, such as Amikacin-liposome is hypothesized that the intracellular acidic pH promotes release of the amikacin after phagocytosis by the macrophage effectively delivering drug to the intracellular pathogens (Schiffelers et al., 2001). Therefore, treatment of intracellular pathogens such as *Mycobacterial* spp. using unilamellar liposomal NMs or NPs and the macrophage as the carrier is improved compared to free drug (Sharma et al., 2004). The majority of pharmaceutical research on nanoparticle DDS has been done in the area of oncology. These nanocarriers can concentrate preferentially in tumor masses, inflammatory sites, and infectious sites by virtue of the enhanced permeability and retention (EPR) effect on the vasculature (Shenoy et al., 2005). Characteristics of the EPR effect include extensive angiogenesis, defective vascular architecture, impaired lymphatic drainage/recovery system, and increased production of a number of permeability mediators. These EPR characteristics are essential components of solid tumors and are not associated with normal tissues or organs. The EPR effect provides an opportunity for more selective targeting of polymer conjugated anticancer drugs to the tumor (Maeda et al., 2000; Vicent and Duncan, 2006; Sengupta et al., 2005). Future generation systems will include biosensing functionalities with in vivo feedback that will permit "smart" drug delivery. We are trying to focus, how one can use NMs for efficient drug delivery system and other potential applications.

7.4 DRUG DELIVERY SYSTEM AND THEIR ROUTES OF ADMINISTRATION

The drug delivery system (DDS) can be defined as the robust delivery systems that encapsulate drugs and biocompatible materials of varying chemistry and molecular weight into NMs and deliver to targeted locations in human and animals. The aim of DDS is to deliver drugs to the right place, at the right concentration, for the right period of time. As drug characteristics differ substantially in chemical composition, molecular size, hydrophilicity, and protein binding, the essential characteristics that identify efficacy are highly complex. Targeted drug delivery to the site of action can be achieved either through passive targeting or by active targeting of the drug. Passive targeting can be achieved by two modes of action, first by enhanced permeability and retention (EPR) effect and, second by localized delivery. Passive targeting exploits the anatomical differences between normal and diseased tissues to deliver the drugs to the required site because the physiology of diseased tissues may be altered in a variety of physiological conditions through the enhanced permeability and retention (EPR) effect (Maeda et al., 2000). This occurs because tumor vasculature is leaky, hence circulating NPs can accumulate more in the tumor tissues than in normal tissues. The EPR effect has been greatly exploited for delivering various therapeutic molecules at the site of action and many studies potentially support this mechanism of passive targeting. A number of passively targeting nano-carriers were developed, the earliest one is Doxil (or Caelyx), a sterically stabilized PEGylated liposome that encapsulates doxorubicin. Doxil has shown good drug retention in the liposomal formulation with enhanced circulation time and is up to 6 times more effective in comparison with free doxorubicin (Gullotti and Yeo, 2009). It was approved for the treatment of advanced ovarian cancer, metastatic breast cancer and AIDS-related Kaposi's sarcoma.

Another approach known as localized delivery is the direct intratumor delivery of anticancer agents using NPs, which can be used in the treatment of local cancers such as prostate, head and neck cancers. Recently, Sahoo et al. have demonstrated that transferrin (Tf) conjugated paclitaxel (Tx)-loaded biodegradable NPs are more effective in demonstrating the antiproliferative effect of the drug than its solution or with unconjugated Tx-loaded NPs. In human prostate cancer cell line (PC3), the IC50 with Tf-conjugated NPs was fivefold lower than that with unconjugated NPs (TX-NPs) or the drug in solution.

A DDS can enhance a drug pharmacokinetics and cellular penetration. Moreover, obstacles arising from low drug solubility, degradation, fast clearance rates, nonspecific to□icity, and inability to cross biological barriers may also be addressed by a DDS. To be useful, a DDS is required to be biocompatible with processes in the body as well as with the drug to be delivered. An example of a DDS is a liposome. Liposomes are closed bilayered phospholipids first designed 40 years ago (Torchilin 2005a, b). Table 7.4 lists examples of FDA approved nanoparticle DDS. These were collectively designed to be taken up and delivered by mononuclear phagocytes (MP). Liposomes, solid lipids nanoparticles, dendrimers, polymers, silicon or carbon materials, and magnetic nanoparticles are the examples of nanocarriers that have been tested as DDS (Figure 7.1).

The goals of drug delivery systems are (1) to overcome the inherent limitations associated with bio-macromolecular therapeutics, which include a short plasma half-life, poor stability and potential immunogenicity, and (2) to maximize the therapeutic activity while minimizing the toxic side effects of drugs (Qiu and Bae, 2006). Current drug-delivery systems are effective at releasing drugs in a controlled fashion to produce a high local concentration, however, the scope is limited to targeting tissues rather than individual cells. At the expense of lower drug loading capacity, the NMs enhances the ability of drug-delivery carriers to cross cell membranes, reduces the risk of undesired clearance from the body through the liver or spleen and minimizes their uptake by the reticuloendothelial system (Svenson and Tomalia, 2005; Qiu and Bae, 2006). Smaller NM particles have greater surface area-to-volume ratios, which increase the particle's dissolution rate, enabling them to overcome solubility-limited bioavailability (Rabinow, 2004). NM Particle size is extremely important for the biological properties and, hence for effective functioning of nano-scale drug-delivery systems (Vinogradov et al., 2002).

Using traditional drug delivery approaches, including the oral and injection methods, to deliver the therapeutic agents (TA) at the targeted site could be useless and toxic (Davis and Illum, 1998). In the common oral doses, these TA are often destroyed during intestinal transit or inadequately absorbed and therefore become ineffective. Moreover, the

TABLE 7.4 Nanoparticles as Drug Delivery Carriers

Nanoparticle	Approx. Size (nm)	Carried therapeutic agent	Examples of potential targeted therapeutic application	Advantage	Reference
Ceramic Nanoparticles	<100	Proteins, DNA, anticancer therapeutic agents, high molecular weight compounds	Photodynamic therapy, liver therapy, diabetes therapy	Can be easily prepared, water-soluble, and stable in biological environment	Roy et al., 2003b, Roy et al., 2003a, Dey, 2005, Cherian et al., 2000
Carbon Nanotubes	2.5–100	Small Anticancer agents (cisplatin, Doxorubicin, etc.), proteins, RNA, DNA or genes	Cancer therapy, CNTs recombined ricin A chain protein toxin (RAT) for tumor targeting, Gene therapy	Can easily penetrate cell membranes providing tunnels for carriers to deliver therapeutic agents into the cytoplasm and, oftenly into the nucleus.	Borowiak-Palen et al., 2009; Branton et al., 2009; Kong et al., 2009; Zhang et al., 2011

TABLE 7.4 (Continued)

Nanoparticle	Approx. Size (nm)	Carried therapeutic agent	Examples of potential targeted therapeutic application	Advantage	Reference
Dendrimers	<10	DNA, anticancer therapeutic agents, antibacterial therapeutic agents, antiviral therapeutic agents, high molecular weight compounds	Tumors therapy, bacterial infection treatment, HIV therapy	Can be modified to carry hydrophobic or hydrophilic drug	Kobayashi et al., 2001; Quintana et al., 2002; Latallo et al., 2005; Chen and Cooper, 2002, Boas and Heegaard, 2004, Witvrouw et al., 2000; Rojo and Delgado, 2004
Liposomes	50–100	Proteins, DNA, anticancer therapeutic agents	Tumors therapy, HIV therapy, vaccine delivery	Effective in reducing system toxicity and can stay longer in targeted tissue	Goren et al., 1996; Versluis et al., 1998; Lasic et al., 1999; Slepushkin et al., 1996; Rao and Alving, 2000
Metallic Nanoparticles	<50	Anticancer therapeutic agents, proteins, DNA	Cancer therapy	Extremely small size with vast surface area to carry large dose	Wang et al., 2004; Priyabrata et al., 2005

	Size	Cargo	Applications	Properties	References
Polymeric biodegradable nanoparticles	10–1,000	Plasmid DNA, proteins, peptides, low molecular weight compounds	Brain tumor therapy, bone healing, vaccine adjuvant, coating gut suture, restenosis, inflamed colonic mucosa, diabetes therapy	Sustain localized drug therapeutic agent for weeks	Olivier et al., 1999; Kreuter et al., 2003; Labhasetwar et al., 1999; Raghuvanshi et al., 2001; Cohen et al., 2000; Guzman et al., 1996; Panyam et al., 2002; Lamprecht et al., 2001; Al Khouri et al., 1986; Watnasirichaikul et al., 2000
Polymeric Micelles	<100	DNA, anticancer therapeutic agents, proteins	Solid tumors therapy, antifungal treatment	Have hydrophobic core, and so they are suitable carriers for water-insoluble drug	Yokoyama et al., 1990; Kataoka et al., 2001; Rapoport et al., 2003; Yu et al., 1998

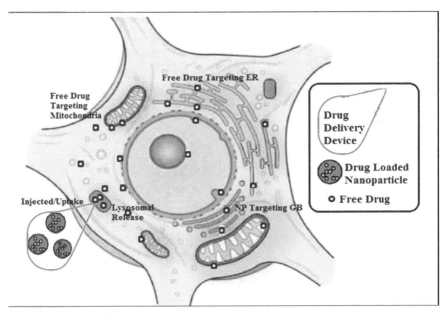

FIGURE 7.1   Cytoplasmic drug delivery concept of nanoparticles and free drugs.

uncontrolled level of these TA could cause concentration spikes, thereby harming the body. However, attaching these TA to NPs may circumvent most of these delivery challenges and produce smart pills. Nanoparticles drug delivery systems, due to their diminutive size, can penetrate across barriers through small capillaries into individual cells to allow efficient drug accumulation at the targeted locations in the body (Unezaki et al., 1996; Hobbs et al., 1998). By doing so, the TA toxicity is reduced, the drug side effects are decreased and the treatment efficacy is enhanced.

Most of the current research is mainly focusing on using NPs as drug delivery carriers for challenging, in most cases, lifelong diseases such as cancer, HIV and diabetes (see Table 7.1). Also, because anticancer drugs tend to disperse to the entire body destroying both the cancer and the normal cells, traditional chemotherapy in some cases might not be successful. Targeting cancer cells using NPs loaded with anticancer agents is a promising tactic that could meet these challenges.

In addition to oral administration (Langer, 2000), the use of nanoparticles for nasal (Illum et al., 2001) and ophthalmic delivery of drugs (Bourlais et al., 1998; Campos et al., 2001) has been investigated. Nanoparticles have enabled crossing the blood–brain barrier that represents an insurmountable obstacle for a large number of drugs, including antibiotics, antineoplastic agents, and a variety of central nervous system active drugs, especially neuropeptides (Kreuter, 2001; Schroeder et al., 1998). Furthermore, nanosize carriers of such molecules as vitamins, vitamin A (Jenning et al., 2000) and vitamin E (Dingler et al., 1999), have potential applications in dermatology and cosmetics. Vaccination against fast communicable diseases is yet another area of investigation that is being optimistically affected by the development of NP-based DDS. A novel cancer vaccine delivery system using NPs have been developed. Liposome polycation pDNA (LPD) NPs were used to carry a strong peptide antigen to immunize mice. This immunization method seems to be more effective than using traditional vaccines, which exploit either live attenuated or killed bacteria or viruses.

Targeted DDS can convey drugs more effectively and conveniently than those of the past, increase patient compliance, extend the product life cycle, provide product differentiation and reduce healthcare costs. In addition, novel DDS would offer protection and improve the pharmacokinetics of easily degradable peptides and proteins that often have short half-lives in vivo (Lee, 2002; Parveen and Sahoo, 2006). Therefore, the development of techniques that could selectively deliver drugs to the pathological sites is currently one of the most important areas of nano-scale DDS.

Through rational design, nano-scale DDS can be manufactured to combine desirable modules, both biological and synthetic, for various applications, including implantable, inhalable, injectable, oral, topical and transdermal drug delivery. Nano-scale DDS can be localized or systemic. Parenteral drug delivery is typically achieved using liposomal or polymeric carriers (Langer, 1998). There is no universal platform, as each system has its advantages and disadvantages, therefore we cannot rely on a single mode of nano-scale DDS. Since liposomes have a high drug-carrying capacity, their release profiles are more difficult to regulate. In some cases, nanoparticles other than liposomes are more efficient drug carriers due to their better stability and possess more useful control release properties. In contrast, polymeric DDS can be synthesized to generate specific molecular weights and compositions, but their drug-carrying capacity is relatively low (Duncan, 2003). Therefore, a universal general drug-delivery platform may never be realized, hybrid DDS that incorporate the fruits of various technological approaches will be tailored to address the needs of specific applications.

7.5   NANOMEDICINE: NEXT GENERATION MEDICINE

Nanomedicine may be defined as the monitoring, repair, construction, and control of human biological systems at the molecular level, using engineered NMs, nano-devices and nanostructures. It can also be regarded as another implementation of nanotechnology in the field of medical science and diagnostics. The field of nanomedicine is currently in its early phase of sharp growth and still a considerable number of years away from maturity. There are several issues to be discussed and incorporated into the nanomedicine: (1) detection of molecular changes responsible for disease pathogenesis, (2) disease diagnosis and imaging, (3) drug delivery and therapy, (4) multifunctional systems for combined therapeutic and diagnostic applications, (5) vehicles to report the in vivo efficacy of a therapeutic agent, and finally (6) the most important issues is the proper distribution of drugs and other therapeutic agents within the patient's body.

Indeed, there have been a rapidly increasing number of nanoscale vehicles with distinct physical and biochemical properties that have been developed, and a growing number of these vehicles are currently under clinical evaluation. This surge has been fuelled by better approaches of molecular assembly and the design of safer and more effective NMs. The most common nanotech platforms today include polymeric nanoparticles, nanoshells, micelles, liposomes, dendrimers, nucleic acid-based nanoconstructs, engineered viral nanoparticles, magnetic nanoparticles, silicon oxide nanoparticles, and quantum dots (Ferrari, 2005). The surface engineering of these nanoparticles to yield them "stealth" may lead to a prolonged residence in blood and increase their efficacy, and the functionalization of these particles with ligands such as antibodies, peptides, nucleic acid aptamers, carbohydrates, small molecules can differentially target their delivery or update by a subset of cells, further increasing their specificity and efficacy (Vlerken and Amiji, 2006; Farokhzad et al., 2006).

Controlled release polymeric nanoparticles represent an effective nanocarrier platform (Brannon-Peppas and Blanchette, 2004). Over the past five decades controlled release polymer technology has impacted virtually every branch of medicine, including ophthalmology, pulmonary, pain medicine, endocrinology, cardiology, orthopedics, immunology, neurology and dentistry. Several examples of these systems in clinical practice today are Atridox, Lupron Depot, Gliadel, Zoladex, Trelstart Depot and Sandostatin

LAR. The first liposome drug delivery system to gain FDA approval in 1995 was Doxil (doxorubicin liposomes) for the treatment of AIDS associated Kaposi's Sarcoma. Other examples of lipsome drug delivery systems in clinical practice today include DaunoXome (daunorubicin liposomes), DepotDur (morphine liposomes) and Ambisome (amphotericine B liposomes). An example is Abraxane which is a 130 nm paclitaxel decorated albumin which was FDA approved in 2005 as second line treatment for patients with breast cancer (Harries et al., 2005). Abraxane concentrates in tumour in part through the passive enhanced permeability and retention (EPR) effect and in part through the trans-endothelial transport mechanisms via the albumin binding protein gp60. Clinical studies have shown that Abraxane almost doubles the therapeutic response rate, increases time to disease progression and increases overall survival in patients with breast cancer (Harries et al., 2005).

Several nanoscale imaging platforms have also been introduced and are currently under pre-clinical or clinical investigation. The most characterized system is iron oxide NPs for use in conjugation with MRI for enhanced resolution imaging. Combidex which is in late stage clinical evaluation is a formulation of iron o☐idenanoparticles which has been shown to be significantly more sensitive for detection of prostate cancer lymph node metastasis as compared to conventional MRI (Harisinghani et al., 2003). Fluorescent nanocrystal quantum dots and nanoshells represent another promising platform for molecular or cellular imaging applications. Preclinical studies have demonstrated that quantum dots may be effective for intra-Operative sentinel lymph node mapping (Kim et al., 2004) and targeted nanoshells have been shown to be effective for combined imaging and therapy in animal models of cancer (Loo et al., 2005). The next generation of nanoscale imaging modalities will likely have enhanced surface characteristics for improved biocompatibility and targeting moieties for enhanced sensitivity and specificity.

In addition to novel engineering approaches, the elucidation of the biochemical and molecular abnormalities that underlie a disease state are critical to the development of more effective nanoscale diagnostic and therapeutic modalities. The wealth of knowledge that has been gained in tumor vascular biology for example, including a better understanding of the tumor microenvironment has been of critical importance to the development of more effective nanoscale systems for cancer therapy (Maeda, 2001).

The application of biodegradable NMs for the development of nanomedicines is one of the most successful ideas. Nanocarriers composed of biodegradable polymers undergo hydrolysis in the body, producing biodegradable metabolite monomers, such as lactic acid and glycolic acid. Some researchers have reported a minimal systemic toxicity associated with using of PLGA for drug delivery or biomaterial applications. Such nanoparticles are biocompatible with tissue and cells (Panyam and Labhasetwar, 2003). Drug-biodegradable polymeric nanocarrier conjugates used for drug delivery are stable in blood, non-toxic, and non-thrombogenic. They are also non-immunogenic as well as non-proinflammatory, and they neither activate neutrophils nor affect reticuloendothelial system (Rieux et al., 2006).

7.6   NANOMEDICINE AND NANOTOXICOLOGY: HEAD OR TAIL?

There is emerging concern that nanosized particles merit a more rigorous assessment of their potential effects on health, because their surface area and toxicity may be significantly greater than those of larger particles. However, specific mechanisms and pathways through which nanobiomaterials may exert their toxic effects remain unknown. A recent report described unusual redox features of single-walled carbon nanotubes (SWCNTs) in the presence of physiologically relevant redox agents (Zheng and Diner, 2004).

There are several reports on potential toxic effects of nanoparticles on different types of cells in vitro (Brown et al., 2001; Li et al., 2003; Shvedova et al., 2003; Sayes et al., 2005; Cui et al., 2005; Jia et al., 2005; Monteiro-Riviere et al., 2005), but only a few on in vivo toxicity of nanobiomaterials (Oberdorster et al., 2004; Muller et al., 2005). Aspiration of SWCNTs elicited an unusual inflammatory response in the lungs of e□posed mice. The inflammatory and fibrogenic responses were accompanied by a detrimental decline in pulmonary function and enhanced susceptibility to infection (Shvedova et al., 2005). Transition metals such as iron and nickel are particularly effective as catalysts of o□idativestress in cells, tissues, and biofluids. Therefore, the inflammatory responses caused by a combined presence of carbon nanotubes with metals can be particularly damaging. Carbon nanotubes are not effectively recognized by macrophages. Thus, neither engulfment nor oxidative/nitrosative activation of macrophages was triggered by carbon nanotubes. As a result, phagocytosis is probably not involved in elimination of non-functionalized carbon nanotubes from the lung. Due to a number of factors which may affect the toxicity of nanoparticles, their estimation is rather difficult and, thus, to☐icologicalstudies of each new DDS formulation are needed. However, with respect to their size, one can make some generalizations – smaller particles have a greater surface area, thus, they are more reactive and, in consequence, more toxic (Ai et al., 2011). It is generally accepted that nanoparticles with a hydrodynamic diameter of 10–100 nm have optimal pharmacokinetic properties for in vivo applications. Smaller nanoparticles are subjects to tissue extravasations and renal clearance whereas larger are quickly opsonized and removed from the bloodstream via the macrophages of the reticuloendothelial system (Cole et al., 2011).

7.7 GOLD NANOPARTICLES: TARGETED DELIVER FOR CANCER THERAPY

The advances in cancer therapy are progressing quickly both in terms of new agents against cancer and new ways of delivering both old and new agents. Hopefully this progress can move us away from near-toxic doses of non-specific agents. Targeted delivery for cancer therapy can be achieved through, (1) enhanced permeability and retention, (2) tumor-specific targeting, (3) targeting through angiogenesis, (4) targeting tumor vasculature, and (5) delivery of specific agents like gold and silver nanoparticles.

Recently, rod-shaped gold nanoparticles have emerged as precisely tunable plasmonic nanomaterials that may be synthesized in bulk, have narrow size distributions, optical absorption coefficients 104-fold to 106-fold higher than conventional organic fluorochromes. The long precedence of gold nanoparticles in clinical rheumatoid arthritis therapies make gold NRs appealing new candidates for nanoantenna-based photo thermal ablation and a wide array of other biomedical applications. Already, gold NRs have been used for a diversity of biological purposes, including multiple☐ed in vitro detection, two-photon fluorescence imaging, and photo thermal heating of tumor and bacterial cell targets.

A critical advantage in treating cancer with advanced, non-solution-based therapies is the inherent leaky vasculature present serving cancerous tissues. The defective vascular architecture, created due to the rapid vascularization necessary to serve fast-growing cancers, coupled with poor lymphatic drainage allows an enhanced permeation and retention effect (EPR effect) (Sledge and Miller, 2003; Teicher, 2000). Tumor-activated prodrug therapy uses the approach that a drug conjugated to a tumor-specific molecule will remain inactive until it reaches the tumor. One such type of target is monoclonal antibodies which were first shown to be able to bind to specific tumor antigens. Some currently available targeted cancer therapy using antibodies are Rituximab, Trastuzumab, Gemtuzumabozogamicin, Alemtuzumab, Ibritumomab tiuxetan and Gefitinib. Targeting the tumor vasculature is a strategy that can allow targeted delivery to a wide range of tumor types (Eatock et al., 2000; Chen et al., 2001; Reynolds et al., 2003). Antiangiogenic drug Avastin (Genentech) targets vascular endothelial growth factor (VEGF) which is a powerful angiogenesis stimulating protein that also causes tumor blood vessels to become more permeable.

There are frequent biomedical applications of gold nanoparticles (AuNPs) as delivery, diagnostic, and therapeutic agents for novel cancer therapies. The gold nanoparticles in clinical rheumatoid arthritis therapies make gold NRs appealing new candidates for nanoantenna-based photothermal abla-tion and a wide array of other biomedical applications. Already, gold NRs have been used for a diversity of biological purposes, including multiplexed in vitro detection (Yu et al., 2007), two-photon fluorescence imaging (Wang et al., 2005), and photothermal heating of tumor and bacterial cell targets (Skirtach et al., 2008; Tong et al. 2007; Huff et al., 2007).

7.8   COMPUTATIONAL AND BIOINFORMATICS INSIGHTS INTO THE NANOBIOMATERIALS

Although many strategies have been applied to advance nanotechnology but computational technologies, bioinformatics, information science, and molecular simulations have been gaining terrain as key methodologies to nanobiotechnology and nanoinformatics research. In recent years, compu-tational molecular design has become an increasingly important field in researching new nanobiomaterials (Rickman and LeSar, 2002), these results from increased calculational capacity and the consolidation of a number of methodologies of computational chemistry (Gao, 2001). Computational chemistry is a powerful tool to design, model, simulate, and visualize nano-materials (Gao, 2001) and NPs, such as dendrimers (Belting and Wittrup, 2009; Selim and Lee, 2009), metallic NPs (Lee et al., 2009c; Murali Mohan et al., 2009; Prasad and Jha, 2010), nanocapsules (Fan and Hao, 2009), nano-spheres (Lee et al., 2009a), and quantum dots (Kang et al., 2009). These NPs are being used in nanomedicine as carriers, sensors, and systems for early diagnosis of diseases (Belting and Wittrup, 2009; Kang et al., 2009).

Peptide enabled synthesis of nanobiomaterials (NMs) offer the ability to control size distributions, morphologies, and provide a range of surface functionalities. Since, peptides provide biological specificity and multi-functionality into inorganic or carbon-based NMs, while at the same time controlling nanoparticle (NP) structures and properties (Dickerson et al., 2008; Chen and Rosi, 2010). At present, though, the effect of the primary and secondary peptide structure on peptide–NMs interaction is not well understood and lacks precise characterization with regard to peptide binding interactions. So to get complete insight into peptide–NMs interactions through computational modeling, bioinformatics tools are very versatile to understand these binding and interaction phenomenon. Some researchers have exploited antibody–antigen binding interactions and epitope mapping techniques to explore the conformational space of several different peptides assembled on a nanoparticle surface and the ability of an antibody to recognize the appropriate peptide conformation (Slocik et al., 2002). Additionally, they have showed that the positive antibody response to Au, Ag, or ZnS nanoparticles encapsulated by a histidine-rich peptide epitope differed from the non-binding response of TiO_2 and Ag2S with the same peptide epitope. These e□amples highlight the influence of different NP surfaces and lattice structures on the secondary and tertiary structure of peptides, but also the importance of possessing many different peptide epitope conformations on the NP surface. Conceptually, computational modeling can provide a tremendous amount of molecular and atomic level detail into peptide–NMs interactions that can be used to support and explain the experimental data. However, given the development and refinement of molecular mechanics, design of new force field parameters, and the advancement and parallelization of supercomputers, there has been a tremendous amount of molecular and atomic level detail generated for peptide–NM interactions with metal-binding peptides. These computations have been empirical and largely underrepresented by the scientific community till now.

The study of nano-scale DDS is not straightforward process, it requires further consideration and comprehensive analysis. The targeted drug techniques such as Molecular dynamics, Brownian motion, and stochastic approaches such as Monte Carlo simulation are helpful to capture the motion and movement of NMs. In recent years, in vitro release profile of drug from controlled release platform has been combined with the state of art Computational fluid dynamics (CFD) simulation, molecular dynamics simulations (Fermeglia and Pricl, 2009; Gates et al., 2005), molecular mechanics (Shapiro et al., 2008) and quantum mechanics (Gates and Hinkley, 2004; Shapiro et al., 2008) to predict the spatial and temporal variation of the drug transport in the living tissues. Besides the use of these computational methods at the atomic level, it is necessary to implement new approaches in the use of quantitative structure–activity relationship (QSAR) studies in this field (Shapiro et al., 2008).

It's the recent advancement in the bioinformatics and computational science that made computational modeling very promising for targeted drug delivery application. Since, the surface property of functionalized NMs would play a crucial role to dictate the efficiency of the targeted drug delivery by providing targeted selectivity. Computational modeling tool will lead to insights of the dynamic delivery process, thus facilitate better design of nanoparticles.

7.9   NOVEL APPLICATIONS AND THEIR LIMITATIONS

There is a rapid growth in the field of nanobiomaterials (NMs) that are revolutionizing therapy, imaging and early diagnosis of various diseases, but some inherent problems exists in this field that should be pin-pointed with utter concern. Dendrimers-based polymers can be harnessed as effective carriers of many pharmaceuticals, such as polyamidoamines (PAMAMs), polyamines, polyamides (polypeptides), poly (aryl ethers), polyesters, carbohydrates and DNA, in most cases PAMAM dendrimers are used (Turnbull and Stoddart, 2002; Li et al., 2004). One of the earliest antitumor drug delivery using dendrimers was achieved by complexing the anticancer drug cisplatin to the surface groups of a G-4 carboxylate-terminated PAMAM dendrimer. These conjugates exhibited slower release, higher accumulation in solid tumors and lower toxicity in comparison with free cisplatin (Malik et al., 1999). Currently, researchers are also investigating using dendrimers as potential gene-delivery vehicles because dendrimers form compact polycations under physiological conditions. Different functionalized PAMAM dendrimers, poly (propylene imine) dendrimers and partially hydrolyzed PAMAM dendrimers have been effectively used as DNA-delivery systems (Sato et al., 2001).

Polymeric micelles represent a micellar system as drug carrier provides a set of unsurpassable advantages over other methods (Jones and

Leroux, 1999). Enhancing drugs' solubility using micelle-forming surfactants results in increased water solubility of a poorly soluble drug. They also improve drugs' bioavailability by enhancing their permeability across physiological barriers. They have been well demonstrated as effective drug carriers. Drugs, such as diazepam, indomethacin, adriamycin, anthracycline antibiotics and polynucleotides were effectively solubilised by polymeric micelles and demonstrated superior properties and lower toxicity. PEG-phosphatidylethanolamine (PEG-PE) micelles can efficiently incorporate a number of sparingly soluble substances like paclitaxel, tamoxifen, campothesin, porphyrine and vitamins.

Coated or uncoated NPs have a tendency to accumulate in the liver in varying amounts. Therefore, a detailed mechanism for exclusion from the body needs to be properly addressed. Thus, deposition of drug-bearing NPs in the lungs may offer the potential for sustained drug action and release throughout the lumen of the lungs, not only in the deep lung or alveolar region, where macrophage clearance occurs. As a result of these limitations, NPs are not presently being explored commercially or clinically as vehicles for drug delivery in the lungs. However, the utility of NPs for drug release is severely limited because of their low inertia, which causes them to be predominantly exhaled from the lungs after inspiration (Heyder and Rudolf, 1994; Heyder et al., 1986). Moreover, their small size leads to particle–particle aggregation, making physical handling of NPs difficult in liquid and dry powder forms, this is a common practical problem that must be overcome before using NPs for oral drug delivery (Hinds, 1998; Kabbaj and Phillips, 2001).

In addition, the mechanism of endocytosis and degradation pathways needs to be addressed because these are still poorly understood, despite their primary importance for clinical transition. However, a number of different classes of NPs with different physico-chemical properties account for the adverse biological responses (Fadeel and Garcia-Bennett, 2010; Kunzmann et al., 2011). In case of slowly degradable or nondegradable NPs used for drug delivery may show persistence and accumulation at the site of action, ultimately resulting in chronic inflammatory response (Jong and Borm, 2008). NPs are attributed qualitatively different physico-chemical characteristics from bulky materials, which may lead to changes in body distribution, passage across the BBB, and triggering of blood coagulation pathways (Jong and Borm, 2008). Cationic NPs including Gold (Au) and polystyrene have been shown to cause hemolysis and blood clotting, and usually anionic particles are rather non-toxic (Gupta and Gupta, 2005). Recent studies with carbon-derived NMs showed that platelet aggregation was induced by both single and multiwalled carbon nanotubes (Radomski et al., 2005). Choi and his investigation team have demonstrated that the nonmodified cadmium telluride QDs induced lipid peroxidation in the cells (Choi et al., 2007). Similarly, Cho and their group have also showed naked QDs to be cyto-toxic by induction of reactive oxygen species (ROS), resulting in damage to plasma membranes, mitochondria and nucleus (Cho et al., 2007). Silicon oxide exposure resulted in an increased ROS levels and reduced glutathione levels indicating an increased in oxidative stress (Lin et al., 2006). Chang and his colleagues have reported that silica NPs are toxic at high dosages as revealed by reduction in cell viability and by lactate dehydrogenase release from the cells indicating membrane damage (Chang et al., 2007). NPs could also cause mitochondrial damage, uptake through olfactory epithelium, platelet aggregation and cardiovascular effects. These effects require a novel way of handling the toxicology of NPs (Morimoto et al., 2010).

Furthermore, there is an urgent need for the development of safety guide-lines by the government regarding the environmental effects, safety and tox-icological issues and the potential effects on the health of people, those who are manufacturing and those who are consuming the NMs.

7.10 FUTURE OF NANOBIOMATERIALS: ESPECIALLY IN HEALTH CARE

Medical diagnosis with proper and efficient delivery of therapeutic mole-cules is the medical fields where nanoBmaterials (NMs) have found practi-cal implementations. However, there are several other intriguing proposals for practical applications of nano-mechanical tools into the fields of imag-ing, medical research and clinical practice.

The therapeutic applications of NPs are diverse, ranging from cancer therapeutics, antimicrobial actions, vaccine delivery, gene delivery and site-specific targeting to avoid the undesirable side effects of the current therapeutics. Many chemotherapeutic drugs such as carboplatin, paclitaxel, doxorubicin and etoposide, etc., have been successfully loaded onto NPs and these nanoparticulate systems are very potent against various cancers as demonstrated by the studies of various research groups. Some of the NPs that can be effectively used for therapeutics can be seen in Table 7.5.

TABLE 7.5 Nanoparticles (NPs) as Therapeutic Agents

Type of nanobiomaterial (NMs)	Encapsulant	Indicator	Therapeutic improvement	Reference
Chitosan NP (CNP)	siRNA	Ovarian cancer	Increased selective intratumoral delivery and significant inhibition of tumor growth compared to controls	Hee-Dong et al., 2010
Cetyl alcohol/polysorbate NPs	Paclitaxel	Brain tumor	Higher brain and tumor cell uptake, thus leading to greater cytotoxicity, also effective towards P-glycoprotein expressing tumor cells	Koziara et al., 2004
Chitosan-alginate NPs	Carboplatin	Retinoblastoma	Enhanced antiproliferative activity and cytotoxicity of NPs in comparison with native carboplatin	Parveen et al., 2010
Gold NPs (AuNPs)	-	Various cancers	Effective as radiation sensitizers for cancer therapy	Chithrani et al., 2010
Lipid nanocapsules	Etoposide	Glioma	Greater cytotoxicity. Can overcome P-glycoprotein dependent multidrug resistance	Lamprecht and Benoit 2006
P (4-vinylpyridine) particles	-	Antimicrobial agent	These particles can be used to inhibit bacterial growth for various bacteria as biocolloids	Ozay et al., 2010
PLGA NPs	Paclitaxel	Various cancers	Effective in chemotherapeutic and photothermal destruction of cancer cells	Cheng et al., 2010
Poly (3-hydroxybutyrate-co-3-hydroxyoctanoate) NPs	DOX	Various cancers	Effective in selective delivery of anticancer drug to the folate receptor-Overexpressed cancer cells	Zhang et al., 2010
Polyisohexylcyanoacrylate NPs	DOX	Hepatocellular Carcinoma	Higher antitumor efficacy than native doxorubicin and can overcome multiple drug resistance phenotype	Barraud et al., 2005

Molecular diagnostics has also been greatly benefited by the advent of AuNPs which promises increased sensitivity and specificity, multiple□ing capability and short turnaround times. AuNP-based colorometric assays also show great potential in point-of-care testing assays. Aptamer-conjugated NPs can also be used for the collection and detection of multiple cancer cells (Medley et al., 2008; Corsi et al., 2009; Escosura-Muñiz et al., 2009; Farias et al., 2009; Smith et al., 2006; Smith et al., 2007; Valanne et al., 2005). Cancer is a debilitating disease and early and accurate detection of cancer is often a bottleneck that is responsible for its delayed treatment complications. To address these limitations, various types of NPs are being developed for effective diagnostics. A brief overview of NMs used in diagnostics is presented in Table 7.6.

The development of the effective nano-scale DDS does not only mean the e□ecutionof delivery, but also the positive confirmation of the site-specific delivery of the drug. Consequently, the ability to track and image the fate of any nanomedicine from the systemic to the subcellular level becomes essential. Therefore, NMs can be successfully exploited to improve the utility of fluorescent markers for medical imaging particularly in the imaging of tumors and other diseases in vivo. Recently, fluorescent silica NPs (FSNPs), which are a new class of engineered optical probes consisting of silica NPs loaded with fluorescent dye, have also garnered immense interest in cancer imaging. Water-soluble, functionalized QDs that are highly stable against o□idation,manifest stable fluorescent properties and also offer new prospects for live cells, in vivo imaging and diagnostics. Magnetic iron oxide NPs also have attracted extensive interest as novel contrast agents for biomedical imaging due to their capability of deeP-tissue imaging, noninvasiveness and low toxicity. Dynamic magnetomotion of magnetic NPs (MNPs) detected with magnetomotive optical coherence tomography (MM-OCT) also represents a new methodology for contrast enhancement and therapeutic interventions in molecular imaging. There are various types of functionalized NPs for imaging of diseased cells enlisted in Table 7.7.

Miniaturized nanoscale sensors like QDs, nanocrystals, and nanobarcodes can sense and monitor biological signals such as the release of proteins or antibodies in response to cardiac or inflammatory events. Therefore, NMs can help in revealing the mechanisms involved in various cardiac diseases (Guccione et al., 2004). NMs have future medical applications in the field of nanodentistry, by which it is possible to maintain near-perfect oral health through the use of nanobiomaterials and nanorobotic (Shi et al., 1999; West

TABLE 7.6 Nanoparticles (NPs) as Diagnostic Agents

Type of nanobiomaterial (NMs)	Diagnostic strategy	Advantages	Reference
Aptamer conjugated NPs	Aptamer-conjugated magnetic NPs can be used for selective targeting cell extraction and aptamerconjugated fluorescent NPs can be used for sensitive cancer detection	Enables the collection and detection of multiple cancer cells	Smith et al., 2007
AuNPs	The selectivity and specific affinity of aptamers is combined with spectroscopic advantages of AuNPs to detect diseased cells	For sensitive detection of cancer cells. Can easily differentiate between different types of target and control cells-based on the aptamer	Medley et al., 2008
AuNPs	Identification is-based on the reaction of cell surface proteins with specific antibodies conjugated with AuNPs	Rapid identification and quantification of tumor cells	de la Escosura-Muñiz et al., 2009
Magnetofluorescent particle systems	These bimodal contrast agents allows detection of cancer cells	Noninvasive diagnosis of breast cancer	Corsi et al., 2009
Fluorescent europium (III)-chelate-doped NP	Highly fluorescent europium (III)-chelate-doped NP labels, together with high affinity monoclonal antibodies (antihexon) coated on label particles and micro titration wells provides a sensitive adenovirus immunoassay	Has potential in sensitive screening of viral analytes	Valanne et al., 2005
Semiconductor fluorescent QDs	These fluorescent biomarkers are analyzed by their resulting fluorescence and thus enables efficient cancer diagnostics	Enables fast and precise cancer diagnostics	Farias et al., 2009
Semiconductor QDs	Intense stable fluorescence enables the detection of cancer biomarkers	Useful for molecular diagnostics of cancer	Smith et al., 2006

TABLE 7.7 NPs as Imaging Agents

Type of nanobiomaterial (NMs)	Imaging strategy	Advantages	Reference
AuNPs	NP bioconjugates coated with dithiol bearing hetero-bifunctional PEG (polyethylene glycol), and cancer-specific monoclonal antibody F-19 can be used to label sections of healthy and cancerous pancreatic tissue.	These NPs can be used for human brain endothelial cell imaging	Craig et al., 2010
AuNPs	Surface functionalized AuNPs with prostatespecific membrane antigen (PSMA) RNA aptamer that binds to PSMA enables specific imaging of prostate cancer cells that expresses the PSMA protein	Multifunctional NPs that that enables combined prostate cancer imaging by computed tomography (CT) and anticancer therapy	Kim et al., 2010
Carboxyl-functionalized silica-coated QDs	The stable fluorescent property of QDs enables specific imaging	Monodisperse and stable in aqueous solution, provides specific targeting and are easy for bio-conjugation. Can serve as efficient targeting probes for cell imaging	Qian et al., 2010
CNPs	Tumor targeted CNPs containing dual imaging agents (near-infrared fluorescent dye, Cy5.5 (20), and gadolinium (Gd (III)) ions) was designed as dual-modality cancer imaging agents	Effective as an optical/MR (magnetic resonance) dual imaging agent for cancer treatment	Nam et al., 2010
Fluorescent silica NPs	Silica NPs can be loaded with fluorescent dyes for sensitive imaging of cancer cells. In addition, for targeting cancer cells, these can be conjugated to specific biomolecules overexpressed on cancer cells	Useful for cancer targeting and imaging	Santra 2010
Multifunctional super paramagnetic iron oxide NPs	Folate provides specific targeting, and DOXloaded super paramagnetic iron oxide NPs serve as a therapeutic agent as well as MRI contrast agent	Promising candidate for treating liver cancer as well as monitoring the cancer using MRI	Maeng et al., 2010

Polymer-Ag@SiO$_2$ Hybrid			
Fluorescent NPs	Cationic surface and suitable size allows the nanocomposites to be rapidly internalized into cells, thus effective in cellular imaging	These nanoparticles show cytocompatibility and bright fluorescence and thus is especially useful for efficient cellular imaging	Tang et al., 2010
Poly (alkyl cyanoacrylate) NPs	The fluorescent rhodamine B-tagged poly (alkyl cyanoacrylate) amphiphilic copolymer nanoparticles enables specific human brain endothelial cell imaging	These NPs can be used for human brain endothelial cell imaging	Brambilla et al., 2010
QD-loaded micelles	Lipid conjugated QDs together with Herceptin enhances tumor cell uptake and thus can be used for simultaneous tumor therapy and imaging	Can be used for targeting, imaging and treatment of cancer in the early stages	Nurumabi et al., 2010
Streptavidin NPs	Biotinylated anti-Her2 Herceptin antibody to provide tumor targeting, whereas a biotinylated DOTA chelator labeled with ^{111}In and a biotinylated Cy5.5 fluorophore to a streptavidin NP provides specific imaging of the tumor	Streptavidin NPs were effective for multimodality imaging of tumor in mice by fluorescence and nuclear detection	Liang et al., 2010

and Halas, 2000). Nanobiomaterials can act as new and effective constituents of bone materials, because bone is also made up of nanosized organic and mineral phases. NMs proposed as ideal scaffolds for cell growth should be biocompatible, osteoinductive, osteoconductive, integrative, porous, and mechanically compatible with native bone to fulfill their desired role as bone implants and substitutes. Greater in vitro osteoblast adhesion has also been observed on helical rosette nanotube-coated titanium compared with uncoated titanium, because these helical rosette nanotubes mimic the dimensions of the nanostructure of the bone components (Chun et al., 2004).

The newly designed nanodevices and nanomachines can have a paradigm-shifting impact in the treatment of the dreaded diseases. These machines have three key elements meant for sensing, decision making, and carrying out the intended purpose. For instance, abciximab, a chimeric mouse-human monoclonal antibody used to lessen the chance of heart attack in people who need percutaneous coronary intervention, can be considered as an example of a simple nanomachine. It has sensors that bind to the GP2b3a receptor and also has an effector that inhibits the receptor through steric hindrance. Thus, by inhibiting the ability of the GP2b3a receptor to bind fibrinogen, abciximab changes platelet behavior, impeding platelet aggregation and activation (Kong et al., 2005). Such nano-tools and nano-devices might be very helpful, and become a reality in the near future. One function of nano-devices in medical sciences could be the replacement of defective or incorrectly functioning cells, such as the respirocyte. This artificial red blood cell is theoretically able to provide oxygen and can do it even more effectively than an erythrocyte. It could replace defective natural red cells in blood circulation. Thus the next application phase of nano-machines could be providing metabolic support in the event of impaired circulation. It has also been postulated that nano-machines could distribute drugs within the patient's body. Such nano-constructions could deliver medicines to particular sites, making more adequate and precise treatment possible (Fahy, 1993a, b; Triggle, 1999). For example, Nanorobots operating in the human body, could monitor levels of different compounds and store that information in internal memory. They could be used to rapidly examine a given tissue location, surveying its biochemistry, biomechanics, and histometric characteristics in greater detail. This would help in better disease diagnosing (Freitas, 1996a; Lampton, 1995). The use of nano-devices would give the additional benefits of reduced intrusiveness, increased patient comfort, and greater fidelity of results, since the target tissue can be examined in its active state in the actual host environment.

KEYWORDS

- bioinformatics
- drug delivery system
- nanobiomaterials
- nanoinformatics
- nanomedicine
- nanoparticles
- nanotoxicology

REFERENCES

Afergan, E., Epstein, H., Dahan, R., Koroukhov, N., Rohekar, K., & Danenberg, H. D., (2008). Delivery of serotonin to the brain by monocytes following phagocytosis of liposomes. *J. Control Re.l, 132*, 84–90.

Agnihotri, S. A., & Aminabhavi, T. M., (2006). Novel interpenetrating network chitosan-poly (ethylene oxide-*g*-acrylamide) hydrogel microspheres for the controlled release of capecitabine. *Int J Pharm, 324*, 103–115.

Ai, J., Biazar, E., Montazeri, M., Majdi, A., Aminifard, S., Safari, M., et al., (2011). Nanotoxicology and nanoparticle safety in biomedical designs. *Int J Nanomedicine, 6*, 1117–1127.

Albertorio, F., Hughes, M. E., Golovchenko, J. A., & Branton, D., (2009). Base dependent DNA-carbon nanotube interactions: Activation enthalpies and assembly-disassembly control. *Nanotechnol, 20*, 395101.

Arayachukeat, S., Wanichwecharungruang, S. P., & Tree-Udom, T., (2011). Retinyl acetate-loaded nanoparticles: Dermal penetration and release of the retinyl acetale. *Int J Pharm, 404*, 281–288.

Arias, J. L., Lopez-Viota, M., Delgado, A. V., & Ruiz, M. A. (2010). Iron/ethylcellulose (core/shell) nanoplatform loaded with 5-fluorouracil for cancer targeting. *Colloids Surf B Biointerfaces, 77*, 111–116.

Arsawang, U., Saengsawang, O., Rungrotmongkol, T., Sornmee, P., Wittayanarakul, K., Remsungnen, T., et al., (2011). How do carbon nanotubes serve as carriers for gemcitabine transport in a drug delivery system? *J. Mol. Graph. Model, 29*, 591–596.

Asmatulu, R., Zalich, M. A., Claus, R. O., & Riffle, J., (2005). Synthesis, characterization and targeting of biodegradable magnetic nanocomposite particles by external magnetic fields. *J. Magn. Magn. Mater, 292*, 108–119.

Bajpai, A. K., & Gupta, R., (2011). Magnetically mediated release of ciprofloxacin from polyvinyl alcohol based super paramagnetic nanocomposites. *J. Mater. Sci. Mater. Med, 22*, 357–369.

Balogh, L., Swanson, D. R., Tomalia, D. A., Hagnauer, G. L., & McManus, A. T., (2001). Dendrimer-silver complexes and nanocomposites as antimicrobial agents. *Nano. Lett, 1*, 18–21.

Bangham, A. D., Standish, M. M., & Watkins, J. C., (1965). Diffusion of univalent ions across the lamellae of swollen phospholipids. *J. Mol. Biol*, *13*, 238–252.

Barraud, L., Merle, P., Soma, E., Lefrançois, L., Guerret, S., Chevallier, M., et al., (2005). Increase of doxorubicin sensitivity by doxorubicin-loading into nanoparticles for hepatocellular carcinoma cells in vitro and in vivo. *J Hepatol*, *42*, 736–743.

Beg, S., Rizwan, M., Sheikh, A. M., Hasnain, M. S., Anwer, K., & Kohli, K., (2011). Advancement in carbon nanotubes: basics, biomedical applications and toxicity. *J. Pharm. Pharmacol*, *63*, 141–163.

Belting, M., & Wittrup, A., (2009). Macromolecular drug delivery: Basic principles and therapeutic applications. *Molec. Biotechnol*, *43*, 89–94.

Bhirde, A. A., Patel, V., Gavard, J., Zhang, G., Sousa, A. A., Masedunskas, A., Leapman, R. D. et al., (2009). Targeted killing of cancer cells in vivo and in vitro with EGF-directed carbon nanotube-based drug delivery. *ACS Nano*, *3*, 307–316.

Bilensoy, E., Sarisozen, C., Esendagl, G., Dogan, L. A., Aktas, Y., Sen, M., & Mangan, A. N., (2009). Intravesical cationic nanoparticles of chitosan and polycaprolactone for the delivery of Mitomycin C to bladder tumors. *Int. J. Pharm*, *371*, 170–176.

Birrenbach, G., & Speiser, P., (1976). Polymerized micelles and their use as adjuvants in immunology. *J. Pharm. Sci. 65*, 1763–1766.

Bodnar, M., Hartmann, J. F., & Borbely, J. (2005). Preparation and characterization of chitosan-based nanoparticles. *Biomacromolecules*, *6(5)*, 2521–2527.

Bourlais, C. L., Acar, L., Zia, H., Sado, P. A., Needham, T., & Leverge, R., (1998). Ophthalmic drug delivery systems-recent advances. *Prog. Retin. Eye Res*, *17*, 33–58.

Brambilla, D., Nicolas, J., Droumaguet, B. L., Andrieu☐,K., Marsaud, V., Couraud, P. O., et al., (2010b). Design of fluorescently tagged poly (alkyl cyanoacrylate) nanoparticles for human brain endothelial cell imaging. *Chem. Commun*, *46*, 2602–2604.

Brannon-Peppas, L., & Blanchette, J. O., (2004). Nanoparticle and targeted systems for cancer therapy. *Adv. Drug Deliv. Rev.*, *56*, 1649–1659.

Brown, D. M., Wilson, M. R., MacNee, W., Stone, V., & Donaldson, K., (2001). Size dependent pro-inflammatory effects of ultrafine polystyrene particles: A role for surface area and oxidative stress in the enhanced activity of ultra fines. *Toxicol. Appl. Pharmacol. 175*, 191–199.

Bruchez, M., Moronne, M., Gin, P., Weiss, S., & Alivisatos, A. P., (1998). Semiconductor nanocrystals as fluorescent biological labels. *Science*, *281*, 2013–2016.

Caminade, A. M., Laurent, R., & Majoral, J. P., (2005). Characterization of dendrimers. *Adv. Drug Deliv. Rev.*, *57*, 2130–2146.

Campos, A. M. D., Sanchez, A., & Alonso, M. J., (2001). Chitosan nanoparticles: A new vehicle for the improvement of the delivery of drugs to the ocular surface. Application to cyclosporin A. *Int. J. Pharm.*, *224*, 159–168.

Chan, W. C., & Nie, S., (1998). Quantum dot bioconjugates for ultrasensitive nonisotopic detection. *Science. 281*, 2016–2018.

Chang, J. S., Chang, K. L., Hwang, D. F., & Kong, Z. L., (2007). *In vitro* cytotoxicitiy of silica nanoparticles at high concentrations strongly depends on the metabolic activity type of the cell line. *Environ. Sci. Technol.*, *41*, 2064–2068.

Chauhan, A. S., Diwan, P. V., Jain, N. K., & Tomalia, D. A., (2009). Une☐pected *in vivo* anti-inflammatory activity observed for simple surface functionalized poly (amidoamine) dendrimers. *Biomacromolecules. 10*, 1195–1202.

Chen, C. L., & Rosi, N. L. (2010). Peptide-based Methods for the Preparation of Nanostructured Inorganic Materials. *Angew. Chem. Int. Ed. 49*, 1924–1942.

Chen, J. F., Ding, H. M., Wang, J. X., & Shao, L., (2004). Preparation and characterization of porous hollow silica nanoparticles for drug delivery application. *Biomaterials, 25(4)*, 723–727.

Chen, Q. R., Zhang, L., Gasper, W., & Mixson, A. J., (2001). Targeting tumor angiogenesis with gene therapy. *Molecular Genetics and Metabolism, 74*, 120–127.

Chen. Z., Pierre, D., He, H., Tan, S., Pham-Huy, C., Hong, H., & Huang. J., (2011). Adsorption behavior of epirubicin hydrochloride on carboxylated carbon nanotubes. *Int. J. Pharm., 28(405)*, 153–161.

Cheng, F. Y., Su, C. H., Wu, P. C., & Yeh, C. S., (2010). Multifunctional polymeric nanoparticles for combined chemotherapeutic and near-infrared photo thermal cancer therapy *in vitro* and *in vivo. Chem. Commun, 46*, 3167–3169.

Chithrani, D. B., Jelveh, S., Jalali, F., Prooijen, M., Allen, C., Bristow, R. G. et al., (2010). Gold nanoparticles as radiation sensitizers in cancer therapy. *Radiat. Res., 173*, 719–728.

Cho, M., Cho, WS., Choi, M., Kim, S. J., Han, B. S., Kim, S. H., Kim, H. O. et al., (2009). The impact of size on tissue distribution and elimination by single intravenous injection of silica nanoparticles. *Toxicol. Lett., 189*, 177–183.

Cho, S. J., Maysinger, D., Jain, M., Roder, B., Hackbarth, S., & Winnik, F. M., (2007). Long-term exposure to CdTe quantum dots causes functional impairments in live cells. *Langmuir 23*, 1974–1980.

Choi, A. O., Cho, S. J., Desbarats, J., Lovric, J., & Maysinger, D., (2007). Quantum dotinduced cell death involves Fas upregulation and lipid peroxidation in human neuroblastoma cells. *J. Nanobiotechnol, 5*, 1–13.

Chun, A. L., Moralez, J. G., Fenniri, H., & Webster, T. J., (2004). Helical rosette nanotubes: a more effective orthopedic implant material. *Nanotechnol. 15*, S234–S239.

Cole, A. J., Yang, V. C., & David, A. E., (2011). Cancer theranostics: the rise of targeted magnetic nanoparticles. *Trends Biotechnol, 29*, 323–332.

Corsi, F., De, P. C., Colombo, M., Allevi, R., Nebuloni, M., Ronchi, S. et al., (2009). Towards ideal magneto fluorescent nanoparticles for bimodal detection of breast-cancer cells. *Small, 5*, 2555–2564.

Craig, G. A., Allen, P. J., & Mason, M. D., (2010). Synthesis, characterization, and functionalization of gold nanoparticles for cancer imaging. *Methods Mol. Biol., 624*, 177–193.

Crescenzo, D. A., Velluto, D., Hubbell, J. A., & Fontana, A., (2011). Biocompatible dispersions of carbon nanotubes: a potential tool for intracellular transport of anticancer drugs. *Nanoscale. 3*, 925–928.

Cui, D., Tian, F., Ozkan, C. S., Wang, M., & Gao, H., (2005). Effect of single wall carbon nanotubes on human HEK293 cells. *Toxicol. Lett., 155*, 73–85.

Cui, Y., Wei, Q., Park, H., & Lieber, C. M., (2001). Nanowire nanosensors for highly sensitive and selective detection of biological and chemical species. *Science., 293*, 1289–1292.

D'Emanuele, A., & Attwood, D., (2005). Dendrimer-drug interactions. *Adv. Drug Deliv. Rev., 57*, 2147–2162.

Davis, S., & Illum, L., (1998). Drug delivery systems challenging molecules. *Inter. J. Pharm., 176*, 1–8.

Dev, A., Binulal, N. S., Anita, A., Nair, S. V., Furuike, T., Tamura, H., & Jayakumar. R., (2010). Preparation of poly (lactic acid)/chitosan nanoparticles for anti-HIV drug delivery applications. *Carbohydr. Polym., 80*, 833–838.

Dey, S., (2005). Inorganic biohybrid nanoparticles for targeted drug delivery. Proceedings of 6th Annual. Cambridge Health Institute.

Dhar, S., Liu, Z., Thomale, J., Dai, H., & Lippard, S. J., (2008). Targeted single-wall carbon nanotube-mediated Pt (IV) prodrug delivery using folate as a homing device. *J. Am. Chem. Soc., 27(130)*, 11467–11476.

Dickerson, M. B., Sandhage, K. H., & Naik, R. R., (2008). Protein- and peptide-directed syntheses of inorganic materials. *Chem. Rev., 108*, 4935–4978.

Dingler, A., Blum, R. P., Niehus, H., Muller, R. H., & Gohla, S., (1999). Solid lipid nanoparticles (SLN/Lipopearls) a pharmaceuticaland cosmetic carrier for the application of vitamin E in dermal products. *J. Microencapsul, 16*, 751–767.

Dobrovolskaia, M. A., & McNeil, S. E., (2007). Immunological properties of engineered nanomaterials. *Nat. Nano. 2*, 469–478.

Duncan, R., (2003). The Dawning Era of Polymer Therapeutics. *Nat. Rev. Drug Discovery., 3*, 347.

Eatock, M. M., Scha¨tzlain, A., & Kaye, S. B., (2000). Tumor vasculature as a target for anti-cancer therapy. *Cancer Treatment Reviews., 26*, 191–204.

Escosura-Muñiz, A. D. L., Sánchez-Espinel, C., Díaz-Freitas, B., González-Fernández, A., Maltez-da, C. M., & Merkoçi, A., (2009). Rapid identification and quantification of tumor cells using an electrocatalytic method based on gold nanoparticles. *Anal Chem., 81*, 10268–10274.

Fadeel, B., & Garcia-Bennett, A. E., (2010). Better safe than sorry: Understanding the toxicological properties of inorganic nanoparticles manufactured for biomedical applications. *Adv. Drug Deliv. Rev., 62*, 362–374.

Fahy, G. M., (1993b). Molecular nanotechnology and its possible pharmaceutical implications. In, *2020* Visions: Health Care Information Standards and Technologies, Bezold, C., Halperin, J. A., Eng, J. L., (Eds.), US Pharmacopoeia Convention: Rockville. *152–159.*

Fan, D., & Hao, J., (2010). Magnetic aligned vesicles. *J. of Colloid and Interface Science. 342(1)*, 43–48.

Farias, P. M., Santos, B. S., & Fontes, A. (2009). Semiconductor fluorescent quantum dots: Efficient biolabels in cancer diagnostics. *Methods Mol. Biol., 544*, 407- 419.

Farokhzad, O. C., Cheng, J., Teply, B. A., Sherifi, I., Jon, S., Kantoff, P. W., et al., (2006). Targeted nanoparticle-aptamer bioconjugates for cancer chemotherapy *in vivo. Proc. Natl. Acad. Sci., USA. 103*, 6315–6320.

Farokhzad, O. C., Karp, K. J., & Langer, R., (2006). Nanoparticle-aptamer bioconjugates for cancer targeting. *Expert Opin. Drug Deliv. 3*, 311–324.

Fermeglia, M., & Pricl, S. (2009). Multiscale molecular modeling in nanostructured material design and process system engineering. *Computers and Chemical Engineering. 33*, 1701–1710.

Ferrari, M. (2005). Cancer nanotechnology: Opportunities and challenges. *Nat. Rev. Cancer., 5*, 161–171.

Fielding, R. M., Lewis, R. O., & Moon-McDermott, L., (1998). Altered tissue distribution and elimination of amikacin encapsulated in unilamellar, low-clearance liposomes (MiKasome). *Pharm. Res., 15*, 1775–1781.

Figuerola, A., Corato, R. D., Manna, L., & Pellegrino, T., (2010). From iron oxide nanoparticles towards advanced iron-based inorganic materials designed for biomedical applications. *Pharmacol. Res., 62*, 126–143.

Freitas, R. A. (1996a). The future of computers. *Analog., 116*, 57–73.

Fritzen-Garcia, M. B., Zanetti-Ramos, B. G., Oliveira, C. S. D., Soldi, V., & Pasa, A. A., (2009c). Creczynski-Pasa, T. B. Atomic force microscopy imaging of polyurethane nanoparticles onto different solid. *Mater Sci. Eng., 29*, 405–409.

Gaihre, B., Khil, M. S., Lee, D. R., & Kim, H. Y., (2009). Gelatin-coated magnetic iron oxide nanoparticles as carrier system: drug loading and in vitro drug release study. *Int. J. Pharm.*, *365*, 180–189.

Gao, H., (2001). Modelling strategies for nano-and biomaterials. In: European White Book on Fundamental Research in Materials Science: Germany, *144*–148.

Gates, T., & Hinkley, J., (2003). Computational materials: Modeling and simulation of nano-structured materials and systems. 44th AIAA/ASME/ASCE/AHS/ASC Structures, Structural Dynamics, and Materials Conference Norfolk, Virginia.

Gates, T., Odegard, G., Frankland, S., & Clancy, T., (2005). Computational materials: Multiscale modeling and simulation of nanostructured materials. *Composites Science and Technology*. *65*, 2416–2434.

Gensini, G. F., Conti, A. A., & Lippi, D. (2007). The 150th anniversary of the birth of Paul Ehrlich, chemotherapy pioneer. *J. Infect. 54(3)*, 221–224.

Giuberti, C. D. S, Reis, D. OE. C., Rocha, T. G. R., Leite, E. A., Lacerda, R. G., Ramaldes, G. A., et al., (2011). Study of the pilot production process of long-circulating and pH-sensitive liposomes containing cisplatin. *J. Liposome Res.*, *21*, 60–69.

Gratton, S. E. A., Ropp, P. A., Pohlhaus, P. D., Luft, J. C., Madden, V. J., Napier, M. E., et al., (2008). The effect of particle design on cellular internalization pathways. *Proc. Natl. Acad. Sci., USA. 105(33)*, 11613–11618.

Gref, R., Minamitake, Y., Peracchia, M. T., Trubetskoy, V., Torchilin, V., & Langer, R., (1994). Biodegradable long-circulating polymeric nanospheres. *Science.*, *263*, 1600–1603.

Guccione, S., Li, K. C., & Bednarski, M. D., (2004). Vascular-targeted nanoparticles for molecular imaging and therapy. *Methods Enzymol.*, *386*, 219–236.

Gupta, A. K., & Gupta, M., (2005). Synthesis and surface engineering of iron oxide nanoparticles for biomedical applications. *Biomaterials.*, *26*, 3995–4021.

Han, L., Huang, R., Liu, S., Huang, S., & Jiang, C. H., (2010). Peptideconjugated PAMAM for targeted doxorubicin delivery to transferrin receptor over expressed tumors. *Mol. Pharm.*, *7*, 2156–2165.

Harisinghani, M. G., Barentsz, J., Hahn, P. F., Deserno, W. M., Tabatabaei, S., van de Kaa, C. H., et al., (2003). Noninvasive detection of clinically occult lymph-node metastases in prostate cancer. *N. Engl. J. Med.*, *348*, 2491–2499.

Harries, M., Ellis, P., & Harper, P., (2005). Nanoparticle albumin-bound paclitaxel for metastatic breast cancer. *J. Clin. Oncol. 23*, 7768–7771.

Hawker, C. J., & Wooley, K. L., (2005). The Convergence of Synthetic Organic and Polymer Chemistries. *Science. 309*, 1200–1205.

Hee-Dong, H., Mangala, L. S., Lee, J. W., Shahzad, M. M., Kim, H. S., Shen, D. Y., et al., (2010). Targeted gene silencing using rgd-labeled chitosan nanoparticles. *Clin. Cancer Res.*, *16(15)*, 3910–3922.

Heyder, J., Gebhart, J., Rudolf, G., Schiller, C., & Stahlhofen, W., (1986). Deposition of particles in the human respiratory tract in the size range 0. 005–15 μm. *J. of Aerosol Sci.*, *17(5)*, 811–825.

Heyder, J., & Rudolf, G., (1984). Mathematical models of particle deposition in the human respiratory tract. *J. Aerosol Sci.*, *15*, 697–707.

Hinds, W. C., (1998). *Aerosol Technology: Properties, Behaviour and Measurement of Airborne Particles,* Wiley: New York.

Hobbs, K., Monsky, W., Yuan, F., Roberts, W., Griffith, L., Torchilin, P., & Jain, R., (1998). Regulation of transport pathways in tumor vessels: Role of tumor type and microenvironment. *Proc. Natl. Acad. Sci. USA. 95*, 4607–4612.

Hofheinz, D., Gnad-Vogt, U., Beyer, U., & Hochhaus, A., (2005). Liposomal encapsulated anti-cancer drugs. *Anticancer Drugs.*, *16*(7), 691–707.

Hua, M. Y., Liu, H. L., Yang, H. W., Chen, P. Y., Tsai, R. Y., Huang, C. Y., et al., (2011). The effectiveness of a magnetic nanoparticle-based delivery system for BCNU in the treatment of gliomas. *Biomaterials*, *32*, 516–527.

Hua, M. Y., Yang, H. W., Chuang, C. K., Tsai, R. Y., Chen, W. J., Chuang, K. L., et al., (2010). Magnetic-nanoparticle modified paclitaxel for targeted therapy for prostate cancer. *Biomaterials*, *31*, 7355–7363.

Huff, T. B., Tong, L., Zhao, Y., Hansen, M. N., Cheng, J. X., & Wei, A., (2007). Hyperthermic effects of gold nanorods on tumor cells. *Nanomedicine-UK.* 2, 125–132.

Illum, L., Jabbal-Gill, I., Hinchcliffe, H., Fisher, A. N., & Davis, S. S., (2001). Chitosan as a novel nasal delivery system for vaccines. *Adv. Drug Deliv. Rev.*, *51*, 81–96.

Jang, W. D., Selim, K. M. K., Lee, C. H., Kang, I. K. (2009). Bioinspired application of dendrimers: From bio-mimicry to biomedical applications. *Prog. Polymer. Sci.*, *34*, 1–23.

Jenning, V., Gysler, A., Schafer-Korting, M., & Gohla, S. H., (2000). Vitamin a loaded solid lipid nanoparticles for topical use: Occlusive properties and drug targeting to the upper skin. *Eur. J. Pharm. Biopharm*, *49*, 208–211.

Jia, G., Wang, H., Yan, L., Wang, X., Pei, R., & Yan, T., (2005). Cytotoxicity of carbon nanomaterials: single-wall nanotube, multi-wall nanotube, and fullerene. *Environ. Sci. Technol*, *39*, 1378–1383.

Jingting, C., Huining, L., & Yi, Z., (2011). Preparation and characterization of magnetic nanoparticles containing Fe_3O_4-dextran-anti-*b*-human chorionic gonadotropin, a new generation choriocarcinoma-specific gene vector. *Int. J. Nanomedicine*, *6*, 285–294.

Jong, W. H. D., & Borm, P. J., (2008). Drug delivery and nanoparticles: applications and hazards. *Int J Nanomedicine.* 3, 133–149.

Kabbaj, M., & Phillips, N. C., (2001). Anticancer activity of mycobacterial dna: Effect of formulation as chitosan nanoparticles. *J. Drug Targetin,* 9, 317–328.

Kale, S. N., Jadhav, A. D., Verma, S., Koppikar, S. J., Kaul-Ghanekar, R., Dhole, S. D., & Ogale, S. B., (2012). Characterization of biocompatible $NiCo_2O_4$ nanoparticles for applications in hyperthermia and drug delivery. *Nanomedicine*, *8*, 452–459.

Kang, W., Chae, J., Cho, Y., Lee, J., & Kim, S., (2009). Multiplex imaging of single tumor cells using quantum-dot-conjugated aptamers. Weinheiman der Bergstrasse: Germany.

Katz, E., & Willner, I., (2004). Integrated nanoparticle-biomolecule hybrid systems: synthesis, properties, and applications. *Angew. Chem. Int. Edn. Engl*, *43*(45), 6042–6108.

Kayser, O., Lemke, A., & Hernandez-Trejo, N., (2005). The impact of nanobiotechnology on the development of new drug delivery systems. *Curr. Pharm. Biotechnol*, *6*, 3–5.

Kempe, H., & Kempe, M., (2010). The use of magnetite nanoparticles for implant-assisted magnetic drug targeting in thrombolytic therapy. *Biomaterials*, *31*, 9499–9510.

Kim, D., Jeong, Y. Y., & Jon, S., (2010). A Drug-Loaded Aptamer-Gold Nanoparticle Bioconjugate for Combined CT Imaging and Therapy of Prostate Cancer. *ACS Nano.* 4(7), 3689–3696.

Kim, S., Lim, Y. T., Soltesz, E. G., De Grand, A. M., Lee, J., Nakayama, A., (2004). Near-infrared fluorescent type II quantum dots for sentinel lymph node mapping. *Nat. Biotechnol*, *22*, 93–97.

Kong, D. F., & Goldschmidt-Clermont, P. J., (2005). Tiny solutions for giant cardiac problems. *Trends Cardiovasc. Med.*, *15*, 207–211.

Koziara, J. M., Lockman, P. R., Allen, D. D., & Mumper, R. J., (2004). Paclitaxel nanoparticles for the potential treatment of brain tumors. *J. Control Release. 99*, 259–269.

Kreuter, J. (2001). Nanoparticulate systems for brain delivery of drugs. *Adv. Drug. Deliv. Rev., 47*, 65–81.

Kreuter, J., Ramge, P., Petrov, V., Hamm, S., Alyautdin, R., Briesen, H., et al., (2003). Direct evidence that polysorbate-80-coated poly (butylcyanoacrylate) nanoparticles deliver drugs to CNS via specific mechanisms requiring prior binding of drug to the nanoparticles. *Pharm. Res. 20*, 409–416.

Kumar, R. M. N., (2000). Nano and microparticles as controlled drug delivery devices. J. Pharm. *Pharmaceut. Sci., 3*, 234–258.

Kunzmann, A., Andersson, B., Thurnherr, T., Krug, H., Scheynius, A., & Fadeel, B., (2011). Toxicology of engineered nanomaterials: focus on biocompatibility, biodistribution and biodegradation. *Biochim. Biophys. Acta., 1810*, 361–373.

Lai, J., Lu, Y., Yin, Z., Hu, F., & Wu, W., (2010). Pharmacokinetics and enhanced oral bioavailability in beagle dogs of cyclosporine A encapsulated in glyceryl monooleate/poloxamer 407 cubic nanoparticles. *Int. J. Nanomedicine. 5*, 13–23.

Lambert, G., Fattal, E., & Couvreur, P., (2001). Nanoparticulate systems for the delivery of antisense oligonucleotides. *Adv. Drug. Deliv. Rev. 47*, 99–112.

Lamprecht, A., & Benoit, J. P., (2006). Etoposide nanocarriers suppress glioma cell growth by intracellular drug delivery and simultaneous P-glycoprotein inhibition. *J. Control Release. 112*, 208–213.

Lampton, C., (1995). Nanotechnology promises to revolutionize the diagnosis and treatment of diseases. *Genetic Eng. News. 4*, 23.

Langer, R., (1998). Drug Delivery and Targeting. *Nature. 392*, 5.

Langer, R., (2000). Biomaterials in drug delivery and tissue engineering: one laboratory's experience. *Acc. Chem. Res., 33*, 94–101.

Langer, R., & Folkman, J., (1967). Polymers for the sustained release of proteins and other macromolecules. *Nature., 263*, 797–800.

Latallo, J., Candido, K., Cao, Z., Nigavekar, S., Majoros, I., Thomas, T., et al., (2005). Nanoparticles targeting of anticancer drug improves therapeutic response in animal model of human epithelial cancer. *Cancer Res., 65*(12), 5317–5324.

Lee, H. J., (2002). Protein drug oral delivery: the recent progress. *Arch. Pharm. Res., 25*, 572–584.

Lee, H., Yang, H., & Holloway, P., (2009a). Functionalized CdS nanospheres and nanorods. *Physica. B. Condensed Matter, 404*(22), 4364–4369.

Lee, W., Piao, L., Park, C., Lim, Y., Do, Y., Yoon, S., et al., (2010). Facile synthesis and size control of spherical aggregates composed of Cu_2O nanoparticles. *J. Colloid Interface Sci., 342*(1), 198–201.

Li, N., Sioutas, C., Cho, A., Schmitz, D., Misra, C., & Sempf, J., (2003). Ultrafine particulate pollutants induce oxidative stress and mitochondrial damage. *Environ. Health Perspect. 111*, 455–460.

Li, Y., Tseng, Y. D., Kwon, S. Y., D'Espaux, L., Bunch, J. S., McEuen, P. L., et al., (2004). Controlled assembly of dendrimer-like DNA. *Nat. Mater., 3*, 38–42.

Liang, M., Liu, X., Cheng, D., Liu, G., Dou, S., Wang, Y., et al., (2010). Multimodality Nuclear and Fluorescence Tumor Imaging in Mice Using a Streptavidin Nanoparticle. *Bioconjug. Chem., 21*(7), 1385–1388.

Lin, W., Huang, Y. W., Zhou, X. D., & Ma, Y., (2006). *In:vitro* toxicity of silica nanoparticles in human lung cancer cells. *Toxicol. Appl. Pharmacol., 217*, 252–259.

Liu, T. Y., Chen, S. Y., Liu, D. M., & Liou, S. C., (2005). On the study of BSA-loaded calcium-deficient hydroxyapatite nano-carriers for controlled drug delivery. *J. Control Rel., 107*(1), 112–121.

Lo, C. L., Lin, K. M., & Hsiue, G. H., (2005). Preparation and characterization of intelligent core-shell nanoparticles based on poly (D, L-lactide)-g-poly (N-isopropyl acrylamide-co-methacrylic acid). *J. Control Release., 104*(3), 477–488.

Loo, C., Lowery, A., Halas, N., West, J., & Drezek, R., (2005). Immuno targeted nano-shells for integrated cancer imaging and therapy. *Nano. Lett., 5,* 709–711.

Lope, AL., Reins, RY., McDermott, A. M., Trautner, B. W., & Cai, C., (2009). Antibacterial activity and cytotoxicity of PEGylated poly (amidoamine) dendrimers. *Mol. Biosyst., 5,* 1148–1156.

Luo, X., Matranga, C., Tan, S., Alba, N., & Cui, X. T., (2011). Carbon nanotube nanoreservior for controlled release of anti-inflammatory dexamethasone. *Biomaterials., 32,* 6316–6323.

Maeda, H., (2001). The enhanced permeability and retention (EPR) effect in tumor vasculature: the key role of tumor-selective macromolecular drug targeting. *Adv. Enzyme Regul., 41,* 189–207.

Maeda, H., Wu, J., Sawa, T., Matsumura, Y., & Hori, K., (2000). Tumor vascular permeability and the EPR effect in macromolecular therapeutics: A review. *J. Control Release., 65,* 271–284.

Maeng, J. H., Lee, D. H., Jung, K. H., Bae, Y. H., Park, I. S., Jeong, S., et al., (2010).Multifunctional doxorubicin loaded superparamagnetic iron oxide nanoparticles for chemotherapy and magnetic resonance imaging in liver cancer. *Biomaterials. 31,* 4995–5006.

Mainardes, R. M., Khalil, N. M., & Gremião, M. P. D., (2010). Intranasal delivery of zidovudine by PLA and PLA–PEG blend nanoparticles. *Int. J. Pharm., 395,* 266–271.

Malik, N., Evagorou, E. G., & Duncan, R., (1999). Dendrimer-platinate: A novel approach to cancer chemotherapy. *Anticancer Drugs, 10,* 767–776.

Medley, C. D., Smith, J. E., Tang, Z., Wu, Y., Bamrungsap, S., & Tan, W., (2008). Gold nanoparticle-based colorimetric assay for the direct detection of cancerous cells. *Anal. Chem., 80,* 1067–1072.

Meng, X., Seton, H. C., Lu, L. T., Prior, I. A., Thanh, N. T., & Song, B., (2011). Magnetic CoPt nanoparticles as MRI contrast agent for transplanted neural stem cells detection. *Nanoscale., 3,* 977–984.

Menjoge, A. R., Kannan, R. M., & Tomalia, D. A., (2010). Dendrimer based drug and imaging conjugates: design considerations for nanomedical applications. *Drug Discov. Today, 15,* 171–187.

Mohan, M. Y., Vimala, K., Thomas, V., Varaprasad, K., Sreedhar, B., Bajpai, S., et al., (2010). Controlling of silver nanoparticles structure by hydrogel networks. *J. Colloid Interface Sci., 342*(1), 73–82.

Monteiro-Riviere, N. A., Nemanich, R. J., Inman, A. O., Wang, Y. Y., & Riviere, J. E., (2005). Multi-walled carbon nanotube interactions with human epidermal keratinocytes. *Toxicol. Lett., 155,* 377–384.

Morimoto, Y., Kobayashi, N., Shinohara, N., Myojo, T., Tanaka, I., & Nakanishi, J., (2010). Hazard assessments of manufactured nanomaterials. *J. Occup. Health, 52,* 325–334.

Muller, J., Huaux, F., Moreau, N., Misson, P., Heilier, J. F., & Delos, M., (2005). Respiratory toxicity of multi-wall carbon nanotubes. *Toxicol. Appl. Pharmacol, 207,* 221–231.

Nam, T., Park, S., Lee, S. Y., Park, K., Choi, K., Song, I. C., et al., (2010). Tumor targeting chitosan nanoparticles for dual-modality optical/MR cancer imaging. *Bioconjug. Chem, 21,* 578–582.

Nanjwade, B. K., Singh, J., Parikh, K. A., & Manvi, F. V., (2010). Preparation and evaluation of carboplatin biodegradable polymeric nanoparticles. *Int. J. Pharm.*, *385*, 176–180.

Nasongkla, N., Shuai, X., Ai, H., Weinberg, B. D., Pink, J., Boothman, D. A., et al., (2004). cRGD-functionalized polymer micelles for targeted doxorubicin delivery. *Angew. Chem. Int. Edn. Engl.*, *43*(46), 6323–6327.

Natali, S., & Mijovic, J., (2010). Dendrimers as drug carriers: dynamics of PEGylated and methotrexate-loaded dendrimers in aqueous solution. *Macromolecules.*, *43*, 3011–3017.

Nevozhay, D., Kañska, U., Budzyñska, R., & Boratyñski, J., (2007). Current status of research on conjugates and related drug delivery systems in the treatment of cancer and other diseases. *Postepy. Hig. Med. Dosw.*, *61*, 350–360.

Nurunnabi, M., Cho, K. J., Choi, J. S., Huh, K. M., & Lee, Y. K., (2010). Targeted near-IR QDs-loaded micelles for cancer therapy and imaging. *Biomaterials*, *31*, 5436–5444.

Oberdorster, G., Sharp, Z., Atudorei, V., Elder, A., Gelein, R., & Kreyling, W., (2004). Translocation of inhaled ultrafine particles to the brain. *Inhal. Toxicol.*, *16*, 437–445.

Ozay, O., Akcali, A., Otkun, M. T., Silan, C., Aktas, N., & Sahiner, N., (2010). P (4-VP) based nanoparticles and composites with dual action as antimicrobial materials. Colloids Surf. B. *Biointerfaces*, *79*(2), 460–466.

Paavola, A., Kilpeläinen, I., Yliruusi, J., & Rosenberg, P., (2000). Controlled release injectable liposomal gel of ibuprofen forepidural analgesia. *Int. J. Pharm.*, *199*, 85–93.

Panyam, J., & Labhasetwar, V., (2003). Biodegradable nanoparticles for drug and gene delivery to cells and tissue. *Adv Drug Deliv. Rev.*, *55*, 329–347.

Park, J., Fong, P. M., Lu, J., Russell, K. S., Booth, K. J., Saltzman, W. M.,et al., (2009). PEGylated PLGA nanoparticles for the improved delivery of doxorubicin. *Nanomedicine*, *5*, 410–418.

Parveen, S., & Sahoo, S. K., (2006). Nanomedicine: clinical applications of polyethylene glycol conjugated proteins and drugs. *Clin. Pharmacokinet. 45*, 965–988.

Parveen, S., Mitra, M., Krishnakumar, S., & Sahoo, S. K., (2010). Enhanced antiproliferative activity of carboplatin-loaded chitosan-alginate nanoparticles in a retinoblastoma cell line. *Acta Biomater*, *6*, 3120–3131.

Paul, W., & Sharma, C., (2001). Porous hydroxyapatite nanoparticles for intestinal delivery insulin. *Trends Biom. Artif. Organs*, *14*(2), 37–38.

Piddubnyak, V., Kurcok, P., Matuszowicz, A., Glowala, M., Fiszer-Kierzkowska, A., Jedlinski, Z., et al., (2004). Oligo-3-hydroxybutyrates as potential carriers for drug delivery. *Biomaterials. 25*(22), 5271–5279.

Pinto-Alphandary, H., Andremont, A., & Couvreur, P., (2000). Targeted delivery of antibiotics using liposomes and nanoparticles: research and applications. *Int. J. Antimicrob. Agents. 13*, 155–168.

Prajapati, R. N., Tekade, R. K., Gupta, U., Gajbhiye, V., & Jain, N. K., (2009). Dendrimermediated solubilization, formulation development and *in vitro-in vivo* assessment of piroxicam. *Mol Pharm*, *6*, 940–950.

Prajapati, V. K., Awasthi, K., Gautam, S., Yadav, T. P., Rai, M., Srivastava, O. N., et al., (2011). Targeted killing of *Leishmania donovani* in vivo and in vitro with amphotericin B attached to functionalized carbon nanotubes. *J Antimicrob. Chemother. 66*, 874–879.

Prasad, K., & Jha, A. (2010). Biosynthesis of CdS nanoparticles: An improved green and rapid procedure. *J Colloid Interface Sci.*, *342*(1), 68–72.

Priyabrata, M., Resham, B., & Debabrata, M., (2005). Gold nanoparticles bearing functional anti-cancer drug and anti-angiogenic agent: A "2 in 1" system with potential application in therapeutics. *J. Biomd. Nanotech*, *1*(2), 224–228.

Qian, J., Wang, Y., Gao, X., Zhan, Q., Xu, Z., & He, S., (2010). Carboxyl-functionalized and bio-conjugated silica-coated quantum dots as targeting probes for cell imaging. *J. Nanosci. Nanotechnol, 10*.

Qiu, L. Y., & Bae, Y. H., (2006). Polymer architecture and drug delivery. *Pharm. Res., 1*, 1–30.

Rabinow, B. E., (2004). Nanosuspensions in drug delivery. *Nat. Rev. Drug Discov., 3*, 785–796.

Radomski, A., Jurasz, P., Alonso-Escolano, D., Drews, M., Morandi, M., Malinski, T., et al., (2005). Nanoparticle-induced platelet aggregation and vascular thrombosis. *Br. J. Pharmacol., 146*, 882–893.

Raghuvanshi, R., Mistra, A., Talwar, G., Levy, R., & Labhasetwar, V., (2001). Enhanced immune response with combination of alum and biodegradable nanoparticles containing tetanus toxoid. *J. Micoencapsul., 18*, 723–732.

Rejinold, N. S., Chennazhi, K. P., Nair, S. V., Tamura, H., & Jayakumar, R., (2011). Biodegradable and thermo-sensitive chitosan-g-poly (N-vinylcaprolactam) nanoparticles as a 5-fluorouracil carrier. *Carbohydr. Polym, 83*, 776–786.

Reynolds, A. R., Moghimi, S. M., & Hodivala-Dilke, K., (2003). Nanoparticle-mediated gene delivery to tumor vasculature. *Trends in Molecular Medicine., 9*, 2–4.

Rickman, J., & LeSar, R., (2002). Free-energy calculations in materials research. *Annu. Rev. Mater. Res., 32*, 195–217.

Rieu☐,A. D., Fievez, V., Garinot, M., Schneider, Y. J., & Préat, V., (2006). Nanoparticles as potential oral delivery systems of proteins and vaccines: a mechanistic approach. *J Control Release., 116*, 1–27.

Robert, L., & Nicholas, A. P., (2003). Advances in Biomaterials, Drug Delivery, and Bionanotechnology. *Bioengineering Food and Natural Products., 49*(12), 2990–3006.

Roney, C., Kulkarni, P., Arora, V., Antich, P., Bonte, F., Wu, A. M., et al., (2005). Targeted nanoparticles for drug delivery through the blood-brain barrier for Alzheimer's disease. *J. Controlled Release. 108*, 193.

Salata, O. V., (2004). Applications of nanoparticles in biology and medicine. *J. Nanobiotechnol. 2*, 3.

Santra, S., (2010). Fluorescent silica nanoparticles for cancer imaging. *Methods Mol. Biol., 624*, 151–162.

Saraog, G. K., Gupta, P., Gupta, U. D., Jain, N. K., & Agrawal, G. P., (2010). Gelatin nanocarriers as potential vectors for effective management of tuberculosis. *Int. J. Pharm., 385*, 143–149.

Sayed, F. N., Jayakumar, O. D., Sudakar, C., Naik, R., & Tyagi, A. K., (2011). Possible weak ferromagnetism in pure and M (Mn, Cu, Co, Fe and Tb) doped $NiGa_2O_4$ nanoparticles. *J. Nanosci. Nanotechnol, 11*, 3363–3369.

Sayes, C. M., Gobin, A. M., Ausman, K. D., Mendez, J., & West, J. L., Colvin, V. L., (2005). Nano-C (60) cytotoxicity is due to lipid peroxidation. *Biomaterials, 26*, 7587–7595.

Schiffelers, R. M., Koning, G. A., Hagen, T. L., Fens, M. H., Schraa, A. J., Janssen, A. P., et al., (2003). Anti-tumor efficacy of tumor vasculature targeted liposomal doxorubicin, *J. Control. Release, 91*, 115–122.

Schroeder, U., Sommerfeld, P., Ulrich, S., & Sabel, B. A., (1998). Nanoparticle technology for delivery of drugs across the blood–brain barrier. *J. Pharm. Sci., 87*, 1305–1307.

Sengupta, S., Eavarone, D., Capila, I., Zhao, G., Watson, N., Kiziltepe, T., et al., (2005). Temporal targeting of tumor cells and neovasculature with a nanoscale delivery system. *Nature, 436*, 568–572.

Shapiro, B., Bindewald, E., Kasprzak, W., & Yingling, Y., (2008). Protocols for the *in silico* design of RNA nanostructures. *Methods Molec. Biol.*, *474*, 93.

Sharma, A., Pandey, R., Sharma, S., & Khuller. G. K., (2004). Chemotherapeutic efficacy of poly (DL-lactide-co-glycolide) nanoparticle encapsulated antitubercular drugs at sub-therapeutic dose against experimental tuberculosis. *Int. J. Antimicrob. Agents*, *24*, 599–604.

Shenoy, D., Little, S., Langer, R., & Amiji, M., (2005). Poly (ethylene oxide)-modified poly (beta-amino ester) nanoparticles as a pH-sensitive system for tumor targeted delivery of hydrophobic drugs: part 2. *In vivo* distribution and tumor localization studies. *Pharm. Res.*, *22*, 2107–2114.

Shi, H., Tsai, W. B., Garrison, M. D., Ferrari, S., & Ratner, B. D., (1999). Template imprinted nanostructured surfaces for protein recognition. *Nature.*, *398*, 593–597.

Shvedova, A. A., Castranova, V., Kisin, E. R., Schwegler-Berry, D., Murray, A. R., &Gandelsman, V. Z., (2003a). E posure to carbon nanotube material: assessment of nanotube cytotoxicity using human keratinocyte cells. *J. Toxicol. Environ. Health.*, *66*, 1909–1926.

Shvedova, A. A., Kisin, E. R., Mercer, R., Murray, A. R., Johnson, V. J., & Potapovich, A. I., (2005). Unusual inflammatory and fibrogenic pulmonary responses to single walled carbon nanotubes in mice. *Am. J. Physiol. Lung Cell Mol. Physiol.*,*283*, L698–L708.

Singh, P., Gupta, U., Asthana, A., & Jain, N. K., (2008). Folate and Folate-PEG-PAMAM dendrimers: synthesis, characterization, and targeted anticancer drug delivery potential in tumor bearing mice. *Bioconjugate Chem.*, *19*, 2239–2252.

Skirtach, A. G., Karageorgiev, P., De Geest, B. G., Pazos-Perez, N., Braun, D., & Sukhorukov, G. B., (2008). Nanorods as wavelength-selective absorption centers in the visible and near-infrared regions of the electromagnetic spectrum. *Adv. Mater.*, *20*, 506–510.

Sledge, G., & Miller, K., (2003). Exploiting the hallmarks of cancer: the future conquest of breast cancer. *European Journal of Cancer*, *39*, 1668–1675.

Slocik, J. M., Moore, J. T., & Wright, D. W., (2002). Monoclonal antibody recognition of histidine-rich peptide encapsulated nanoclusters. *Nano Lett.*, *2*, 169–173.

Smith, A. M., Dave, S., Nie, S., True, L., & Gao, X., (2006). Multicolor quantum dots for molecular diagnostics of cancer. *Expert Rev. Mol. Diagn*, *6*, 231–244.

Smith, J. E., Medley, C. D., Tang, Z., Shangguan, D., Lofton, C., & Tan, W., (2007). Aptamer-conjugated nanoparticles for the collection and detection of multiple cancer cells. *Anal. Chem*, *79*, 3075–3082.

Smolensky, E. D., Park, H. Y, Berquó, T. S., & Pierre, V. C., (2011). Surface functionalization of magnetic iron oxide nanoparticles for MRI applications–effect of anchoring group and ligand exchange protocol. *Contrast Media Mol. Imaging*, *6*, 189–199.

Soppimath, K. S., Aminabhavi, T. M., Kulkarni, A. R., & Rudzinski, W. E., (2001). Biodegradable polymeric nanoparticles as drug delivery devices. *J. Control. Release.*, *70*, 1–20.

Sun, Y., Mayers, T., & Xia, Y., (2002). Template-engaged replacement reaction: A one-step approach to large scale synthesis of metal nanostructure with hollow interior. *Nano. Lett.*, *2*, 481–485.

Sunderland, C. J., Steiert, M., Talmadge, J. E., Derfus, A. M., & Barry, S. E., (2006). Targeted nanoparticles for detecting and treating cancer. *Drug Dev. Res.*, *67*, 70–93.

Svenson, S., & Tomalia, D. A., (2005). Dendrimers in biomedical applications – reflections on the field. *Adv. Drug Deliv. Rev.*, *57*, 2106– 2129.

Tang, F., He, F., Cheng, H., & Li, L., (2010). Self-assembly of conjugated polymerag SiO_2 hybrid fluorescent nanoparticles for application to cellular imaging. *Langmuir, 26*(14), 11774–11778.

Tanis, I., & Karatasos, K., (2009b). Association of a weakly acidic anti-inflammatory drug (Ibuprofen) with a poly (amidoamine) dendrimer as studied by molecular dynamics simulations. *J Phys. Chem., 113*, 10984–10993.

Teicher, B. A., (2000). Molecular targets and cancer therapeutics: Discovery, development and clinical validation. *Drug Resistance Updates, 3*, 67–73.

Tong, L., Zhao, Y., Huff, T. B., Hansen, M. N., Wei, A., & Cheng, J. X., (2007). Gold nanorods mediate tumor cell death by compromising membrane integrity. *Adv. Mater., 19*, 3136–3141.

Tong, Q., Li, H., Li, W., Chen, H., Shu, X., Lu, X., & Wang, G., (2011). In: vitro and in vivo anti-tumor effects of gemcitabine loaded with a new drug delivery system. *J. Nanosci. Nanotechnol, 11*, 3651–3658.

Torchilin, V. P., (2005a). Lipid-core micelles for targeted drug delivery. *Curr. Drug. Deliv, 2*, 319–327.

Tosh, D. K., Yoo, L. S., Chinn, M., Hong, K., Kilbey, M. S., Barrett, M. O., et al., (2010). Polyamidoamine (PAMAM) dendrimer conjugates of "clickable" agonists of the A3 adenosine receptor and coactivation of the P2Y14 receptor by a tethered nucleotide. *Bioconjug. Chem., 21*, 372–384.

Triggle, D. J., (1999). Rho Chi lecture. Pharmaceutical sciences in the next millennium. *Ann. Pharmacother, 33*, 241–246.

Tripisciano, C., Costa, S., Kalenczuk, R. J., & Borowiak-Palen, E., (2010). Cisplatin filled multiwalled carbon nanotubes–a novel molecular hybrid of anticancer drug container. *Eur. Phys. J. B. 75*, 141–146.

Tripisciano, C., Kraemer, K., Taylor, A., & Borowiak-Palen, E., (2009). Single-wall carbon nanotubes based anticancer drug delivery system. *Chem. Phys. Lett., 478*, 200–205.

Turkova, A., Roilides, E., & Sharland, M., (2011). Amphotericin B in neonates: deoxycholate or lipid formulation as first-line therapy - is there a 'right' choice? *Curr. Opin. Infect. Dis., 24*, 163–171.

Turnbull, W. B., & Stoddart, J. F., (2002). Design and synthesis of glycodendrimers. *J. Biotechnol. 90*, 231–255.

Turos, E., Shim, J. Y., Wang, Y., Greenhalgh, K., Reddy, G. S., Dickey, S., et al., (2007). Antibiotic-conjugated polyacrylate nanoparticles: New opportunities for development of anti-MRSA agents. *Bioorg. Med. Chem. Lett., 17*, 53–56.

Ueno, Y., Futagawa, H., Takagi, Y., Ueno, A., & Mizushima, Y., (2005). Drug-incorporating calcium carbonate nanoparticles for a new delivery system. *J. Control. Rel., 103*(1), 93–98.

Unezaki, S., Maruyama, K., Hosoda, J., Nagae, I., Koyanagi, Y., Nakata, M., et al., (1996). Direct measurement of the extravasation of polyethyleneglycol-coated liposomes into solid tumor tissue by in vivo fluorescence microscopy. *Int. J. Pharm., 144*, 11–17.

Valanne, A., Huopalahti, S., Soukka, T, Vainionpää, R., Lövgren, T., & Härmä, H., (2005). A Sensitive adenovirus immunoassay as a model for using nanoparticle label technology in virus diagnostics. *J. Clin. Virol., 33*, 217–223.

Van den Hoven, J. M., Van Tomme, S. R., Metselaar, J. M., Nuijen, B., Beijnen, J. H., & Storm, G., (2011). Liposomal drug formulations in the treatment of rheumatoid arthritis. *Mol. Pharm, 8*, 1002–1015.

VanVlerken, L. E., & Amiji, M. M., (2006). Multi-functional polymeric nanoparticles for tumor-targeted drug delivery. *Expert Opin. Drug Deliv.*, *3*, 205–216.

Vicent, M. J., & Duncan, R., (2006). Polymer conjugates: Nano sized medicines for treating cancer. *Trends Biotechnol*, *24*, 39–47.

Vinogradov, S. V., Bronich, T. K., & Kabanov, A. V., (2002). Nanosized cationic hydrogels for drug delivery: preparation, properties and interactions with cells. *Adv. Drug Deliv. Rev.*, *54*(1), 135–147.

Waite, C. L., & Roth, Ch., (2009). M. PAMAM-RGD conjugates enhance siRNA delivery through a multicellular spheroid model of malignant glioma. *Bioconjug Chem.*, *20*, 1908–1916.

Wang, H., Huff, T. B., & Zweifel, D. A. (2005). *In: vitro* and *in vivo* two-photon luminescence imaging of single gold nanorods. *Proc. Natl. Acad. Sci. USA.* *102*, 15752–15756.

Wangler, C., Moldenhauer, G., Eisenhut, M., Haberkorn, U., & Mier, W., (2008). Antibody-dendrimer conjugates: the number, not the size of the dendrimers, determines the imunoreactivity. *Bioconjug. Chem.*, *19*, 813–820.

Weng, X., Wang, M, Ge, J., Yu, S., Liu, B., Zhong, J., & Kong, J., (2009). Carbon nanotubes as a protein toxin transporter for selective HER2-positive breast cancer cell destruction. *Mol BioSysts*, *5*, 1224–1231.

West, J. L., & Halas, N. J., (2000). Applications of nanotechnology to biotechnology commentary. *Curr. Opin. Biotechnol*, *11*, 215–217.

Wilson, B., Samanta, M. K., Santhi, K., Kumar, K. P., Ramasamy, M., & Suresh, B., (2010). Chitosan nanoparticles as a new delivery system for the anti-Alzheimer drug tacrine. *Nanomedicine.*, *6*, 144–152.

Wu, W., Chen, B., Cheng, J., Wang, J., Xu, W., Liu, L., et al., (2010). Biocompatibility of Fe3O4/DNR magnetic nanoparticles in the treatment of hematologic malignancies. *Int J. Nanomedicine*, *5*, 1079–1084.

Xiong, F., Mi, Z., & Gu, N., (2011). Cationic liposomes as gene delivery system: transfection efficiency and new application. *Pharmazie*, *66*, 158–164.

Yallapu, M. M., Othman, S. F., Curtis, E. T., Gupta, B. K., Jaggi, M., & Chauhan, S. C., (2011). Multi-functional magnetic nanoparticles for magnetic resonance imaging and cancer therapy. *Biomaterials*, *32*, 1890–1905.

Yang, J., Park, S. B., Yoon, H. G., Huh, Y. M., & Haam, S., (2006). Preparation of poly e-caprolactone nanoparticles containing magnetite for magnetic drug carrier. *Int. J. Pharm.*, *324*, 185–190.

Yu, C., Nakshatri, H., & Irudayaraj, J., (2007). Identity profiling of cell surface markers by multiplex gold nanorod probes. *Nano. Lett.*, *7*, 2300–2306.

Yukihara, M., Ito, K., Tanoue, O., Goto, K., Matsushita, T., Matsumoto, Y., et al., (2011). Effective drug delivery system for duchenne muscular dystrophy using hybrid liposomes including gentamicin along with reduced toxicity. *Biol. Pharm. Bull.*, *34*, 712–716.

Zhang, C., Zhao, L., Dong, Y., Zhang, X., Lin, J., & Chen, Z., (2010). Folate-mediated poly (3-hydroxybutyrate-co-3-hydroxyoctanoate) nanoparticles for targeting drug delivery. *Eur. J. Pharm Biopharm.* *76*(1), 10–16.

Zhang, D., Pan, B., Wu, M., Wang, B., Zhang, H., Peng, H., et al., (2011). Adsorption of sulfamethoxazole on functionalized carbon nanotubes as affected by cations and anions. *Environ. Pollut*, *159*, 2616–2621.

Zhang, W., Zhong, Z. Z., & Zhang, Y., (2011). The application of carbon nanotubes in target drug delivery systems for cancer therapies. *Nanoscale Research Letters*, *6*, 555.

Zheng, M., & Diner, B. A., (2004). Solution redox chemistry of carbon nanotubes. *J. Am. Chem. Soc.*, *126*, 15490–15494.

PART III

NANOBIOMATERIALS AS CARRIERS IN CANCER THERAPEUTICS

CHAPTER 8

ADVANCES OF NANOTECHNOLOGY IN CANCER THERAPY

URMILA JAROULIYA,[1] RAJ K. KESERVANI,[2]
and RAJESH K. KESHARWANI[3]

[1]School of Studies in Biotechnology, Jiwaji University, Gwalior (M.P.), 474011, India

[2]Faculty of Pharmaceutics, Sagar Institute of Research and Technology-Pharmacy, Bhopal–462041, India, E-mail: rajksops@gmail.com

[3]Department of Biotechnology, NIET, Nims University, Jaipur-303121, India

CONTENTS

ABSTRACT

Cancer is a leading cause of deaths worldwide, so nanotechnology is currently under intense development for applications in cancer imaging, molecular diagnosis and targeted therapy. Nanotechnology refers to the interactions of cellular and molecular components and engineered materials-typically, clusters of atoms, molecules, and molecular fragments into incredibly small particles-between 1 and 100 nm. A large number of nanoparticle delivery systems have been developed for cancer therapy and currently they are in the preclinical stages of development. In the last several decades, nanoparticles such as liposomes, gelatin nanoparticles, micelles has been studied and developed primarily for use in novel drug-delivery systems. A recent explosion in engineering and technology has led to the development of many new nanoparticles, including quantum dots, nanoshells, gold nanoparticles, paramagnetic nanoparticles, and carbon nanotubes. When linked with biotargeting ligands, such as monoclonal antibodies, peptides or small molecules, these nanoparticles are used to target malignant tumors with high affinity and specificity. These developments raise exciting opportunities for personalized oncology in which genetic and protein biomarkers are used to diagnosis and treatment of cancer on molecular basis in individual patients. In this chapter, we will discuss the available preclinical and clinical nanoparticles technology platforms and their impact for cancer therapy.

8.1   INTRODUCTION

As the death toll from infectious disease has declined in the world, cancer has become the second-ranking cause of death, led only by heart disease. Current estimates project that one person in three in the United States will develop the cancer, and that one in five will die from cancer. From immunologic view point, cancer cells can be viewed as altered self-cells that have escaped normal growth regulating mechanism. In most organs and tissues of a mature animal, a balance is usually maintained between cell renewal and cell death. The various types of mature cells in the body have a given life span, as these cells die, new cells are generated by the proliferation and differentiation of various types of stem cells. Under normal circumstances, the production of new cell is also regulated that the number of any particular type of cells remains constant. Occasionally, though, cell arises that no longer responds to normal growth-control mechanisms. These cells give rise to clones of cells that can expand to a considerable size, producing a tumor.

Word *cancer* was first applied to the disease by Hippocrates (460–370 B.C.), the Greek philosopher, who used the words *carcinos* and *carcinoma* to refer to non-ulcer forming and ulcer forming tumors. The words refer to a crab, probably due to the external appearance of cancerous tumors, which have branch-like projections that resemble the claws of a crab (Mansoori et al., 2007). Cancer is an inherently biological disease, in which cell replication-one of the hallmarks of life-fails to be regulated by the usual mechanisms. It is a major cause of morbidity and mortality in developed countries. It is projected that the number of new cases of all cancers worldwide will be 15.4 million in the year 2020. In India the number of projected cancer cases will be 182,602 and 1,148,757 by the year 2015 and 2020. Despite the intensive efforts within the last 50 years, it has only been possible to slightly, but not substantially, lower the mortality of cancer.

At generic level, cancer is a genetic disease because it can be traced to alterations within specific genes, but in most cases, it is not an inherited disease. In an inherited disease, the genetic defect is present in the chromosome of a parent and is transmitted to the zygote. In contrast, the genetic alternation that leads to most cancers arises in the DNA of a somatic cell during the lifetime of an affected individual. Because of these genetic changes, cancer cell proliferate uncontrollably, producing malignant tumors that invade surrounding healthy tissue. While at the cellular level, cancer involves the simultaneous occurrence of two general categories of malfunctions. The first category causes the *replication of a cell to become permanently enabled* due to a natural or carcinogen-induced genetic mutation, chromosome translocation or gene amplification (genetic instability). The second category is also due to genetic mutations, and causes the *apoptosis complex*, also known as the suicide complex, to become permanently disabled (Figure 8.1). As stated, both of these problems must occur in the same cell, at the same time, in order to cause cancer. Under normal circumstances, the cells carefully control their divisions using apoptosis complex activated by the *p53 tumour suppressor protein*. There are other mechanisms triggering the apoptosis complex, including receptor mediated death, which is dependant to chemical messengers, especially tumor necrosis factors (Souhami et al., 2004; Fu et al., 2005). But, when both of these mechanisms malfunction,

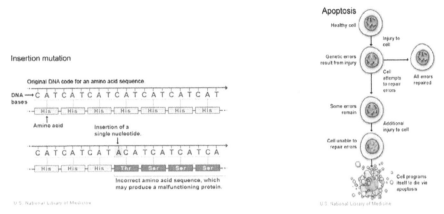

FIGURE 8.1 A single nucleotide base inserted (or removed) out of sequence can cause the entire subsequent chain of amino acids, to be incorrect. When this happens in the gene sequence that codes for the protein (s) responsible for apoptosis or restricting cell division, the cell permanently loses the ability to carry out that particular function.

the body has no other option. As the uncontrolled cell division continues, a cluster of fairly unspecialized cells committed to dividing develops and becomes larger and larger. In addition, the cluster of cells releases chemicals to promote *abnormal capillary growth* into the tumor. This kind of a cell cluster is known as a *malignant tumor*, and can severely damage the surrounding tissue as it sucks up essential nutrients and displaces healthy cells. Eventually, when the tumor grows large enough, some of the tumor cells can find their way into the bloodstream, forming tumors in other parts of the body. This latter phenomenon is known as *metastasis*. It effectively multiplies the cancer as well as its effects, and eventually will prove fatal to the patient. Many other mutations can also be cancerous.

Historically, chemistry has been one of the most effective tools for treating cancer cells: chemotherapy-treatment with cytotoxic chemicals – kills cancer cells. But most chemotherapeutics also kill healthy cells. Making drugs that differentiate between cancer and normal cells is difficult, and when it works, it may not work for long. Cancer cells replicate rapidly, so they evolve rapidly and are extraordinarily quick at developing drug resistance. With a new generation of nanotech drugs, researchers are fighting cancer by approaching it as a physics problem-a problem of mass transport and fluid mechanics. They've already achieved some success, but the drugs have introduced a new series of challenges unique to the physics of nanomaterials.

At the present time, there are three basic approaches for treating cancer: surgery, radiotherapy, and chemotherapy. However, all of these procedures have considerable side effects and often are not sufficient for the curative treatment of metastatic cancer. The tumor-specific drugs, such as the tyrosinekinase inhibitors, have now been in use for several years with good results in certain types of cancer (e.g., Imatinib for chronic myeloid leukemia) (Druker et al., 2001). The antibodies are also increasingly being used in oncology, although the antibody therapy sometimes has adverse systemic effects and often is expensive. For a subgroup of breast cancer patients, for example, the Her2-targeting antibodies are used successfully in treatment (Murphy et al., 2012). However, the approval of some previous indications for the antibody therapy has even been withdrawn recently (the British National Institute for Health and Clinical Excellence refused Avastin® (Bevacizumab) in colorectal carcinomas) (NICE, 2010). There has not yet been a real breakthrough in cancer therapy, in general. According to the estimates by the US National Cancer Institute, nanomedicine will prove to be trailblazing in the future prevention, diagnostic investigation, and treatment of cancer (Tiwari et al., 2012). Nanotechnology has already found a use in many medical specialties, for example, in otorhinolaryngology (Durr et al., 2012). It is widely used in the different disciplines, from imaging to regenerative medicine. Before being used for medical purposes, however, those substances have to be investigated closely to determine their effects on the organism. This field of research is called nanoto□icology. It investigates the effects of both the nanoparticles that occur naturally (in the environment) and those that are created by industry and traffic.

8.1.1 CANCER NANOTECHNOLOGY

Cancer nanotechnology is out coming as a new area of multidisciplinary research that includes biology, chemistry, engineering, and medicine, and is expected to lead to major advances in cancer detection, diagnosis, and treatment. The traditional, well known, forms of cancer treatment are chemotherapy and radiation. Although these treatments are helpful in the fight against cancer, they are infamous for destroying normal cells and harming other organs in the process. Researchers are confident that nanoparticles technology will allow them to bring cancer treatment to another level. The term "nano" is derived from the Greek word "nanos" which means small and it is used as the prefix for 1 billionth part (10^{-9}). Nanoparticles are typically smaller in size and ranges from 1 to 100 nm, but in the scientific community advocate that in terms of size, nanoparticles extend up to 1000 nm, comparable to large biological molecules such as enzymes, receptors, and antibodies. With the size of about one hundred to ten thousand times smaller than human cells, these nanoparticles can offer unrivalled interactions with biomolecules both on the surface of and inside the cells, which may revolutionize cancer diagnosis and treatment. Nanoparticles have unique properties, including minimal renal filtration, high surface-to-volume ratios enabling modification with various surface functional groups that home, internalize, or stabilize, and may be constructed from a wide range of materials used to encapsulate or solubilize therapeutic agents for drug delivery or to provide unique optical, magnetic, and electrical properties for imaging and remote activation. Recent advances in nanoscience and nanotechnology have led to the development of nanomaterials for molecular and cellular imaging, cancer therapy, and integrated nanodevices for cancer detection and screening (Jain, 2005; Nie et al., 2007; Sengupta and Sasisekharan, 2007; Wang et al., 2007). Nanoparticles not only provide the specific imaging information in cancer patients but also selectively deliver anticancer drugs to tumor sites. Nanoparticles, the metallic nanoparticles respond resonantly to the magnetic field which varies with time so they transfer enough toxic thermal energy to the tumor cells as hyperthermic agents to destroy the tumor. Presently, there is limited knowledge of suitable biomarkers for imaging, selection of the imaging target and contrast enhance materials, and the chemistry required to assemble the bioactive imaging probe.

There is numerous hurdles faced in developing cancer-specific imaging agents, such as (i) delivery of the probe to the targeted tissue/tumor, (ii) biocompatibility and toxicity, (iii) stability of the probe and effective signal enhancement (iv) adequate imaging methods and strategies. During chemotherapy, pharmacologically active cancer drugs reach the tumor tissue with poor specificity and induce dose-limiting to□icities.Nanoparticle drug delivery may provide a more efficient, less harmful solution to overcome these problems.

The commonly studied nanoparticles (NPs) include quantum dots, carbon nanotubes, paramagnetic NPs, liposomes, gold NPs, polymeric, lipid and silver NPs. The well-studied nanoparticles include quantum dots (Cai et al., 2007b), carbon nanotubes (Liu et al., 2007b), paramagnetic nanoparticles (Thorek et al., 2006), nanoshells (Park et al., 2004), gold nanoparticles (Huang et al., 2007b), and many others (Ferrari, 2005, Grodzinski et al., 2006).

Quantum dots (QDs) a tiny light-emitting particles on the nanometer scale, are emerging as a new class of fluorescent probes for biomolecular and cellular imaging (Gao X et al., 2007). Iron oxide-based super para-magnetic nanoparticles display magnetism only when an e□ternal field is applied, leading to their direct application in clinical cancer imaging-based on magnetic resonance. These are widely explored for numerous biomedical applications in cancer imaging, drug delivery, tissue repair, deto□ification of biological fluids, and hyperthermia (Yang et al., 2011). Gold nanoparticles absorb high levels of ionizing radiation, boosting the impact of radiotherapy treatments that work by damaging DNA in tumor cells.

Carbon-based NPs, especially carbon nanotubes, have found increasing interest from the point-of-view of biomedical applications such as photo thermal therapy and drug delivery (Bianco et al., 2005). Nanoshells are adjustable core/shell nanoparticles that strongly absorb in the near-infrared (NIR) region where light transmits deeply into tissue. When injected sys-temically, these particles have been shown to accumulate in the tumor due to the enhanced permeability and retention (EPR) effect and induce photo thermal excision of the tumor when irradiated with an NIR laser (Mortan et al., 2010).

The application of nanotechnology is rapidly progressing, and has tre-mendous potential to make a revolutionary impact in healthcare, with intense effects on current treatment on pattern of various diseases. Thus, this unique technology might be a ray of hope in treating a complicated disease like can-cer, a disease that accounted for up to 7,021,000 deaths in 2007 worldwide, and is the second leading global killer (12.5% of deaths). Scientists have made a continuous effort over the past few decades for cancer treatment that consist of doses of compounds that are non-specific and highly to□ic. Also, the inability of conventional diagnosis tools to detect cancer in an early and potentially curable stage further hinders effective treatment options, and thus by the time cancer is detected it may be too late to prevent metastasis to other organs in the body. Due to the unique physical and chemical properties of different nano-carriers these makes a possible solution for many of the drawbacks associated with existing cancer treatments. The rapid advances in the development of nanotechnology-based materials, explaining the toxicity of nanoparticles is also crucial. The nanoparticles are undergoing extensive study to determine the role that they might play in cancer detection and treatment. Hence, in this chapter, we seek to explore the biomedical properties of nanoparticles (quantum dots, nanoshells, paramagnetic NPs, gold NPs and carbon nanotubes), with particular prominence on their use as therapeutic platforms in cancer biology.

8.2   QUANTUM DOTS

The quantum dot (QD) is defined as an artificially structured system with the capacity to load electrons. It has special physical and chemical properties that differentiate it from other naturally occurring biogenic and anthropogenic nanoparticles. QDs are one type of nanoparticles (NPs) having three characteristic properties: semiconductors, zero-dimension, and strong fluorescence. Although semiconductor QDs are single crystals with diameters of a few nanometers, their sizes be precisely controlled by the duration, temperature, and ligand molecules during the synthetic processes. The well-controlled synthetic process yields QDs with composition and size-dependent absorption and emission (Liu et al., 2013). The specific properties of QDs enable them with wide applications in chemistry, chemical biology and biomedicine. In the past decade, QDs have been broadly applied in fluorescence resonance energy transfer (FRET) analysis, gene technology, fluorescent labeling of cellular proteins, cell tracking, in vivo animal imaging and tumor biology investigation. Their unique optical properties, such as high brightness, long-term stability, simultaneous detection of multiple signals and tunable emission spectra, make them appealing as potential diagnostic and therapeutic systems in cancer biology (Han et al., 2013). These properties are most important for improving the sensitivity of molecular imaging and quantitative cellular analysis by 1–2 orders of magnitude. In this section, we will discuss the QDs in tumor imaging and therapy.

Although relatively fewer researches on QDs for tumor therapy were reported, it is possible that QDs have the potentialities for tumor therapy due to their large surface areas available for the modification of functional groups or therapeutic agents such as anti-cancer drugs. Recent advances have led to multifunctional nanoparticle probes that are highly bright and stable under complex in-vivo conditions. A new structural design involves encapsulating luminescent QDs with amphiphilic block copolymers, and linking the polymer coating to tumor-targeting ligands and drug-delivery functionalities. Polymer-encapsulated QDs are essentially nontoxic to cells and small animals, but their long-term in-vivo toxicity and degradation need more careful studies. Nonetheless, bioconjugated QDs have raised new possibilities for ultrasensitive and multiplexed imaging of molecular targets in living cells and animal models. QDs are nearly spherical semiconductor particles with diameters on the order of 2–10 nanometers, containing roughly 200–10, 000 atoms. The semiconducting nature and the size dependent fluorescence of these nanocrystals have made them very attractive for use in optoelectronic devices, biological detection, and also as fundamental prototypes for the study of colloids and the size-dependent properties of nanomaterials.

QDs synthesis was first described in 1982 by Efros and Ekimov, who grew nanocrystals and microcrystals of semiconductors in glass matrices. After that, a wide variety of synthetic methods have been devised for the preparation of QDs in different media, including aqueous solution, high-temperature organic solvents, and solid substrates. Colloidal suspensions of QDs are commonly synthesized through the introduction of semiconductor precursors under conditions that thermodynamically favor crystal growth, in the presence of semiconductor-binding agents, which function to kinetically control crystal growth and maintain their size within the quantum-confinement size regime.

For use in biology and medicine, QD probes most frequently take the form shown in Figure 8.2, with an inorganic semiconductor core surrounded by a monolayer of ligands and an amphiphilic polymer coat that is linked to biomolecules. QD cores are most commonly prepared from cadmium selenide (CdSe), a binary semiconductor with size-dependent band gap energy that can be tuned to emit light of any color throughout the visible spectrum. CdSe QDs have been thoroughly studied and can be produced in large quantities with fluorescence emission efficiencies as high as 90% at room temperature (Pan et al., 2005). These crystalline CdSe cores are synthesized and capped with a protective zinc sulfide (ZnS) shell in a high temperature organic solvent (Murray et al., 1993; Hines and Guyot-Sionnest, 1996; Dabbousi et al., 1997), and are coated with a monolayer of nonpolar coordinating ligands such as trioctylphosphine oxide (TOPO).

After synthesis, the nonpolar, hydrophobic QDs are transferred to an aqueous phase, a nontrivial process that is crucial to the generation of high-quality QDs. Amphiphilic polymers, such as polyethylene glycol (PEG)-modified

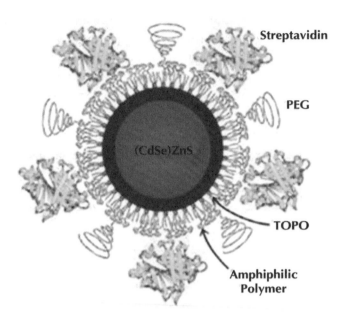

FIGURE 8.2 Quantum dot probes for biological imaging.

lipids and octylamine-modified polyacrylic acid (Dubertret et al., 2002), have been used to produce water-soluble. After transfer to water, QDs may be cross-linked to a variety of biologically active molecules such as antibodies, enzymes, nucleic acids, small molecule inhibitors, or biologically inert PEG. Many coupling schemes have been used to generate cross-links, such as electrostatic adsorption (Mattoussi et al., 2000), covalent bond formation (Chan and Nie, 1998), and biomolecular bridging via streptavidin-biotin binding (Goldman et al., 2002). These methods give rise to QD probes that are highly stable and have e□cellentaffinity toward their targets, although it is difficult to control the number and orientation of biomolecules attached to a single QD. Quantum dots (QDs) have several advantages and unique applications, (i) QDs have large molar e□tinctioncoefficients on the order of $0.5–5 \times 106\ M^{-1}cm^{-1}$ (Leatherdale et al., 2002) and have efficiency of 10–50 times more in adsorbing photons than organic dyes at the same excitation photon flu□,leading a significant improvement in probe brightness. (ii) QDs are several thousand times more stable against photo bleaching than dye molecules, allowing extended imaging and quantitative biomarker studies of cells and tissue specimens. (iii) QDs have size and composition-tuneable fluorescence emission from visible to infrared wavelengths and one light source can be used to e☐citemultiple colors of fluorescence emission. This leads to very large stokes spectral shifts (measured by the distance between the excitation and emission peaks) that can be used to further improve detection sensitivity. This factor becomes especially important for clinical tissue studies and in-vivo animal imaging due to the high auto fluorescence background often seen in complex biomedical specimens. Organic dye signals are often buried by strong tissue auto fluorescence, whereas QD signals can be readily separated from the background by wavelength-resolved or spectral imaging (Gao et al., 2004).

Imaging of tumors presents a unique challenge not only because of the urgent need for sensitive and specific imaging agents of cancer, but also because of the unique biological ascribed intrinsic to cancerous tissue. Blood vessels are abnormally formed during tumor-induced angiogenesis, having irregular architectures and wide endothelial pores. These pores are large enough to allow the extravasation of large macromolecules up to ~400 nm in size, which accumulate in the tumor microenvironment due to a lack of effective lymphatic drainage (Jain, 2001, 1999; Maeda et al., 2000; Matsumura, 1986). This "enhanced permeability and retention" effect (EPR effect) has inspired the development of a variety of nano therapeutics and nano particulates for the treatment and imaging of cancer. QDs hold great promise for this disease mainly due to their intense fluorescent signals, multiplexing capabilities and physiochemical properties, which could allow a high degree of sensitivity and selectivity in cancer imaging with multiple antigens, which has promising advantages intumor imaging and therapy.

8.3 NANOSHELLS

Nanoshell is a type of spherical nanoparticle consisting of a dielectric core which is covered by a thin metallic shell, these nanoshells involve a quasi-particle called plasmon which is a collective excitation or quantum plasma oscillation where the electrons simultaneously oscillate with respect to all the ions. The discovery of the nanoshell was made by Professor Naomi J. Halas and her team at Rice University in 2003. Nanoshells were the first nanoparticle used to demonstrate photo thermal cancer therapy. Nanoshell particles are highly functional materials with tailored properties, which are quite different than either of the core or of the shell material. Indeed, they show modified and improved properties than their single component counter parts or nanoparticles of the same size. Therefore, nanoshell particles are preferred over nanoparticles. Their properties can be modified by changing either the constituting materials or core-to-shell ratio. The term nanoshell is used specifically because thickness of the shell is 1–20 nm and the properties of shell materials (metal or semiconductor) having thickness in nanometers, become important when they are coated on dielectric cores to achieve higher surface area. Nanoshell materials can be synthesized practically using any material, like semiconductors, metals and insulators. Usually dielectric materials such as silica and polystyrene are commonly used as core because they are highly stable. Sometimes they are referred as core shell or core@ shell particles also. Thicker shells can also be prepared, but their synthesis is restricted mainly to achieve some specific goal, such as providing thermal stability to core particles.

Synthesis of nanoshells can be useful for creating novel materials with different morphologies, as it is not possible to synthesize all the materials in desired morphologies. Core particles of different morphologies such as rods, wires, tubes, rings, cubes, etc. can be coated with thin shell to get desired morphology in core shell structures (Ohmari et al., 2006). They are chemically inert and water-soluble, therefore they can be useful in biological applications. Numerous techniques have been developed to synthesize nanoshell particles. Preparation of nanoshell particles involves multistep synthesis procedure. It requires highly controlled and sensitive synthesis protocols to ensure complete coverage of core particles with the shell material. There are various methods to fabricate core shell structures, for example, precipitation (Ocana et al., 1991), grafted polymerization (Okaniwa et al., 1998), micro emulsion, reverse micelle (Huang et al., 1999), sol–gel condensation (See et al., 2005), layer-by-layer adsorption technique (Caruso et al., 2001). Although several methods have been established, it is still difficult to control the thickness and homogeneity of the coating. If the reaction is not controlled properly, it eventually leads to aggregation of core particles, formation of separate particles of shell material or incomplete coverage. However, with the new emerging techniques it is now possible to deposit homogeneous coatings. One such method is to coat the surface of colloids with appropriate primer (coupling agent) to enhance coupling of the shell material with the core (Ung et al., 1998). Also, the surface of the core material can be charged, and shell material can be adsorbed on its surface by electrostatic attraction (Zhang et al., 2004).

Nanoshell particles can be synthesized in a variety of combinations such as (core-shell) dielectric–metal (Imhof et al., 2001), dielectric–semiconductor (Monterio et al., 2002), dielectric-dielectric (Caruso et al., 1998 and 1999), semiconductor–metal (Kim et al., 2005), metal–metal, semiconductor–semiconductor (Hines et al., 1996), semiconductor–dielectric (Gerion et al., 2001), metal–dielectric (Kim et al., 2004), dye–dielectric (Ethiraj et al., 2005). Core shell particles can be assembled and further used for creation of another class of novel materials like colloidal crystal or quantum bubbles. It is indeed possible to create unique core shell structures having multishells (Chen et al., 2004). Multishell particles can be visualized as core particles having a number of shells around them. Also the core particles can be coated with a shell to obtain a single nanoshell. Further, these combinations of core and shell can be repeated again to get multishells. These structures show tuneable optical properties from the visible to infrared region of the electromagnetic spectrum by choosing different combinations of core and shell. Nanoshells can be synthesized on polysacchride (PS) core and the core can be removed easily either by calcinations or by dissolution. These particles are synthesized for a variety of purposes like providing chemical stability to colloids (Smith et al., 1996), enhancing luminescence properties, engineering band structures, biosensors, drug delivery.

A schematic diagram showing a variety of core shell particles is depicted in Figure 8.3. Surface of the core particle can be modified using bifunctional molecules and then small particles can be anchored on it (Figure 8.3a). Nanoparticles grow around the core particle and form a complete shell (Figure 8.3b). In some cases, a smooth layer of shell material can be deposited directly on the core by co-precipitation method (Figure 8.3c). Small core particles such as gold or silver (10–50 nm) can be uniformly encapsulated with silica (Figure 8.3d). Also a number of colloidal particles can be encapsulated inside a single particle (Figure 8.3e). Core particles can be removed either by calcinations or by dissolving them in a proper solvent. This gives rise to hollow particles also known as quantum bubbles (Figure 8.3f). Concentric shells also can be grown on core particles to form a novel structure known as multishell or nanomatryushka (named after the Russian doll, Figure 8.3g).

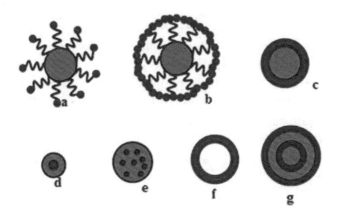

FIGURE 8.3 Variety of core shell particles. *a*, Surface-modified core particles anchored with shell particles. *b*, More shell particles reduced onto core to form a complete shell. *c*, Smooth coating of dielectric core with shell. *d*, Encapsulation of very small particles with dielectric material. *e*, Embedding number of small particles inside a single dielectric particle. *f*, Quantum bubble. *g*, Multishell particle.

There well known material Silica and gold is used for nanoshell formation. Silica (SiO2) (dielectric constant ~4.5) is a popular material to form core shell particles because of its extraordinary stability against coagulation. It has non-coagulating nature due to its very low value of Hamaker constant, which defines the Vander Waal forces of attraction among the particles and the medium. It is also chemically inert, optically transparent and does not affect redox reactions at core surfaces. For various purposes it is desirable that particles remain well dispersed in the medium, which can be achieved by coating silica on them to form an encapsulating shell (Lide, 1993). Whereas Gold nanoshells are spherical particles with diameters typically ranging in size from 10 to 200 nm (Figure). They are composed of a dielectric core covered by a thin gold shell. As novel nanostructures, they possess a remarkable set of optical, chemical and physical properties, which make them ideal candidates for enhancing cancer detection, cancer treatment, cellular imaging and medical biosensing.

Coating of colloidal particles with shells offers the most simple and versatile way of modifying their surface chemical (Matijevic, 1993), reactive (Mulvaney et al., 1997), optical, magnetic (Caruso, 1999) and catalytic properties (Hanprasopwattana et al., 1996). Silica particles coated with gold shell have been studied for fascinating optical properties. CdSe nanoparticles coated with CdS or ZnTe and CdTe nanoparticles coated with CdSe have been studied for enhancement in their luminescent properties. Similarly, semiconductor particles such as ZnS doped with Mn can be embedded inside a single silica particle. Such doped semiconductor nanoparticles are known to be highly efficient fluorescent materials. Nanoshell materials have received considerable attention in recent years because of potential applications associated with them. These materials have been synthesized for a variety of applications like fluorescent diagnostic labels, catalysis, avoiding photo degradation, enhancing photoluminescence, creating photonic crystals, preparation of bio conjugates, chemical and colloidal stability (Graph et al., 2002).

Nanoshells have gained considerable attention in clinical and therapeutic applications (Hirsch et al., 2003). By carefully choosing the core-to-shell ratio, it is possible to design novel nanoshell structures, which either absorb light or scatter it effectively (Loo, 2004). Strong absorbers can be used in photothermal therapy, while efficient scatterers can be used in imaging applications (Figure 8.4).

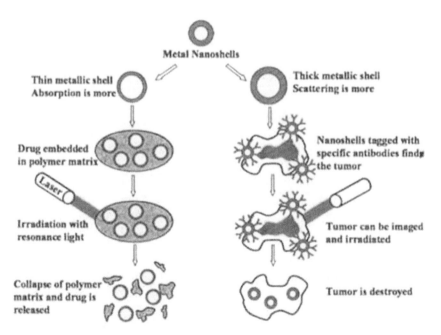

FIGURE 8.4   Drug delivery and imaging system-based on nanoshells.

Core shell (mostly gold nanoshells) particles conjugated with enzymes and antibodies can be embedded in a matrix of the polymer (West et al., 2000). These polymers, such as Nisopropylacrylamide (NIPAAm), and acrylamide (AAm), have a melting temperature which is slightly above body temperature. When such a nanoshell and polymer matrix is illuminated with resonant wavelength, nanoshells absorb heat and transfer to the local environment. This causes collapse of the network and release of the drug, as shown in Figure 8.3(a). In core shell particles-based drug delivery systems either the drug can be encapsulated or adsorbed onto the shell surface (Sparnacci et al., 2002). The shell interacts with the drug via a specific functional group or by electrostatic stabilization method. When it comes in contact with the biological system, it directs the drug. In imaging applications, nanoshells can be tagged with specific antibodies for diseased tissues or tumors. When these nanoshells are inserted in the body, they get attached to diseased cells and can be imaged (Figure 8.3b). Once the tumor has been located, it is irradiated with resonance wavelength of the nanoshells. This leads to localized heating of the tumor and it is destroyed. The power required for destroying diseased cells is almost half that required to kill healthy cells. The usual methods of tumor treatment, such as chemotherapy or radiotherapy have various side effects like substantial loss of hair, lack of appetite, diarrhea, etc. The process of attacking the tumor, also leads to the loss of many healthy cells. Nanoshells offer an effective and relatively safer strategy to cure this disease.

8.4   GOLD NANOPARTICLES

Gold nanoparticles are appearing as promising agents for cancer therapy and are being investigated as drug carriers, photo thermal agents, contrast agents and radio-sensitizers. It is metallic nanoparticles which have abundant use in the field of biotechnology and biomedicine because they have large surface bio conjugation with molecular probes and they also have many optical properties which are mainly concerned with localized Plasmon resonance (PR). Properties of gold nanoparticles are different from its bulk form because bulk gold is yellow solid and it is inert in nature while gold nanoparticles are wine red solution and are reported to be anti-oxidant. Inter particle interactions and assembly of gold nanoparticles networks play key role in the determination of properties of these nanoparticles (Deb et al., 2011). Gold nanoparticles e□hibit various sizes ranging from 1 nm to 8 μm and they also unveil different Shapes such as spherical, sub-Octahedral, octahedral, decahedral, icosahedral multiple twined, multiple twined, irregular shape, tetrahedral, nanotriangles, nanoprisms, hexagonal platelets and nanorods. GNPs are suitable for the delivery of the drugs to cellular destinations due to their ease of synthesis, functionalization, small size, biocompatibility, high atomic number (high-Z) and the ability to bind targeting agents (Figure 8.5).

Among all these shapes triangular shaped nanoparticles show attractive optical properties as compared to the spherical shaped nanoparticles. Some of the active substance presents in the plant extract used to synthesize the gold nanoparticles and it is an important biosynthesis technique to purify gold nanoparticles and to investigate their uses in medical science. Due to its efficient and targeted drug delivery to tumor site gold nanoparticles have been widely used in the field of radiation medicine as radiation enhancer (Ganesh Kumar et al., 2012). As a nanomaterial it have various applications in biomolecular ultrasensitive detection, killing cancer cells by hyperthermal treatment, labeling of cells and proteins and delivering therapeutic agents within cells.

Fluorescent nanoparticles or nanoprobes-based on gold nanoparticles have good biocompatibility for molecular imaging of many enzymes and metabolites which is necessary for cellular functions in cancer. Gold nanorods

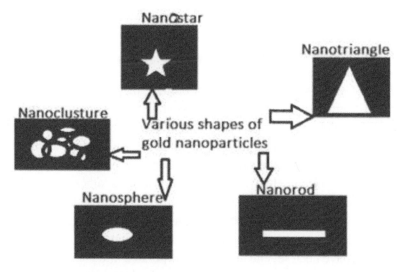

FIGURE 8.5   Various shapes of gold nanoparticles.

have unique anisotropic geometry which enables them to get tunable absorption in both visible and near infrared (NIR) regions and make them suitable for potential applications in the fields of biosensing, gene delivery and photo thermal therapy. Major advantage of gold nanoparticles is that they are non-cytoto☐icand most important benefit is regarding their surfaces, as they have large surface area due to which their surfaces are readily available for modification with targeting molecules or specific biomarkers and applicable in biomedical purposes. Due to the unique optical and electronic properties of gold nanoparticles they have been widely used in the color indicating probes in the development of analytical techniques which are used for the sensing of various analytes.

Colloidal gold nanoparticles have also gain much attention due to their easy preparation through chemical methods and they can easily be imported to the tissues and cells because of their very small size which is equal to the biological molecules like DNA and proteins (Amjadi et al., 2014). The gene gun gold nanoparticles have been also extensively used for epidermal delivery of DNA vaccines and this method is one of the best methods to deliver DNA vaccine to the affected site. Coated walls of gold nanocages with temperature-sensitive polymer were used as drug carrier which releases their effectors with interaction of near-infrared irradiations.

GNPs are the colloidal suspension of gold particles. Generally, GNPs are produced in a liquid ("liquid chemical methods") by reduction of chloroauric acid ($H[AuCl_4]$). After dissolving $H[AuCl_4]$, reducing agent is added and the solution is stirred continuously. This causes Au^{3+}ions to be reduced to neutral gold atoms. As more and more of these gold atoms form, the solution becomes supersaturated, and gold gradually starts to precipitate in the form of sub-nanometer particles. The rest of the gold atoms stick to the existing particles, and, if the solution is stirred vigorously, the particles will be fairly uniform in size. In order to prevent the aggregation of particles, stabilizing agent that sticks to the nanoparticle surface is added. The size of GNPs is depends on the salt concentration, temperature and rate of addition of reactants and by varying the salt concentration and temperature, GNPs of 1–100 nm size can be achieved. GNPs can also be synthesized by Brust method in organic liquids (like toluene), in which chlorauric acid reacts with tetraoctylammonium bromide (TOAB) solution in toluene and sodium borohydride is added as an anti-coagluant and a reducing agent (Brust et al., 1994). There have been several changes incorporated into the basic methods that resulted into an array of techniques to synthesize and manipulate these nanoparticles to satisfy the needs of a specific research objective (Hung et al., 2007 and Bhattacharya et al., 2003). In another method chemical reduction using L-Tryptophan as a reducing agent for ionic gold and polyethylene glycol was used to produce $AuCl_4^-$ ions to provide higher stability and uniformity in size, shape, and particle distribution (Akbarzadeh et al., 2009). Photochemical found in various plant sources have been used as a means of developing a more economical and environmentally friendly synthetic pathway in the formation of gold nanoparticles. According to the principles of "green chemistry," these methods employ the use of nontoxic chemicals, less energy consumption, renewable materials, and environmentally benign solvents to minimize the use, disposal, and health effect of hazardous chemicals. One "green" method that has been employed in the formation of gold nanoparticles uses the phytochemicals and polyphenols in Darjeeling black tea leaves, with water acting as a favorable solvent at room temperature. The phytochemicals in the black tea reduce $HAuCl_4$ to its salt and stabilize the aggregation of the gold atoms as the nanoparticle is formed. In addition to the "green" benefits of using black tea, the size of the nanoparticle is influenced by the concentration of the tea, and the absorbance and size of the gold nanoparticles formed can be easily determined using UV-Vis spectrometry abiding that an increase in λma□correlates to an increase in the size of the nanoparticle (Sharma et al., 2012 and Haiss et al., 2007). Other "green" methods that have been studied and employed include the use of Elettaria cardamomum (cardamom) and cinnamon in the synthesis of gold nanoparticles, as well as Syzygium aromaticum (cloves) in the formation of copper nanoparticles and table sugar in the formation of silver nanoparticles (Nune et al., 2009; Pattanayak et al., 2013; Chanda et al., 2011 and Subhankari et al., 2013). Generally the GNPs is in the spherical shape but it is also synthesized in various other shapes affecting their physical and biochemical properties. Gold nanocages (having six to eight surfaces) and the nanorods have also been used in biomedical applications for cancer imaging and photo thermal therapy (Huang et al., 2010). In a study by Vekilov (2011), the GNPs were grown in a lysozyme crystal which could be useful as a bifunctional molecule for specific catalytic activity. Various methods are being originated to synthesize GNPs with functional moieties to enhance their affinity to biological molecules and use as drug-carriers into the cells with increased specificity (Erik et al., 2012). GNPs functionalized with targeted specific biomolecules that can effectively destroy cancerous cells (Figure 8.6) (Duncan et al., 2010). Taking advantage of their unique properties, most studies of

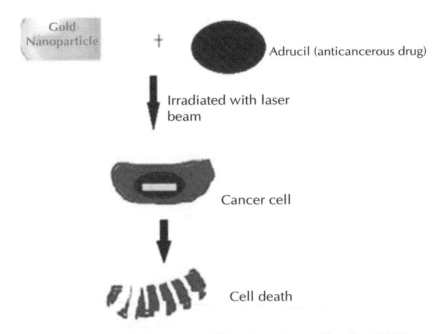

FIGURE 8.6   Gold nanoparticles in drug delivery: Targeting specific cells with higher loading efficiency, targeted delivery and efficient release of drugs.

gold nanoparticles-based cancer therapy have used photo thermal therapy for the destruction of cancer cells or tumor tissue, which may be potentially useful in the clinical setting. When irradiated with focused laser pulses of suitable wavelength, targeted gold nanospheres, nanorods, nanoshells, and nanocages can kill bacteria and cancer cells (Zharov et al., 2006b).

GNPs functionalized with fluorescently labeled heparin which have been recently used for the targeted detection and apoptotic killing of metastatic cancer cells. The idea used in the study is the over-expression of heparin-degrading enzymes by metastatic cancer cells is that, when attached to GNPs fluorescence of heparin is quenched and upon cleavage by heparinase/heparanase the fluorescence is regained and cancer cells can be detected. Also, it was shown that heparin binds to RGD peptide over-expressed in cancer cells inducing apoptosis. These fGNPs could thus be useful for both diagnosis and treatment of cancer.

GNPs have the ability to exhibit different surface plasmon resonances, when placed close to each other, hence they have been shown to differentiate between the normal and the cancerous cells when conjugated to anti-epidermal growth factor receptor antibodies as a biomarker agent. Biocompatible GNPs with two functional domains have been used for delivery of drugs into cells without being internalized (Kim et al., 2009). These domains included a hydrophobic alkane thiol interior and ahydrophilic shell composed of a tetraethylene glycol (TEG) unit terminated with a zwitterionic head group. These functionalized particles minimize non-specific binding with bio-macromolecules. It was demonstrated that hydrophobic dyes/drugs can be stably entrapped in hydrophobic pocket of GNPs and released into the cell by membrane-mediated diffusion without uptake of the carrier nanoparticle. The small size of these nanocarriers coupled with their biocompatible surface functionality provides longer circulation life time and preferential accumulation in tumor tissues due to the enhanced permeability and retention effect. Additionally, the non-interacting nature of their monolayer makes these systems highly amenable for targeting strategies.

GNPs have also been used to the co-administration of protein drugs due to their ability to cross the cell membranes because of their interaction to the cell surface lipids. It was investigated that the absorbance wavelength (in the visible range) of gold nanosphere is less and is not optimal for in vivo applications (Zharov et al., 2005). GNPs have so many unique properties that are advantageous for the destruction of cancer cells or tumor tissue, which may be potentially useful in cancer treatment.

8.5   MAGNETIC NANOPARTICLES

Magnetic nanoparticles seem to hold the greatest potential of success in medicine because they have a wide variety of materials that are being used in cancer treatment. They are already established in clinical use for magnetic resonance imaging (MRI). Magnetic nanoparticles can be guided non-invasively (drug delivery) and have the ability to be heated (hyperthermia) (Alexiou et al., 2010) by external magnetic fields. A combination of diagnostic investigation and therapy, as "theranostics," is, thus, possible. Therefore, magnetic nanoparticles have excellent potential to improve the treatment of cancer. Over the past decade metallic NPs have been extensively studied for their potential in biomedical applications (Thanh et al., 2010; Fadeel et al., 2010). Compared with other NPs, metallic NPs have proven to have unique chemical and physical properties-based on their quantum-size which lead to a range of interesting biomedical applications. There are many different kinds of chemical methods for synthesizing magnetic nanoparticles. The most commonly used are precipitation-based approaches, either by co-precipitation or reverse micelle synthesis. They can be easily synthesized with a high level of control of their size, shape and composition. Commonly studied metallic NPs include metal oxides such as iron oxide (Gupta et al., 2005), copper oxide (Ren et al., 2009), zinc oxide and aluminium oxide (Ansari et al., 2011). It is essential that metallic NPs are carefully surface engineered before introduction into biological environments due to their inherent instability and associated toxicity (Fisichella et al., 2009). In biomedicine, particularly, iron oxide nanoparticles are being used. However, the magnetic force on these particles decreases very rapidly with the increasing distance. As a result, it is hard to accumulate magnetic nanoparticles in tumors deep within the body by using an external magnetic field, making therapy difficult. Besides such implantation, it is also imaginable that the magnets can be put in preformed body cavities, which would make it possible to concentrate magnetic nanoparticles in deeper regions as well. The particular focus of this section is on the use of iron oxide nanostructures.

8.5.1   IRON OXIDE METALLIC NANOPARTICLES

Magnetic iron oxide nanoparticles (MNPs) have been the focus of vast scientific interest due to their potential for numerous applications in nanomedicine (Figure 8.7) (Acar et al., 2005). This includes the magnetic bioseparation, targeted therapy, drug delivery, biological detection and imaging. Magnetic separation techniques possess the advantage of rapid, high efficacy, and cost-effectiveness. Also, they have been shown to be highly efficient as supports in heterogeneous catalytic reactions received to the high specific area and magnetic restoration. MNPs possess large surface area to volume ratios due to their nano-size, low surface charge at physiological pH and they aggregate easily in solution due to their inherent magnetic nature. In some cases an unwanted aggregation may decrease the long-term stability of products leading to large nanoparticle clusters which are undesirable for medical application, so the surface of Iron oxide nanoparticles should be coated with a variety of different moieties that can eliminate or minimize their aggregation under physiological conditions. The amphiphilic polymeric surfactants such as poloxamers, poloxamines and poly (ethylene glycol) (PEG) derivatives are usually used for coating the surface of Iron oxide

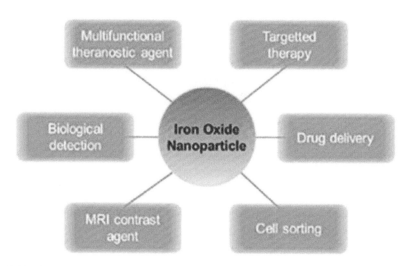

FIGURE 8.7   Showing uses of iron oxide nanoparticles in biomedicine.

nanoparticles, since they can minimize or eliminate opsonization of Iron oxide nanoparticles. Among them, PEG is the most used chemical material, which confers on Iron oxide nanoparticles several important properties such as high solubility and stability in aqueous solutions, biocompatibility, and prolonged blood circulation time. More importantly, the functional groups of modified PEG allow for bioconjugation of various ligands or therapeutic agents to Iron oxide nanoparticles.

In addition to PEG coating, other materials such as antibiofouling poly (TMSMA-r-PEGMA) hyaluronic acid (HA) layers and carboxyl functionalized poly (amidoamine) (PAMAM) dendrimers of generation 3 (G3) (Shi et al., 2007) have also been used to coat the surface of Iron oxide nanoparticles for either increasing circulation time in the blood or delivering peptides at high efficiency. Recently, we have developed a new class of super paramagnetic iron particles that have uniform sizes ranging from 5–30 nm and can be further functionalized through surface coating with amphiphilic triblock polymers, which provide functional groups for conjugating tumor-targeting biomolecules such as peptides or antibodies (Xiang–Hong Peng et al., 2008).

In addition, magnetic Iron oxide nanoparticles can also be encapsulated in liposomes to create magnetoliposomes. Degradation of iron oxide into free ions in physiological environments (Shubayev et al., 2009) increases the production of free radicals in cells, which may cause cell death (Hoslins et al., 2012). Therefore, these particles are commonly coated with organic macromolecules such as polyacrylic acid (PAA) (Mak, 2005), Dextran (Li et al., 2013) and polyethyleneimine (PEI) (Wang et al., 2009) or silica (Santra et al., 2001), carbon (Mendes et al., 2014) and precious metals (e.g., gold or silver) (Mandal et al., 2005). Common problems with chemically synthesized hybrid structures are the multiplicity of synthesis steps, incomplete understanding of the fundamentals of particle formation and, finally, a broad range of structures with ill-controlled configurations are produced (Chen et al., 2009).

The iron compounds are predominantly used because of their biocompatibility. They show the lowest toxicity and are even used therapeutically for iron substitution (Spinowitz et al., 2008). The nanoparticles are coated in order to prevent agglomeration, ensure stability, and provide a positive effect on biodistribution. A wide variety of materials, including fatty acids, polyethyleneglycol (PEG), de□tran, and chitosan may be used (Veiseh et al., 2010). Magnetic nanoparticles can transport various different substances and molecules, such as chemotherapeutic agents, antibodies, nuclear acids, radionuclide, etc. In principle, this approach can be used for any tumor, irrespective of its size, differentiation, or site. These targeted macromolecules could, bind to surface antigens over expressed by certain cell types. These moieties increase the uptake of the nanoparticles by cancer cells and increase the therapeutic payload to these cells. Detection capability of cancerous cells can be obtained by imaging the coating of fluorescent dyes near infrared region (Wang et al., 2011), Gd chelates to provide T1-weighted MR contrast (Bae et al., 2010), positron emission tomography (PET) tracers for MRI/PET dual imaging (Yang et al., 2011), or gold nanoparticles for MRI/surface enhanced Raman scattering (SERS) imaging (Yigit et al., 2011). Other iron oxide nonmaterial in biomedicine is super paramagnetic and ferromagnetic iron oxide nanoparticles.

8.5.2   SUPER PARAMAGNETIC IRON OXIDE NANOPARTICLES

Iron oxide nanoparticles are typically classed as super paramagnetic or ferromagnetic-based on their size. Although the majority of studies in nanomedicine are-based on super paramagnetic NPs, some studies have also highlighted the potential of ferromagnetic. The term super paramagnetism is assigned to the magnetic phenomena observed in fine magnetic particle systems exhibiting close similarities to atomic paramagnetism. Freeman et al., were the first to describe the concept of the use of magnetism in medicine in the 1970s (Freeman et al., 1960). Super paramagnetism occurs when the size of the nanoparticulate is so small (20 nm or less) resulting in a single-domain nanoparticle whose magnetic anisotropy barrier is lower than the thermal energy, and in consequence, the orientation of the magnetic moment of the particle is unstable due to the thermal agitation, like the atomic moments in a paramagnet. Unlike their paramagnetic counterparts, super paramagnetic materials retain no remnant magnetization upon the removal of an external magnetic field (Bate, 1980). Super paramagnetic iron oxide nanoparticles (SPIONs) are small synthetic α-Fe_2O_3 (hematite), γ-Fe_2O_3 (maghemite) or Fe_3O_4 (magnetite) particles with a core diameter ranging from 10 nm to 20 nm. SPIONs have been exploited as targeted magnetic resonance contrast agents for MRI, which improve diagnosis of progressive diseases like cancer in their early phases (Amstad et al., 2009). From a drug delivery viewpoint, targeting of cancer is the most pursued area, with emphasis on delivery of radio therapeutics and chemotherapeutics (Moghimi et al., 2001). However, increasing applications of SPIONs have also been used in the areas of cell death by using local hyperthermia, gene delivery, and delivery of antibodies and peptides to their site of action.

8.5.3 Ferromagnetic Iron Oxide Nanoparticles

Ferromagnetic iron oxide NPs offer some attractive possibilities in biomedicine. First of all, they have controllable sizes ranging from a few nanometers up to tens of nanometers, which locates them at dimensions that are smaller than or comparable to those of a virus (20–450 nm), a cell (10–100 µm), a protein (5–50 nm) or a gene (2 nm wide and 10–100 nm long). This demonstrates that they may enter to a biological entity of interest. These nanomaterials can be coated with bio molecules for binding to or interacting with a biological entity, in order to produce a controllable addressing or tagging tool (Pankhurst et al., 2003). Moreover, these magnetic NPs can be influenced by external magnetic fields. In electromagnetic fields, energy can be transferred from the exciting fields to the nanoparticle, thus, ferromagnetic iron oxide NPs can be designed to resonantly react to a time-changing magnetic fields. Ferromagnetic nanoparticles possess stronger magnetic potential than their SPION counterparts which is attractive for external magnetic guidance and increased contrast-ability (Hilger et al., 2005; Hergt et al., 2006). These are made available in biomedicine due to the special physical properties of ferromagnetic iron oxide NPs. Commonly ferromagnetic iron oxide NPs are synthesized via co-precipitation, like SPIONS size, shape and crystallinity can be tailored-based on reaction conditions. While SPIONs are less magnetic they are often preferred due to the relative instability of ferromagnetic, making them difficult to suspend due to aggregation and making them more challenging for biomedical applications.

8.5.4   IRON OXIDE NANOPARTICLES AS VEHICLES FOR CHEMOTHERAPY

Iron oxide NPs are increasingly being investigated for their potential as drug delivery vehicles for chemotherapy (West et al., 2014; Hedgire et al., 2014; Yallapu et al., 2013). Nano-sized formulations of cytotoxic agents have proved to passively target pancreatic adenocarcinomas and promote increased drug efficacy. This is thought to be due to the accumulation via EPR resulting in deeper drug penetration. This factor combined with the rapid diagnostic and treatment platform of this technology results in a system with great potential to act as a localized therapy with reduced dosages, thus minimizing harsh side effects and improve clinical outcomes for patients with pancreatic cancer.

There are many chemical methods that are applied for the adjoining of therapeutic, targeting and imaging carrier molecules with NP surfaces. These can be classified into covalent linkage strategies (direct nanoparticle conjugation, covalent linker chemistry, click chemistry) and physical interactions (hydrophilic/hydrophobic, electrostatic and affinity interactions). Physicochemical properties, functional groups found on the NPs coating and ligands are factors which influence the choice of chemistry. The primary aim is to bind the targeting, therapeutic or imaging moiety without making changes in its functionality. Functionality in such assemblies is dictated by the nature of the ligand (e.g., conformation of biomolecules) and the manner in which it is attached. For instance, if an antibody is conjugated to the NP but its recognition site is shielded by proteins, it might lose its potency to bind or reach to a target. The surface of iron oxide NPs has been modified with anticancer drugs such as do□orubicin(Do□)(Jain et al., 2005), Catechin– De□tran(Vittorio et al., 2014) and Paclita□el (Jain et al., 2005). Catechin-dextran conjugated Endorem NPs increased the intracellular concentration of the drug compared with the free drug. Endorem is an FDA approved dextran coated iron oxide NP. The Catechin-dextran-Endorem formulation induced apoptosis in 98% of human pancreatic cancer cell line (MIA PaCa-2) placed under a magnetic field. The findings suggest that conjugation of catechin–dextran with Endorem enhances the anticancer activity of this drug and provides a novel means for targeted drug delivery to tumor cells driven by magnetic fields. Often the surface of iron oxideNPs for chemotherapy will also contain a targeting agent which can identify the receptors over-expressed on the external surface of cancer cells (Hwu et al., 2009; Guo et al., 2013). Recently, Lee et al., designed urokinase plasminogen activator receptor (uPAR)-targeted magnetic iron oxide nanoparticles (IONPs) carrying gemcitabine as a chemotherapy drug for targeted delivery into uPAR-expressing tumor and stromal cells. The novel formulation was prepared by conjugating IONPs with the amino-terminal fragment peptide of the receptor-binding domain of uPA. uPA is a naturally occurring ligand of uPAR, and gemcitabine via a lysosomally cleavable tetrapeptide linker. These theranostic nanoparticles were designed to enable intracellular release of gemcitabine following receptor-mediated endocytosis into tumor cells while providing contrast enhancement in magnetic resonance imaging (MRI) of tumors the results demonstrated the pH- and lysosomal enzyme-dependent release of gemcitabine, preventing the drug from enzymatic degradation while imaging inside the tumor was possible.

8.6   CARBON NANOTUBES

Carbon nanotubes (CNTs) opened new frontiers in the field of nanotechnology and nanoscience. CNTs have attracted tremendous attention due to their unique properties as one of the most promising nanomaterials for a variety of biomedical applications. In comparison with other nanomaterials, CNTs appear to be more dynamic in their biological application CNTs are hollow, tubular and carbon graphite nanomaterials with nanometer-size and axial symmetry, also have high surface area, ultralight weight this gives it unique physical and chemical properties and have applications in developing electronic devices and sensors to nanocomposite materials of high strength and low weight that can be used in the diagnosis and treatment of cancer and to deliver drugs directly to targeted cells and tissues (Bhirde et al., 2010; Sahoo et al., 2011). There are two types of nanotubes: single-walled carbon nanotubes (SWCNTs) and multiwalled carbon nanotubes (MWCNTs), which differ in the arrangement of their carbon layers. SWCNTs consist of a single cylindrical carbon layer with a diameter in the range of 0.4–2 nm (Klumpp et al., 2006), and its synthesis depends on the temperature, at higher temperature the diameter of SWCNTs is larger. In contrast, MWCNTs are usually made from several cylindrical carbon layers with diameters in the range of 1–3 nm for the inner tubes and 2–100 nm for the outer tubes (Bekyarova et al., 2005).

The distance between each layer of a MWNT is about 0.34 nm. Due to the carbon arrangement, their structure varies and can be metallic or semiconducting. The structure of SWCNTs is organized according to armchair, zigzag, chiral, or helical arrangements and the structure of MWCNTs can be divided into two types according to the arrangement of the graphite sheets. One is a "Russian-doll"-like structure where the graphite sheets are arranged in concentric layers and the other is a parchment-like model where the single sheet of graphite is rolled around itself (Danailov et al., 2002) (Figure 8.8). CNTs exhibit different electrical properties: SWNT can be either semi-conducting or metallic while MWNT are only semi-conducting. There are several methods to produce CNT including arc discharge, laser ablation and chemical vapor deposition. CNTs act as sensing materials in pressure, flow, thermal, gas, optical, mass, position, stress, strain, chemical, and biological sensors.

8.6.1   CELLULAR DELIVERY OF CNTS

The exact mechanism of cellular delivery of CNTs into the cell is still not well understood. Because of their needle-like shapes, CNTs might be able to penetrate cellular membrane and pass into the cellular components without causing apparent cell damage (Klumpp et al., 2006).

There has been great interest in the mechanism of cellular uptake of CNTs in the literature, and different methods have been investigated to elucidate this concept. Labeling CNTs with fluorescent materials, such as quantum dots,

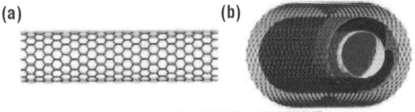

(a)   **(b)**

FIGURE 8.8   Structures of (a) SWCNT and (b) MWCNT.

enables researchers to track the movement of CNTs. Additionally, detection of CNTs by non-labeling methods such as transmission electronic microscopy or atomic force microscopy has also been undertaken. The unique structural, mechanical and electronic properties of CNT were initially exploited in the field of materials science. Soon after this new types of nanostructures started to be used in interaction mainly with proteins aiming to develop efficient biosensors. The combination of nanotubes with proteins and other natural products including nucleic acids and polysaccharides paved the way for the compatibility of such materials with biological systems. CNT are practically insoluble in any type of solvent and only the recent development of strategies for linking chemical moieties on the tubes has facilitated their applicability.

Because of their tube structure, carbon nanotubes can be made with or without end caps, meaning that without end caps the inside where the drug is held would be more accessible. Drug delivery through carbon nanotubes arises some problem like the lack of solubility, clumping, and half-life (Pastorin and Giorgia, 2009). These nanotubes function with a larger inner volume to be used as the drug container, large aspect ratios for numerous functionalization attachments, and the ability to be readily taken up by the cell. Also, drug encapsulation in CNTs has been shown to enhance water disseminately, better bioavailability, sustained half-life, increased cell penetration and uptake and reduced toxicity, all of which are currently novel but undeveloped ideas.

8.6.2 CELLULAR UPTAKE OF CNTS

There is number of methods have been used by various research groups to investigate how the anticancer drug-loaded CNTs recognize the cancer cell. In a method given by Xiao et al. (2010), the surface of the CNT was coated with a particular antibody that is having affinity for the target cancer cell and they observed that the attachment of antibodies tithe CNT surface does not alter the antibody specificity for the target cell but also the antibody can successfully deliver anticancer drug-loaded CNTs to the site of action. In an experiment, a SWCNTs functionalized by PEG and rituxan (the monoclonal antibody against CD20, found primarily on B cells), selectively targeted the CD20 cell surface receptor on B cells that binds to T cells (Big et al., 2011). Coating of CNT with functional groups, (such as antibodies) achieve cell specificity and the movement of CNTs in the living organism is tracked by quantum dots, which have the ability to generate fluorescence when exposed to certain wavelengths of light, and have been functionalized to the walls of CNTs (Shi et al., 2009). Uptake of large proteins (such as streptavidin, staphylococcal protein A, or bovine serum albumin) that are attached to CNTs, are taken up via endocytosis, whereas CNTs attached to small molecules (such as ammonium, methotrexate, or amphotericin B) enter cells by a diffusion process (Raffa et al., 2010). Both the CNTs have different mechanism of cell penetration. SWCNTs have the ability to be internalized into cells, whereas MWCNTs are excluded from the interior of the cell. There is various mode of drug release have been reported from CNTs to the targeted cell. Once the drug is loaded into the both ends sealed CNTs, it will be evacuated from the tube into the cellular environment and the molecules cleaved off intracellularly. Various factors such as pH, temperature and reducing agents affect the cleavage of drug intracellularly (Samorri et al., 2010). In an experiment by Kam et al. (2005), CNTs surface was attached to the small interfering RNA (siRNA) through the disulfide bond and delivered to the targeted cell, there by the endocytotic entry of the CNT-siRNA, the disulfide bond was cleaved off by the thiolreductase and siRNA was released. Happening of this process requires acidic pH and temperature for the release of drugs. Another potential therapy for cancer treatment is thermal extirpation, it is non-invasive and harmless to normal cells and is very effective for the removal of tumor cells. In an experiment by Gannon et al. (2007) showed that both in vitro and in vivo, cancer cells are destroyed thermally when CNTs were exposed to radiofrequency. In another findings resulted that the heat shock protein works as an endogenous marker of thermal stress and is induced by elevated temperature (typically in excess of 43°C). Researchers have shown that when cancer cells are exposed to laser alone, heat shock protein expression is induced adjacent to the incident laser and gradually reduces more distal to it. In contrast, when the laser is exposed to CNTs, the heat generated from these materials is high enough to increase the surface temperature of the cells and cause embolitic necrosis. As a result, heat shock protein induction is seen in deeper tissue, indicating that CNTs can be used to extend the depth of thermal therapy. The CNTs could act as a protective shell that protects the biological environment against oxidation and toxicity of the other nanoparticles. Thus, the nanostructures would be able to target the malignant tumor site specifically and selectively to excise the tumor cells, leaving nearby healthy tissue relatively unharmed, providing a far less invasive alternative to surgery.

8.7 CONCLUSION

Nanotechnology is a broad word that comprises an assortment of sub-disciplines in biology, biotechnology, engineering, chemistry, and physics and also it has large potential in detection and treatment of cancer in its incipient stage. It is having the unique physical and chemical properties that give them advantages as drug delivery carriers, or "nano-carriers," and diagnosis probes and this potential arises due to the ability of NP entering inside the cells and access to the chromosomes/DNA molecules. Certain nano structures like a paramagnetic particle, gold particles, nanotubes, nanoshells and quantum dots are prospective structures that would help in detecting and treatment of cancers. Still there are many challenges that are to be met before use of nanoparticles, which becomes a reality.

KEYWORDS

- cancer therapy
- carbon nanotubes
- drug delivery
- gold nanoparticles
- magnetic particles
- nanoparticles
- nanoshells
- nanotechnology
- quantum dots

REFERENCES

Acar, H. Y. C., Garaas, R. S., Syud, F., Bonitatebus, P., & Kulkarni, A. M., (2005). Super paramagnetic nanoparticles stabilized by polymerized PEGylated coatings. *J. Magn. Magn. Mater, 293*(1), 1–7.

Akbarzadeh, A., Zare, D., Farhangi, A., Mehrabi, M. R., Norouzian, D., Tangestaninejad, S., et al., (2009). Synthesis and characterization of gold nanoparticles by tryptophane. *Am. J. Appl. Sci., 161*, 83–91.

Alexiou, C., Tietze, R., Schreiber, E., & Lyer, S., (2010). Nanomedicine: Magnetic nanoparticles for drug delivery and hyperthermia–new chances for cancer therapy. Bundesgesundheitsblatt Gesundheitsforschung Gesundheitsschutz, *53*, 839–845.

Ali Mansoori, G., Pirooz Mohazzabi., Percival McCormack., & Siavash Jabbari., (2007). Nanotechnology in cancer prevention, detection and treatment: bright future lies ahead. *World Review of Sci. Tech. and Sust. Develop., 4*, 226–257.

Amjadi, M., & Farzampour, L., (2014). Fluorescence quenching of fluoroquinolones by gold nanoparticles with different sizes and its analytical application. *J. Luminesc. 145*, 263–268.

Amstad, E., Zurcher, S., Mashaghi, A., Wong, J. Y., & Textor, M., (2009). Surface functionalization of single super paramagnetic iron oxide nanoparticles for targeted magnetic resonance imaging. *Small, 5*(11), 1334–1342.

Ansari, S. A., & Hussain, Q., (2011). Immobilization of kluyveromyces lactis β galactosidase on concanavalin-A layered aluminium oxide nanoparticles its future aspects in biosensor applications. *J. Mol. Catal. B Enzymatic., 70*(3–4), 119–126.

Ansari, S. A., Hussain, Q., Qayyum, S., & Azam, A., (2011). Designing and surface modification of zinc oxide nanoparticles for biomedical applications. *Food Chem. Toxicol. 49*(9), 2107–2115.

Bae, K. H., Kim, Y. B., Lee, Y., Hwang, J., Park, H., & Park, T. G., (2010). Bioinspired synthesis and characterization of gadolinium-labeled magnetite nanoparticles for dual contrast T1- and T2-weighted magnetic resonance imaging. *Bioconj. Chem., 21*, 505–512.

Bekyarova, E., Ni, Y., & Malarkey, E. B., (2005). Applications of carbon nanotubes in biotechnology and biomedicine. *J Biomed Nanotechnol, 1*, 3–17.

Bhattacharya, S., & Srivastava, A., (2003). Synthesis of gold nanoparticles stabilized by metal-chelator and the controlled formation of close-packed aggregates by them. *Proc. Indian Acad. Sci. Chem. Sci., 115*, 613–619.

Bhirde, V., Patel, J., & Gavard., (2009). "Targeted killing of cancer cells *in vivo* and *in vitro* with EGF-directed carbon bimodal detection of breast-cancer cells. *Small., 5*, 2555–2564.

Brust, M., Walker, M., Bethell, D., Schiffrin, D. J., & Whyman, R. J., (1994). Synthesis of thiol derivatized gold nanoparticles in a two phase liquid-liquid system. *J. Chem. Soc. Chem. Commun, 7*, 801–802.

Cai, W., Hsu, A. R., Li, Z. B., & Chen, X., (2007). Are quantum dots ready for *in vivo* imaging in human subjects? *Nanoscale Res. Lett., 2*(6), 265–281.

Caruso, F., Caruso, R. A., & Möhwald, H., (1998). Nanoengineering of inorganic and hybrid hollow spheres by colloidal templating. *Science., 282*, 1111–1114.

Caruso, F., Caruso, R. A., & Möhwald, H., (1999). Production of hollow microspheres from nanostructured composite particles. *Chem. Mater., 11*, 3309–3314.

Caruso, F., Spasova, M., Salgueiriño-Maceira, V., & Liz-Marzán, L. M., (2001). Multilayer assemblies of silica-encapsulated gold nanoparticles on decomposable colloid templates. *Adv. Mater., 13*, 1090–1094.

Chan, W. C. W., & Nie, S. M., (1998). Quantum dot bioconjugates for ultrasensitive nonisotopic detection. *Science, 281*, 2016–2018.

Chanda, Nripen., (2011). An Effective Strategy for the Synthesis of Biocompatible Gold Nanoparticles using Cinnamon Phytochemicals for Phantom CT Imaging and Photo acoustic Detection of Cancerous Cells. *Pharmaceu. Res., 28*(2), 279–291.

Chen, W., Bian, A., Agarwal, A., Liu, L., & Shen, H., (2009). Nanoparticle superstructures made by polymerase chain reaction: Collective interactions of nanoparticles and a new principle for chiral materials. *Nano Lett., 9*(5), 2153–2159.

Chen, Z., Wang, Z. L., Zhan, P., Zhang, J. H., Zhang, W. Y., Wang, H. T., & Ming, N. B., (2004). Preparation of metallodielectric composite particles with multishell structure. *Lang., 20*, 3042–3046.

Dabbousi, B. O., Rodriguez, Viejo. J., & Mikulec, F. V., (1997b). (CdSe) ZnS core-shell quantum dots: Synthesis and characterization of a size series of highly luminescent nanocrystallites. *J. Phy. Chem., 101*, 9463–9475.

Danailov, D., Keblinski, P., Nayak, S., & Ajayan, P. M., (2002). Bending properties of carbon nanotubes encapsulating solid nanowires. *J. Nanosci. Nanotech., 2*(5), 503–507.

Deb, S., Patra, H. K., Lahiri, P., Dasgupta, A. K., Chakrabarti, K., & Chaudhuri, U., (2011). Multistability in platelets and their response to gold nanoparticles. *Nanomed. Nanotechnol Biol. Med., 7*, 376–384.

Druker, B. J., Talpaz, M., Resta, D. J., Peng, B., Buchdunger, E., Ford, J. M., et al., (2001). Efficacy and safety of a specific inhibitor of the BCR-ABL tyrosine kinase in chronic myeloid leukemia. *N. Engl. J. Med., 344*, 1031–1037.

Dubertret, B., Skourides, P., & Norris, D. J., (2002). *In: vivo* imaging of quantum dots encapsulated in phospholipid micelles. *Science, 298*, 1759–1762.

Duncan, B., Kim, C., & Rotello, V. M., (2010). Gold nanoparticle platforms as drug and biomacromolecule delivery systems. *J. Contr. Release 148*, 122–127.

Durr, S., Tietze, R., Lyer, S., Lyer, S., & Alexiou, C., (2012). Nanomedicine in otorhinolaryngology–future prospects. *Laryngorhinootologie, 91*, 6–12.

Efros, A. L., & Efros, A. L., (1982). Interband absorption of light in a semiconductor sphere. *Sov. Phys. Semicond., 16*, 772–775.

Ekimov, A. I., & Onushchenko, A. A., (1982). Quantum size effect in the optical-spectra of semiconductor microcrystals. *Sov. Phys. Semicond, 16*, 775–778.

Erik, C. Dreaden., Lauren, A. Austin., Megan, A. Mackey., & Mostafa, A., (2012). El-Sayed. Size matters: gold nanoparticles in targeted cancer drug delivery. *Ther. Deliv., 3*(4), 457–478.

Ethiraj, A. S., Hebalkar, N., Sainkar, S. R., & Kulkarni, S. K., (2005). Photoluminescent core shell particles of organic dye in silica. Adsorption and encapsulation of fluorescent probes in nanoparticles. *J. Lumin., 114*, 15–23.

Fadeel, B., & Garcia-Bennett, A. E., (2010). Better safe than sorry: Understanding the toxicological properties of inorganic nanoparticles manufactured for biomedical applications. *Adv. Drug Deliv. Rev., 62*(3), 362–374.

Fisichella, M., Dabboue, H., Bhattacharya, S., Saboungi, M. L., & Salvetat, J. P., (2009). Mesoporous silica nanoparticles enhance MTT formazan exocytosis in He-La cells and astrocytes. *Toxicol. In Vitro, 23*(4), 697–703.

Freeman, M. W., Arrott, A., & Watson, J. H. L., (1960). Magnetism in medicine. *J. Appl. Phy., 31*, S404.

Fu, X. S., Hu, C. A., Chen, J., Wang, J., & Liu, K. J. R., (2005). 'Cancer genomic, proteomics and clinic applications', in: Dougherty, E., Shmulevich, I., Chen, J., & Wang, J., (Eds.), Genomic Signal Processing and Statistics, European Applied Signal Processing Inc., Hindawi Publishing Corporation, New York, pp. 267–408.

Ganeshkumar, M., Sastry, T. P., Sathish, Kumar. M., Dinesh, M. G., Kannappan, S., & Suguna, L., (2012). Sun light mediated synthesis of gold nanoparticles as carrier for 6-mercaptopurine: Preparation, characterization and toxicity studies in zebra fish embryo model. *Mater. Res. Bull., 47*, 2113–2119.

Gannon, C. J., Cherukuri, P., & Yakobson, B. I., (2007). Carbon nanotube-enhanced thermal destruction of cancer cells in a noninvasive radiofrequency field. *Cancer, 110*, 2654–2665.

Gao, X., & Dave, S. R., (2007). Quantum dots for cancer molecular imaging. *Adv. Exp. Med. Biol., 620*, 57–73.

Gao, X. H., Cui, Y. Y., & Levenson, R. M., (2004). *In: vivo* cancer targeting and imaging with semiconductor quantum dots. *Nat. Biotechnol. 22*, 969–976.

Gerion, D., Pinaud, F., Williams, S. C., Parak, W. J., Zanchet, D., Weiss, S., & Alivisatos, A. P., (2001). Synthesis and properties of biocompatible water-soluble silica-coated CdSe/ZnS semiconductor quantum dots. *J. Phys. Chem. B. 105*, 8861–8871.

Goldman, E. R., Balighian, E. D., & Mattoussi, Avidin, H., (2002). A natural bridge for quantum dot-antibody conjugates. *J. Am. Chem. Soc., 124*, 6378–6382.

Graph, C., & Van Blaaderen, A., (2002). Metallodielectric colloidal core-shell particles for photonic applications. *Lang., 18*, 524–534.

Grodzinski, P., Silver, M., & Molnar, L. K., (2006). Nanotechnology for cancer diagnostics: promises and challenges. *Expert. Rev. Mol. Diagn., 6*, 307–318.

Guo, Y., Zhang, Z., Kim, D. H., Li, W., & Nicolai, J., (2013). Photothermal ablation of pancreatic cancer cells with hybrid iron-oxide core gold-shell nanoparticles. *Int. J. Nanomedicine, 8*, 3437–3446.

Gupta, A. K., & Gupta, M., (2005). Synthesis and surface engineering of iron oxide nanoparticles for biomedical applications. *Biomat., 26*(18), 3995–4021.

Haiss, Wolfgang., (2007). Determination of Size and Concentration of Gold Nanoparticles from UV-Vis Spectra. *Analy. Chem., 79*(11), 4215–4221.

Han, S., Xia, T., Li, Q., Guo, J., & Lu, P., (2013). Application of functional quantum dots in cancer diagnosis and therapy: A review. Sheng Wu Gong Cheng Xue Bao., *29*(1), 10–20.

Hedgire, S. S., Mino-Kenudson, M., Elmi, A., Thayer, S., & Fernandez-del, Castillo. C., (2014). Enhanced primary tumor delineation in pancreatic adenocarcinoma using ultra small super paramagnetic iron oxide nanoparticle-ferumoxytol: An initial experience with histopathological correlation. *Int. J. Nanomed., 9*, 1891–1896.

Hergt, R., Dutz, S., Muller, R., & Zeisberger, M., (2006). Magnetic particle hyperthermia: nanoparticle magnetism and materials development for cancer therapy. *J Phys. Conden. Matt. 18*: S2919–S2934.

Hilger, I., Hergt, R., & Kaiser, W. A., (2005). Use of magnetic nanoparticle heating in the treatment of breast cancer. *IEE Proc Nanobiotechnol. 152*(1), 33–39.

Hines, M. A., & Guyot-Sionnest, P., (1996). Synthesis and characterization of strongly luminescing ZnS-capped CdSe nano,crystals. *J Phys Chem., 100*, 68–71.

Hirsch, L. R., Jackson, J. B., Lee, A., Halas, N. J., & West, J. L., (2003). A whole blood immunoassay using gold nanoshells. *Anal. Chem., 75*, 2377–2381.

Hoskins, C., Cuschieri, A., & Wang, L., (2012). Cytotoxicity of polycationic iron oxide nanoparticles: Common endpoint assays and alternative approaches for improved understanding of cellular response mechanism. *J. Nanobiotechn. 10*, 15.

Huang, X., Jain, P. K., & El-Sayed, I. H., (2007). Gold nanoparticles: interesting optical properties and recent applications in cancer diagnostics and therapy. *Nanomed., 2*, 681–693.

Huang, H., Remsen, E. E., Kowalewski, T., & Wooley, K. L., (1999). Nanocages derived from shell cross-linked micelle templates. *J. Am. Chem. Soc., 121*, 3805–3806.

Hung, L., & Leel, A. P., (2007). Microfluidic devices for the synthesis of nanoparticles and biomaterials. *J. Med. Biol. Eng. 27*, 1–6.

Hwu, J. R., Lin, Y. S., Josephrajan, T., Hsu, M. H., & Cheng, F. H., (2009). Targeted paclitaxel by conjugation to iron oxide and gold nanoparticles. *J. Am. Chem. Soc., 131*(1), 66–68.

Imhof, A., (2001). Preparation and characterization of titania-coated polystyrene spheres and hollow titania shells. *Lang., 17*, 3579–3585.

Jain, K. K., (2005). Role of nanobiotechnology in developing personalized medicine for cancer. Technol. *Cancer Res. Treat., 4*, 645–650.

Jain, R. K. (2001). Delivery of molecular medicine to solid tumors: lessons from in vivo imaging of gene expression and function. *J. Control. Rel. 74*, 7–25.

Jain, R. K., (1999). Transport of molecules, particles, and cells in solid tumors. *Annu. Rev. Biomed. Eng. 1*, 241–263.

Kam, N. W., & Liu, Z., Dai, H., (2005). Functionalization of carbon nanotubes via cleavable disulfide bonds for efficient intracellular delivery of siRNA and potent gene silencing. *J. Am. Chem. Soc., 127*, 12492–12493.

Kim, H., Achermann, M., Balet, L. P., Hollingsworth, J. A., & Klimov, V. I., (2005). Synthesis and characterization of Co/CdSe shell nanocomposites: Bifunctional magnetic–optical nanocrystals. *J. Am. Chem. Soc., 127*, 544–546.

Kim, S. W., Bae, D. S., Shin, H., & Hong, K. S., (2004). Optical absorption behavior of platinum core-silica shell nanoparticles layer and its influence on the reflection spectra of a multi-layer coating system in the visible spectrum range. *J. Phys. Condens. Mater., 16*, 3199–3206.

Klumpp, C., Kostarelos, K., Prato, M., & Bianco, A., (2006). Functionalized carbon nanotubes as emerging nanovectors for the delivery of therapeutics. *Biochem. Biophy. Acta, 1758*, 404–412.

Leatherdale, C., Woo, W., & Mikulec, F., (2002). On the absorption cross section of CdSe nanocrystal quantum dots. *J. Phys. Chem. B. 106*, 7619–7622.

Li, L., Mak, K. Y., Shi, J., Leung, C. H., & Wong, C. M., (2013). Sterilization on dextran-coated iron o☐idenanoparticles: Effects of autoclaving, filtration UV irradiation, and ethanol treatment. *Microelectronic Eng., 111*, 310–313.

Liu, Z., Cai, W., & He, L., (2007). In: vivo biodistribution and highly efficient tumor targeting of carbon nanotubes in mice. *Nat. Nanotechnol., 2*, 47–52.

Lide, D. R. (Ed.), (1993). CRC Handbook of Chemistry and Physics, 74th edn.

Longfei, Liu., Qingqing, Miao & Gaolin, Liang., (2013). Quantum Dots as Multifunctional *Materials for Tumor Imaging and Therapy. Mater., 6*, 483–499.

Maeda, H., Wu, J., Sawa, T., Matsumura, Y., & Hori, K., (2000). Tumor vascular permeability and the EPR effect in macromolecular therapeutics: a review. *J. Control. Release., 65*, 271–284.

Mak, S. Y., & Chen, D. H., (2005). Binding and sulfonation of poly(acrylic acid) on iron oxide nanoparticles: a novel, magnetic, strong acid cation nano-adsorbent. *Macromol. Rapid Comm., 26*(19), 1567–1571.

Mandal, M., Kundu, S., Ghosh, S. K., Panigrahi, S., & Sau, T. K., (2005). Magnetite nanoparticles with tunable gold or silver shell. *J. Colloid Interface Sci., 286*(1), 187–194.

Matijevic, E., (1993). Preparation and properties of uniform size colloids. *Chem. Mater., 5*, 412–426.

Matsumura, Y., & Maeda, H., (1986). A New Concept for Macromolecular Therapeutics in Cancer- Chemotherapy - Mechanism of tumoritropic accumulation of Proteins and The Antitumor Agent SMANCS. *Cancer Res., 46*, 6387–6392.

Mattoussi, H., Mauro, J. M., & Goldman, E. R., (2000). Self-assembly of CdSe-ZnS quantum dot bioconjugates using an engineered recombinant protein. *J. Am. Chem. Soc., 122*, 12142–12150.

Mendes, R. G., Koch, B., Bachmatiuk, A., El-Gendy, A. A., & Krupskaya, Y., (2014). Synthesis and toxicity characterization of carbon coated iron oxide nanoparticles with highly defined size distributions. *Biochim Biophys Acta., 1840*(1), 160–169.

Moghimi, S. M., Hunter, A. C., & Murray, J. C., (2001). Long-circulating and target-specific nanoparticles: theory to practice. *Pharmacol. Rev., 53*(2), 283–318.

Monteiro, O. C., Esteves, A. C. C., & Trindade, T., (2002). The synthesis of SiO2@CdS nanocomposites using single molecule precursors. *Chem. Mater., 14*, 2900–2904.

Morton, J. G., Day, E. S., Halas, N. J., & West, J. L., (2010). Nanoshells for photothermal cancer therapy. *Methods Mol. Biol., 624*, 101–117.

Mulvaney, P., Giersig, M., Ung, T., & Liz-Marzán, L. M., (1997). Direct observation of chemical reactions in silica-coated gold and silver, nanoparticles. *Adv. Mater., 9*, 570–575.

Murphy, C. G., & Morris, P. G., (2012). Recent advances in novel targeted therapies for HER2-positive breast cancer. *Anti-Cancer Drugs., 23*, 765–776.

Murray, C. B., Norris, D. J., & Bawendi, M. G., (1993). Synthesis and characterization of nearly monodisperse CdE (E=S, Se, Te) semiconductor nanocrystallites. *J. Am. Chem. Soc., 115*, 8706–8715.

NICE rejects other cancer drugs. Dt Medical Sheet, (2010). Available at http://www.arzteblatt. de/nachtfricten/43932/nice-lehnt_weithre-krebsmedikamente-ab. Accessed on October 7, 2016.

Nie, S., Xing, Y., & Kim, G. J., (2007). Nanotechnology applications in cancer. *Ann. Rev. Biomed. Eng., 9*, 257–288.

Nune., & Satish, K., (2009). Green Nanotechnology from Tea: Phytochemicals in Tea as Building Blocks for Production of Biocompatible Gold Nanoparticles. *J. Mat. Chem., 19*, 2912–2920.

Ocana, M., Hsu, W. P., & Matijevic, E., (1991). Preparation and properties of uniform coated colloidal particles. *Lang., 7*, 2911–2916.

Ohmari, M., & Matijevic, E., (2006). Preparation and properties of uniform coated inorganic colloidal particles, 8. Silica on iron. *J. Colloid Interface Sci., 160*, 288–292.

Okaniwa, M., (1998). Synthesis of poly (tetrafluoroethylene)/poly (butadiene) coreshell particles and their graft copolymerization. *J. Appl. Poly. Sci., 68*, 185–190.

Pan, D. C., Wang, Q., & Jiang, S. C., (2005). Synthesis of extremely small CdSe and highly luminescent CdSe/CdS core-shell nanocrystals via a novel two-phase thermal approach. *Adv. Mater., 17*, 176–179.

Pankhurst, Q. A., Connolly, J., Jones, S. K., & Dobson, J., (2003). Applications of magnetic nanoparticles in biomedicine. *J Phys D: Appl. Phys., 36*, 167–181.

Pastorin, Giorgia., (2009). Crucial Functionalizations of Carbon Nanotubes for Improved Drug Delivery: A Valuable Option?. *Pharma. Res., 26*(4), 746–769.

Pattanayak, Monalisa & P. L. Nayak., (2001). Green Synthesis of Gold Nanoparticles using Elettaria Cardamomum (ELAICHI) Aqueous Extract. *World, 2*(1), 1–5.

Raffa, V., Ciofani, G., Vittorio, O., Riggio, C., & Cuschieri, A., (2010). Physicochemical properties affecting cellular uptake of carbon nanotubes. *Nanomed. 5*, 89–97.

Ren, G., Hu, D., Cheng, E. W., Vargas-Reus, M. A., & Reip, P., (2009). Characterisation of copper oxide nanoparticles for antimicrobial applications. *Int. J. Antimicrob. Agents., 33*(6), 587–590.

Sahoo, N. G., Bao, H., & Pan, Y., (2011). "Functionalized carbon nanomaterials as nanocarriers for loading and delivery of a poorly water-soluble anticancer drug: a comparative study" *Chem. Commun, 47*(18), 5235–5237.

Samori, C., li-Boucetta, H., & Sainz, R., (2010). Enhanced anticancer activity of multi-walled carbon nanotube-methotrexate conjugates using cleavable linkers. *Chem. Commun. (Camb). 46*, 1494–1496.

Santra, S., Tapec, C., Theodoropoulou, N., & Dobson, J. Hebard, A., (2001). Synthesis and characterization of silicacoated iron oxide nanoparticles in microemulsion: The effect of non-ionic surfactants. *Lang., 17*(10), 2900–2906.

See, K. H., Mullins, M. E., Mills, O. P., & Heiden, P. A., (2005). A reactive core-shell nanoparticle approach to prepare hybrid nanocomposites: Effects of processing variables. *Nanotechnol., 16*, 1950–1959.

Sengupta, S., & Sasisekharan, R., (2007). Exploiting nanotechnology to target cancer. *Br. J. Cancer. 96*, 1315–1319.

Sharma, R. K., Gulati, S., & Mehta, S., (2012). Preparation of Gold Nanoparticles using Tea: A Green Chemistry Experiment. *J. Chem. Educ.*, *89*(10), 1316–1318.

Shi, X., Thomas, T. P., & Myc, L. A., (2007). Synthesis, characterization, and intracellular uptake of carboxyl-terminated poly(amidoamine) dendrimer-stabilized iron oxide nanoparticles. *Phys. Chem. Chem. Phys., 9*, 5712–5720.

Shi, D. L., Cho, H. S., & Huth, C., (2009). Conjugation of quantum dots and Fe_3O_4 on carbon nanotubes for medical diagnosis and treatment. *Appl. Phys. Lett., 95*, 223–702.

Shubayev, V. I., Pisanic, T. R., & Jin, S., (2009). Magnetic nanoparticles for theragnostics. *Adv. Drug Deliv. Rev., 61*(6), 467–477.

Smith, J. N., Meadows, J., & Williams, P. A., (1996). Adsorption of polyvinylpyrrolidone onto polystyrene lattices and the effect on colloid stability. *Lang., 12*, 3773–3778.

Souhami, R. L., Tannok, I., Hohernberger, P., & Horiot, J. C., (2002). Oxford Textbook of Oncology, Benjamin Lewin, Genes VIII, Prentice-Hall, p. 889.

Sparnacci, K., Laus, M., Tondelli, L., Magnani, L., & Bernardi, C., (2002). Core-shell microspheres by dispersion polymerization as drug delivery systems. *Macromol. Chem. Phys., 203*, 1364–1369.

Subhankari, Ipsa. & Nayak, P. L., (2013). Synthesis of Copper Nanoparticles using Syzygium Aromaticum (Cloves) Aqueous Extract by using Green Chemistry. *World., 2*(1), 14–17.

Thanh, N. T. K., & Green, L., (2010). Functionalization of nanoparticles for biomedical applications. *Nano Today. 5*, 213–230.

Thorek, D. L., Chen, A. K., & Czupryna, J., (2006). Super paramagnetic iron oxide nanoparticle probes for molecular imaging. *Ann Biomed Eng., 34*, 23–38.

Tiwari, M. (2012). Nano cancer therapy strategies. *J. Cancer Res. Ther., 8*, 19–22.

Ung, T., Liz-Marzan, L. M., & Mulvaney, P., (1998). Controlled method for silica coating of silver colloids. Influence of coating on the rate of chemical reactions. *Lang., 14*, 3740–3748.

Veiseh, O., Gunn, J. W., & Zhang, M., (2010). Design and fabrication of magnetic nanoparticles for targeted drug delivery and imaging. *Adv. Drug. Deliv. Rev., 62*, 284–304.

Vekilov, P. G., (2011). Gold nanoparticles: Grown in a crystal. *Nat. Nanotech. 6*, 82–83.

Vittorio, O., Voliani, V., Faraci, P., Karmakar, B., & Iemma, F., (2014). Magnetic catechin dextran conjugate as targeted therapeutic for pancreatic tumor cells. *J. Drug Target. 22*(5), 408–415.

Wang, M. D., Shin, D. M., & Simons, J. W., (2007). Nanotechnology for targeted cancer therapy. *Expert Rev Anticancer Ther., 7*, 833–837.

Wang, P., Yigit, M. V., & Medarova, Z., (2011). Combined small interfering RNA therapy and in vivo magnetic resonance imaging in islet transplantation. *Diabetes, 60*, 565–571.

West, D. L., White, S. B., Zhang, Z., Larson, A. C., & Omary, R. A., (2014). Assessment and optimization of electroporation-assisted tumoral nanoparticle uptake in a nude mouse model of pancreatic ductal adenocarcinoma. *Int. J. Nanomed, 9*, 4169–4176.

West, J. L., & Halas, N. J., (2000). Application of nanotechnology to biotechnology. *Curr. Opin. Biotechnol, 11*, 215–217.

Xiang-Hong Peng., Ximei Qian., Hui Mao., Andrew Y Wang., Zhuo (Georgia) Chen., Shuming Nie., et al., (2008). Targeted magnetic iron oxide nanoparticles for tumor imaging and therapy. *Int. J. Nanomed, 3*(3) 311–321.

Xiao, Y., Gao, X. G., & Taratula, O., (2010). Anti-HER2 IgY antibody-functionalized single-walled carbon nanotubes for detection and selective destruction of breast cancer cells. *BMC Cancer. 95*, 351.

Yallapu, M. M., Ebeling, M. C., Khan, S., Sundram, V., & Chauhan, N., (2013). Novel curcmin-loaded magnetic nanoparticles for pancreatic cancer treatment. *Mol. Cancer Ther. 12*(8), 1471–1480.

Yang, X., Hong, H., & Grailer, J. J., (2011). cRGD-functionalized, DOX-conjugated, and Cu-labeled superparamagnetic iron oxide nanoparticles for targeted anticancer drug delivery and PET/MR imaging. *Biomat, 32*, 4151–4160.

Yigit, M. V., Zhu, L., & Ifediba, M. A., (2011). Noninvasive MRI-SERS imaging in living mice using an innately bimodal nanomaterial. *ACS Nano., 5*, 1056–1066.

Zhang, J., Liu, J., Wang, S., Zhan, P., Wang, Z., & Ming, N., (2004). Facile methods to coat polystyrene and silica colloids with metal. *Adv. Funct. Mater., 14*, 1089–1095.

Zharov, V. P., Galitovskaya, E. N., & Johnson, C., (2005). Synergistic enhancement of selective nanophotothermolysis with gold nanoclusters: potential for cancer therapy. *Lasers Surg Med., 37*, 219–226.

CHAPTER 9

NANOBIOMATERIALS FOR CANCER DIAGNOSIS AND THERAPY

CECILIA CRISTEA,[1] FLORIN GRAUR,[2] RAMONA GĂLĂTUŞ,[3] CĂLIN VAIDA,[4] DOINA PÎSLĂ,[4] and ROBERT SĂNDULESCU[1]

[1]Analytical Chemistry Department, Faculty of Pharmacy, Iuliu Haţieganu University of Medicine and Pharmacy, 4 Pasteur St., 400012, Cluj-Napoca, Romania, E-mail: ccristea@umfcluj.ro

[2] Surgical Department, Faculty of Medicine, Iuliu Hatieganu University of Medicine and Pharmacy, 8 Victor Babes St., 400012, Cluj-Napoca, Romania

[3]Basis of Electronics Department, Faculty of Electronics, Telecommunication and Information Technology, Technical University of Cluj-Napoca, 28 Memorandumului St., 400114 Cluj-Napoca, Romania

[4]Research Center for Industrial Robots Simulation and Testing, Technical University of Cluj-Napoca, 28 Memorandumului St., 400114 Cluj-Napoca, Romania

CONTENTS

"The purpose of medicine is to prevent significant disease, to decrease pain and to postpone death... Technology has to support these goals-if not, it may even be counterproductive."

—*Joel J. Nobel*

9.1 NANOBIOMATERIALS INVOLVED IN CANCER DIAGNOSIS AND THERAPY: A SHORT INTRODUCTION

The third millennium encounters a very provocative challenge: there are more and more cancer patients, as a direct consequence of continuous increase of the world population, higher average life span and treatment for competing co-morbidities. This makes cancer one of the main deaths causes nowadays, due to a complex set of uncontrollable natural and artificial factors. In response to that, the scientific community made huge efforts in interdisciplinary and transdisciplinary approaches to fight against this disease, aiming to improve survival and life expectancy. Doctors (oncologists, surgeons, geneticists) together with chemists, physicists, engineers and programmers created clusters of Research and Development delivering innovation, many times at the border of current human knowledge. New techniques have been developed from whole body treatments up to intracellular level in an effort to understand and overcome the complexity of cancer. An overview of some recent cancer statistic data (Znaor et al., 2013; Ferlay et al., 2015; Siegel et al., 2015) revealed few general conclusions: cancer is growing in incidence and while advanced countries are able to improve survival rates and quality of life, developing countries still struggle to find feasible strategies to fight cancer. Unfortunately, economics will always play an important role in the medical systems whereas the scientific community will always search for better and more cost efficient ways do defeat cancer.

In fact, when one or multiple subsets of genes undergo alterations, activation of oncogenes or inactivation of tumor suppressor genes occur, the following steps being malignant proliferation of cancer cells, tissue infiltration, and dysfunction of organs (Braicu et al., 2009; Diaconu et al., 2013).

In the last decade, the number of clinical trials has increased continuously and it is likely that this trend will continue in the future due to the creation of new nanobiomaterials and progress in material sciences. The clinical trials are focusing on cancer treatment entrapping the cancer drugs in nanocarriers, therefore most nanobiomaterials that have entered clinical trial, must have the property of switching functionality.

Surgery, radiation and chemotherapy are still the main pillars for cancer therapy, even though cancer treatment has evolved significantly within the last years. Unfortunately, most current therapies for cancers are not curative, they do prolong life but only for a few months or years.

Advances in cancer management depend on early cancer diagnosis, prevention of recurrence and drug selection-based on genetic profiling and personalized treatment. On every stage, the novel nanobiomaterials could have the highest impact. Early stage cancer diagnosis is desirable to prevent metastasis, but remains challenging because clinical symptoms usually appear only in advanced cancer stages. An important goal is to perform this early stage detection by using non-invasive or minimal invasive techniques, requiring a fast and accurate response. The personalized medicine, which is probably the future of medicine, will use the personalized drug and specific drug delivery assemblies in order to attack and destroy only the malignancies without the healthy cells.

Nanotechnology offers many potential benefits to medical applications including the early detection of cancer, disease targeting, increased biocompatibility, and multifunctionality encompassing both imaging and therapeutic capabilities, allowing simultaneous disease treatment and monitoring (Wang et al., 2013).

Nanotechnology is a multidisciplinary field of science intended to create materials and technology at length scales between 1 and 100 nm and to use those nanoscale materials as building blocks for novel structures and devices. Due to their size, nanomaterials show unique properties and functions (Avti, 2011).

Regarding their application in biomedical field, nanomaterials are anticipated to revolutionize the cancer diagnosis and therapy. Several approaches have been used for application of functionalized nanoparticles for targeting tumors, imaging technology for tumor visualization, allowing the early diagnosis of cancer. Other applications in the biomedical field are drug delivery, where intelligent nanocarriers could improve the therapy efficacy by transporting anticancer drugs to a well-established place where the release of the drugs take place.

Natural or synthetic materials are used, such as polymers, carbon-based nanomaterials (nanotubes, graphene, fullerene), quantum dots, superparamagnetic iron oxide, and their composites. They are used for accurate diagnosis and effective drugs administration in cancer.

There are major advantages when using nanotechnology-based diagnosis techniques such as:

- **Fast screening, rapid testing even in the doctor's office**. This could lead to rapid intervention with medication without needing another visit to the doctor's office or lab for testing.
- **Early diagnosis, if possible, in early stage of cancer**. This could lead to early intervention, stopping the disease earlier and increasing the survival rate.

Bionanomaterials, whatever as they are natural or synthetic materials, are originated from discoveries and technologies from several research areas such as medicine, pharmaceutics, biology, tissue engineering, material science and chemistry (Atala, 2010). In spite of being used in drug delivery, implants, tissue regeneration, the bio and nanomaterials have some drawbacks (Graur et al., 2011). The need for innovative and biocompatible nanobiomaterials with improved properties represents a challenge for scientists. For instance, biopolymers have many advantages compared with cationic gene delivery carriers such as target-specific release of genetic materials and their applications in gene therapy for the treatment of incurable diseases such as cancer. For instance, Lim (Lim et al., 2014) presented some applications of carbon-based materials used as carriers for drug delivery. For the transportation of water insoluble drugs and some biomaterials such as antibodies, aptamers or nucleic acid to the cancer cell, several types of carbon allotropes as carbon nanotubes, graphene oxide, and nanodiamonds (NDs) were used.

The systems-based on gold nanoparticles (AuNPs) are applied in drug delivery and molecular imaging (Jeong et al., 2014), but also for immuno-sensors design (Florea et al., 2013; Taleat et al., 2013; Florea et al., 2015). Early diagnosis to any disease condition is vital to ensure that early treatment is started and perhaps resulting in a better chance of cure. Due to its versatility, AuNPs suffered many surface modifications being involved in applications like the delivery of small drugs as well as biomaterials such as DNAs and siRNAs (Lee et al., 2014).

The early stage cancer diagnose, when the curative treatment and the cancer patients outcome improvement is possible, represent one of the major challenges at this moment (Peng et al., 2010). In order to improve diagnosis, many new methods and techniques have been developed and an area of considerable interest is the development of biocompatible nanoparticles for molecular targeted diagnosis and treatment.

Some nanobiomaterials and nanodevices involved in diagnosis, monitoring and treatment for cancer are (Graur et al., 2010; Graur, 2015):

- *Quantum dots* (QDs) are mostly used for biomarker screening and medical imaging, for example, the detection of primary tumor such as ovarian cancer, breast cancer, prostate cancer and pancreatic cancer, as well as regional lymph nodes and distant metastases (Peng et al., 2010). Quantum dots have great advantages over organic dyes and allow tracking activities over a period of time. Because are synthesized from heavy metals, their use is limited (toxicity);

- *Gold nanoparticles* loaded with substances are attracted to the target cells, and reacts to light stimuli;

- *Magnetic nanoparticles modified with anti-tumor antibodies* were developed against cancer cells. Nanosized contrast agents for magnetic resonance imaging (MRI) and computed tomography (CT), activated by laser, during CT or MRI scanning, and leading to selective destruction of tumor cells, play an important role in cancer diagnosis and treatment;

- *Genosensors used for DNA detection* are highly important for the diagnosis of infectious diseases, genetic mutations, and clinical medicine;

- *Nanoparticles carrying cytostatics* into solid tumors and neoformation blood vessels;

- *Video capsules* with nanocameras being swallowed were developed using nanotechnology;

- *Polyethylene nanostructured silicon membranes* used for therapeutic substances administration and also, as biosensors (Popat et al., 2003);

- *Silicon nanoparticles* have been discovered to have the ability to encapsulate DNA (Choi et al., 2008);

- *High-resolution microscopes* (atomic force microscopy, near-fields scanning optical microscopy) providing images of bi- or three-dimensional biological structures (DNA, molecules, proteins) (Aleksandr Noy et al., 1997; Lehenkari et al., 2000; Edidin, 2001).

9.2 ABOUT MALIGNANCIES: THE TRIGGERING MECHANISM OF CANCER

- Environmental factors may be responsible for the development of certain cancers. These environmental factors are triggers because their exposure is associated with the development of a certain type of cancer. Some of these environmental factors are: X-rays, ultraviolet rays, tobacco, and viruses. Environmental factors are not independent of cancer genes, individual genetic differences cause an individual susceptibility to the carcinogenic effects of environmental factors (Does et al., 2016).
- Some researches established that hypoxia drives cancer progression by promoting genomic instability and that inactivation of apoptosis is essential for tumor-cell survival during this process. Hypoxia is associated with the maintenance of stem-cell-like phenotypes and increased invasion, angiogenesis and metastasis in cancer patients. Also hypoxia leads to increased aggressiveness and tumor resistance to chemotherapy and radiation (Shaduri et al., 2013).
- The period between the fixation of a carcinogen to chromosomal DNA and the appearance of a population of neoplastic cells can be divided in the following stages: initiation, promotion and progression. In initiation stage the carcinogen actions on DNA, inducing a lesion, which can be repaired or reproduced. The initiation stage continues with promotion. In this stage chronic genetic alteration determine the neoplastic transformation and the appearance of cells that are capable of autonomous growth. Progression is characterized by genetic changes, gene alterations and rearrangements, and the tumor is phenotypically characterized by a rapid proliferation, invasive and metastasizing properties (Baba et al., 2007).
- Tumors need oxygen and nutrients to grow, therefore angiogenesis is an essential process in their development. To ensure with blood all tumor cells, it is necessary to develop new blood vessels, this process is called angiogenesis. Angiogenesis is mediated by growth factors synthesized by the tumor cells, which stimulates the formation of capillaries, up-regulation of the activity of angiogenic factors is not sufficient itself for angiogenesis, negative regulators of vessel growth need to be also down-regulated. More than a dozen different proteins have been identified as angiogenic activators, including endothelial growth factor, angiogenin, tumor necrosis factor and other. Treatments acting

by inhibiting angiogenesis were developed knowing the factors that stimulate angiogenesis (Nishida et al., 2006).

- Despite research on cancer had led to new therapies over the last years, the triggering mechanism of carcinogenesis is still unknown. There are many theories trying to explain some parts of the whole mechanism.

- The *cancer-stem-cell (CSC)* concept was first demonstrated in the study of leukemia (Lapidot et al., 1994; Dick et al., 2005). It is widely accepted that the CSC sub-population of cancer cells plays a determinant role in initiation, progression and recurrence of cancer. In this situation one question remain: are neoplasms arising from normal stem cells due to maturation arrest or due to transformation of mature cells into CSC?

- The *somatic mutation theory* of carcinogenesis has been dominant since early 20th century. Somatic mutations in DNA are present in cancer cell genomes. Studies have found oncogenes that, helped by tumor suppressor genes, can reproduce some aspects of cancer progression (Hameroff, 2004).

- The *viral/microbial theory of cancer* regards viruses/microbes as potential triggers of a neoplastic process. It is well known that some oncoviruses, for example, human papillomavirus, hepatitis B and hepatitis C virus, Epstein-Barr virus, etc. are indeed associated with increased incidence of human cancers (Thompson et al., 2004; Yu et al., 2011).

- The *embryonic theory of cancer* is based on the similarity between embryogenesis and carcinogenesis, firstly proved by John Beard (Beard, 1911). The main idea in this theory is that certain fetal cells/atavistic genes give rise to malignancies.

- Approximately 35, 000 genes in the human genome have been associated with cancer. These genes can be classified in:

- *proto-oncogenes*, protein products that normally increase cell division or inhibit normal cell death,

- *tumor suppressors* that normally prevent cell division or cause cell death and

- *DNA repair genes*, which prevent mutations to lead to cancer (Does et al., 2016).

Cancer cells behave as independent cells, growing without control to form tumors. A series of steps are made to become a tumor: hyperplasia → dysplasia → anaplastic cells → metastasis cells. The first three modifications are made locally.

9.3   TYPES OF BIONANOMATERIALS APPLIED IN EARLY DIAGNOSIS, TREATMENT, THERAPY MONITORING, AND DRUG DELIVERY

Nanobiomaterials were used as nanocarriers for drug delivery or as enhancement factors for sensitivity or selectivity in point-of-care (POC) devices.

Metallic and magnetic nanoparticles, micelles, carbon nanotubes, graphene, fullerenes, dendrimers, quantum dots (QDs), and nanofibers were synthesized or prepared from organic and inorganic materials and act as nanocarriers (Cai et al., 2007; 2007) (Figure 9.1). They have shown great potential in cancer therapy by enhancing the performance of medicines and reducing systemic side effect in order to gain therapeutic efficiency.

The biosensors are analytical devices that associate a bioelement (antibodies, aptamer, miRNA, DNA, enzymes or tissues) with transducers, capable of transforming the biological event into a measurable signal. Several types of transducers are available allowing the design and development of electrochemical, optical, mass or piezoelectric biosensors.

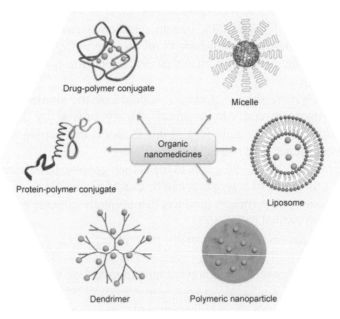

FIGURE 9.1   Organic and inorganic nanobiomaterials for cancer diagnosis and therapy (Reprinted from Tang, L., & Cheng, J., (2013). "Nonporous silica nanoparticles for nanomedicine application." *Nano Today*, 8(3), 290–312. © 2013. With permission from Elsevier.)

Immunosensors, being a class of biosensors, are analytical devices coupling the immunochemical reaction with a transducer. They belong to the most important classes of affinity biosensors which are-based on the specific recognition of antigens by antibodies (Ab) to which they form a stable complex that is formed also in immunoassays. The immunosensors are preferred to perform an early diagnosis, because these assays are able to measure peptides and proteins *directly* in the sample, without any sample pretreatment or any separation (Luppa et al., 2001). The direct detection can be performed making use of the specific interaction of proteins and Ab, aptamers and micro RNA (miRNA).

Nanobiomaterials were incorporated in the design of biosensors in order to increase sensitivity, or for multiple ing. When micro fluidic devices are associated the biosensors, they may have a potential use in clinical analysis.

Cancer biomarkers, like any other biomarker, are generally defined as "a characteristic that is objectively measured and evaluated as an indicator of normal biological processes, pathogenic processes, or pharmacological responses to a therapeutic intervention" (Leteurtre et al., 2004). They are used as an indicator for a neoplastic process detected after their aberrant e pression in biological fluids (serum, saliva) or cells and tissues. Their abnormal levels appear from early stages of cancer, they could be use in early detection improving the survival rate of patients. Generally, there are several biomarkers over expressed for the same type of cancer. Therefore, in a panel for simultaneous detection of several significant cancer biomarkers is usually required in order to avoid false positive results.

Examples of assays using different types of nanobiomaterials for cancer biomarkers will be presented (Cristea et al., 2015).

9.3.1 METALLIC NANOPARTICLES

Metallic nanoparticles have been used for sensing applications, drug delivery, photothermal therapy, cell tracking, due to their amazing properties (Giljohann et al., 2010). The mostly used are gold and silver nanoparticles, which can be easily prepared by chemical or electrochemical methods.

9.3.1.1 Gold Nanoparticles

Due to their facile synthesis techniques, enhanced optical properties, and relatively good biocompatibility gold nanoparticles (AuNPs) have been used lately for biomedical applications.

One of the easier ways of producing AuNPs was reported by McFarland et al.(McFarland et al., 2004) by using tetrachlorauric acid and trisodium citrate dihydrate as a reducing agent. Smaller nanoparticles (2 to 6 nm) could be produced by Brust method (Brust et al., 1994), in which the reaction of a chlorauric acid solution with tetraoctylammonium bromide solution in toluene and in the presence of an anti-coagulant and a reducing agent, respectively takes place. Modern approaches like the presence of ultrasonic waves (Baigent et al., 1980) or electrochemistry (Guo et al., 2007) were also reported. Although the synthesis of AuNPs makes great progresses, the size control, their monodispersion, morphology, and surface chemistry are still a great challenge (Daniel et al., 2004).

The commonly synthesized AuNPs could have spherical shapes, or could be as nanorods, nanoshells, and nanocages (Huang et al., 2010). By controlling the size, shape, composition, and structure of the nanoparticle, their optical properties such as surface plasmon absorption and scattering, or near-infrared (IR) fluorescence, could be enhanced (Lee et al., 2005; Jain et al., 2006). Due to the fact that AuNPs can convert absorbed light into heat (Link et al., 2000) they are used in photothermal therapy in cancer. 0

Gold nanoshells have been used as agents capable of assuring the contrast in optical coherence tomography (OCT) in vivo and for tumor therapy by near-IR (NIR) photothermal ablation combining diagnostic and treatment applications (Gobin et al., 2007).

Other interesting application of AuNP relies on its use as sensitivity enhancer in biosensing. An important step in the design of electrochemical-based biosensor includes finding the proper aptamer immobilization and the right amount, the maintenance of target accessibility and achieving a useful detection signals. AuNPs are excellent candidates for bioconjugation, owing to the fact that they are biocompatible and bind readily to a range of biomolecules. The immobilization of the nanoparticles on the electrode surfaces increases the surface area and promotes the adsorption capability of the electrode.

Florea et al. (2013) by binding aptamer immobilized on graphite and gold screen printed electrodes (SPE) modified with AuNPs, have developed an electrochemical assays-based on a Mucin 1 protein (MUC1). The loosely packed aptamers are self-assembled onto gold surface. After immobilization of proteins to the sensor surface an increase in charge transfer resistance is observed from impedance measurements. In the second approach, the electrochemical response of methylene blue (MB) is studied in the presence of different concentrations of MUC1 protein. The estimated detection limits (LODs) of the MUC1 protein was 0.95 ng mL^{-1} at AuNPs-modified gold screen printed electrodes (SPE).

Other electrochemical aptasensor-based on MB-labeled hairpin aptamers for MUC1 detection was described (Ma et al., 2013). The detection was-based on the conformational changes of the labeled aptamer after binding the protein or on a dual signal amplification strategy by using poly (o-phenylenediamine)-AuNPs film (for aptamer immobilization) and AuNPs functionalized silica/MWCNTs core-shell nanocomposites as tracing tag for detection aptamer (Chen et al., 2015).

A 3D label-free immunosensor for protein serum prostate-specific antigen (PSA) is based on crumpled graphene-gold nanocomposites. The graphene sheets decorated with gold nanoparticles were prepared by aerosol spray pyrolysis. A limit of detection (LOD) of 0.59 ng mL^{-1} and a linear range of 0–10 ng mL^{-1} were achieved, due to the synergistic effect of graphene and AuNPs, with increased biocompatibility, a higher amount of bound antigen and fast electron transfer (Zhou et al., 2010).

The multiple biomarker detection (CEA, epidermal growth factor receptor, hEGFR and cancer antigen 15–3, CA15–3) was possible by using an interdigitated electrode-based capacitive sensor-based on AuNPs amplification. CEA and hEGFR could be successfully detected in the concentration range of 20–1000 pg mL^{-1}, while CA15–3 was detected in the range of 10–200 U mL^{-1}, with LOD of 20 pg mL^{-1} and 10 U mL^{-1}, respectively (Altintas et al., 2014).

Many aptamer-metallic nanomaterials conjugate targeted to assemble on the surface of a specific type of cancer cell through the recognition of the aptamer to its target on the cell membrane surface have been reported. For example, by using cancer cell aptamer-conjugated AuNPs, Kim et al. (Kim et al., 2010) reported PSA aptamer-conjugated multifunctional AuNPs for combined computed tomography (CT) and cancer therapy. In this study, AuNPs were first functionalized with an anti-PSA and then loaded with doxorubicin (DOX) (Savla et al., 2011) developed a tumor targeted, pH-responsive quantum dot-mucin1 aptamer-doxorubicin (QD-MUC1-DOX) conjugate for the chemotherapy of ovarian cancer. In their design, QD was conjugated with a DNA aptamer specific for mutated MUC1 over e⬚pressed in many cancer cells, including ovarian carcinoma. DOX was attached to QD via a pH-sensitive hydrazone bond in order to provide stability of the complex for systemic circulation and drug release in the acidic environment inside cancer cells. The results show that this bond was stable at neutral and slightly basic pH and underwent rapid hydrolysis in mildly acidic pH. The QD-MUC1-DOX conjugate was successfully applied for in vivo imaging and treatment and showed a higher cytotoxicity than free DOX in multidrug-resistant cancer cells, preferentially accumulating in ovarian tumor.

9.3.1.2   Silver Nanoparticles

The first step in obtaining silver nanoparticles (AgNPs) of different sizes is to synthesize them by the chemical reduction of silver nitrate. AgNPs can also be synthesized by using biological methods. As was reported by Shahverdi et al. (2007) and Nanda (Nanda et al., 2009), AgNPs can be made by reduction of aqueous Ag ion with the culture supernatants of bacteria such as *Klebsiella pneumoniae* or *Staphylococcus aureus*. Furthermore, many others reported the use of fungus to produce AgNPs (Fayaz et al., 2010; Dar et al., 2013). Other methods of synthesis consist in evaporation-condensation, which could be carried out using a tube furnace at atmospheric pressure, laser ablation of metallic bulk materials in solution, etc. From chemical point of view chemical reduction is the most frequently applied method for the preparation of AgNPs as stable, colloidal dispersions in water or organic solvents. Commonly used reducing agents are borohydride, citrate, ascorbate and elemental hydrogen. Also, AgNPs can be prepared inside micro emulsion. Metal clusters formed at the interface are stabilized, due to their surface coated with stabilizer molecules occurring in the nonpolar aqueous medium (Abou El-Nour et al., 2010).

The availability of AgNPs has ensured a rapid adoption in medical practice, their application can be divided into diagnostic and therapeutic uses for cancer. Lin et al. (2011) reported AgNP-based Surface-Enhanced Raman Spectroscopy (SERS) in non-invasive cancer detection. This approach is highly promising and may prove to be an indispensable tool for the future.

Regarding therapeutics, one of the most well documented and commonly used applications of AgNPs is in wound healing. Compared with other silver compounds, many studies have demonstrated the superior efficacy of AgNPs in healing time, as well as achieving better cosmetic after healing (Kwan et al., 2011). For oncology, Tse et al. (2011) presented a novel method to selectively destroy cancer cells. Human epidermoid cancer cell line was targeted with folated silver-dendrimer composite nanodevices and the labeled cancer cells were destroyed by the micro bubbles generated through increased uptake of laser light energy by AgNPs. AgNPs display plasmonic properties, depending on its shape, size, and the dielectric medium that surrounds them. In fact, its properties in the dielectric medium that can be exploited and make it an ideal candidate (Prabhu et al., 2012) for biosensing. Nanosilver biosensors can quantify a large number of proteins that normal biosensors find hard to detect. This unique advantage that nanosilver has can be used for detecting various diseases in the human body including cancer (Zhou et al., 2011).

9.3.2 MAGNETIC NANOPARTICLES

Magnetic nanoparticles (MNP) have as main constituent ferrous or ferric oxide. The main properties of magnetic particles for medical applications are their biocompatibility, injectability, lack of toxicity, and high-level accumulation in the target tissue or organ (Ito et al., 2005). Magnetic particles are attracted to high magnetic flux density, this feature is used for drug targeting and bioseparation including cell separation and purification. Currently, magnetic nanoparticles are used as contrast agents for magnetic resonance imaging and heating mediators for cancer thermotherapy (hyperthermia). Furthermore, a novel application of magnetic nanoparticles as platform of immobilization for immunosensor development has been proposed.

Magnetic nanoparticles are used in an experimental cancer treatment called magnetic hyperthermia (Rabias et al., 2010) in which the fact that nanoparticles heat when they are placed in an alternative magnetic field is used.

Affinity ligands such as hEGF, folic acid, aptamers, lectin, etc. can be attached to the magnetic nanoparticle surface in order to place them in the vicinity of tumors, enabling thus the magnetic nanoparticles to accumulate in specific tissues or cells (Kralj et al., 2013). This strategy is used in cancer research to target and treat tumors in combination with magnetic hyperthermia or nanoparticle-delivered cancer drugs.

Another potential treatment of cancer includes attaching magnetic nanoparticles to free-floating cancer cells, allowing them to be captured and carried out of the body. The treatment has been tested in the laboratory on mice and will be evaluated in survival studies (Scarberry et al., 2008).

Recently, very small super paramagnetic iron oxide particles (particle size 4–8 nm), have been developed as a new class of MRI contrast agents. Their properties such as charge and nature of the coating material determine the stability, biodistribution, metabolism, pharmacokinetics, and pharmacodynamics of ultra small super paramagnetic iron oxides with sizes smaller than 50 nm.

Other applications of these MNP include cellular labeling and separation. By tagging the iron oxide nanoparticles with antibody, cells expressing a specific ligand in any diseased state could be identified. These iron oxide-labeled cells can then be separated by a process called magnetophoresis (Winoto-Morbach et al., 1994) and could be used for a variety of clinical applications such as detection of metastases, metastatic lymph nodes, inflammatory diseases, and degenerative diseases (Na et al., 2009; Li et al., 2013; Khalkhali et al., 2015).

During recent years, there has been an increased interest regarding the use of magnetic nanoparticles as contrast agents, such as dextran magnetite in MRI (Magin et al., 1986; Gillis et al., 1987). Monoclonal antibodies (mAbs) specifically targeted to cancer cells could serve as a possible tool for active targeting. The MRI contrast agent, mAbs-magnetite, which was prepared by covalently linking polyethylene glycol-coated magnetite to a mAbs specific for a human glioma cell surface antigen was prepared and tested. When mAbs-magnetite was injected intravenously into tumor-bearing nude mice, magnetite nanoparticles accumulated in the tumor tissue 24 or 48 h after the mAbs-magnetite injection, and a 50% decrease in the T2 signal intensity of the tumor was observed (Suzuki et al., 1996).

Hyperthermia is a promising approach to cancer therapy, and various methods inducing hyperthermia, such as the use of hot water, capacitive heating, and induction heating among others, have been employed. Back in 70s, Gordon et al. (1979) first proposed the concept of inducing intracellular hyperthermia using dextran magnetite nanoparticles.

A significant development in the area of biomedical applications of MNP has been their use in gene delivery and gene therapy. A viral vector carrying the appropriate gene is attached to the coating of the MNP and when approaching the target site, gene transfection and expression occur, rectifying genetic disorders. Attractive targets for gene therapy include the epithelial surfaces of the lungs and the gastrointestinal tract and endothelial cells lining the blood vessels. Magnetic transfection (or magnetofection) has also aimed at expanding into non-viral transfection of DNA, siRNA, and other biomolecules or indeed in studying specific genes involved in disease pathways (Issa et al., 2013).

Magnetic nanoparticles could also be used for the early detection and diagnosis of cancer being building blocks of electrochemical immunoassays.

An interesting configuration of an electrochemical immunoassay using magnetic nanoparticles and carbon-based SPE could be designed to be used as disposable device. At the surface of the nanoparticles could be attached an ELISA like sandwich assay with a primary antibody recognizing the antigen (like MUC1 or CA125) followed by the secondary and the third, labeled antibody that will transform the substrate into a recognizable product (by reduction or oxidation). Their high selectivity and sensitivity as well as the possibility to miniaturize these systems have made possible the utilization of electrochemical immunosensors for in vivo analyses. Some important tumor markers, such as CEA, carcinoma antigen 125 (CA125), alpha-fetoprotein (AFP), prostate specific antigen (PSA), CA15–3 and human chorionic gonadotropin (hCG) have been widely detected for the diagnosis of breast, epithelial ovarian tumors, endometriosis and prostate cancer (Diaconu et al., 2013). They are suitable for assessing the activity and complications of the disease and for monitoring therapy in the tumor prevention stage, as well as for the follow-up examination during therapy. In the management of cancer patients, they are also used in noninvasive tests for relapse detection (Magdelenat et al., 1992). Thus, the detection of tumor marker levels in human serum is absolutely necessary in clinical assay (Lin et al., 2005).

A quick and reproducible electrochemical-based immunosensor technique, using magnetic core/shell particles coated with self-assembled multilayer of gold, has been developed. Magnetic particles that are structured from Au/Fe_3O_4 core-shells were prepared and aminated after a reaction between gold and thiourea, and additional multilayered coatings of gold nanoparticles were assembled on the surface of the core/shell particles. The CEA antibody was immobilized on the modified magnetic particles, which were then attached on the surface of solid paraffin carbon paste electrode by an e□ternalmagnetic field. In this case the sensitivity and response features of this immunoassay were significantly affected by the surface area and the biological compatibility of the multilayered gold. The linear range for the detection of CEA was from 0.005 to 50 ng mL^{-1} and the LOD was 0.001 ng mL^{-1}. The immunosensor was approximately 500 times more sensitive than that of the traditional enzyme-linked immunosorbent assay for CEA detection (Li et al., 2010).

Marrazza et al. (2012) developed an immunoassay-based on a sandwich format which starts with a primary monoclonal antibody anti-HER2 is coupled to protein A modified magnetic beads. The modified beads are then used to capture the protein from the sample solution and a sandwich assay is performed by adding a secondary monoclonal antibody anti-HER2 labeled with biotin is added. The enzyme alkaline phosphatase (AP) conjugated with streptavidin and its substrate (1-naphthyl-phosphate) are then used for the electrochemical detection by differential pulse voltammetry (DPV) and applied to the serum samples from hospital patient analyses.

Following the same principle, Florea et al. (2015) developed two single-use bioassays based on a sandwich format in which aptamers or antibodies were coupled, respectively to streptavidin or protein G modified magnetic beads (Figure 9.2). The bioreceptor modified beads were used to capture the MUC1 protein from the sample and the sandwich assay was performed by the addition of a labeled secondary aptamer or antibody. The AP and 1-naphthyl phosphate were then used for the electrochemical detection by DPV. Using the optimized conditions, a linear range from 0 to 0.28 nM was

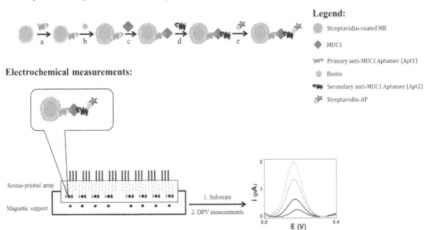

FIGURE 9.2 Scheme of aptamer-based bioassay for MUC1 detection: (a) incubation of streptavidin modified magnetic beads with biotinylated primary aptamer (Apt1), (b) blocking step with biotin, (c) incubation with MUC1 protein, (d) incubation with secondary biotinylated aptamer (Apt2), (e) incubation with streptavidin-alkaline phosphatase. Electrochemical measurements: screen-printed arrays were placed on a magnetic support and modified-magnetic beads were deposited on the surface of each working electrode of the array (Reprinted from Florea, A., Ravalli, A., Cristea, C., Sandulescu, R., & Marrazza, G., (2015). "An optimized multiplexed bioassay for sensitive Mucin1 detection in serum samples." *Electroanal, 27*(3), 638–647. © 2015 John Wiley and Sons. With permission.)

obtained, with the LOD of 0.19 nM for the antibody-based and 0.07 nM for the aptamer-based sandwich assay in MUC1 buffered solutions. The results also showed that the aptamer-based approach exhibited higher selectivity for MUC1, allowing the detection of the protein in real samples, complex matrices such as blood and plasma.

The developed aptasensor for MUC1 detection was applied on serum samples obtained from cancer patients, providing promising perspectives for clinical applications.

9.3.3 SILICA BASED NANOMATERIALS

In the field of nanotechnology, silica-based nanoparticles (SiNPs) have a dominant role because of their fundamental characteristics, such as: size (generally from 5 to 1000 nm), unique optical properties, high specific surface area, low density, adsorption capacity, capacity for encapsulation, biocompatibility and low toxicity. Besides all these properties they exhibit unique properties amenable for in vivo applications (Barbé et al., 2004; Slowing et al., 2008), such as hydrophilic surface, versatile silane chemistry for surface functionalization, good biocompatibility, ease of large-scale synthesis, and low cost of NP production.

These features lead to the widely use of SiNPs as an inert solid supporting or entrapping matrix. SiNPs offer considerable advantages and have opened new avenues of biomedical research in numerous leading edge applications, such as biosensors (Tallury et al., 2008), enzyme supporters, controlled drug release and delivery (Vivero-Escoto et al., 2010) and cellular uptake. SiNPs are used in medical imaging and its applications as contrast agents have an important auxiliary role. These are used to encapsulate contrast agent particles, such as gold (Viarbitskaya et al., 2011), silver, iron oxide, organic dyes and quantum dots (Bitar et al., 2012).

SiNPs offer a promising alternative to organic drug delivery systems and exhibit interesting properties, such as highly controllable size and shape. Nonporous SiNPs have numerous biomedical applications for the delivery of drugs, proteins, and genes and for molecular imaging. However, there are some drawbacks regarding their use in clinical applications: the need for improving drug loading, spatial and temporal control of drug release, highly efficient targeting of disease sites, scalable manufacturing, long-term stability, and well-understood biocompatibility and potential toxicity.

By modifying the properties of SiNPs with organic functional groups, the new organo-silica hybrid NPs could provide more sophisticated silica-based nanomedicines with highly controllable drug loading and responsive drug release (Tang et al., 2013).

9.3.4 CARBON BASED NANOMATERIALS

Different type of carbon nanostructures become popular due to their specific structures, properties, and the possibility of being used for many applications. A wide variety of carbon-based materials are available, such as: nanoparticles, nanodiamonds, nanoonions, peapods, nanofibers, nanorings, fullerenes and nanotubes, which have been extensively used in analytical and biomedical applications. The basic structure of fullerenes and nanotubes is composed of a layer of sp^2-bonded carbon atoms. This configuration, which resembles that of graphene, is responsible for their good electrical conductivity and their ability to form charge-transfer complexes when in contact with electron donor groups (Barnes et al., 2007). This configuration is responsible for the development of strong van der Waals forces that significantly hamper the solubility and dispersion of carbon-based nanoparticles. In order to avoid these problems, different pretreatment methods have been proposed, such as the addition of polar groups (oxygen, hydroxyl, phenyl and polyvinylpyrrolidone). The surface defects could also affect the stability, the mechanical and electrical properties of carbon nanostructures (Săndulescu et al., 2015).

Carbon nanotubes are cylinders of one several coaxial graphite layers with a diameter in the order of nanometers. They can be classified into two general categories-based on their structure: single-walled carbon nanotubes (SWCNTs) with a single cylindrical carbon wall and multi-walled carbon nanotubes (MWCNTs) with multiple walls cylinders nested within other cylinders (Lacerda et al., 2006).

CNTs are also cylindrical graphene structures with unique physical and chemical properties such as high mechanical strength, electrical conductivity, and thermal conductivity. Semiconducting SWCNTs display near-IR fluorescence. The IR spectrum between 900 and 1300 nm is an important optical window for biomedical applications because of its lower optical absorption (greater penetration depth of light) and small self-fluorescent background. Additionally, CNTs display good photo stability and good electric conductivity being used lately not only for drug delivery, but also for expanding the electrodes surfaces in biosensors's design. Thanks to their unique electronic, thermal, and structural characteristics, they can offer a promising approach for gene and drug delivery for cancer therapy (Tanaka et al., 2004; Lacerda et al., 2006).

Heating of organs and tissues by placing multifunctional nanomaterials at tumor sites is emerging as an art of tumor treatment by "nanothermal therapy" (Sharma et al., 2008) and it is applicable for all types of nanobiomaterials. CNTs are good candidates to kill cancer cells via local hyperthermia, due to their thermal conductivity and optical properties. A research showed that oligonucleotides could be translocated into cell nucleus by nanotubes and cause cell death with continuous near-infrared radiation (NIR) because of excessive local heating of SWCNT in vitro (Kam et al., 2005). The CNTs could be functionalized with tumor-specific ligands and antibodies, like folic acid and monoclonal antibody acting as targeting agents for many tumors. As an example, Zhang et al. (2009) developed a highly effective drug delivery system triggered by pH change, via CNTs coated with a polysaccharide material and then modified with folic acid. The encapsulated do[]orubicin was only released from the modified nanotubes at a low pH, resulting in nuclear DNA damage and the Hela cell proliferation inhibition. In order to ensure their future clinical applications more investigations must be performed for the pharmacokinetics, biodistribution, toxicity, and activity of the novel functional carbon nanotubes in vivo.

Graphene is another material that shows great perspectives in the future of analytical chemistry, consisting in one atom thick sheet of six-member rings of sp^2 hybridized carbon, that provide a surface area that is nearly twice as large as that of SWCNT (Pumera et al., 2010). This material present high mechanical strength, high elasticity, high thermal conductivity, but also presents a major disadvantage: its poor dispersion in aqueous medium. Stable dispersions of graphene sheets have been achieved using as dispersing agent amphiphilic polymers, alkyl amines and hydrophilic carboxyl groups, among others (Bustos-Ramírez et al., 2013). The functionalization of graphene is considered an important route for improving its dispersion. Therefore, different authors have tried producing graphene oxide, a graphite derivative with hydroxyl, epoxy and carboxyl groups covalently attached to its layers, a better dispersion in some solvents being obtained. A wide variety of methods are available for graphene fabrication, the first being published in 2004 (Novoselov et al., 2004) and consisting in the mechanical exfoliation (repeated peeling) of small patches of highly ordered pyrolytic graphite. Many other fabrication methods (Săndulescu et al., 2015) have been developed, including unzipping MWCNT to form graphene ribbons, and spray deposition of graphene oxide-hydrazine dispersions, etc.

Fullerenes are spheroidal carbon nanostructures, with exceptional physical, photochemical, and agonists and antagonists for tumor targeting (Bosi et al., 2003). The water-soluble gadolinium metallofullerenes had a prolonged blood circulation time, up to 48 h, and delayed clearance by the excretory system (Mikawa et al., 2001). Further advances have resulted in enhanced proton relativities obtained by manipulating the pH effect (Toth et al., 2005). More studies regarding the surface tailoring of fullerene formulations need to be achieved for active tumor targeting, and investigated for their in vivo safety aspects, in addition, all of the findings need to be tested at clinically applicable MRI (Nazir et al., 2014).

All carbon-based nanomaterials were widely used for the construction of biosensors using different bioelements such as antibodies, aptamers, DNA or miRNA in order to develop devices capable of early diagnosis of cancer. This is due to their large surface area, high conductivity, and easy chemical modification, desirable properties for use in electrochemical immunosensors.

Rusling's group (Yu et al., 2006) developed nanostructured electrodes composed of densely packed films of upright SWCNT forests and used them for ultrasensitive detection of PSA with a LOD of 4 pg mL^{-1} for PSA spiked into undiluted calf serum.

An example of multiplexing electrochemical immunosensor was developed for ultrasensitive detection of PSA and Interleukin 8 (IL-8). Disposable carbon-based SPEs array was used as detection platform and multi-labeled nanoprobe consisting in HRP and goat-antirabbit IgG (secondary antibody, Ab_2) onto MWCNTs. By using the carbon-based SPEs array and the universal nanoprobe, it was possible to detect as low as 5 pg mL^{-1} of PSA and 8 pg mL^{-1} of IL-8 simultaneously (Wan et al., 2011).

The association between two carbon-based nanomaterials, such as graphene and carbon nanospheres (CNSs) labeled with horseradish peroxidase-secondary antibodies (HRP-Ab_2) for sensitive detection of cancer biomarker AFP was developed by Lin's group (Du et al., 2010). By introducing the multibioconjugates of HRP-Ab_2-CNSs onto the electrode surface through "sandwich" immunoreactions the sensitivity was enhanced. After that, functionalized graphene sheets used for the biosensor platform increased the surface area to capture a large amount of primary antibody (Ab_1), thus amplifying the detection response. The proposed immunosensor respond to 0.02 ng mL^{-1} AFP with a linear calibration range from 0.05 to 6 ng mL^{-1}. This amplification strategy is a promising platform for clinical screening of cancer biomarkers and point-of-care diagnostics.

More recently, Liu et al. (2014) developed an AuNPs-ionic liquid (IL) functionalized reduced graphene oxide (rGO) immunosensing platform (IL-rGO-AuNPs) for simultaneous electrochemical detection of CEA and AFP. This sandwich type electrochemical immunosensor is-based on IL-rGO-AuNPs nanocomposite platform, using chitosan (CS) coated polystyrene beads nanoparticles (PBNPs) or cadmium hexacyanoferrate nanoparticles (CdNPs) loaded with AuNPs as distinguishable signal tags. The electrochemical signals were simultaneously measured by DPV at different peak potentials because of the presence of PBNPs and CdNPs. The immunosensor exhibited LODs of 0.01ng mL^{-1} for CEA and 0.006 ng mL^{-1} for AFP, respectively. The results provided in clinical serum samples were consistent with those obtained by an ELISA method, indicating the immunosensing platform as a possible application for the simultaneous determination of CEA and AFP in clinical diagnostics.

Although electrochemical immunosensors-based on carbon nanomaterials may offer several advantages for enhancing the capabilities of current analytical methodologies by permitting rapid and highly accurate analysis, this field is still new and there are many points to be addressed. Researches should also be made on exploring the novel biosensor design in complex matrices and solve the problem of nanomaterial aggregation in these media as well as the interferents issue.

9.3.5 POLYMERIC NANOCOMPOSITES

Biological and synthetic polymers were used for the development of polymeric nanobiomaterials. Polysachharides such as alginate, chitosan, hyaluronic acid and their derivatives were used as biological polymers besides some proteins (collagen or fibrin). Among synthetic polymers, those exhibiting biocompatibility and other relevant properties for clinical use (promoting cell adhesion and tissue regeneration), such as: poly (glycolic acid), poly (lactic acid), poly (hydroxyl butyrate) are preferred. For biosensing applications different modified polypyrrole derivatives are also in use as well as polyethyleneimine, polythiophene, polyaniline, etc.

Comparing the main classes of polymers it could be concluded that synthetic polymers in general have better mechanical strength than biological polymers and can be synthetically manipulated to allow biological degradation.

Polymeric nanocomposites are hybrids in which nanomaterials are used as fillers to improve the polymer's surface properties or in bulk solution. These hybrid materials, whether of biological or chemical origin, have gained much attention due to their role in improving the physicochemical properties of the polymer matrix.

Polymeric nanoparticles with less than 1 μm diameter are prepared from natural or synthetic polymers. Depending on the methods of preparation, nanoparticles with different properties and different release characteristics can be obtained by forming matrix-type or reservoir-type structure, named nanospheres or nanocapsules (Bajpai et al., 2008). Due to their specificity of action of drugs by changing their tissue distribution and pharmacokinetics (Fonseca et al., 2002), they have been considered the most promising carriers for drug delivery.

Polymeric nanoparticles can be formulated for passive delivery to the lymphatic system, brain, arterial walls, lungs, liver, spleen, or made for long-term systemic circulation. In this regard, they could play an important role in delivering antitumor drugs in a targeted manner to the malignant tumor cells, thereby reducing the systemic toxicity and increasing their therapeutic efficacy. An important feature of polymeric nanoparticles as drug carriers is that they are amenable to surface functionalization for active targeting to tumor tissues or cells and for stimulus-responsive controlled release of drug. Active targeting of nanoparticles to action site is-based on the pathological state of tumor tissues, such as the angiogenesis and the over expressed receptors. Thus, varieties of researchers have focused on formulating multifunctional nanoparticles to improve the effectiveness of drug delivery and therapy (Qiao et al., 2010).

Another interesting carrier system, polymeric micelles, originally composed of an amphiphylic block, have become an increasingly exciting field in research due to its significant therapeutic potential (Meng et al., 2015). As an example, the design of a self-assembled aptamer-micelle nanostructure able to enhance the binding ability of the aptamer moiety was reported by Wu et al. (2010). This nanostructure have many advantages: greatly improved binding affinity, low off rate once on the cell membrane, rapid targeting ability, high sensitivity, and effective drug delivery. To prove the potential detection/delivery application of this aptamer-micelle in biological living systems, the authors mimicked a tumor site in the blood stream by immobilizing tumor cells onto the surface of a flow channel device. The nanostructure aptamer-micelles showed great dynamic specificity in flow channel systems that mimic drug delivery in the blood system, shown high potential for cancer cell recognition and for in vivo drug delivery applications.

A pH-sensitive drug delivery system capable of releasing drugs-based on polypyrrole nanoparticles (PPyNPs), was synthesized via a simple one-step micro emulsion technique and was reported by Zare's group (Samanta et al., 2015). These nanoparticles being highly monodisperse, are stable in solution over a month, and have good drug loading capacity (~15 wt %). PPyNPs can be tuned to release drugs at both acidic and basic pH by varying the pH value, the charge of the drug, as well as by adding small amounts of charged amphiphiles. Such a nanoparticle-hydrogel composite drug delivery system is very promising for cancer treatment requiring controlled, localized, and sustained drug release.

PPyNPs could also be used as organic photothermal agents. They are biocompatible, present high photothermal efficiency, and low cost. The near infrared absorption and photothermal effect of two different sizes PPyNPs were compared in a study by Wang et al. (2014) which concluded that the PPyNPs with small size can kill the cancer cells effectively under the irradiation of 808 nm laser with a low-power density. These findings may provide better information for the application of the PPyNPs nanoparticles on the photothermal ablation of cancer cells.

9.3.6 BIOLOGICAL NANOMATERIALS

Other nanobiomaterials used for drug delivery and tumor targeting use the biological ones, represented by lipoprotein-based nanomaterials and peptide nanoparticles.

Lipoproteins are the assembly of proteins and lipids into spherical nanostructures involved in the transport of water-insoluble lipids in the blood. They contain a polar core of triglyceride and cholesteryl esters surrounded by phospholipid monolayer shell containing apolipoprotein and unesterified cholesterol. Depending on their density there are different types of lipoproteins with various diameters (ranging from 5–15 nm in the first case to 30–80 in the last one): high-density (HDL), low-density (LDL), intermediate density, very-low-density lipoprotein. The main advantage of lipoprotein-based nanoparticles is represented by their completely biodegradable nature, their biocompatibility and stability in blood flow. The hydrophobic core of the plasma-derived LDL could be used for incorporating lipophilic drugs for drug delivery to tumor target sites expressing LDL receptor (Rensen et al., 2001).

Peptide-based nanoparticles are the small peptide sequences having the ability of self-assembling or binding with various other nanomaterials. The self-assembly of peptides may lead to various structure formations such as oligomeric coiled-α-coilheli□(trimeric and pentameric) (Raman et al., 2006). Further, these peptide nanoparticles can be functionalized with other targeting moieties like QDs as contrast agents for fluorescence imaging, AuNPs as contrast agents for electron microscopy, and/or magnetic nanoparticles as probes for MRI, binding of specific oligonucleotide sequences for effective gene delivery and therapy, incorporation of small hydrophobic molecules such as vitamins and lipophilic drugs (Lehner et al., 2013).

9.4 NANOROBOTS USED IN DRUG DELIVERY AND CANCER THERAPY

Robotics evolved throughout the history, being in a very general sense, *advanced tools which improve the efficiency and outcome of a certain process*, being linked with every major technological progress in human evolution. Naming just a few exceptional achievements in medical robotics, one has to mention the da Vinci robotic which is a high performance system for minimally invasive surgery (Crisan et al., 2013; Ng et al., 2014), with applications also in oncology (Yuh et al., 2016), the Magellan intravascular robotic system (Baumann et al., 2015) and multiple robotic devices designed for brachytherapy (Pisla et al., 2014) (a technique which aims to treat cancer by placing radioactive seeds directly inside the malignant tumors) where the efficiency of these system enabled the definition of standards for the development of new solutions.

Going from macro to nano levels one encounters a totally new world where robotics or nanorobots change their structure from metallic components and motors to enzymes, polymers, proteins assembled in complex nanostructures aiming to serve as drug delivery agents capable of targeting specific molecular structures inside the body. A comprehensible definition of nanorobots is proposed by Freitas in his book (Freitas, 1999): "a computer-controlled robotic device constructed of nanoscale components to molecular precision, usually microscopic in size."

Nanorobots or nanobots are still tools positioned at the frontier of human knowledge, but current progress in science shows that in the next years the first successful applications in medicine will be applied to cure different diseases.

9.4.1 NANOROBOTS WITH SENSORS FOR MEDICAL APPLICATIONS

A joint effort of research centers from Australia and USA propose a nanorobot architecture which integrates biosensors and actuators enabling target identification and drug delivery at specific locations (Cavalcanti et al., 2008) The foreseen range of applications for such devices is wide and combine early and accurate diagnosis and smart drug delivery, whereas the most relevant target cells are all forms of cancer, aneurysm development (by measuring the levels of nitric oxide synthase (Mohamed et al., 2012)), early detection of Alzheimer disease by monitoring the levels of amyloid-β (Bloom, 2014), smart chemotherapy, efficient neurotransmitters, and so on. The design of the nanorobots proposed by the authors integrates sensory functions and motion capability as it can be seen in Figure 9.3, achieved in a real time nanorobot simulator which allows the complete analysis of the nanorobot behavior.

The nanorobot has a dynamic real-time sensing and actuation system which can adapt and take decisions-based on information gathered from the sensors (Figure 9.4). The authors claim that in the near future the manufacturing strategies development will enable the actual development of such complex nanosystems which could in time provide a revolution in medicine.

9.4.2 SYNTHETIC MICROMOTORS

Important efforts are made towards the development of biologically inspired or non-biological nanodevices as nanorobots with self-actuation/navigation inside the human body. At micro/nano scale the motion and propulsion is achieved in totally different ways, as the moving device has to overcome the viscous drag to be able to propel itself inside a fluid, while it requires a

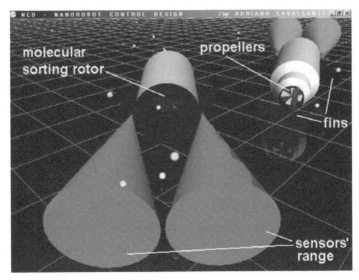

FIGURE 9.3 Nanorobot design (Reprinted with permission from Cavalcanti, A., Shirinzadeh, B., Freitas, R., A., & Hogg, T., "Nanorobot architecture for medical target identification." *Nanotechnology, 19*(1), 015103. 2008. http://iopscience.iop.org/article/10.1088/0957-4484/19/01/015103/meta © IOP Publishing.)

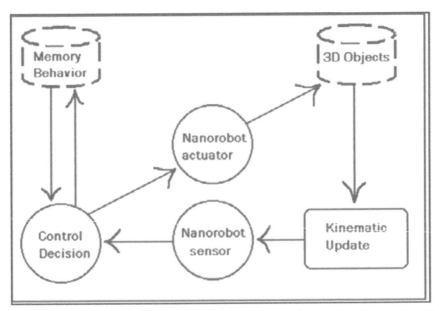

FIGURE 9.4 Adaptive real time interactive sensing and actuation (Reprinted with permission from Cavalcanti, A., Shirinzadeh, B., Freitas, R., A., & Hogg, T., "Nanorobot architecture for medical target identification." *Nanotechnology, 19*(1), 015103. 2008. http://iopscience.iop.org/article/10.1088/0957-4484/19/01/015103/meta © IOP Publishing.)

sufficiently high power to achieve a significant motion with respect to the Brownian one (Hong et al., 2010). Most of the techniques used to create autonomous micro/nanoscale motions are summarized in Hong et al. (2010): asymmetric mechanical method, surface tension, polymerization (with only a few results), catalytic self-electrophoresis (Paxton et al., 2004; Nourhani et al., 2015), where PtAu submicron rods are used, self-thermophoresis, self-diffusiophoresis and osmotic propulsion (in which the unbalanced concentration is generated by the reaction between the colloidal particle – the motor and the bath of particles will create the osmotic pressure imbalance which will determine the motion of the motor (Córdova-Figueroa et al., 2008)).

The department of Nanoengineering, University of California (USA), has successfully completed a series of in vivo tests on mouse models where they developed a zinc-based micromotor, evaluating the behavior of the device inside the stomach, using the gastric acid as propulsion (Gao et al., 2015). Figure 9.5a illustrates the development of the micromotor and its behavior inside the mouse stomach, showing its self-propulsion and tissue

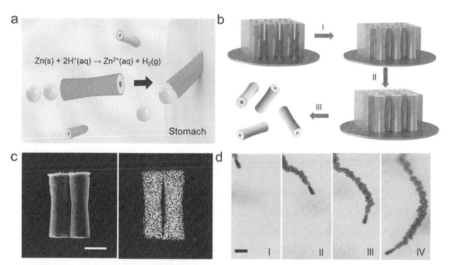

FIGURE 9.5 Preparation and characterization of PEDOT/Zn micromotors. (a) Schematic of the in vivo propulsion and tissue penetration in mouse stomach. (b) Preparation of PEDOT/Zn micromotors (c) Scanning electron microscopy (SEM) image (left) of the PEDOT/Zn micromotors and energy-dispersive X-ray spectroscopy (EDX) data (right) of elemental Zn in the micromotors. Scale bar, 5 μm. (d) Time lapse images (1 s intervals, I-IV) Scale bar, 20 μm (From Gao W, Dong R, Thamphiwatana S, et al. Artificial Micromotors in the Mouse's Stomach: A Step toward in Vivo Use of Synthetic Motors. ACS Nano. 2015;9(1):117-123. doi:10.1021/nn507097k. http://pubs.acs.org/doi/pdf/10.1021/nn507097k. Reproduced with permission from the American Chemical Society.)

penetration of acid-driven poly (3,4-ethylenedioxythiophene) (PEDOT)/ zinc (Zn). Figure 9.5b illustrates the fabrication of the micromotors using Cyclopore polycarbonate membrane templates with microconical pores, in a three steps procedure: (I) deposit of the PEDOT tube, (II) the inner zinc layer deposition galvanostatically and (III) membrane dissolution with the release of the micromotors. Figure 9.5c displays microscopic imaging of the micromotors, on the left having a Scanning Electron Microscopy view, showing the 20 μm long and 5 μm wide devices, while on the right, using energy-dispersive X-ray spectroscopy the presence of zinc is demonstrated. Figure 9.5d illustrates the self-propulsion of the micromotors in an acid simulated environment (pH 1.2) with images taken at 1 second intervals.

9.4.3 NANOROBOTS-BASED ON DNA

A robotic DNA device which is capable to interact with selected cells to deliver signaling molecules to the cell surface was reported by Douglas et al. (2012). The nanorobot has a hexagonal barrel shape, with two domains covalently attached in the rear by single-stranded scaffold hinges and capable of non-covalent fastening in the front by staples modified with DNA aptamer-based locks. The lock itself can be stabilized in a dissociated state using its own antigen key. The lock duplex has a length of 24 bp, having in the nonaptamer strand a 18-up to 24-base thyamine spacer. The nanorobot can act as a carrier of gold nanoparticles and antibody Fab' fragments. Targeting a response to proteins, the lock mechanism opens in response to binding antigen keys. The structure has 12 payload attachment sites in the middle, these being staple strands with 3' extensions complementary to the linker sequence of the cargo. Following the nanorobot folding and purification two types of cargo were loaded: 5-nm gold particles covalently attached to 5'-thiol-modified linkers and various Fab' antibody fragments covalently attached to 5'-amine-modified linkers using a HyNic/4FB coupling kit. The validation of the nanorobot was performed by using antibody fragments against human leukocyte antigen HLA-A/B/C and the robot was able to open and bind to the cell surface (using its antibody payload) when the proper combination of antigen keys was encountered. The outstanding results of this DNA robot has encouraged the research team to schedule first human trials, with promise of a potential cure for leukemia.

As a glimpse into the possible future of nanorobots, the diamondoid nanorobots have to be mentioned (Freitas, 2013). The mechanosynthesis (molecular positional fabrication) is the formation of covalent chemical bonds by applying accurate mechanical forces to build atomically precise structures. An elected material for diamondoid nanorobots is the crystalline allotrope of carbon, pure diamond, which has properties like: extreme hardness, high thermal conductivity, chemical inertness, low frictional coefficient, wide electronic bandgap being the strongest and stiffest material known at ordinary pressures. Hand in hand with technology progress, the future development of nanofactories is foreseen as major step in the development of medical diamondoid nanorobots.

9.5 PHOTONICS APPROACHES AND MEDICAL IMAGING

Recent advances in optical techniques have created a great range of possibilities for early diagnosis and therapeutics, focused on major cancer incidences. Tools-based on light physics helps to observe and manipulate objects on a scale from several nanometers to centimeters. Optical methods can be combined smoothly with other techniques due to high compatibility with electronics and can be used for multiple functionalities such as morphological, chemical, and mechanical movement. Photonics is the only available technology to cover time scale from 10^{-15} to 10^6 s, using the ultrashort light pulses of femtosecond lasers (which are extremely fast and stable, permitting analysis or manipulation with high precision in time) (Popp, 2014).

9.5.1 PHOTONIC TECHNOLOGY

Photonic technology can complement existing approaches and offer advantages such as:
- non-invasive or minimally invasive techniques for in live cell imaging and treatments with therapeutic laser systems;
- good resolution, filling the gaps between the macro, micro and nanoscales, in two or three dimensions, single molecule resolution with high sensitivity screening and monitoring tool for in vitro or in vivo detection (with performances depending on markers and labels used);
- adaptation for specific applications that can produce results in real time;

- device capabilities with multiplexing-based techniques (frequency, spectral);
- offering long-term and stable monitoring of molecular processes;
- at cellular level there are innovative microscopes and endoscopes that help us to inspect and understand cell processes, tissues and model organisms;
- augmented workplace (computer-generated image providing a composite view of the real world) in surgery for minimal invasive tumor removal or for pathology diagnostics;
- automated real time processing of large amount of data: real time detection in 3-D (30 frames), higher throughput and high-content screening (Bryan, 2010).

The important methods and technologies-based on photonic approaches are given in the following subsections.

Spectroscopy
- Absorption (THz, microwaves, IR, UV-Vis),
- Emission (one-photon/multi-photon fluorescence, fluorescence resonance energy transfer – FRET, fluorescence recovery after photobleaching – FRAP, fluorescence lifetime imaging microscopy-FLIM, single-atom spin-Flip spectroscopy – FLIP),
- Elastic and inelastic light scattering (Rayleigh, Mie scattering, Raman, coherent anti-Stokes Raman scattering – cARS, stimulated Raman scattering – SRS, surface-enhanced Raman scattering – SERS, tip enhanced Raman scattering – TERS).

Light Microscopy
- Fluorescence microscopy (observing self-fluorescence or with fluorescence labels, for example, proteins), including sub-diffraction techniques (optical nanoscopy) and 3D imaging techniques (e.g., confocal and multi-photon excitation microscopy),
- Raman microscopy (based on scattering phenomena such as CARS, SRS, TERS),
- Other contrast methods: phase contrast microscopy, digital holographic microscopy.

Multimodal Approaches
- Molecular imaging in combination with nanotechnology (nanobiophotonics),
- Lab on a chip for point-of-care diagnosis.

These tools or techniques are able to process large numbers of samples using methods adapted from other fields, such as high-throughput, array-based techniques for cell-based assays and rapid imaging and metrology tools for in vivo studies. At this moment, the area of the body affected by the disease must be accessible to photons, (no deeper than a few centimeters beneath accessible body surfaces). Alternatively metabolic products in body fluids such as saliva, tears and urine can be investigated by photonic methods and can provide valuable information. For instance Cell-CT platform generate the image cells with Optical Projection Tomography. It is an automated 3D imaging technology from VisionGate (Seattle, WA) that can analyze any body fluid with cellular content in suspension by computing the three-dimensional internal structure of cells-based on stain absorption densities, marker color, and simultaneous optional fluorescence or any microscopy contrast mode (Nelson, 2003). The '*lab on a chip*' ultra-sensitive biosensors technology, that results by combining microfluidics with photonics, can measure little amounts of substances in small sample volumes in low concentration, and make it possible to assess patients rapidly at the point-of-care (Luka et al., 2015).

The photonic technology is already relative widely applied in *laboratory testing* for diagnosis (using for instance innovative miniaturized laboratory tools-based on plasmonic phenomena). The main drawback is that the in vitro analysis of samples e☐tractedfrom the body finds cancer typically at a late stage in its development, when the cancer has already metastasized.

The photonic technology use for *in-vivo* diagnosis and treatment is quite limited to few techniques such as photodynamic therapy (PDT) and to fluorescence endoscopy to detect tumor lesions during surgery (Muthusamy et al., 2009). A novel promising technique-based on photo-acoustic phenomena (sound-and-light-based imaging) is "one of the most quickly growing bioimaging modalities that fundamentally impact clinical applications," said Professor Dr. Vasilis Ntziachristos from Technical University Munich and Helmholtz Center Munich, founder of Photoacoustic Journal. The optical fiber endoscope is also a helpful technique to perform in vivo cancer diagnosis (Katzir, 1993). An optical diagnostic procedure-based on laser-induced fluorescence (LIF) has been developed and reported by Vo-Dinh team at Duke University (Vo-Dinh et al., 2003) for direct *in vivo* cancer diagnosis without requiring biopsy. Endogenous fluorescence of normal and malignant tissues was measured directly using a fiber optic probe inserted through an endoscope. The differential normalized fluorescence (DNF) technique is used to enhance small but consistent spectral differences between the normal and malignant tissues.

The market for optical healthcare equipment include both laboratory and surgical microscopes, fixed and flexible endoscopes (with cameras and video modules) and medical imaging-based systems (computed radiography, optical coherence tomography, fluorescence diagnostic and therapeutic laser systems). New diagnostic and treatment markets will emerge from innovative optical technologies, especially by the combination of existing photonic and non-photonic imaging methods, such as positron emission tomography (PET) (RadiologyInfo.org). Some efficient approaches,-based on appropriate combination of existing methods and technologies, such as digital microscopy with Raman spectroscopy, to gain both morphological and molecular information in concert with automated image and pattern recognition, are nowadays used. 3D reconstruction and stereology with numerical and computational image analysis are important model-driven methods for cancer diagnosis and monitoring, using imaging devices. These methods are enabling the efficient acquisition of only the necessary image information.

An understanding of cell processes on the molecular level is also needed for the concept of *personalized medicine*. The available genetic information, such as that obtained by innovative optical imaging systems, is used for diagnostic and treatment adapted to the needs of individual patients (Conde et al., 2012). Biochips and optical methods for gene sequencing offer new routes to better diagnosis, thereby opening ways to optimize the treatment of cancer. Many photonic and imaging methods can identify cancer cells only when the eye of a trained clinician analyzes the visual images. The cancer detection is often not specific or sensitive enough to lead to a conclusive result: it can be difficult to select a biopsy sample by the usual *visual inspection*, especially in the early stages of cancer or if the affected area is small. The standard biopsy methods cannot always scan the whole of the extracted tissue sample, which could lead to an erroneous finding, especially for inhomogeneous types of cancer (such as astrocytomas). The photonic tools and methods can improve specificity and sensitivity for detecting cancer earlier and more gently and to allow their use for screening, staging, grading and delineation of tumors for accurate removing of the tumors. At Harvard University the coherent anti-Stokes Raman scattering (CARS) and stimulated Raman scattering (SRS) are used to visualize the margins of cancerous brain tumors.

The nowadays trend is to move *"subjective" cancer analysis into the "objective" territory*. For instance the Block's commercially available Laser Scan analyzer provides highly sensitive and selective spectral signatures for biological tissues (such as prostate cancer), with much greater power than a conventional Fourier-transform IR (FTIR). The Laser Scan use the mid-infrared (mid-IR) laser-based absorption spectroscopy and a tunable quantum-cascade laser (QCL) source to rapidly scan over a 5 μm spectral region anywhere within its 6 to 12 μm wavelength operating range – an area of the spectrum known as the fingerprint region. But also a combination of FTIR microscope, computer algorithms and a synchrotron IR source can be successfully used to analyze spectroscopic images of tissue and to develop classification schemes for IR chemical imaging of prostate and breast cancers (University of Illinois, Urbana–Champaign). Other methods such as quantum-cascade laser-based absorption spectroscopy can illuminate tissue samples with enough spectral power density to potentially differentiate between different forms of cancer or precancerous tissue using "objective" spectroscopic trace results rather than "subjective" image inspection done by a trained medical specialist.

Another common technique is to attach *"marker"* molecules to cancer cells to make them more visible. New and more versatile markers were developed, but the process is expensive and time-consuming and is justified where marker-free methods are not suitable. Marker-free imaging techniques such as infrared and Raman microscopy, coherent anti-Stokes Raman scattering microscopy (CARS), stimulated Raman scattering (SRS) and self-fluorescence-based methods can detect cancer quickly and reliably but they must be further developed and improved, in particular at an application-specific level. To meet rising demand, a high degree of automation is implemented in order to reduce reliance on human effort at all stages, including sampling, recording and evaluation of large amount of data. This implies image guided biopsy and computer-aided diagnostics (Leondes, 2005).

To meet the requirements of biomedical researchers, the next generation of photonic devices must offer higher spatial resolution (submicrometer to a few nanometers), multifunctional or multimodal tools, combining different optical, biological and medical techniques and providing readout for multiple parameters. Another trends guide the research activities toward the miniaturization of the point-of-care applications, and the improvement of the signal-to-noise ratio.

9.5.2 NANOPHOTONICS

Today, with the advent of nanotechnology, the use of light has reached its own dimension where light-matter interactions take place at wavelength and sub-wavelength scales and where the physical/chemical nature of nanostructures controls the interactions. This is the field of nanophotonics which allows for the exploration and manipulation of light in and around nanostructures, single molecules, and molecular complexes. Nanophotonics deals with the interaction of light with matter at a nanometer scale, providing challenges for fundamental research and opportunities for new technologies. It involve the study of new optical interactions, materials, fabrication techniques, and architectures, including the exploration of natural and synthetic or artificially engineered structures such as photonic crystals, holey optical fibers, quantum dots, sub wavelength structures, and plasmonics (Conde et al., 2012). The nanophotonics for *diagnostics* includes: nanoplasmonics such as plasmons on surfaces (waveguides and optical fibers (Cristea et al., 2014; Cennamo et al., 2015; Gupta et al., 2015)) and on nanoparticle (El-Sayed et al., 2005; Giannini et al., 2011), Raman Spectroscopy-based systems, nanophotonics bioimaging in non-optical techniques like MRI, fluorescence-based systems, photo-acoustic imaging, two-photon luminescence (Durr et al., 2007), quantum dots (QD) for in-vivo imaging (Masuhara et al., 2006). Nanophotonics *for therapy* includes: photonics photodynamic therapy and photo-thermal therapy (Conde et al., 2012). The "objective" method of diagnosing cancer is about engineer nanoparticles and dyes, to light up cancer cells for easy identification. Biomedical applications of plasmonic nanoparticles (Rai, 2011) imply the use of gold *nanostars and nanoshell* markers. These markers are infiltrating cancer cells and lighting up these tissues in conjunction with multiple imaging modalities, such as optical molecular imaging. Some nanoparticles provide a photo-thermal effect not only to identify cancer cells but to kill them at their source. The research group led by Prof. Liesbet Lagae (IMEC-Leuven, Belgium) has developed a biosensor-based on asymmetric gold nanostructures that combine *nanorings and nanodiscs*. By exploiting surface plasmon resonance (SPR) for these structures, a change in spectral response upon binding with specific cancer markers in human blood is observed. The width of the resonance peak and the shift in resonance for spectral response is identified more easily than with traditional *nanosphere-based* cancer diagnosis methods. A new nanoparticle design-based on Au/Ag alloy *nanocages* was reported by researchers from Ji-Xin Cheng group at Purdue University (West Lafayette, Indiana) and Washington University (Saint Louis, Missouri). The new imaging approach uses a phenomenon called "three-photon luminescence" (3PL). The results are hundreds of times brighter than conventional fluorescent dyes and ten times brighter than other previous reported experimental results using *gold nanospheres and nanorods* (Tong et al., 2011). The researchers in the Prof. Shuming Nie group at Emory University (Atlanta, Georgia) are able to use multicolor and multiplexing capabilities of semiconductor quantum dots (QDs) staining method, conjugated to antibodies, for molecular and cellular mapping of tumor homogeneity in human prostate cancer specimens classification.

A *"hypermodal"* imaging technique,-based on a new nanoparticles design was developed by researchers group led by Jonathan Lovell, PhD (Buffalo University). The nanoparticle is composed by two components: an *"upconversion"* core that glows blue when is exposed to near-infrared light, and an outer material of porphyrin-phospholipids (PoP) that wraps around the core. In the future, patients could receive a single injection of the nanoparticles to have all six types of imaging (Rieffel et al., 2015):

- computed tomography (CT) scanning,
- positron emission tomography (PET) scanning,
- photo-acoustic imaging,
- fluorescence imaging,
- upconversion imaging,
- Cerenkov luminescence imaging.

With hexamodal imaging, the doctors will obtain a much clearer picture of patients' organs and tissues to diagnose disease and identify the boundaries of tumors using just one contrast agent.

Going a step further implies to *kill the cancer* cells by combining gold nanoparticles, the binding properties of cells and lasers. The gold nanoparticles are nontoxic and can be injected into the human body and can detect and kill cancer in a noninvasive manner. Many cancer cells have a protein, known as epidermal growth factor receptor (EFGR), all over their surface. By conjugating, or binding, the gold nanoparticles to an antibody for EGFR (anti-EGFR) the nanoparticles can attach themselves to the cancer cells in order to destroy them by using photo-thermal effect. The tumors can be destroyed after only a few minutes (Chen et al., 2010). Preliminary successes of photo-thermal therapy in cancer studies was reported by El-Sayed at UCSF School of Medicine and by Hsian-Rong Tseng group at the University of California, Los Angeles (UCLA) (Tseng, 2010, Wang et al., 2010). Surgical removal of a tissue sample is now the standard for diagnosing cancer. Such procedures, known as biopsies, are accurate but offer only a snapshot of the tumor at a single moment in time. *Monitoring a tumor* for weeks or months after the biopsy and tracking its growth and the response to treatment, would be much more valuable using monitoring implants (Trafton, 2009) or trapping rare circulating tumor cells (CTCs) from the blood of cancer patients (Ignatiadis et al., 2012). The CTC-based technique can be used to characterize tumors from patients with lung and prostate cancer. Also CTC can serve as early indicators of disease and therapeutic efficacy. It allows the possibility to monitor the changes of the cancer cells through the process of metastasis. Several methods have been developed to detect CTCs from blood samples, including photo-acoustic spectral detection (Strohm et al., 2014) or CTC-chip-based on magnetism principles and microfluidics (Li et al., 2013).

In photo-acoustic detection, laser-induced optical effects in individual cells – such as absorption, photo-acoustic or photo-thermal phenomena, fluorescence, and elastic scattering (Rayleigh) and inelastic (Raman) scattering – are detected with optical or non-optical (e.g., ultrasound) detectors.

The first generation chip-based on magnetism principles and microfluidics is called CTC-chip and was developed at the Massachusetts General Hospital Center for Engineering in Medicine and the Mass General Cancer Center. The silicon chip with microfluidic is capable of trapping rare circulating tumor cells (CTCs) from the blood of cancer patients. The second generation is called the HB-(herringbone) Chip. It provides more comprehensive and easily accessible data from captured tumor cells and can also process larger volume blood samples, increasing the ability to find rare CTCs and growth in culture than the first CTC-chip. The integrated photonics solutions for lab-on-a-chip technology offers advantages as small size, low power consumption, fle☐ibledesign and efficiency and reliability of batch fabrication. The possibility of integration with electronics (at front-end, and on signal processing levels), with micromechanical, micro-optical and microfluidic components permits to develop comple☐ multi-modal devices capabilities. Modern waveguide-based set-ups use a variety of read-outs, such as surface plasmon resonance with non-functionalized or functionalized transducer mechanism for label-free detection. The complex photonics lab-on-a-chip combined photonic components, grating coupler, resonant mirror, Young interferometers, and Mach-Zehnder chips, optical resonators and special micro-structured waveguides (El-Sayed et al., 2005), conventional and micro-structured fibers known as photonic crystal fibers (PCF) (U et al., 2012).

9.6 FINAL REMARKS AND PERSPECTIVES

Being such a dynamic field of research, material science will provide in the future amazing and innovative nanobiomaterials to be use in diagnosis and therapy of cancer. Due to their outstanding properties given by their dimensions, several nanobiomaterials were intensively investigated for diagnosis, imaging and cancer therapy. There are already approved and commercialized nanomaterials known as nanomedicines for the treatment and detection of cancer, while many other are in clinical trials. Before being accepted in clinical use more long-term toxicity tests and ways of removal from the body must be performed. But still, nanobiomaterials are excellent candidates for drug delivery having many advantages: they improve the solubility of poorly soluble drugs, they could circulate into blood stream for a longer time before being recognized by macrophages, and they could be designed to release the drug at the desired organ or tissue. Nanobiomaterials being at the same size like biomolecules (such as antibodies and nucleic acids) could be easily functionalized with biomolecules, enabling them to target certain tissues or cells. They present also interesting physical properties such as optical, necessary for imaging. Electrochemical and optical immunoassay for early diagnosis of cancer is another promising field, which in the future will be helpful in clinical diagnosis. The limiting factor because of which the electrochemical immunoassay are not yet commercial available relies on the lack of internal validation, the long-term stability of the developed assays and the length and complicated (sometimes) protocols used for immune/aptasensors incorporating nanobiomaterials. A future trend in the development of smart materials and personalized drugs will focus on several issues: smaller dimensions in order to easily excrete by renal system, target specificity allowing the accumulation of nanobiomaterials only in the tumor or other specific targets, good biocompatibility and lower long-term toxicity, the use of natural polymers, biomolecules for creating hybrid materials, versatility for using those materials for diagnosis purposes but also for monitoring the therapy efficacy and finally the healing of the cancer. Nanorobots will probably be able of imaging, drug delivery and releasing multiple drugs at the desired environment in a predictable manner.

ACKNOWLEDGEMENTS

The authors are grateful for the financial support from the Romanian National Authority for Scientific Research UEFISCDI for projects no. PN-II-ID-PCE-2011–3-0355, PN-II-PT-PCCA-2013-4-0647 and PN-II-RU-TE-2014-4-0992.

KEYWORDS

- **carbon based nanomaterials**
- **drug delivery**
- **medical imaging**
- **nanoparticles**
- **nanophotonics**
- **nanorobots**
- **sensors**

REFERENCES

Abou El-Nour, K. M. M., Eftaiha, A. A., Al-Warthan, A., & Ammar, R. A. A. (2010). "Synthesis and applications of silver nanoparticles." *Arabian Journal of Chemistry. 3*(3), 135–140.

Al-Khafaji, Q. A. M., Harris, M., Tombelli, S., Laschi, S., Turner, A. P. F., & Mascini, M., (2012). "An Electrochemical Immunoassay for HER2 Detection." *Electroanalysis, 24*(4), 735–742.

Aleksandr Noy, Dmitri, V., Vezenov, A., & Lieber, C. M., (1997). "Chemical Force Microscopy." *Annual Review of Materials Science, 27*(1), 381–421.

Altintas, Z., Kallempudi, S. S., & Gurbuz, Y., (2014). "Gold nanoparticle modified capacitive sensor platform for multiple marker detection." *Talanta, 118*, 270–276.

Atala, A., (2010). Principles of regenerative medicine. San Diego, *Academic Press.*

Baba, A. I., & Câtoi, C., (2007). Comparative oncology, the publishing house of the romanian academy: Bucharest.

Baigent, C. L., & Müller, G., (1980). "A colloidal gold prepared with ultrasonics." *Experientia. 36*(4), 472–473.

Bajpai, A. K., Shukla, S. K., Bhanu, S., & Kankane, S., (2008). "Responsive polymers in controlled drug delivery." *Progress in Polymer Science, 33*(11), 1088–1118.

Barbé, C., Bartlett, J., Kong, L., Finnie, K., Lin, H. Q., Larkin, M., et al., (2004). "Silica Particles: A Novel Drug-Delivery System." *Advanced Materials, 16* (21), 1959–1966.

Barnes, T. M., Van De Lagemaat, J., Levi, D., Rumbles, G., Coutts, T. J., Weeks, C. L. et al., (2007*b*). "Optical characterization of highly conductive single-wall carbon-nanotube transparent electrodes." *Physical Review, 75* (23), 235410.

Baumann, F., Gandhi, R., Pena, C., & Katzen, B., (2015). "Robotics for endovascular interventions: a need?" *Italian Journal of Vascular and Endovascular Surgery. 22*(2), 81–86.

Beard, J., (1911). The Enzyme Treatment of Cancer and its Scientific Basis, *New Spring Press.*

Bitar, A., Ahmad, N. M., Fessi, H., & Elaissari, A., (2012). "Silica-based nanoparticles for biomedical applications." *Drug Discov Today, 17*(19–20), 1147–1154.

Bloom, G. S., (2014). "Amyloid-beta and tau: the trigger and bullet in Alzheimer disease pathogenesis." *JAMA Neurol, 71*(4), 505–508.

Bosi, S., Da Ros, T., Spalluto, G., & Prato, M, (2003). "Fullerene derivatives: an attractive tool for biological applications." *Eur J Med Chem, 38*(11–12), 913–923.

Braicu, C., Burz, C., Berindan-Neagoe, I., Balacescu, O., Graur, F., Cristea, V. et al., (2009). "Hepatocellular Carcinoma: Tumorigenesis and Prediction Markers." *Gastroenterology Research, 2*(4), 191–199.

Brust, M., Walker, M., Bethell, D., Schiffrin, D. J., & Whyman, R., (1994). "Synthesis of thiol-derivatised gold nanoparticles in a two-phase Liquid-Liquid system." *Journal of the Chemical Society, Chemical Communications.* (7), 801–802.

Bryan, N., (2010). Chapter 8: Optical Imaging. Introduction to the Science of Medical Imaging. Cambridge.

Bustos-Ramírez, K., Martínez-Hernández, A., Martínez-Barrera, G., Icaza, M., Castaño, V., & Velasco-Santos, C., (2013). "Covalently Bonded Chitosan on Graphene O☐ide via Redox Reaction." *Materials. 6*(3), 911.

Cai, W., & Chen, X., (2007). "Nanoplatforms for targeted molecular imaging in living subjects." *Small. 3*(11), 1840–1854.

Cai, W., Hsu, A. R., Li, Z. B., & Chen, X., (2007). "Are quantum dots ready for *in vivo* imaging in human subjects?" *Nanoscale Res Lett, 2*(6), 265–281.

Cavalcanti, A., Shirinzadeh, B., Freitas, R., A., & Hogg, T., (2008). "Nanorobot architecture for medical target identification." *Nanotechnology, 19*(1), 015103.

Cennamo, N., Pesavento, M., Lunelli, L., Vanzetti, L., Pederzolli, C., Zeni, L. et al., (2015). "An easy way to realize SPR aptasensor: A multimode plastic optical fiber platform for cancer biomarkers detection." *Talanta, 140*, 88–95.

Chen, J., Glaus, C., Laforest, R., Zhang, Q., Yang, M., Gidding, M. et al., (2010). "Gold nanocages as photothermal transducers for cancer treatment." *Small, 6*(7), 811–817.

Chen, X., Zhang, Q., Qian, C., Hao, N., Xu, L., & Yao, C., (2015). "Electrochemical aptasensor for mucin 1 based on dual signal amplification of poly(o-phenylenediamine) carrier and functionalized carbon nanotubes tracing tag." *Biosens Bioelectron., 64*, 485–492.

Choi, J., Wang, N. S., & Reipa, V., (2008). "Conjugation of the photoluminescent silicon nanoparticles to streptavidin." *Bioconjug Chem. 19*(3), 680–685.

Conde, J., Rosa, J., Lima, J. C., & Baptista, P. V., (2012). "Nanophotonics for Molecular Diagnostics and Therapy Applications." *International Journal of Photoenergy.*

Córdova-Figueroa, U. M., & Brady, J. F., (2008). "Osmotic Propulsion: The Osmotic Motor." *Physical Review Letters, 100* (15), 158303.

Crisan, N., Neiculescu, C., Matei, D. V., & Coman, I., (2013). "Robotic retroperitoneal approach–a new technique for the upper urinary tract and adrenal gland." *Int J Med Robot, 9*(4), 492–496.

Cristea, C., Florea, A., Galatus, R., Bodoki, E., Sandulescu, R., Moga, D. et al. (2014). Innovative Immunosensors for Early Stage Cancer Diagnosis and Therapy Monitoring. The International Conference on Health Informatics: ICHI 2013, Vilamoura, Portugal on 7–9 November, 2013. Zhang, Y.-T. Cham, *Springer International Publishing, 47*–50.

Cristea, C., Florea, A., Tertiş, M., & Săndulescu, R., (2015). Immunosensors. Biosensors - Micro and Nanoscale Applications. *Rinken, T., InTech.*

Daniel, M. C., & Astruc, D., (2004). "Gold nanoparticles: assembly, supramolecular chemistry, quantum-size-related properties, and applications toward biology, catalysis, and nanotechnology." *Chem Rev., 104*(1), 293–346.

Dar, M. A., Ingle, A., & Rai, M., (2013). "Enhanced antimicrobial activity of silver nanoparticles synthesized by Cryphonectria sp. evaluated singly and in combination with antibiotics." *Nanomedicine, 9*(1), 105–110.

Diaconu, I., Cristea, C., Harceaga, V., Marrazza, G., Berindan-Neagoe, I. et al., (2013). "Electrochemical immunosensors in breast and ovarian cancer." *Clin Chim Acta., 425*, 128–138.

Dick, J. E., Lapidot, T., (2005). "Biology of Normal and Acute Myeloid Leukemia Stem Cells." *International Journal of Hematology, 82*(5), 389–396.

Does, A., Johnson, N. A., & Thiel, T., (2016). "Rediscovering Biology-Molecular to Global Perspectives." from https://www.learner.org/courses/biology/support/textbook_full.pdf.

Douglas, S. M., Bachelet, I., & Church, G. M. (2012). "A Logic-Gated Nanorobot for Targeted Transport of Molecular Payloads." *Science. 335*(6070), 831–834.

Du, D., Zou, Z., Shin, Y., Wang, J., Wu, H., Engelhard, M. H., et al., (2010). "Sensitive immunosensor for cancer biomarker based on dual signal amplification strategy of graphene sheets and multienzyme functionalized carbon nanospheres." *Anal Chem., 82*(7), 2989–2995.

Durr, N. J., Larson, T., Smith, D. K., Korgel, B. A., Sokolov, K., & Ben-Yakar, A., (2007). "Two-Photon Luminescence Imaging of Cancer Cells Using Molecularly Targeted Gold Nanorods." *Nano Lett., 7*(4), 941–945.

Edidin, M., (2001). "Near-field scanning optical microscopy, a siren call to biology." *Traffic, 2*(11), 797–803.

El-Sayed, I. H., Huang, X., & El-Sayed, M. A., (2005). "Surface plasmon resonance scattering and absorption of anti-egfr antibody conjugated gold nanoparticles in cancer diagnostics: Applications in Oral Cancer." *Nano Lett., 5*(5), 829–834.

Fayaz, A. M., Balaji, K., Girilal, M., Yadav, R., Kalaichelvan, P. T., & Venketesan, R., (2010). "Biogenic synthesis of silver nanoparticles and their synergistic effect with antibiotics: a study against gram-positive and gram-negative bacteria." *Nanomedicine, 6*(1), 103–109.

Ferlay, J., Soerjomataram, I., Dikshit, R., Eser, S., Mathers, C., Rebelo, M. et al., (2015). "Cancer incidence and mortality worldwide: sources, methods and major patterns in GLOBOCAN 2012." *Int J Cancer, 136*(5), E359–386.

Florea, A., Ravalli, A., Cristea, C., Sandulescu, R., & Marrazza, G., (2015). "An optimized multiplexed bioassay for sensitive Mucin1 detection in serum samples." *Electroanal, 27*(3), 638–647.

Florea, A., Taleat, Z., Cristea, C., Mazloum-Ardakani, M., & Săndulescu, R., (2013). "Label free MUC1 aptasensors based on electrodeposition of gold nanoparticles on screen printed electrodes." *Electrochemistry Communications, 33*, 127–130.

Fonseca, C., Simoes, S., & Gaspar, R., (2002). "Paclitaxel-loaded PLGA nanoparticles: preparation, physicochemical characterization and *in vitro* anti-tumoral activity." *J Control Release, 83*(2), 273–286.

Freitas, R. A., (1999). Volume I: Basic Capabilities. Nanomedicine Georgetown, TX, Landes Bioscience.

Freitas, R. A., (2013). Diamondoid nanorobotics. nanorobotics: Current Approaches and Techniques. Mavroidis, C., Ferreira, A. New York, NY, Springer New York, 93–111.

Gao, W., Dong, R., Thamphiwatana, S., Li, J., Gao, W., Zhang, L. et al., "(2015). Artificial Micromotors in the Mouse's Stomach: A Step toward *in Vivo* Use of Synthetic Motors." *ACS Nano, 9*(1), 117–123.

Giannini, V., Fernández-Domínguez, A. I., Heck, S. C., & Maier, S. A., (2011). "Plasmonic Nanoantennas: Fundamentals and Their Use in Controlling the Radiative Properties of Nanoemitters." *Chem Rev, 111*(6), 3888–3912.

Giljohann, D. A., Seferos, D. S., Daniel, W. L., Massich, M. D., Patel, P. C., & Mirkin, C. A., (2010). "Gold nanoparticles for biology and medicine." *Angew Chem Int Ed Engl, 49*(19), 3280–3294.

Gillis, P., & Koenig, S. H., (1987). "Transverse relaxation of solvent protons induced by magnetized spheres: Application to ferritin, erythrocytes, and magnetite." *Magnetic Resonance in Medicine, 5*(4), 323–345.

Gobin, A. M., Lee, M. H., Halas, N. J., James, W. D., Drezek, R. A., & West, J. L., (2007). "Near-infrared resonant nanoshells for combined optical imaging and photothermal cancer therapy." *Nano Lett., 7*(7), 1929–1934.

Gordon, R. T., Hines, J. R., & Gordon, D. (1979). "Intracellular hyperthermia. A biophysical approach to cancer treatment via intracellular temperature and biophysical alterations." *Med Hypotheses. 5*(1), 83–102.

Graur, F., (2015). Nano-drugs therapy for hepatocellular carcinoma. Nano Based Drug Delivery. Naik, J. Zagreb, Croatia, *IAPC Publishing, 165–178.*

Graur, F., Elisei, R., Szasz, A., Neagoş, H. C., Mureşan, A., Furcea, L. et al., (2011). Ethical Issues in Nanomedicine. International Conference on Advancements of Medicine and Health Care through Technology, 29th August–2nd September 2011, Cluj-Napoca, Romania. Vlad, S., Ciupa, R. V. Berlin, Heidelberg, Springer Berlin Heidelberg. 9–12.

Graur, F., Pitu, F., Neagoe, I., Katona, G., & Diudea, M., (2010). "Applications of nanotechnology in medicine." *Academic Journal of Manufacturing Engineering, 8*(4), 36–42.

Guo, S., Wang, E., (2007). "Synthesis and electrochemical applications of gold nanoparticles." *Analytica Chimica Acta., 598*(2), 181–192.

Gupta, B. D., Srivastava, S. K., & Verma, R., (2015). Fiber Optic Sensors Based on Plasmonics, World Scientific Publishing.

Hameroff, S. R., (2004). "A new theory of the origin of cancer: Quantum coherent entanglement, centrioles, mitosis, and differentiation." *Biosystems, 77*(1–3), 119–136.

Hong, Y., Velegol, D., Chaturvedi, N., & Sen, A., (2010). "Biomimetic behavior of synthetic particles: from microscopic randomness to macroscopic control." *Physical Chemistry Chemical Physics, 12*(7), 1423–1435.

Huang, X., & El-Sayed, M. A., (2010). "Gold nanoparticles: Optical properties and implementations in cancer diagnosis and photothermal therapy." *Journal of Advanced Research, 1*(1), 13–28.

Ignatiadis, M., Sotiriou, C., & Pantel, K. (2012). Minimal Residual Disease and Circulating Tumor Cells in Breast Cancer, *Springer-Verlag Berlin Heidelberg.*

Issa, B., Obaidat, I. M., Albiss, B. A., & Haik, Y., (2013). "Magnetic nanoparticles: Surface effects and properties related to biomedicine applications." *Int J Mol Sci., 14*(11), 21266–21305.

Ito, A., Shinkai, M., Honda, H., & Kobayashi, T., (2005). "Medical application of functionalized magnetic nanoparticles." *Journal of Bioscience and Bioengineering, 100*(1), 1–11.

Jain, P. K., Lee, K. S., El-Sayed, I. H., & El-Sayed, M. A., (2006). "Calculated Absorption and Scattering Properties of Gold Nanoparticles of Different Size, Shape, and Composition: Applications in Biological Imaging and Biomedicine." *The Journal of Physical Chemistry B., 110*(14), 7238–7248.

Jeong, E. H., Jung, G., Hong, C. A., & Lee, H., (2014). "Gold nanoparticle (AuNP)-based drug delivery and molecular imaging for biomedical applications." *Arch Pharm Res., 37*(1), 53–59.

Kam, N. W., O'Connell, M., Wisdom, J. A., & Dai, H., (2005). "Carbon nanotubes as multifunctional biological transporters and near-infrared agents for selective cancer cell destruction." *Proc Natl Acad Sci USA. 102*(33), 11600–11605.

Katzir, A., (1993). Chapter 6: Endoscopy. Lasers and Optical Fibers in Medicine. San Diego, Academic Press. 156–179.

Khalkhali, M., Sadighian, S., Rostamizadeh, K., Khoeini, F., Naghibi, M., Bayat, N. et al., (2015). "Synthesis and characterization of dextran coated magnetite nanoparticles for diagnostics and therapy." *Bioimpacts, 5*(3), 141–150.

Kim, D., Jeong, Y. Y., & Jon, S., (2010). "A drug-loaded aptamer-gold nanoparticle bioconjugate for combined CT imaging and therapy of prostate cancer." *ACS Nano. 4*(7), 3689–3696.

Kralj, S., Rojnik, M., Kos, J., & Makovec, D., (2013). "Targeting EGFR-overexpressed A431 cells with EGF-labeled silica-coated magnetic nanoparticles." *Journal of Nanoparticle Research., 15*(5), 1–11.

Kwan, K. H., Liu, X., To, M. K., Yeung, K. W., Ho, C. M., & Wong, K. K., (2011). "Modulation of collagen alignment by silver nanoparticles results in better mechanical properties in wound healing." *Nanomedicine, 7*(4), 497–504.

Lacerda, L., Bianco, A., Prato, M., & Kostarelos, K., (2006). "Carbon nanotubes as nanomedicines: from toxicology to pharmacology." *Adv Drug Deliv Rev, 58*(14), 1460–1470.

Lapidot, T., Sirard, C., Vormoor, J., Murdoch, B., Hoang, T., Caceres-Cortes, J. et al., (1994). "A cell initiating human acute myeloid leukaemia after transplantation into SCID mice." *Nature., 367*(6464), 645–648.

Lee, H., & Kim, Y. H., (2014). "Nanobiomaterials for pharmaceutical and medical applications." *Arch Pharm Res., 37*(1), 1–3.

Lee, K. -S., & El-Sayed, M. A., (2005b). "Dependence of the Enhanced Optical Scattering Efficiency Relative to That of Absorption for Gold Metal Nanorods on Aspect Ratio, Size, End-Cap Shape, and Medium Refractive Index." *The Journal of Physical Chemistry, 109*(43), 20331–20338.

Lehenkari, P. P., Charras, G. T., Nykanen, A., & Horton, M. A., (2000). "Adapting atomic force microscopy for cell biology." *Ultramicroscopy. 82*(1–4), 289–295.

Lehner, R., Wang, X., Marsch, S., & Hunziker, P., (2013). "Intelligent nanomaterials for medicine: Carrier platforms and targeting strategies in the context of clinical application." *Nanomedicine: Nanotechnology, Biology and Medicine, 9*(6), 742–757.

Leondes, C. T., (2005). Medical Imaging Systems Technology – Analysis and Computational Methods, World Scientific Publishing Company.

Leteurtre, E., Gouyer, V., Rousseau, K., Moreau, O., Barbat, A., Swallow, D. et al., (2004). "Differential mucin expression in colon carcinoma HT-29 clones with variable resistance to 5-fluorouracil and methotrexate." *Biol Cell, 96*(2), 145–151.

Li, J., Gao, H., Chen, Z., Wei, X., & Yang, C. F., (2010). "An electrochemical immunosensor for carcinoembryonic antigen enhanced by self-assembled nanogold coatings on magnetic particles." *Anal Chim Acta., 665*(1), 98–104.

Li, L., Jiang, W., Luo, K., Song, H., Lan, F., Wu, Y., & Gu, Z., (2013). "Super paramagnetic Iron Oxide Nanoparticles as MRI contrast agents for Non-invasive Stem Cell Labeling and Tracking." *Theranostics, 3*(8), 595–615.

Li, P., Stratton, Z. S., Dao, M., Ritz, J., & nHuang, T. J., (2013). "Probing circulating tumor cells in microfluidics." *Lab Chip, 13*(4), 602–609.

Lim, D. J., Sim, M., Oh, L., Lim, K., & Park, H., (2014). "Carbon-based drug delivery carriers for cancer therapy." *Arch Pharm Res., 37*(1), 43–52.

Lin, J., & Ju, H. (2005). "Electrochemical and chemiluminescent immunosensors for tumor markers." *Biosens Bioelectron. 20*(8), 1461–1470.

Lin, J., Chen, R., Feng, S., Pan, J., Li, Y., Chen, G. et al. (2011). "A novel blood plasma analysis technique combining membrane electrophoresis with silver nanoparticle-based SERS spectroscopy for potential applications in noninvasive cancer detection." *Nanomedicine., 7*(5), 655–663.

Link, S., & El-Sayed, M. A., (2000). "Shape and size dependence of radiative, non-radiative and photothermal properties of gold nanocrystals." *International Reviews in Physical Chemistry., 19*(3), 409–453.

Liu, N., & Ma, Z., (2014). "Au-ionic liquid functionalized reduced graphene oxide immunosensing platform for simultaneous electrochemical detection of multiple analytes." *Biosens Bioelectron, 51*, 184–190.

Luka, G., Ahmadi, A., Najjaran, H., Alocilja, E., Derosa, M., Wolthers, K. et al., (2015). "Microfluidics Integrated Biosensors: A Leading Technology towards Lab-on-a-Chip and Sensing Applications." *Sensors., 15*(12), 30011–30031.

Luppa, P. B., Sokoll, L. J., & Chan, D. W. (2001). "Immunosensors--principles and applications to clinical chemistry." *Clin Chim Acta. 314*(1–2), 1–26.

Ma, F., Ho, C., Cheng, A. K. H., & Yu, H. Z., (2013). "Immobilization of redox-labeled hairpin DNA aptamers on gold: Electrochemical quantitation of epithelial tumor marker mucin 1." *Electrochimica Acta., 110*, 139–145.

Magdelenat, H., Gerbault-Seureau, M., Laine-Bidron, C., Prieur, M., & Dutrillaux, B., (1992). "Genetic evolution of breast cancer: II. Relationship with estrogen and progesterone receptor expression." *Breast Cancer Res Treat., 22*(2), 119–127.

Magin, R. L., Wright, S. M., Niesman, M. R., Chan, H. C., & Swartz, H. M. (1986). Liposome delivery of NMR contrast agents for improved tissue imaging.

Masuhara, H., Kawata, S., & Tokunaga, F., (2006). Nano Biophotonics, Science and Technology Proceedings of the 3rd International Nanophotonics Symposium Handai, Suita Campus of Osaka University, Osaka, Japan.

Mcfarland, A. D., Haynes, C. L., Mirkin, C. A., Van Duyne, R. P., & Godwin, H. A., (2004). "Color My Nanoworld." *Journal of Chemical Education, 81*(4), 544A.

Meng, H. M., Fu, T., Zhang, X. B., & Tan, W., (2015). "Cell-SELEX-based aptamer-conjugated nanomaterials for cancer diagnosis and therapy." *National Science Review., 2*(1), 71–84.

Mikawa, M., Kato, H., Okumura, M., Narazaki, M., Kanazawa, Y., Miwa, N., et al., (2001). "Paramagnetic Water-Soluble Metallofullerenes Having the Highest Relaxivity for MRI Contrast Agents." *Bioconjug Chem., 12*(4), 510–514.

Mohamed, S. A., Radtke, A., Saraei, R., Bullerdiek, J., Sorani, H., Nimzyk, R. et al., (2012). "Locally different endothelial nitric oxide synthase protein levels in ascending aortic aneurysms of bicuspid and tricuspid aortic valve." *Cardiol Res Pract, 165957*(10), 18.

Muthusamy, V. R., & Chang, K. J., (2009). Photodynamic Therapy (PDT), The Best-Validated Technique. Endoscopic Therapy for Barrett's Esophagus. Sampliner, E. R. Totowa, NJ, Humana Press. 131–154.

Na, H. B., Song, I. C., & Hyeon, T., (2009). "Inorganic Nanoparticles for MRI Contrast Agents." *Advanced Materials, 21*(21), 2133–2148.

Nanda, A., & Saravanan, M., (2009). "Biosynthesis of silver nanoparticles from Staphylococcus aureus and its antimicrobial activity against MRSA and MRSE." *Nanomedicine. 5*(4), 452–456.

Nazir, S., Hussain, T., Ayub, A., Rashid, U., &Macrobert, A. J., (2014). "Nanomaterials in combating cancer: therapeutic applications and developments." *Nanomedicine., 10*(1), 19–34.

Nelson, A. C., (2003). Optical Projection Imaging System and Method for Automatically Detecting Cells Having Nuclear and Cytoplasmic Densitometric Features Associated with Disease. United States Patent 6,519,355. http://www.visiongate3d.com/cell-ct.

Ng, A. T., & Tam, P. C., (2014). "Current status of robot-assisted surgery." *Hong Kong Med J. 20*(3), 241–250.

Nishida, N., Yano, H., Nishida, T., Kamura, T., & Kojiro, M., (2006). "Angiogenesis in Cancer." *Vascular Health and Risk Management, 2*(3), 213–219.

Nourhani, A., Crespi, V. H., Lammert, P. E., & Borhan, A., (2015). "Self-electrophoresis of spheroidal electrocatalytic swimmers." *Physics of Fluids., 27*(9), 092002.

Novoselov, K. S., Geim, A. K., Morozov, S. V., Jiang, D., Zhang, Y., Dubonos, S. V. et al., (2004). "Electric Field Effect in Atomically Thin Carbon Films." *Science. 306*(5696), 666–669.

Paxton, W. F., Kistler, K. C., Olmeda, C. C., Sen, A., St. Angelo, S. K., Cao, Y. et al., (2004). "Catalytic Nanomotors, Autonomous Movement of Striped Nanorods." *J Am Chem Soc., 126*(41), 13424–13431.

Peng, C. W., & Li, Y., (2010). "Application of Quantum Dots-based Biotechnology in Cancer Diagnosis: Current Status and Future Perspectives." *Journal of Nanomaterials., 11.*

Pisla, D., Cocorean, D., Vaida, C., Gherman, B., Pisla, A., & Plitea, N., (2014). Application Oriented Design and Simulation of an Innovative Parallel Robot for Brachytherapy. ASME 2014 International Design Engineering Technical Conferences and Computers and Information in Engineering Conference, Buffalo, New York, USA.

Popat, K. C., Johnson, R. W., & Desai, T. A., (2003b). "Characterization of vapor deposited poly (ethylene glycol) films on silicon surfaces for surface modification of microfluidic systems." *Journal of Vacuum Science & Technology, 21*(2), 645–654.

Popp, J., (2014). Handbook of Biophotonics, John Wiley and Sons.

Prabhu, S., & Poulose, E. K., (2012). "Silver nanoparticles: mechanism of antimicrobial action, synthesis, medical applications, and toxicity effects." *International Nano Letters., 2*(1), 1–10.

Pramod, K., Avti, S. C. P., & Balaji Sitharaman., (2011). *Nanobiomaterials: Current Status and Future Prospects. Nanobiomaterials Handbook.* Group, T. A. F., 1–24.

Pumera, M., Ambrosi, A., Bonanni, A., Chang, E. L. K., & Poh, H. L., (2010). "Graphene for electrochemical sensing and biosensing." *TrAC Trends in Analytical Chemistry, 29*(9), 954–965.

Qiao, W., Wang, B., Wang, Y., Yang, L., Zhang, Y., & Shao, P., (2010). "Cancer Therapy Based on Nanomaterials and Nanocarrier Systems." *Journal of Nanomaterials.*

Rabias, I., Tsitrouli, D., Karakosta, E., Kehagias, T., Diamantopoulos, G., Fardis, M. et al., (2010). "Rapid magnetic heating treatment by highly charged maghemite nanoparticles on Wistar rats exocranial glioma tumors at microliter volume." *Biomicrofluidics, 4*(2).

Radiology info. Org. "Positron Emission Tomography–Computed Tomography," from http://www.radiologyinfo.org/en/info. cfm?pg=pet.

Rai, M., (2011). *Metal Nanoparticles in Microbiology*, Springer.

Raman, S., Machaidze, G., Lustig, A., Aebi, U., & Burkhard, P., (2006). "Structure-based design of peptides that self-assemble into regular polyhedral nanoparticles." *Nanomedicine, 2*(2), 95–102.

Rensen, P. C., De Vrueh, R. L., Kuiper, J., Bijsterbosch, M. K., Biessen, E. A., & Van Berkel, T. J. (2001). "Recombinant lipoproteins: lipoprotein-like lipid particles for drug targeting." *Adv Drug Deliv Rev., 47*(2–3), 251–276.

Rieffel, J., Chen, F., Kim, J., Chen, G., Shao, W., Shao, S. et al., (2015). "Hexamodal imaging with porphyrin-phospholipid-coated upconversion nanoparticles." *Adv Mater., 27*(10), 1785–1790.

Samanta, D., Meiser, J. L., & Zare, R. N., (2015). "Polypyrrole nanoparticles for tunable, pH-sensitive and sustained drug release." *Nanoscale. 7*(21), 9497–9504.

Săndulescu, R., Tertiş, M., Cristea, C., & Bodoki, E., (2015). New Materials for the Construction of Electrochemical Biosensors. Biosensors-micro and nanoscale aplications. Rinken, T., Intech., 1–36.

Savla, R., Taratula, O., Garbuzenko, O., & Minko, T., (2011). "Tumor targeted quantum dot-mucin 1 aptamer-doxorubicin conjugate for imaging and treatment of cancer." *J Control Release, 153*(1), 16–22.

Scarberry, K. E., Dickerson, E. B., Mcdonald, J. F., & Zhang, Z. J., (2008). "Magnetic Nanoparticle–Peptide Conjugates for *in Vitro* and *in Vivo* Targeting and Extraction of Cancer Cells." *J Am Chem Soc., 130*(31), 10258–10262.

Shaduri, M., Bouchoucha, M. *Life-Cycling of Cancer: New Concept. Cancer Treatment - Conventional and Innovative Approaches*. Rangel, L., InTech. 2013.

Shahverdi, A. R., Fakhimi, A., Shahverdi, H. R., & Minaian, S., (2007). "Synthesis and effect of silver nanoparticles on the antibacterial activity of different antibiotics against Staphylococcus aureus and Escherichia coli." *Nanomedicine, 3*(2), 168–171.

Sharma, R., & Chen, C. J., (2008). "Newer nanoparticles in hyperthermia treatment and thermometry." *Journal of Nanoparticle Research, 11*(3), 671–689.

Siegel, R. L., Miller, K. D., & Jemal, A., (2015). "Cancer statistics." *CA Cancer J Clin. 65*(1), 5–29.

Slowing, Ii, Vivero-Escoto, J. L., Wu, C. W., & Lin, V. S., (2008). "Mesoporous silica nanoparticles as controlled release drug delivery and gene transfection carriers." *Adv Drug Deliv Rev., 60*(11), 1278–1288.

Strohm, E. M., Berndl, E. S. L., & Kolios, M. C., (2014). Circulating Tumor Cell Detection Using Photoacoustic Spectral Methods. Photons Plus Ultrasound: Imaging and Sensin, Proc. of SPIE.

Suzuki, M., Honda, H., Kobayashi, T., Wakabayashi, T., Yoshida, J., & Takahashi, M., (1996). "Development of a target-directed magnetic resonance contrast agent using monoclonal antibody-conjugated magnetic particles." *Noshuyo Byori., 13*(2), 127–132.

Taleat, Z., Ravalli, A., Mazloum-Ardakani, M., & Marrazza, G., (2013). "CA 125 Immunosensor Based on Poly-Anthranilic Acid Modified Screen-Printed Electrodes." *Electroanalysis, 25*(1), 269–277.

Tallury, P., Payton, K., & Santra, S., (2008). "Silica-based multimodal/multifunctional nanoparticles for bioimaging and biosensing applications." *Nanomedicine, 3*(4), 579–592.

Tanaka, T., Shiramoto, S., Miyashita, M., Fujishima, Y., & Kaneo, Y., (2004). "Tumor targeting based on the effect of enhanced permeability and retention (EPR) and the mechanism of receptor-mediated endocytosis (RME)." *Int J Pharm, 277*(1–2), 39–61.

Tang, L., & Cheng, J., (2013). "Nonporous silica nanoparticles for nanomedicine application." *Nano Today, 8*(3), 290–312.

Thompson, M. P., & Kurzrock, R., (2004). "Epstein-Barr virus and cancer." *Clin Cancer Res., 10*(3), 803–821.

Tong, L., & Cheng, J. X. (2011). "Label-free imaging through nonlinear optical signals." *Materials Today, 14*(6), 264–273.

Toth, E., Bolskar, R. D., Borel, A., Gonzalez, G., Helm, L., Merbach, A. E. et al., (2005). "Water-soluble gadofullerenes: toward high-relaxivity, pH-responsive MRI contrast agents." *J Am Chem Soc. 127*(2), 799–805.

Trafton, A., (2009). "Implantable device offers continuous cancer monitoring." *MIT Tech Talk. 53*(25).

Tse, C., Zohdy, M. J., Ye, J. Y., O'donnell, M., Lesniak, W., & Balogh, L., (2011). "Enhanced optical breakdown in KB cells labeled with folate-targeted silver-dendrimer composite nanodevices." *Nanomedicine, 7*(1), 97–106.

Tseng, H. R. (2010). "Self-Assembling Gold Nanoparticles Use Light to Kill Tumor Cells." from http://www.cnsi.ucla.edu/news/item?itemid=1588030.

U. S. D., Fu, C. Y., Soh, K. S., Ramaswamy, B., Kumar, A., & Olivo, M., (2012). "Highly sensitive SERS detection of cancer proteins in low sample volume using hollow core photonic crystal fiber." *Biosens Bioelectron, 33*(1), 293–298.

Viarbitskaya, S., Ryderfors, L., Mikaelsson, T., Mukhtar, E., & Johansson, L. B., (2011). "Luminescence enhancement from silica-coated gold nanoparticle agglomerates following multi-photon excitation." *J Fluoresc, 21*(1), 257–264.

Vivero-Escoto, J. L., Slowing, I. I., Trewyn, B. G., & Lin, V. S. Y., (2010). "Mesoporous Silica Nanoparticles for Intracellular Controlled Drug Delivery." *Small. 6*(18), 1952–1967.

Vo-Dinh, T., Panjehpour, M., Overholt, B. F., Julius, C. E., Overholt, S., & Phan, M. N., (2003). Laser-induced fluorescence for the detection of esophageal and skin cancer. Proceedings of SPIE-The International Society for Optical Engineering.

Wan, Y., Deng, W., Su, Y., Zhu, X., Peng, C., Hu, H. et al., (2011). "Carbon nanotube-based ultrasensitive multiplexing electrochemical immunosensor for cancer biomarkers." *Biosens Bioelectron, 30*(1), 93–99.

Wang, Q., Wang, J., Lv, G., Wang, F., Zhou, X., Hu, J. et al., (2014). "Facile synthesis of hydrophilic polypyrrole nanoparticles for photothermal cancer therapy." *Journal of Materials Science, 49*(9), 3484–3490.

Wang, R., Billone, P. S., & Mullett, W. M., (2013). "Nanomedicine in Action: An Overview of Cancer Nanomedicine on the Market and in Clinical Trials." *Journal of Nanomaterials. 12.*

Wang, S., Chen, K. J., Wu, T. H., Wang, H., Lin, W. Y., Ohashi, M. et al., (2010). "Photothermal Effects of Supramolecularly Assembled Gold Nanoparticles for the Targeted Treatment of Cancer Cells." *Angewandte Chemie International Edition, 49* (22), 3777–3781.

Winoto-Morbach, S., Tchikov, V., & Muller-Ruchholtz, W., (1994). "Magnetophoresis: I. Detection of magnetically labeled cells." *J Clin Lab Anal, 8*(6), 400–406.

Wu, Y., Sefah, K., Liu, H., Wang, R., & Tan, W., (2010). "DNA aptamer–micelle as an efficient detection/delivery vehicle toward cancer cells." *Proceedings of the National Academy of Sciences. 107*(1), 5–10.

Yu, X., Munge, B., Patel, V., Jensen, G., Bhirde, A., Gong, J. D. et al. (2006). "Carbon nanotube amplification strategies for highly sensitive immunodetection of cancer biomarkers." *J Am Chem Soc., 128*(34), 11199–11205.

Yu, Y., Clippinger, A. J., & Alwine, J. C., (2011). "Viral effects on metabolism: Changes in glucose and glutamine utilization during human cytomegalovirus infection." *Trends Microbiol., 19*(7), 360–367.

Yuh, B., Yu, X., Raytis, J., Lew, M., Fong, Y., & Lau, C., (2016). "Use of a mobile tower-based robot The initial Xi robot experience in surgical oncology." *Journal of Surgical Oncology, 113*(1), 5–7.

Zhang, X., Meng, L., Lu, Q., Fei, Z., & Dyson, P. J., (2009). "Targeted delivery and controlled release of doxorubicin to cancer cells using modified single wall carbon nanotubes." *Biomaterials. 30*(30), 6041–6047.

Zhou, F., Lu, M., Wang, W., Bian, Z. P., Zhang, J. R., & Zhu, J. J., (2010). "Electrochemical immunosensor for simultaneous detection of dual cardiac markers based on a poly(dimethylsiloxane)-gold nanoparticles composite microfluidic chip: A proof of principle." *Clin Chem., 56*(11), 1701–1707.

Zhou, W., Ma, Y., Yang, H., Ding, Y., & Luo, X., (2011). "A label-free biosensor based on silver nanoparticles array for clinical detection of serum p53 in head and neck squamous cell carcinoma." *Int J Nanomedicine, 6, 381*–386.

Znaor, A., Van Den Hurk, C., Primic-Zakelj, M., Agius, D., Coza, D., Demetriou, A., et al., (2013). "Cancer incidence and mortality patterns in South Eastern Europe in the last decade: gaps persist compared with the rest of Europe." *Eur J Cancer., 49*(7), 1683–1691.

CHAPTER 10

NANOBIOMATERIALS FOR CANCER THERAPY

KEVIN J. QUIGLEY[1] and PAUL DALHAIMER[1,2]

[1]*Department of Chemical and Biomolecular Engineering, University of Tennessee, Knoxville, TN, 37996, USA, E-mail: pdalhaim@utk.edu*

[2]*Department of Biochemistry, Cellular and Molecular Biology, University of Tennessee, Knoxville, TN 37996, USA*

CONTENTS

ABSTRACT

Biomaterials that have at least one dimension on the nanometer scale (<1 micron) are beginning to be used to encapsulate and deliver anti-cancer drugs to target areas in the body with a special emphasis on solid mass tumors. Current FDA-approved nanobiomaterials for cancer therapy that have well-defined structures include Abraxane, Daunoxome, Doxil, Genexol-PM, MM-398, and Myocet. Several additional yet-similar formulations are currently in use in Europe, China, South Korea, and India. These nanobiomaterials or nanoparticles are celebrated for their increased circulation times, relative improved efficacy and reduced side effects over freely administered anti-cancer drugs. However, major complications remain. In this chapter we review the progress that has been made in nanobiomaterial design and the additional challenges that remain in eradicating cancers using nanotechnologies that circulate in the vasculature before reaching their target.

10.1   INTRODUCTION

The size, shape, surface chemistry, and ability to adapt to a dynamic environment collectively determine the targeting effectiveness and efficacy of nanoparticles in combating tumors (Bae and Park, 2011; Bertrand et al., 2014; Doane and Burda, 2013). These design parameters are constantly being tuned to optimize plasma protein deposition, avoid the mononuclear phagocyte system (MPS), increase localization to the tumor, decrease localization to off-target areas, and penetrate and ultimately cause apoptosis in tumor cells (Blanco et al., 2015; Peer et al., 2007). Nanotechnology allows for myriad formulations of nanoparticles to achieve these goals (Scheinberg et al., 2010). However, a complete cure of even one class of tumor still evades the field. In addition, nanoparticles are expensive and their improvements over naked drugs in the clinic are currently modest (Gill et al., 1996; Northfelt et al., 1998a,b; O'Brien et al., 2004). In this chapter, we focus on the challenges of shrinking solid-mass tumors with an emphasis on the design parameters that aim at optimizing the ability of nanoparticles to navigate the above spatial and temporal sequence of events.

Figure 10.1 shows the sketches of current clinically approved nanoparticles for anti-cancer drug delivery. (A) Sketch of a single human albumin protein with eight bound paclitaxel molecules. This should be the correct stoichiometry-based on the 1:9 mass ratio of paclitaxel:albumin in the Abraxane formulation and the molecular weights of the two molecules. The albumin proteins with bound paclitaxel self-assemble into a spherical nanoparticle having a diameter of 130 nm. (B) Sketch of a Daunoxome

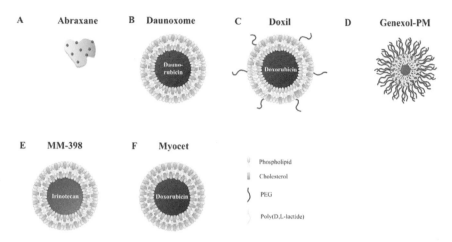

A Abraxane   B Daunoxome   C Doxil   D Genexol-PM

E MM-398   F Myocet

Phospholipid
Cholesterol
PEG
Poly(D,L-lactide)

FIGURE 10.1   Sketches of current clinically approved nanoparticles for anti-cancer drug delivery.

nanoparticle. Daunoxomoe is a phospholipid bilayer of distearoylphosphatidylcholine and cholesterol in a 2:1 molar ratio, which surrounds an interior of daunorubicin. The mean diameter of Daunoxome is ~45 nm. (C) Sketch of a Doxil nanoparticle, which has a phospholipid bilayer of phosphatidylcholine, phosphoethanolamine-PEG2000, and cholesterol in a roughly 8:1:5 molar ratio that surrounds an aqueous phase containing doxorubicin. A subset of the heads of the phospholipids is PEGylated. The diameter of Doxil is ~100 nm. (D) Sketch of a Genexol-PM nanoparticle, which is a spherical micelle comprised of poly (ethylene glycol)-poly (D, L-lactide) copolymers having a paclitaxel hydrophobic core. Genexol-PM has a diameter of ~25 nm. (E) Sketch of a MM-398 nanoparticle which is a phospholipid bilayer vesicle comprised of 1, 2-distearoyl-SN-glycero-3-phosphocholine, cholesterol, and methoxy-terminated PEG in a molar ratio of 227:151:1. MM-398 carries irinotecan and has a diameter of ~110 nm. (F) Sketch of a Myocet nanoparticle. Myocet is composed of egg phosphatidylcholine and cholesterol in a 1:1 ratio. The diameter of Myocet is ~180 nm.

The FDA has approved several nanobiomaterials that can be classified as nanoparticles: Abraxane, Daunoxome, Doxil, Genexol-PM, MM-398, and Myocet (Figure 10.1 A-F). Their targeted cancers, survival benefits, and adverse effects are reviewed in Stylianopoulos and Jain. Of these formulations, Doxil is the most structurally complex as it has a hydrophilic core containing free doxorubicin and a liposome structural scaffold of cholesterol and phospholipids where a subset of the lipids are chemically conjugated to poly (ethylene-glycol): "PEGylated." Daunoxome, MM-398, and Myocet also have hydrophilic cores containing anti-cancer drugs inside of a liposome-cholesterol scaffold, but lack a PEG exterior. Abraxane is a self-assembly of albumin and paclitaxel and Genexol-PM is a spherical micelle of biodegradable PEG-poly (D, L-lactide) with paclitaxel in the hydrophobic center. It is important to note that second generation liposomes, called polymersomes (Discher et al., 1999), have improved upon the stability, circulation time, and targeting of liposomes. Furthermore, the use of liposomes to deliver drugs to tumors is a decades old paradigm. The field of drug delivery has since employed increasingly stable and sophisticated materials such as colloidal nanoparticles made from cross linked polymers, dendrimers, and spherical metal particles to deliver active agents to tumors (Scheinberg et al., 2010). Each of these materials has useful attributes that allow for improvements over liposomes in at least part of the pathway from administration to tumor apoptosis. It is also crucial to understand that the anti-cancer drugs used in the above FDA-approved carriers are also mostly older motifs. Thus, in many ways these nanoparticles are combinations of older yet well-understood technologies. Therefore, it should not be surprising that in many cases statistically significant improvements over naked drugs have not been seen. For example, no increase in overall survival was found for patients having HIV-related Kaposi's sarcoma that were treated with encapsulated drug versus naked doxorubicin, bleomycin, and vincristine (Northfelt et al., 1998A, B). Yet, Doxil showed improved outcomes over topotecan treatments in patients having metastatic ovarian cancer (Gordon et al., 2001). Hopefully, more cases such as the latter will increase as nanoparticle designs improve. In any case, the knowledge gained from safe nanoparticle administration in the clinic cannot be underestimated for the improved design of these technologies.

10.2   SPATIAL AND TEMPORAL OBSTACLES FOR NANOPARTICLES TO OVERCOME TO TREAT TUMORS

10.2.1   PLASMA PROTEIN DEPOSITION

Upon injection, plasma proteins bind nanoparticles forming a corona based on nanoparticle size and surface chemistry (Figure 10.2) (Cedervall et al.,

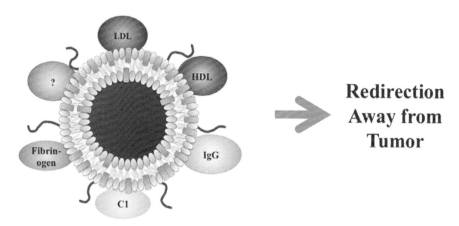

FIGURE 10.2 Plasma proteins bind nanoparticles upon injection. Sketch of a generic nanoparticle with bound plasma proteins. The formation of such a corona could cause nanoparticles to be redirected away from tumor cells.

2007; Lundqvist et al., 2008; Tenzer et al., 2011). The corona consists of immunoglobulin, lipoprotein, complement, acute-phase, coagulation and additional factors. Even slight differences of 10 nm in nanoparticle diameter can have a significant impact on the identity of the bound proteins in these dynamic coronas (Lundqvist et al., 2011; Tenzer et al., 2011). Given that size reproducibility of self-assembling systems such as liposomes is challenging, such variations in nanoparticle diameter could alter the corona and thus the biodistribution of certain nanoparticles. In seminal studies with polystyrene spheres of diameters of 50–100 nm, it was shown that the surface charge (either neutral, positive (NH_2), or negative (COOH)) played an important role in the identity of the protein corona (Lundqvist et al., 2008; Lundqvist et al., 2011; Tenzer et al., 2011). For 100 nm spheres, neutral surface charges tended to attract immunoglobulin and complement factors over positively and negatively charged surfaces (Lundqvist et al., 2008). The same trend was seen for 50 nm beads for the immunoglobulin factors, but positively charged 50 nm spheres had striking increases in lipoprotein binding over the neutral and negatively charged spheres (Lundqvist et al., 2008). Not surprisingly, HDLs, LDLs, and VLDLs bind polymer-coated surfaces stronger than individual lipoproteins (Cornelius et al., 2015).

With the exception of Abraxane, the surface chemistry of FDA-approved nanoparticles consists of Zwitterionic head groups. For example, Doxil liposomes are composed of phosphatidylcholine, phosphoethanolamine-PEG2000, and cholesterol (Barenholz, 2012). The exposed chemical groups should in part determine the identity of the bound proteins and thus the initial reaction of the body to the nanoparticles (Klapper et al., 2015). Many studies have observed that positively charged spherical nanoparticles are rapidly cleared from the bloodstream versus neutral or slightly negatively charged nanoparticles. This is hypothesized to be dependent on the content of the protein corona. The specific binding of proteins can be mitigated through ensuring a delivery vehicle has a near neutral charge. The binding of many serum proteins, such as IgG, can promote aggregation of nanoparticles, which can not only harm drug efficacy but also cause negative clinical side effects. However, increased binding of serum proteins does not necessarily correlate to reduced circulation times, as nonspecific binding of proteins such as albumin has shown to prolong nanoparticle lifetime in blood (Zhu et al., 2009). Interestingly, certain surface chemical modifications such as NH_2 conjugation, can improve the effectiveness of nanoparticles by reducing the levels of coagulation factors in the corona (Yoshida et al., 2015). The mechanisms by which the protein corona affects nanoparticle biodistribution are currently unclear since it is unknown if the bound proteins retain their apo functions. If so, they could re-direct nanoparticles to off-target sites such as macrophages and non-tumor endothelial cells reducing their effectiveness. To date, a comprehensive study identifying the proteins in the corona of a PEGylated liposome such as Doxil seems to be lacking.

Protein deposition can also destabilize self-assembled nanoparticles such as Abraxane, Daunoxome, Doxil, Genexol-PM, MM-398, and Myocet (Wolfram et al., 2014). This can cause the breakdown of the nanoparticle and the premature release of its contents. Liposomes are especially vulnerable to breakdown in circulation because they have relatively low self-association energies. Plasma proteins that have hydrophobic binding domains or bilayer penetrating motifs may be able to destabilize lipid assemblies. Thus, the need for the accurate determination of plasma proteins bound to nanoparticles by mass spectrometry is paramount for improving circulation times.

10.3 AVOIDANCE OF THE MONONUCLEAR PHAGOCYTE SYSTEM (MPS)

The nanoparticles approved for treating cancer in the clinic are spherical and have diameters from ~50–150 nm. This size range is the most widely studied for spherical nanoparticles and is ideal for at least partially evading the MPS (Gustafson et al., 2006; Longmire et al., 2008). As stated, protein deposition plays a significant role in nanoparticle clearance. Coating nanoparticle surfaces with PEG has been shown to increase circulation times and slow clearance by the immune system (Klibanov et al., 1990; Owens and Peppas, 2006, Vllasaliu et al., 2014). This is part of the design philosophy of Doxil. Referred to as "stealth," PEGylated nanoparticles have reduced opsonization and phagocytosis by macrophages and phagocytes, improved protein corona characteristics, and enhanced targeting (Dai et al., 2014). Despite these advantages, small PEGylated particles undergo accelerated blood clearance upon repeated injections – which will likely be needed in the clinic to completely eradicate a tumor – most likely due to the generation of PEG antibodies (Ishida et al., 2007; Klibanov et al., 1990; Schellekens et al., 2013). However, the effects of antibody generation on assembled PEGylated materials such as Doxil are currently unclear (Mima et al., 2015).

It is appreciated that macrophages recognize the size and shape of their targets, with particles having the longest dimension in the range of 2–3 microns exhibiting the highest attachment (Champion and Mitragotri, 2009; Doshi and Mitragotri, 2010; Sharma et al., 2010). When the small axis comes into first contact with a macrophage, elongated particles often enter the cell faster than a spherical particle with the same diameter. Yet shape alone does not account for this circulation benefit, as solid high-aspect ratio nanoparticles are cleared at the same rate as spheres from circulation versus fluid high-aspect ratio nanoparticles (Geng et al., 2007). Also, there appears to be a ma☐imum length of ~8 μm for fluid assemblies such as filomicelles, which have diameters on the order of the discussed FDA-approved nanoparticles (Geng et al., 2007). These benefits have yet to be e☐ploredin the clinic: all of the FDA-approved nanoparticles are spherical. A first high-aspect-ratio nanoparticle would probably be a cylindrical yet short version (length ~200 nm x ~50 nm width) of Genexol-PM. As with nanoparticle size, it will be challenging to prepare samples for the clinic that have consistent aspect ratios – especially for micelles.

It is also becoming appreciated that there are diverse populations of macrophages (Gustafson, 2015). Macrophages are split into two categories-based on their patterns of gene expression: M1 and M2. The M1 phenotype is promoted by Th1 mediators such as LPS and IFN-γ and shows over e☐pres sion of proinflammatory cytokines (Gordon and Taylor, 2005). In contrast,

Th2 mediators drive the M2 phenotype, which activate immunosuppressive expression and PPARγ. These factors promote tissue remodeling and help resolve inflammation. M1 macrophages seem to be responsible for the vast majority of nanoparticle uptake (Gustafson et al., 2015). However, it has been shown recently that M2 polarization increases macrophage uptake of silica nanoparticles (Hoppstadter et al., 2015). Of course, macrophage responses are dictated by the protein corona, which could be unique for each nanoparticle.

10.4 TUMOR LOCALIZATION

The vast majority of nanoparticles localize to the liver and spleen in mice. This is true even if the animal has a solid mass tumor (Christian et al., 2009). Therefore, it is postulated that the elimination of nanoparticles occurs over faster time scales than their accumulation at the tumor by the enhanced permeation and retention (EPR) effect. Additional difficulties in quantifying nanoparticle effectiveness in localizing to the tumor area come from the lack of general guidelines for animal studies as discussed in detail in Dawidczyk et al. (2014).

After reaching the blood vessels surrounding the tumor, nanoparticles must pass through the vessel walls and the interstitial space to reach the tumor cells. Tumor vasculature is highly inconsistent in architecture and connectivity compared to healthy vasculature (Hashizume et al., 2000; Hobbs et al., 1998; Jain, 1995, 1998, 2005; Vakoc et al., 2009). The first step in nanoparticle transport is dictated by the pressure difference between the tumor vasculature and the tumor cells' interstitial fluid (Figure 10.3). This pressure is determined by the permeability of the vessel walls, which is highly variable between different cancer morphologies, and a dysfunctional local lymphatic system characterized by poor drainage and increased fluid flow to the vasculature (Padera et al., 2004). High interstitial pressure often limits transport via nanoparticle diffusion (Boucher et al., 1990). This can be especially difficult for larger nanoparticles to overcome because they have lower diffusion rates than smaller nanoparticles (Perrault et al., 2009; Popovic et al., 2010). It is postulated that nanoparticles that are ~4 nm most likely transfer through intercellular junctions (Hashizume et al., 2000; Hobbs et al., 1998). However, the use of nanoparticles with diameters less than 6 nm is discouraged because these circulating objects are subject to rapid clearance by the

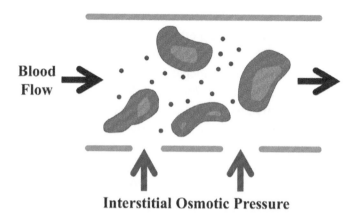

Interstitial Osmotic Pressure

FIGURE 10.3 Transport of nanoparticles from tumor vasculature to the tumor interstitial space. Transport is hampered by the osmotic pressure of the tumor. Nanoparticles are represented as spheres.

kidneys before they reach the tumor as shown with quantum dots (Choi et al., 2007). As the interstitial pressure increases, diffusion is likely to dominate nanoparticle movement from the vessels into the interstitial space. Of course, high interstitial pressure could also drive nanoparticles back into the normal blood flow and to healthy tissues (Boucher et al., 1990).

Transport from the blood vessels to the interstitial space can also be affected by nanoparticle charge as the particle diameter approaches the size of the vessel pores. While negatively charged and neutral particles have increased circulation times, positively charged particles often see greater transvascular transport due to affinity to the negatively charged vessel walls (Campbell et al., 2002; Dellian et al., 2000; Krasnici et al., 2003; Schmitt-Sody et al., 2003; Stylianopoulos et al., 2013; Thurston et al., 1998; Yim et al., 2013). However, many nanocarriers are designed not to cross the barrier from the vasculature to the interstitial space but rather to localize then release drug to diffuse freely into the tumor. Again, pressure barriers may reduce the effectiveness of this strategy.

Effective distribution in the tumor tissue is directly linked to improved patient outcomes. Once inside the interstitial tumor tissue, transport remains limited by diffusion (Jain and Stylianopoulos, 2010). Excessive extracellular fibers hinder the distribution of nanoparticles with diameters larger than 50 nm in contrast to nanoparticles with diameters ~10 nm which generally diffuse homogenously in the tissue (Jain, 1987). Larger nanoparticles are able to extravasate from blood vessels but remain at the edges of the interstitial collagen matrix causing only local effects (Alexandrakis, 2004; Pluen et al., 1999; Pluen et al., 2001; Ramanujan et al., 2002). It will be interesting to see if fle☐iblehigh-aspect ratio nanoparticles can overcome these dynamical issues (Figure 10.4) (Kim et al., 2005). While small particles diffuse rapidly, they are also cleared from the tumor tissue due to poor retention compared to larger particles. Positively and negatively charged particles both see decreased diffusivity due to charged interactions with negatively charged hyaluronic acid, negatively charged sulfated glycosaminoglycan, and positively charged collagen fibers (Netti et al., 2000). Continuing nanoparticle design will need to balance effective transport in the vasculature versus effective transport in the local tumor environment.

10.5 TUMOR APOPTOSIS

The FDA-approved nanoparticles discussed in this chapter carry irinotecan, doxorubicin, or paclitaxel. Irinotecan prevents DNA from unwinding by inhibition of topoisomerase 1, doxorubicin inhibits the progression of topoisomerase II, preventing DNA replication, paclitaxel prevents microtubule disassembly so that the chromosomes are unable to achieve a metaphase spindle configuration. The EPR of solid mass tumors does not permit nanoparticles to deliver cargo uniformly or in high enough quantities to eradicate these tumors (Jain and Stylianopoulos, 2010). This results in the exposure of subpopulations of the tumor to non-lethal amounts of the above drugs, thus promoting resistance. Nanoparticles themselves can cause

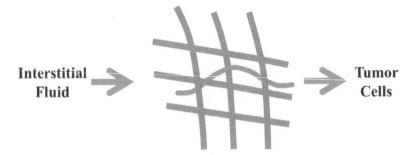

FIGURE 10.4   Different nanoparticle geometries may be advantageous for tumor localization. Sketch of a flexible, elongated nanoparticle diffusing through the interstitial collagen matrix.

apoptosis, usually through the generation of reactive oxygen species (ROS) and the triggering of autophagy, which could sidestep the issue of drug resistance (Stern et al., 2012). However, as with the drugs, current nanoparticles – at the concentrations that are administered – are not internalized by every cell in the tumor at levels that would cause apoptosis.

10.6 IMPROVING NANOPARTICLE DESIGN

Improvements can most certainly be made in nanoparticle circulation time, tumor localization, and tumor shrinkage. Of the current FDA-approved nanoparticles for cancer therapy, Doxil most likely has the longest circulation time because of its size and PEGylation. However, this time is unlikely to be longer than a few hours since only ~5% of PEG-lipid molecules are stable in liposomes after incubation in serum (Bradley et al., 1998). If the shape of the nanoparticle remains spherical, the best chance for increased circulation time is through the conjugation of peptides to the nanoparticle surface.

Foreign objects are recognized as such because they lack "markers of self" that are present on endogenous objects. CD47 has been identified as a marker of self on red blood cells (Oldenborg et al., 2000). CD47 binds the inhibitory receptor signal regulatory protein alpha (SIRPα) beginning a signaling cascade that blocks phagocytosis. Recently, a subunit of CD47 was identified that was conjugated to particle surfaces allowing them to circulate with lower immune system detection (Rodriguez et al., 2013). This is a highly promising avenue to explore in regards to increasing the circulation time of nanoparticles, even for applications beyond tumor targeting. The induction of nonspecific protein binding to serum proteins such as albumin has also been shown to reduce clearance, however many of these modifications to reduce the detection of particles see reduced specificity to target areas. For this reason many researched drugs candidates aim to include a mixed surface of both targeted and "stealth" surface moieties. Also, the conjugation of biomolecules to nanoparticles surface will most certainly modify the content of the corona. However, in the case of the CD47 peptide, it stands to reason that binding by endogenous proteins would be minimal otherwise its function on red blood cells would be abated.

Nanoparticle circulation time can also be extended by geometric modification. Of the modifications that can be made to nanoparticle shape, the augmentation of the aspect ratio seems to be the most promising for increasing the effectiveness of these technologies. High aspect ratio particles such as fle□iblenanorods and cylindrical micelles retain many advantages of small particles due to their comparable diameters while in some cases having the benefit of increased circulation and drug loading (Cai et al., 2007; Kim et al., 2005). The main reason for the extended circulation time of elongated nanoparticles appears to be due to their e□tensionin blood flow. E□tended nanoparticles are able to conform to streamlines around macrophages in in vitro e□periments(Geng et al., 2007). This behavior can be quantified by the Weissenberg number: $W_i = v_{flow}\tau_R/d_{cell}$, which describes the extension of a polymer in flow. A polymer is e□tended in the flow field when $W_i > 1$, and is coiled when $W_i < 1$. Increased circulation times – most likely due to fluid flow properties of e□tendednanoparticles – can potentially lead to lower required doses and fewer necessary injections. In head-to-head comparisons, PEGylated gold nanorods had longer circulation times in the blood and higher accumulation in orthotopic ovarian tumors in 6–8-week-old female *nu/nu* mice over spherical analogs (Arnida et al., 2011). Thus, evidence is emerging that aspect ratios could further improve targeting and tumor penetration (Chauhan et al., 2011).

Current FDA-approved nanoparticles rely on the EPR effect for tumor localization. However, this paradigm is unlikely to yield impressive results. As with the marker-of-self CD47 peptide, several promising peptides have been discovered – usually through phage display – that bind factors upregulated on tumor cells. In a seminal study, RGD peptides were discovered by phage display and used to target doxorubicin to human breast cancer xenografts in nude mice (Arap et al., 1998). Cyclic RGD (cRGD) binds αvβ3 integrin receptors, which are present on endothelial cells and over-expressed on the surface of many solid mass tumors (Desgrosellier and Cheresh, 2010). After binding, cRGD induces endocytosis in a variety of cell types, including HeLa cells (Mickler et al., 2011). PEG-poly (caprolactone) (PCL) micelles functionalized with cRGD showed a 30-fold increase in uptake over pristine controls (Nasongkla et al., 2004). cRGD-functionalized iron oxide particles also reduced cell growth of αvβ3-integrin-e□pressingtumor cells through controlled release of doxorubicin (Nasongkla et al., 2006). Conjugating such peptides to the surfaces of nanoparticles has been an active area of materials design. Promising factors include octreotide, AP peptide, tLyp-1 peptide, and A20 peptide (McGuire et al., 2006; Zhong et al., 2014). Unfortunately only a small number of targeted drugs leave preclinical investigations. Particles that combine advantageous nanoparticle geometries and targeting moieties will hopefully advance the eradication of tumor cells.

Multistage particles could be advantageous for achieving homogeneous distribution once localized in the tumor interstitial space. Several strategies can be used to control release at the tumor site. Stimuli include changes in pH, redox potential, enzyme activity, temperature, light, magnetism, ultrasound, and radiation. Secondary drug release can be accomplished through response to the tumor microenvironment. pH sensitive particles can respond to the commonly acidic microenvironment and particles that decrease in size in response to activated enzymes and undergo hydrolysis. Enzymes can degrade interstitial collagen to facilitate tumor angiogenesis and cell movement. It is important to consider that an increase in complexity in nanoparticle chemistry may necessitate larger diameter particles.

KEYWORDS

- **doxil**
- **enhanced permeation**
- **liposome**
- **micelle**
- **nanoparticle**
- **protein corona**
- **retention**

REFERENCES

Alexandrakis, G., Brown, E. B., Tong, R. T., McKee, T. D., Campbell, R. B., Boucher, Y., et al., (2004). Two-photon fluorescence correlation microscopy reveals the two-phase nature of transport in tumors. *Nat. Med, 10*, 203–207.

Arap, W., Pasqualini, R., & Ruoslahti, E., (1998). Cancer treatment by targeted drug delivery to tumor vasculature in a mouse model. *Science, 279*, 377–380.

Arnida, Janat-Amsbury, M. M., Ray, A., Peterson, C. M., & Ghandehari, H., (2011). Geometry and surface characteristics of gold nanoparticles influence their biodistribution and uptake by macrophages. *European Journal of Pharmaceutics and Biopharmaceutics, 77*, 417–423.

Bae, Y. H., & Park, K. (2011). Targeted drug delivery to tumors: Myths, reality and possibility. *J. Controlled Release, 153*, 198–205.

Barenholz, Y., (2012). Doxil®: the first FDA-approved nano-drug: lessons learned. *J. Controlled Release, 160*, 117–134.

Bertrand, N., Wu, J., Xu, X. Y., Kamaly, N., & Farokhzad, O. C., (2014). Cancer nanotechnology: The impact of passive and active targeting in the era of modern cancer biology. *Advanced Drug Delivery Reviews., 66*, 2–25.

Blanco, E., Shen, H., & Ferrari, M., (2015). Principles of nanoparticle design for overcoming biological barriers to drug delivery. *Nature Biotechnology, 33*, 941–951.

Boucher, Y., Baxter, L. T., & Jain, R. K., (1990). Interstitial pressure gradients in tissue-isolated and subcutaneous tumors: Implications for therapy. *Cancer Res., 50*, 4478–4484.

Bradley, A. J., Devine, D. V., Ansell, S. M., Janzen, J., & Brooks, D. E., (1998). Inhibition of liposome-induced complement activation by incorporated poly(ethylene glycol) lipids. *Arch. Biochem. Biophys, 357*, 185–194.

Cai, S. S., Vijayan, K., Cheng, D., Lima, E. M., & Discher, D. E., (2007). Micelles of different morphologies–advantages of worm-like filomicelles of PEO-PCL in paclitaxel delivery. *Pharmaceutical Research., 24*, 2099–2109.

Campbell, R. B., Fukumura, D., Brown, E. B., Mazzola, L. M., Izumi, Y., Jain, R. K., et al., (2002). Cationic charge determines the distribution of liposomes between the vascular and extravascular compartments of tumors. *Cancer Res., 62*, 6831–6836.

Cedervall, T., Iseult, L., Lindman, S., Berggard, T., Thulin, E., Nilsson, H., et al., (2007). Understanding the nanoparticle-protein corona using methods to quantify exchange rates and affinities of proteins for nanoparticles. *Proc. Natl. Acad. Sci., USA, 104*, 2050–2055.

Champion, J. A., & Mitragotri, S., (2009). Shape induced inhibition of phagocytosis of polymer particles. *Pharmaceutical Research., 26*, 244–249.

Chauhan, V. P., Popovic, Z., Chen, O., Cui, J., Fukumura, D., Bawendi, M. G., et al., (2011). Fluorescent nanorods and nanospheres for real-time *in vivo* probing of nanoparticle shape-dependent tumor penetration. *Angew. Chem. Int. Ed. Engl., 50*, 11417–11420.

Choi, H. S., Wenhao, L., Misra, P., Tanaka, E., Zimmer, J. P., Ipe, B. I., et al., (2007). Renal clearance of quantum dots. *Nat. Biotechnol., 25*, 1165–1170.

Christian, D., Cai, S., Garbuzenko, O. B., Harada, T., Zajac, A. L., Minko, T., et al., (2009). Flexible filaments for in vivo imaging and delivery: persistent circulation of filomicelles opens the dosage window for sustained tumor shrinkage. *Molecular Pharmaceutics., 6*, 1343–1352.

Cornelius, R. M., Macri, J., Cornelius, K. M., & Brash, J. L., (2015). Interactions of Apo lipoproteins A1, A11, B and HDL, LDL, VLDL with polyurethane and polyurethane-PEO surfaces. *Langmuir., 31*, 12087–12095.

Dai, Q., Walkey, C., & Chan, W. C. W., (2014). Polyethylene glycol backfilling mitigates the negative impact of the protein corona on nanoparticle cell targeting. *Angwandte Chemie-International Edition., 53*, 5093–5096.

Dawidczyk, C. M., Kim, C., Park, J. H., Russell, L. M., Lee, K. H., Pomper, M. G., et al., (2014). State-of-the-art design rules for drug delivery platforms: lessons learned from FDA-approved nanomedicines. *J. Controlled Release., 187*, 133–134.

Dellian, M., Yuan, F., Trubestskoy, V. S., Torchilin, V. P., & Jain, R. K., (2000). Vascular permeability in a human tumour xenograft: molecular charge dependence. *Br. J. Cancer 82*, 1513–1518.

Desgrosellier, J. S., & Cheresh, D. A., (2010). Integrins in cancer: biological implications and therapeutic opportunities. *Nat. Rev. Cancer, 10*, 9–22.

Discher, B. M., Won, Y. -Y., Ege, D. S., Lee, J. C. M., Bates, F. S., Discher, D. E., et al., (1999). Polymersomes: tough vesicles made from diblock copolymers. *Science, 284,* 143–146.

Doane, T., & Burda, C., (2013). Nanoparticle mediated non-covalent drug delivery. *Advanced Drug Delivery Reviews, 65,* 607–621.

Doshi, N., & Mitragotri, S., (2010). Macrophages recognize size and shape of their targets. *PLoS One 5,* e10051.

Geng, Y., Dalhaimer, P., Shenshen, C., Tsai, R., Tewari, M., Minko, T., & Discher, D. E., (2007). Shape effects of filaments versus spherical particles in flow and drug delivery. *Nat. Nanotechnol., 2,* 249–255.

Gill, P. S. et al., (1996). Randomized phase III trial of liposomal daunorubicin versus doxorubicin, bleomycin, and vincristine in AIDS-related Kaposi's sarcoma. *J. Clin. Oncol. 14,* 2353–2364.

Gordon, A. N., Fleagle, J. T., Guthrie, D., Parkin, D. E., Gore, M. E., & Lacave, A. J., (2001). Recurrent epithelial ovarian carcinoma: A randomized phase III study of pegylated liposomal doxorubicin versus topotecan. *J. Clin. Oncol., 19,* 3312–3322.

Gordon, S., & Taylor, P. R. (2005). Monocyte and macrophage heterogeneity. *Nat. Rev. Immunol.,* 10, 453–460.

Gustafson, H. H., Holt-Casper, D., Grainger, D. W., & Ghandehari, H., (2015). Nanoparticle uptake: The phagocyte problem. *Nano Today, 10,* 487–510.

Hashizume, H., Baluk, P., Morikawa, S., McLean, J. W., Thurston, G., Roberge, S., et al., (2000). Openings between defective endothelial cells explain tumor vessel leakiness. *Am. J. Pathol., 156,* 1363–1380.

Hobbs, S. K., Monsky, W. L., Yuan, F., Roberts, W. G., Griffith, L., Torchilin, V. P., et al., (1998). Regulation of transport pathways in tumor vessels: Role of tumor type and microenvironment. *Proc. Natl. Acad. Sci.,U. S. A. 95,* 4607–4612.

Hoppstadter, J., Seif, M., Dembek, A., Cavelius, C., Huwer, H., Kraegeloh, A., et al., (2015). M2 polarization enhances silica nanoparticle uptake by macrophages. *Frontiers in Pharmacology, 6,* 55.

Ishida, T., Wang, X., Shimizu, T., Nawata, K., & Kiwada, H., (2007). PEGylated liposomes elicit an anti-PEG IgM response in a T cell-independent manner. *J. Controlled Release, 122,* 349–355.

Jain, R. K., (1987). Transport of molecules in the tumor interstitium: A review. *Cancer Res., 47,* 3039–3051.

Jain, R. K., (1994). Barriers to drug delivery in solid tumors. *Sci., Am. 271,* 58–65.

Jain, R. K., (1998). Determinants of tumor blood flow: a review. *Cancer Res., 48,* 2641–2658.

Jain, R. K., (2005). Normalization of tumor vasculature: an emerging concept in antiangiogenic therapy. *Science., 307,* 58–62.

Jain, R. K., & Stylianopoulos, T., (2010). Delivering nanomedicine to solid tumors. *Nat. Rev. Clin. Oncol.,* 7, 653–664.

Kim, Y., Dalhaimer, P., Christian, D. A., & Discher, D. E., (2005). Polymeric worm micelles as nano-carriers for drug delivery. *Nanotechnology, 16,* S484–S491.

Klapper, Y., Maffre, P., Shang, L., Ekdahl, K. N., Nilsson, B., Hettler, S., et al., (2015). Low affinity binding of plasma proteins to lipid-coated quantum dots as observed by in situ fluorescence correlation spectroscopy. *Nanoscale, 7,* 9980–9984.

Klibanov, A. L., Maruyama, K., Torchilin, V. P., & Huang, L., (1990). Amphipathic polyethyleneglycols effectively prolong the circulation time of liposomes. *FEBS Letters, 268,* 235–237.

Krasnici, S., Werner, A., Eichhorn, M. E., Schmitt-Sody, M., Pahernik, S. A., Sauer, B., et al., (2003). Effect of the surface charge of liposomes on their uptake by angiogenic tumor vessels. *Int. J. Cancer, 105*, 561–567.

Longmire, M., Choyke, P. L., & Kobayashi, H., (2008). Clearance properties of nanosized particles and molecules as imaging agents: considerations and caveats. *Nanomedicine, 3*, 703–717.

Lundqvist, M., Stigler, J., Cedervall, T., Berggard, T., Flanagan, M. B., Lynch, I., et al., (2011). The evolution of the protein corona around nanoparticles: a test study. *ACS Nano, 5*, 7503–7509.

Lundqvist, M., Stigler, J., Elia, G., Lynch, I., Cebervall, T., & Dawson, K. A., (2008). Nanoparticle size and surface properties determine the protein corona with possible implications for biological impacts. *Proc. Natl. Acad. Sci. USA, 105*, 1465–14270.

McGuire, M. J., Samli, K. N., Chang, Y. -C., & Brown, K. C., (2006). Novel ligands for cancer diagnosis: Selection of peptide ligands for identification and isolation of B-cell lymphomas. *Experimental Hematology, 34*, 443–452.

Mickler, F. M., Vachutinsky, Y., Oba, M., Miyata, K., Nishiyama, N., Kataoka, K., et al., (2011). Effect of integrin targeting and PEG shielding on polyplex micelle internalization studied by live-cell imaging. *J. Controlled Release, 156*, 364–373.

Mima, Y., Hashimoto, Y., Shimzu, T., Kiwada, H., & Ishida, T., (2015). Anti-PEG IgM is a major contributor to the accelerated blood clearance of polyethylene glycol-conjugated protein. *Molecular Pharmaceutics, 12*, 2429–2435.

Nasongkla, N., Bey, E., Ren, J., Ai, H., Khemtong, C., Guthi, J. S., et al., (2006). Multifunctional polymeric micelles as cancer-targeted, MRI-ultrasensitive drug delivery systems. *Nano Lett., 6*, 2427–2430.

Nasongkla, N., Shuai, X., Ai, H., Weinberg, B. D., Pink, J., Boothman, D. A., et al., (2004). cRGD-functionalized polymer micelles for targeted doxorubicin delivery. *Angew. Chem. Int. Ed., 43*, 6323–6327.

Netti, P. A., Berk, D. A., Swartz, M. A., Grodzinsky, A. J., & Jain, R. K., (2000). Role of extracellular matrix assembly in interstitial transport in solid tumors. *Cancer Res., 60*, 2497–2503.

Northfelt, D. W., Dezube, B. J., Thommes, J. A., Miller, B. J., Fischl, M. A., Friedman-Kien, A., et al., (1998). Pegylated-liposomal doxorubicin versus doxorubicin, bleomycin, and vincristine in the treatment of AIDS-related sarcoma: results of a randomized phase iii clinical trial. *J. Clin. Oncol., 16*, 2445–2451.

O'Brien, M. E., et al., (2004). Reduced cardiotoxicity and comparable efficacy in a phase iii trial of pegylated liposomal doxorubicin HCl (CAELYX/Doxil) versus conventional doxorubicin for first-line treatment of metastatic breast cancer. *Ann. Oncol., 15*, 440–449.

Oldenborg, P. A., Zheleznyak, A., Fang, Y. F., Lagenaur, C. F., Gresham, H. D., & Lindberg, F. P., (2000). Role of CD47 as a marker of self on red blood cells. *Science, 288*, 2051.

Owens, D. E., & Peppas, N. A., (2006). Opsonization, biodistribution, and pharmacokinetics of polymeric nanoparticles. *Int. J. Pharm., 307*, 93–102.

Padera, T. P., Stoll, B. R., Tooredman, J. B., Capen, D., di Tomaso, E., & Jain, R. K. (2004). Pathology: cancer cells compress intratumour vessels. *Nature 427*, 695.

Peer, D., Karp, J. M, Seungpyo, H., Farokhzad, O. C., Margalit, R., & Langer, R., (2007). Nanocarriers as an emerging platform for cancer therapy. *Nature Nanotechnology, 2*, 751–760.

Perrault, S. D, Walkey, C., Jennings, T., Fischer, H. C., & Chan, W. C. W., (2009). Mediating tumor targeting efficiency of nanoparticles through design. *Nano Letters., 9*, 1909–1915.

Pluen, A., Boucher, Y., Ramanujan, S., McKee, T. D., Gohongi, T., di Tomaso, E., et al., (2001). Role of tumor-host interaction in interstitial diffusion of macromolecules: cranial vs. subcutaneous tumors. *Proc. Natl. Acad. Sci., USA, 98*, 4628–4633.

Pluen, A., Netti, P. A., Jain, R. K., & Berk, D. A., (1999). Diffusion of macromolecules in agarose gels: comparison of linear and globular configurations. *Biophys., J. 77*, 542–552.

Popovic, Z., Liu, W., Chauhan, V. P., Lee, J., Wong, C., Greytak, A. B., et al., (2010). A nanoparticle size series for in vivo fluorescence imaging. *Angew. Chem. Int. Ed. Engl., 49*, 8649–8652.

Ramanujan, S., Pluen, A., McKee, T. D., Brown, E. B., Boucher, Y., & Jain, R. K., (2002). Diffusion and convection in collagen gels: Implications for transport in the tumor interstitium. *Biophys. J. 83*, 1650–1660.

Rodriguez, P. L., Harada, T., Christian, D. A., Pantano, D. A., Tsai, R. K., & Discher, D. E., (2013). Minimal "self" peptides that inhibit phagocytic clearance and enhance delivery of nanoparticles. *Science, 339*, 971–975.

Scheinberg, D. A., Villa, C. H., Escorcia, F. E., & McDevitt, M. R. (2010). Conscripts of the infinite armada: Systemic cancer therapy using nanomaterials. *Nature Reviews Clinical Oncology, 7*, 266–276.

Schellekens, H., Hennink, W. E., & Brinks, V., (2013). The immunogenicity of polyethylene glycol: Facts and fiction. *Pharmaceutical Research, 30*, 1729–1734.

Schmitt-Sody, M., Strieth, S., Krasnici, S., Sauer, B., Schulze, B., Teifel, M., Michaelis, U., et al., (2003). Neovascular targeting therapy: paclitaxel encapsulated in cationic liposomes improves antitumoral efficacy. *Clin. Cancer Res., 9*, 2335–2341.

Sharma, G., Valenta, D. T., Altman, Y., Harvey, S., Xie, H., Mitragotri, S., & Smith, J. W., (2010). Polymer particle shape independently influences binding and internalization by macrophages. *J. Controlled Release, 147*, 408–412.

Stern, S. T., Adiseshaiah, P. P., & Crist, R. M., (2012). Autophagy and lysosomal dysfunction as emerging mechanisms of nanomaterial toxicity. *Particle and Fiber Toxicology, 9*, 20.

Stylianopoulos, T., & Jain, R. K., (2015). Design considerations for nanotherapeutics in oncology. *Nanomedicine: Nanotechnology, Biology, and Medicine, 11*, 1893–1907.

Stylianopoulos, T., Soteriou, K., Fukumura, D., & Jain, R. K., (2013). Cationic nanoparticles have superior transvascular flux into solid tumors: Insights from a mathematical model. *Ann. Biomed. Eng., 41*, 68–77.

Tenzer, S. et al., (2011). Nanoparticle size is a critical physiochemical determinant of the human blood plasma corona: a comprehensive quantitative proteomic analysis. *ACS Nano, 5*, 7155–7167.

Thurston, G., McLean, J. W., Rizen, M., Baluk, P., Haskell, A., Murphy, T. J., et al., (1998). Cationic liposomes target angiogenic endothelial cells in tumors and chronic inflammation in mice. *J. Clin. Invest., 101*, 1401–1413.

Vakoc, B. J., Lanning, R. M., Tyrrell, J. A., Padera, T. P., Bartlett, L. A., Stylianopoulos, T., et al., (2009). Three-dimensional microscopy of the tumor microenvironment in vivo using optical frequency domain imaging. *Nat. Med., 15*, 1219–1223.

Vllasaliu, D., Fowler, R., & Stolnik, S., (2014). PEGylatednanomedicines: Recent progress and remaining concerns. *Expert Opinion on Drug Delivery 11*, 139–154.

Wolfram, J., Suri, K., Yang, Y., Shen, J., Cellia, C., Fresta, M., Zhao, Y., Shen, H., & Ferrari, M., (2014). Shrinkage of pegylated and non-pegylated liposomes in serum. *Colloids and Surfaces B–Biointerfaces, 114*, 294–300.

Yim, H., Park, S. J., Bae, Y. H., & Na, K., (2013). Biodegradable cationic nanoparticles loaded with an anticancer drug for deep penetration of heterogeneous tumors. *Biomaterials, 34*, 7674–7682.

Yoshida, T., Yoshioka, Y., Morishita, Y., Aoyama, M., Tochigi, S., Hirai, T., et al., (2015). Protein corona changes mediated by surface modification of amorphous silica nanoparticles suppress acute toxicity and activation of intrinsic coagulation cascade in mice. *Nanotechnology, 26,* 245101.

Zhong, Y., Meng, F., Deng, C., & Zhong, Z., (2014). Ligand-directed active tumor-targeting polymeric nanoparticles for cancer chemotherapy. *Biomacromolecules 15,* 1955–1969.

Zhu, Y., Li, W., Li, Q., Li, Y., Li, Y., Zhang, X., & Huang, Q., (2009). Effects of serum proteins on intracellular uptake and cytotoxicity of carbon nanoparticles. *Carbon, 47,* 1351–1358.

CHAPTER 11

NANOTECHNOLOGY IN HYPERTHERMIA-BASED THERAPY AND CONTROLLED DRUG DELIVERY

ANAMARIA ORZA[1] and CHRISTOPHER CLARK[2]

[1]*Department of Radiology and Imaging Sciences/Center for Systems Imaging, Emory University School of Medicine Atlanta, Georgia, The United States of America*

[2]*The Movement Group, Inc., Atlanta, Georgia, The United States of America*

CONTENTS

ABSTRACT

In this chapter, an introduction explaining the nexus of nanotechnology and its impact in hyperthermia-based therapies is discussed first, then the mechanism for MNP-based hyperthermia and hyperthermia-based drug delivery is described. And then, an evaluation of certain recent studies on this topic is provided, and finally, the challenges of this treatment and also a future outlook of the prospects in this narrow field are provided.

11.1   INTRODUCTION

Nanotechnology and its engineered nanoparticles have made great strides in the medical field by enabling new promising diagnosis and treatments for various incurable and terminal diseases (Mocan et al., 2011; Orza et al., 2013; Orza et al., 2014; Galanzha et al., 2015).

For instance, Magnetic Nanoparticles (MNPs) have long been classified as having a high utility in cancer diagnosis and treatment (Zhang et al., 2011; Lee et al., 2013; Satpathy et al., 2014). In the realm of treatment, MNPs have been extensively used in (a) MNP-based hyperthermia (Kobayashi, 2011; Soares et al., 2012; Sadhukha et al., 2013; Quinto et al., 2015; Thomas et al., 2015); (b) hyperthermia-based drug release (Meenach et al., 2010; Pradhan et al., 2010; Meenach et al., 2013; Hayashi et al., 2014; Mohammad and Yusof, 2014; Tabatabaei et al., 2015).

The term "hyperthermia" has been confined to the use of heat as a therapy. In 1957, the first report of this kind – describing the utilization of magnetic materials and hyperthermia for cancer treatment – was performed by Gilchest et al. (1957). Along with the introduction of MNPs, the research has evolved and has confirmed that MNP-based hyperthermia allows for increased temperature in tumors, rapid apoptosis and, most profoundly, the destruction of the malignant tissue. This approach has been able to overcome a number of limitations that generally accompany conventional hyperthermia treatment (i.e., limitations such as producing side effects from burns, blisters, and heat penetration to thermal under-dosage in the target region) (Ito et al., 2006).

Moreover, MNP-based hyperthermia shows high effectiveness and selectivity, which is attributed to the capacity of the nanoparticles: (i) to be multi-functionalized with different targeted groups for specific delivery and accumulation to the tumor site (Moros et al., 2010); (ii) to generate heat at specific tumor sites (the oscillating field passes harmlessly through the body), thus allowing a reduction in the side effects (Loynachan et al., 2015); (iii) to deliver the heat using alternating magnetic fields; (iv) to provide heat generation that is much more efficient and homogenous (Ito et al., 2006); (v) to effectively cross the blood–brain barrier and hence serving as promising treatments for brain tumors; and finally, (vi) (Dan et al., 2015) to manipulate and control the delivery of specific drugs (Kaaki et al., 2012; Wang et al., 2013). This last feature offers multifunctional and multi-therapeutic approaches that potentially can address a number of other diseases.

Additionally, the utilization of MNPs in hyperthermia-based drug delivery are increasingly becoming recognized as an attractive alternative treatment for various cancers (Hayashi et al., 2010; Gautier et al., 2012; Gillich et al., 2013; Halupka-Bryl et al., 2014). These alternative strategies prove the efficacy and tolerability of the drug by activating the magnetic nanoparticles with the e☐ternal magnetic field and the activation can be targeted to the tumor site. Two designed photodynamic therapies containing liposomal formulations have shown potential use as anti-tumoral agents (Gillich et al., 2013; Halupka-Bryl et al., 2014). Encapsulation of magnetic liposomes that induce local therapeutic hyperthermia has been recently studied (Hayashi et al., 2010; Dou et al., 2014; Hardiansyah et al., 2014; Di Corato et al., 2015; Staruch et al., 2015). The results of this therapeutic method have led to tumor cell death through apoptosis and necrosis. This is a promising alternative for superficial malignant or premalignant lesions of various cancers.

11.2 THE MECHANISM OF MNP-BASED HYPERTHERMIA AND THE PROMISING IMPLICATIONS FOR THERAPY AND CONTROLLED DRUG DELIVERY

The terms, Specific Absorption Rate (SARS) or Specific Loss Power (SLP), are generally used to describe the transition of magnetic energy into heat (Bornstein et al., 1993; Garaio et al., 2015; Hola et al., 2015). SARS and SLP measure the absorption efficiency of any material to generate heat due to the Amplified Magnetic Field (AMF). MNPs have a superior ability (in comparison to micrometric particles) of being able to obtain higher temperature enhancement rates (Jordan et al., 1993; Hola et al., 2015). The fundamental elements influencing the heating ability of the MNPs are frequency (f), amplitude of the magnetic field (H), particle size and size distributions, and the geometry of the MNPs.

11.2.1   MNP-BASED HYPERTHERMIA

In this procedure, the following is generally performed: MNPs are distributed throughout the tumor site, followed by a flow of heat to the tumor using an external AMF. While there are a number of effects occurring in MNPs, the heat generation mechanism can be attributed to relaxation and hysteresis loss. The relaxation characteristic is of two distinct types: Néel and Brownian relaxation. Heat generation through Néel relaxation is attributed to the rapidly occurring changes in the direction of the magnetic moments relative to the crystal lattice (i.e., the internal dynamics). This Néel process is hindered by energy of anisotropy that tends to orient the magnetic domain in a given direction relative to the crystal lattice. Brownian relaxation is attributed to physical rotation of particles within a medium in which they are placed (external dynamics) and is hindered by the viscosity that tend to counter the movement of particles in the medium (Jeyadevan, 2010).

The thermal losses tend to be generated by the internal (Néel) and external (Brownian) sources of friction that result in a phase lag between applied magnetic field and the direction of the magnetic moments. By using linear response models with known Néel and Brownian relaxation times, one can easily predict the SPL values for MNPs (Rosensweig, 2002). One significant heating mechanism, which governs the heating capacity of a given MNP is-based on the induction of the rapid variation of the magnetic moments. In general, the SPL values increase with the frequency of the applied magnetic field. These values are also proportional to the square of the magnetic field's intensity. SPL values are measured in terms of rise in temperature per unit of time and per gram of magnetic material, all multiplied by the calorific capacity of a sample.

Additional Observations: (1) normally, SPL values depend on parameters such as the MNPs' structure (size), magnetic properties (magnetic anisotropy and temperature dependence of magnetizations) and amplitude (H) and frequency (f) of AMF (Hergt et al., 1998; Gupta and Gupta, 2005; Fortin et al., 2007; Fortin et al., 2008; Shah et al., 2015), and (2) reduction in SPL values with polydispersity of MNPs is observed and this can be due to the decrease in the proportion of particles contributing to the total generation of heat, (3) the heating power of many MNPs also change with the surrounding environment, such as the environment for those particles internalized within the cells, (4) in these cases, the nuclear endosomes and other intracellular components may hinder the movement of the particles and, as a result, lead to a total heat contribution largely attributed to the Néel relaxation (Riviere et al., 2007; Zhao et al., 2009), and (5) accordingly, for intracellular Magnetic Fluid Hyperthermia (MFH), Néel relaxation is usually the major contributor for heat release.

The most effective AMF parameters that have been reported for hyperthermia applications (measured by where the highest SLP was obtained), are a frequency of 500 kHz and with a field amplitude of 10 kA/m. (Hergt and Dutz, 2007)

11.2.2 MNPS FOR HYPERTHERMIA-BASED CONTROLLED DRUG DELIVERY

For this type of procedure, two distinct mechanisms are considered.

Under one mechanism, a drug molecule is attached to a MNP through a linker and upon the application of the AMF, the drug molecule is released due to the heating of the linker molecule attached to the MNP's surface (Derfus et al., 2007). By varying the power of the Electromagnetic Field (EMF) stimulus, it is possible to release multiple drugs in series, or in combinations. This approach can be termed as MNP for hyperthermia-based controlled drug delivery through Bond Breaking (DBB).

Under an alternate mechanism, the release of drugs takes place from within a polymeric matri□ (nano/micro particles/thin films) encapsulated with MNPs on application of AMF/EMF (Kost et al., 1987; Lu et al., 2005) A possible mechanism for this type of hyperthermia-based drug delivery is the formation of crevices or cracks of nanometer scale within a polymeric matrix due to local heat generated by the MNPs, thereby releasing encapsulated drugs. The second type of controlled drug release can be termed as MNP for hyperthermia-based controlled drug delivery through Enhanced Permeability (DEP).

Additional observations are listed as follows: (1) depending on the nature of the polymer, hyperthermia-based drug release can be due to the creation of mechanically forced openings or thermally responsive openings in the case of thermoresponsive polymers (Brazel, 2009); (2) the change in the dimension of nano-crevices is physically reversible upon short-term field exposure (Following longer-term exposure, however, these nano-crevices further enlarge to nanometer-scale cracks that become characterized along the spherical shell structure and ultimately form irreversible deformations); (3) having absorbed sufficient amounts of magnetic energy, the shell ruptures resulting in an increase of both pore volume and surface area (as for the frequency dependency of permeability, the aggregates rotate or oscillate under the action of the magnetic field with elastic deformation of capsule walls); and (4) at higher frequencies, these aggregates will not follow variation of magnetic fields and the deformation will become minimal leading to a decrease in permeability (this is similar to the case of forced oscillation of a pendulum at frequencies above the resonance frequency).

11.3 AN EVALUATION OF RECENT CASE STUDIES USING MNP-BASED HYPERTHERMIA AND HYPERTHERMIA-BASED DRUG RELEASE

By tuning the magnetic properties of the nanoparticles (and by tuning the size, shape, composite materials (such as core shell and dimer like structures) and other properties), one can manipulate the nanoparticles to respond selectively to a wide range of frequencies and oscillating magnetic fields in order to treat different complications. Here, we present (a) the types of nanoparticles used in hyperthermia; (b) the effect of size, size distribution, and the shape of nanoparticles-based therapies along with a detailed overview of the recent studies that have been reported on; (c) MNP-based hyperthermia; and (d) MNP for hyperthermia-based controlled drug delivery. Additionally, as the supporting information Table 11.1 provides a comparison and detailed overview of the recent studies that have been reported-based on size dependent hyperthermia and other properties, as described (viscosity, anistropy of the material and stabilizing ligand).

11.3.1 TYPES OF NANOPARTICLES USUALLY USED IN HYPERTHERMIA AND HYPERTHERMIA-BASED DRUG RELEASE

Recently, the manufacturing of systems using MNPs has flourished exponentially as a field. A number of nanoparticles such as, and their oxides, have been widely used solely for their hyperthermic capabilities. The MNPs are mainly used in hyperthermia and in drug delivery systems and also primarily consist of pure magnetite, paramagnetic nanoparticles (Mn, Fe, Co, Ni, Zn, Gd, Mg) or a composite core consisting of Fe_2O_4-Au, (Wijaya et al., 2007;

TABLE 11.1 Comparison and Detailed Overview of Some of the Recent Studies (2014 and 2015) on MNP-Based Hyperthermia and MNP for Hyperthermia-Based Controlled Drug Delivery

S. No.	Year	Type of Magnetic Nanoparticles	Applications
1	2015	Hyaluronic acid coated Iron Oxide Nanoparticles (IONPs)	Hyperthermia-based therapy
2		Phospholipid-polyethylene glycol (PEG) coating IONPs	Hyperthermia through Bond Breaking
3		IONPs	Hyperthermia-based therapy
4		IONPs	Hyperthermia-based therapy
5		Poly (ethylene glycol)- poly (aspartate) [PEG-p (Asp)] block copolymers-IONP	Hyperthermia Through Enhanced Permeability
6		Magnetic Lipozoms (Thermodox, Celsion, Lawrenceville, NJ)	Hyperthermia Through Enhanced Permeability
7		Magnetic Lipozoms	Hyperthermia-based therapy
8		IONPs	Hyperthermia-based therapy
9		IONPs	Hyperthermia-based therapy
10		MFe_2O_4 (M = Ni, Co, Zn, Mn)	Hyperthermia-based therapy
11		Carbon encapsulated-IONPs	Hyperthermia through Bond Breaking
12		Dextran coated-La0.7Sr0.3MnO3	Hyperthermia-based therapy
13		Lauric Acid-/Albumin-Coated IONP	Hyperthermia-based therapy
14		Hyaluronic acid-modified Fe3O4@Au core/shell nanostars	Hyperthermia-based therapy
15		Poly (vinylpyridine) polymer coated-cubic-IONPs	Hyperthermia through Bond Breaking
16		PEGylated FePt-Fe3O4 composite nanoassemblies	Hyperthermia through Bond Breaking
17		Polymer (poly-N isopropylacrylamide-co-poly glutamic acid)-IONPs	Hyperthermia Through Enhanced Permeability
18		PNIPAM-based Microgels with CoFeO Nanoparticles	Hyperthermia Through Enhanced Permeability
19		Magnetoresponsive virus-mimetic nanocapsules	Hyperthermia Through Enhanced Permeability
20	2014	Magnetic gold nanoshells	Hyperthermia through Bond Breaking
21		Poly (ethylene glycol)-block-poly (4-vinylbenzylphosphonate)-IONPs	Hyperthermia Through Enhanced Permeability

TABLE 11.1   (Continued)

S. No.	Year	Type of Magnetic Nanoparticles	Applications
22		Magnetic liposomes	Hyperthermia Through Enhanced Permeability
23		Cobalt-doped ferrite nanoparticles	Hyperthermia-based therapy
24		Cationic vesicle-IONPs	Hyperthermia Through Enhanced Permeability
25		Thermosensitive liposomes	Hyperthermia Through Enhanced Permeability
26		Poly (N-isopropylacrylamide)-IONPs	Hyperthermia through Bond Breaking
27		$Fe_3O_4/mSiO_2$	Hyperthermia Through Enhanced Permeability
28		P(EO-co-LLA) functionalized Fe3O4@mSiO$_2$	Hyperthermia Through Enhanced Permeability

Guo et al., 2013), MFe_2O_4 (M = Ni, Co, Zn, Mn) (Sabale et al., 2015) and other Fe-, Ni/Zn/Cu-, Li-, Co- and Co/Ni-ferrites (Kim et al., 2005; Fantechi et al., 2014), etc. They were stabilized by a variety of ligands, such as polyethylene glycol-folic acid (Jang et al., 2013; Sadhasivam et al., 2015), liposomes (Clares et al., 2013; Bealle et al., 2014), polyvinyl alcohol (Levy et al., 2008), dextran (Haghniaz et al., 2015), and lauric acid (Pradhan et al., 2007; Zaloga et al., 2015).

The composite nanoparticles were reported as hybrid composite (Sadhasivam et al., 2015) or core shell nanoparticles (Li et al., 2015). From those, the core shell nanoparticles offer a number of therapies over monometallic oxides for hyperthermia-based therapy. Some advantages, as stated in the literature, of using core shell nanomaterials for hyperthermia are: (1) they provide oxidative stability, (2) they enhance hyperthermia, (3) they reduce toxicity, and (4) they increase biocompatibility (Mohammad and Yusof, 2014). In the case of MNP for hyperthermia-based controlled drug delivery, using core shell magnetic nanomaterials is relatively new and recent examples from the literature highlight their uniqueness (Li et al., 2015).

However, the most used hyperthermic agents are-based on iron oxide nanoparticles. Yet, other promising composite materials (based on a combination of magnetic and paramagnetic materials) were reported as having superior properties in hyperthermia. Although extremely high heating performances were reported-based on magnetic composites (e.g., iron cobalt), iron oxide magnetic nanoparticles are still more attractive due to their lower toxicity compared with the composites. It has been well documented that hyperthermia-efficiency depends on properties ranging from the particle size, the size distribution, to the shape and viscosity of fluid at which they are dispersed.

11.3.2 EFFECTS OF SIZE, SIZE DISTRIBUTION, AND THE SHAPE OF NANOPARTICLES

The effects of size and size distributions on hysteresis losses of MNPs in magnetic hyperthermia were investigated by Levy et al. (2008), Hergt and Dutz (2007) and Derfus et al. (2007). The reports of these investigations generally found heating to be mainly caused by Néel relaxation. In addition, studies have also ventured into the effects of nanoparticle shapes (Kawai et al., 2008; Sharma and Chen, 2009; Kakwere et al., 2015) surrounding environment or viscosity (Bae et al., 2006) on hyperthermia heating in the AC magnetic field. The key issue has centered on the particle clustering mechanism in magnetic ferrofluids, as it may change the global magnetic behavior due to dipolar interactions between particles.

Recent theoretical and experimental studies demonstrate that particles are prone to chain formation in the AC magnetic field (Saville et al., 2014). Branquinho's et al. (2013) proved that the SAR of ferrite-based nanoparticles decreases with increasing volume fractions due to the formation of chains. Branquinho's group demonstrated the optimal experimental conditions including the chain size and the particle size which all significantly affected the heating ability of the MNPs.

11.3.3 MNP-BASED HYPERTHERMIA

Based on their multifunctionality, nanoparticles offer the opportunity to be used not only in therapeutic settings but also in the innovative diagnosis of different diseases (Hola, Markova et al., 2015). Therefore, the utilization of these multifunctional nanoparticles will facilitate the finding of solutions to incurable diseases and illnesses. A targeted form of hyperthermia is one that is self-regulated and self-regulation is possible with MNPs that have a Curie temperature at or around the therapeutic temperature range. Following a process similar to conventional hyperthermia, the MNPs are heated once the AC field is applied. Moreover, once the MNPs reach the targeted Curie temperature, the Ms of particles drops to zero and the heating stops. Therefore, if the Curie temperature can be fixed by a calculated selection of particle composition and size, then the hyperthermia can be applied and carefully controlled, preventing it from reaching excessive temperatures. The Curie temperature is also, conclusively, anticipated to change with size and with the composition of MNPs. Therefore, by extending the known compositional dependence of the Curie temperature for the face centered cube (fcc)-phase at high Ni concentrations, one will observe metastable FeNi alloys to have low Curie temperature's in the Ferich region of the phase diagram. Therefore, making the MNPs suitable for self-regulated radio frequency (Rf) heating in cancer hyperthermia (McNerny, Kim et al., 2010).

11.3.4 MNP FOR HYPERTHERMIA-BASED CONTROLLED DRUG DELIVERY

This magnetically stimulated platform for controlled drug release demonstrates the ability to remotely trigger the release of a biomolecule (if necessary from the sequence) from the surface of the MNPs (Fourmy et al., 2015). There are a number of reports in the literature relating to nanoparticles in hyperthermia-based drug release (Mohammad and Yusof, 2014; Li et al., 2015; Sahu et al., 2015; Thomas et al., 2015). The reports can be classified into two types—hyperthermia-based controlled drug release through (a) bond breaking, and (b) enhanced permeability.

11.3.4.1 Hyperthermia Through Bond Breaking (DBB)

The first example of this concept was reported by using radio frequencies in order to release the surface-bonded fluorescein-labeled 18bp (Derfus et al., 2007). Here, the magnetic nanoparticles were linked with the biomolecules through a heat sensitive linker and mixed with matrigel. This mixture was then injected subcutaneously into the mammary fat pad of mice. When applied to the radiofrequency field, the fluorescent biomolecules were released into the surrounding tissue. As a result, this controlled drug release through magnetic stimuli demonstrates the ability of using magnetic nanoparticles as a platform to confirm the release of the biomolecules. Other examples that further confirm this concept releases fluorophorebimane amine from the surface of the SPIONs in the presence of oscillating magnetic fields (Sahu et al., 2015), release of doxorubicine from ferucarbotran nanoparticles surface (Derfus et al., 2007), of human ferritin protein cage from MNPs surface (Fantechi et al., 2014). Click chemistry functionalization methods (drug-nanoparticles surface) were reported as being ideal and result in unprecedented hyperthermia-induced drug release. N'Guyen et al. (2013) employed orthogonal click reactions on MNPs surface and showed unprecedented hyperthermia-induced drug release through magnetically stimulated retro-Diels–Alder (rDA) process. (N'Guyen, Duong et al., 2013)

11.3.4.2 Hyperthermia Through Enhanced Permeability (DEP)

In this case, the magnetic nanoparticle and the drugs are encapsulated in polymer nanoparticles. Hydrogels (Meenach et al., 2010; Lopez-Noriega et al., 2014), polymers (Meenach et al., 2013), and thermoresponsive polymers (Kakwere et al., 2015) were reported as being ideal for such therapies. Using temperature-sensitive poly (N-isopropylacrylamide) hydrogels incorporated with SPIONs, a variety of drugs were released (Chiang et al., 2013; Davaran et al., 2014; Patra et al., 2015) More advanced hydrogels, such as microgels of poly (N-isopropylacrylamide), containing magnetic nanoparticles were demonstrated to have the ability to tune the magnetic and thermoresponsive properties of both the nanoparticles and the microgel (Wong et al., 2008; Backes et al., 2015) Yet other report (McGill et al., 2009) confirms a similar concept by triggering the release of Fluorophorebimane amine from the surface of super paramagnetic iron oxide nanoparticles (SPIONs) in the presence of oscillating magnetic fields. Example demonstrations showing this remote-controlled pulsatile drug release for a number of different drugs and of different "on–off" durations for oscillating magnetic fields was demonstrated from temperature sensitive poly (N-isopropylacrylamide) hydrogels incorporated with SPIONs (Satarkar and Hilt, 2008). Other hydrogel-based DEP systems are poly (N-isopropylacrylamide) microgel-containing MNPs with the ability to tune magnetic and thermoresponsive properties of individual components (NPs and microgels) (Wong et al., 2007; Purushotham and Ramanujan, 2010).

Examples of organic polymers that were used for this purpose are: polylactic acid (Huang et al., 2012), poly (ethylene glycol) ethyl ether (Ghosh et al., 2010; Huang et al., 2012), methacrylate-copoly (ethylene glycol) methyl ether methacrylate (Liu et al., 2008), and Pluronic (Urbina et al., 2008). Additionally, silica-based polymers were also reported as being efficient (Guo et al., 2014; Tao and Zhu, 2014). Moreover, natural polymers (based on bacteria and viruses) have also been found to be suitable for these applications are: magneto responsive virus-mimetic (Fang et al., 2015) and Microbial exopolysaccharides (Mohammad and Yusof, 2014). In addition, layer-by-layer self-assembly approaches for synthesis purposes offer a great opportunity to modulate the nanoparticles' properties in order to create for more efficient therapies.

Additionally, as the supporting information Table 11.1 provides a comparison and detailed overview of the recent studies that have been reported on MNP-based hyperthermia and MNPs for MNP for hyperthermia-based controlled drug delivery. Despite the extensive research on magnetic nanoparticles, additional studies will be needed to explore the development of high-level computational methods for complex biological systems and magnetic hyperthermia therapy.

11.4 CURRENT PERSPECTIVES AND CHALLENGES OF HYPERTHERMIA-BASED THERAPY AND CONTROLLED DRUG DELIVERY

According to the observation of recent research results, major challenges and opportunities still exist for MNPs in the realm of thermotherapy and controlled drug delivery. One of the top challenges is fulfilling the objective of finding new biocompatible super-magnetic materials with tunable properties such as magnetization, suitable and relatively small sizes, different shapes with a good size distribution and with the ability to maximize the Specific Absorption Rate in hyperthermia.

Moreover, prospective opportunities exist for hyperthermia treatment and drug delivery featuring multi-therapeutic modalities. Three examples of nano-constructs are the following: (i) magnetic core polymer shell nanoparticles with an encapsulated drug within a polymeric shell and a targeting agent on the construct's surface, (ii) MNP, drug and polymer nanocomposite with a targeting agent on the construct's surface, and (iii) a core containing drug with a polymer shell encapsulated with MNPs and a targeting agent on the construct's surface. Still, there is a need to validate the potential of these type of constructs for the use in hyperthermia-based therapy and drug release.

Although the shape of the nanoparticles plays an important role in magnetism, minimal effort has been invested in using this approach to design magnetic nanomaterials for hyperthermia.

In order for this manner of treatment to be considered further in order to be applied in the clinic, it is necessary for the drug's pharmokinetics and clearance issues to be well understood (biocompatibility, clearance, etc.). An additional challenge, further study into the systematic dosage and thermal responsive relationship at the targeted sites. Another challenge is to quantify the amount of aggregation of the drug at the targeted site along with the drug's distribution.

KEYWORDS

- **hyperthermia therapy**
- **image guided**
- **magnetic nanoparticles**
- **MNP-based hyperthermia drug delivery**
- **targeting**

REFERENCES

Backes, S., Witt, M. U., Roeben, E., Kuhrts, L., Aleed, S., Schmidt, A. M., et al., (2015). Loading of PNIPAM Based Microgels with CoFeO Nanoparticles and Their Magnetic Response in Bulk and at Surfaces. *Journal of Physical Chemistry B. Phys. Chem., B, 119* (36), 12129–12137.

Bae, S., Sang Won, L., Takemura, Y., Yamashita, E., Kunisaki, J., Zurn, S., et al., (2006). Dependence of Frequency and Magnetic Field on Self-Heating Characteristics of NiFeO Nanoparticles for Hyperthermia. *IEEE Transactions on Magnetics, 42*(10), 3566–3568.

Bealle, G., Lartigue, L., Wilhelm, C., Ravaux, J., Gazeau, F., Podor, R., et al., (2014). Surface decoration of catanionic vesicles with super paramagnetic iron oxide nanoparticles, a model system for triggered release under moderate temperature conditions. *Physical Chemistry Chemical Physics, PCCP. 16*(9), 4077–4081.

Bornstein, B. A., Zouranjian, P. S., Hansen, J. L., Fraser, S. M., Gelwan, L. A., Teicher, B. A., et al., (1993). Local hyperthermia, radiation therapy, and chemotherapy in patients with local-regional recurrence of breast carcinoma. *International Journal of Radiation Oncology, Biology, Physics., 25*(1), 79–85.

Branquinho, L. C., Carrião, M. S., Costa, A. S., Zufelato, N., Sousa, M. H., Miotto, R., et al., (2013). Effect of magnetic dipolar interactions on nanoparticle heating efficiency, *Implications for cancer hyperthermia. Scientific Reports, 3,* 2887.

Brazel C. S., (2009). Magnetothermally-responsive nanomaterials, combining magnetic nanostructures and thermally-sensitive polymers for triggered drug release. *Pharmaceutical Research, 26*(3), 644–656.

Chiang, W. H., Ho, V. T., Chen, H. H., Huang, W. C., Huang, Y. F., Lin, S. C., et al., (2013). Super paramagnetic hollow hybrid nanogels as a potential guidable vehicle system of stimuli-mediated MR imaging and multiple cancer therapeutics. *Langmuir, 29*(21), 6434–6443.

Clares, B., Biedma-Ortiz, R. A., Saez-Fernandez, E., Prados, J. C., Melguizo, C., Cabeza, L., et al., (2013). Nano-engineering of 5-fluorouracil-loaded magnetoliposomes for combined hyperthermia and chemotherapy against colon cancer. European journal of pharmaceutics and biopharmaceutics, *Official Journal of Arbeitsgemeinschaft fur Pharmazeutische Verfahrenstechnik eV. 85*(3 Pt A), 329–338.

Dan, M., Bae, Y., Pittman, T. A., & Yokel R. A., (2015). Alternating magnetic field-induced hyperthermia increases iron oxide nanoparticle cell association/uptake and flux in blood-brain barrier models. *Pharmaceutical Research, 32*(5), 1615–1625.

Davaran, S., Alimirzalu, S., Nejati-Koshki, K., Nasrabadi, H. T., Akbarzadeh, A., Khandaghi, A. A., et al., (2014). Physicochemical characteristics of Fe3O4 magnetic nanocomposites based on Poly(N-isopropylacrylamide) for anti-cancer drug delivery. *Asian Pacific Journal of Cancer Prevention, APJCP. 15*(1), 49–54.

Derfus, A. M., vonMaltzahn, G., Harris, T. J., Duza, T., Vecchio, K. S., Ruoslahti, E., et al., (2007). Remotely Triggered Release from Magnetic Nanoparticles. *Advanced Materials, 19*(22), 3932–3936.

Di Corato, R., Bealle, G., Kolosnjaj-Tabi, J., Espinosa, A., Clement, O., Silva, A. K., et al., (2015). Combining magnetic hyperthermia and photodynamic therapy for tumor ablation with photoresponsive magnetic liposomes. *ACS Nano, 9*(3), 2904–2916.

Dou, Y. N., Zheng, J., Foltz, W. D., Weersink, R., Chaudary, N., Jaffray, D. A., et al., (2014), Heat-activated thermosensitive liposomal cisplatin (HTLC) results in effective growth delay of cervical carcinoma in mice. Journal of controlled release, *Official Journal of the Controlled Release Society, 178,* 69–78.

Fang, J. H., Lee, Y. T., Chiang, W. H., & Hu, S. H., (2015). Magnetoresponsive virus-mimetic nanocapsules with dual heat-triggered sequential-infected multiple drug-delivery approach for combinatorial tumor therapy. *Small, 11*(20), 2417–2428.

Fantechi, E., Innocenti, C., Zanardelli, M., Fittipaldi, M., Falvo, E., Carbo, M., et al., (2014). A smart platform for hyperthermia application in cancer treatment, cobalt-doped ferrite Nanoparticles Mineralized in Human Ferritin Cages. *ACS nano., 8*(5), 4705–4719.

Fortin, J. P., Gazeau, F., & Wilhelm, C., (2008). Intracellular heating of living cells through Neel relaxation of magnetic nanoparticles. European biophysics journal , EBJ. *37*(2), 223–228.

Fortin, J. P., Wilhelm, C., Servais, J., Menager, C., Bacri, J. C., & Gazeau, F., (2007). Size-sorted anionic iron oxide nanomagnets as colloidal mediators for magnetic hyperthermia. *Journal of the American Chemical Society., 129*(9), 2628–2635.

Fourmy, D., Carrey, J., & Gigou▢ V., (2015). Targeted nanoscale magnetic hyperthermia, challenges and potentials of peptide-based targeting. *Nanomedicine (London, England)., 10*(6), 893–896.

Galanzha, E. I., Nedosekin, D. A., Sarimollaoglu, M., Orza, A. I., Biris, A. S., Verkhusha, V. V, et al., (2015). Photoacoustic and photothermal cytometry using photo switchable proteins and nanoparticles with ultrasharp resonances. *Journal of Biophotonics, 8*(8), 687.

Garaio, E., Sandre, O., Collantes, J. M., Garcia, J. A., Mornet, S., & Plazaola, F., (2015). Specific absorption rate dependence on temperature in magnetic field hyperthermia measured by dynamic hysteresis losses (ac magnetometry). *Nanotechnology., 26*(1), 015704.

Gautier, J., Munnier, E., Paillard, A., Herve, K., Douziech-Eyrolles, L., Souce, M., et al., (2012). A pharmaceutical study of doxorubicin-loaded PEGylated nanoparticles for magnetic drug targeting. *International Journal of Pharmaceutics, 423*(1), 16–25.

Ghosh, S., GhoshMitra, S., Cai, T., Diercks, D. R., Mills, N. C., & Hynds, D. L., (2010). Alternating Magnetic Field Controlled, Multifunctional Nano-Reservoirs, Intracellular Uptake and Improved Biocompatibility. *Nanoscale Research Letters., 5*(1), 195–204.

Gilchrist, R. K., Medal, R., Shorey, W. D., Hanselman, R. C., Parrott, J. C., & Taylor, C. B., (1957). Selective inductive heating of lymph nodes. *Annals of Surgery, 146*(4), 596–606.

Gillich, T., Acikgoz, C., Isa, L., Schluter, A. D., Spencer, N. D., & Textor, M., (2013). PEG-stabilized core-shell nanoparticles, impact of linear versus dendritic polymer shell architecture on colloidal properties and the reversibility of temperature-induced aggregation. *ACS nano. 7*(1), 316–329.

Guo, W., Yang, C., Lin, H., & Qu, F., (2014). P(EO-co-LLA) functionalized $Fe_3O_4@mSiO_2$ nanocomposites for thermo/pH responsive drug controlled release and hyperthermia. Dalton transactions (Cambridge, England, 2003). *43*(48), 18056–18065.

Guo, Y., Zhang, Z., Kim, D. H., Li, W., Nicolai, J., Procissi, D., et al., (2013). Photothermal ablation of pancreatic cancer cells with hybrid iron-oxide core gold-shell nanoparticles. *International Journal of Nanomedicine, 8*, 3437–3446.

Gupta, A. K., & Gupta, M., (2005). Synthesis and surface engineering of iron oxide nanoparticles for biomedical applications. *Biomaterials. 26*(18), 3995–4021.

Haghniaz, R., Umrani, R. D., & Paknikar, K. M., (2015). Temperature-dependent and time-dependent effects of hyperthermia mediated by dextran-coated La0. 7Sr0. 3MnO3, in vitro studies. *International Journal of Nanomedicine. 10*, 1609–1623.

Halupka-Bryl, M., Asai, K., Thangavel, S., Bednarowicz, M., Krzyminiewski, R., & Naga-saki, Y., (2014). Synthesis and *in vitro* and *in vivo* evaluations of poly(ethylene glycol)-block-poly(4-vinylbenzylphosphonate) magnetic nanoparticles containing doxorubicin as a potential targeted drug delivery system. *Colloids and Surfaces, B., Biointerfaces, 118*, 140–147.

Hardiansyah, A., Huang, L. Y., Yang, M. C., Liu, T. Y., Tsai, S. C., Yang, C. Y., et al., (2014), Magnetic liposomes for colorectal cancer cells therapy by high-frequency magnetic field treatment. *Nanoscale Research Letters., 9*(1), 497.

Hayashi, K., Nakamura, M., Miki, H., Ozaki, S., Abe, M., Matsumoto, T., et al., (2014), Magnetically responsive smart nanoparticles for cancer treatment with a combination of magnetic hyperthermia and remote-control drug release. *Theranostics., 4*(8), 834–844.

Hayashi, K., Ono, K., Suzuki, H., Sawada, M., Moriya, M., Sakamoto, W., et al., (2010), High-frequency, magnetic-field-responsive drug release from magnetic nanoparticle/organic hybrid based on hyperthermic effect. *ACS applied materials & interfaces., 2*(7), 1903–1911.

Hergt, R., & Dutz, S., (2007). Magnetic particle hyperthermia biophysical limitations of a visionary tumour therapy. *Journal of Magnetism and Magnetic Materials., 311*(1), 187–192.

Hergt, R., Andra W, d'Ambly, C. G., Hilger, I., Kaiser, W. A., Richter, U., et al., (1998). Physical limits of hyperthermia using magnetite fine particles. *Magnetics, IEEE Transactions on., 34*(5), 3745–3754.

Hola, K., Markova, Z., Zoppellaro, G., Tucek, J., & Zboril, R., (2015). Tailored functionalization of iron oxide nanoparticles for MRI, drug delivery, magnetic separation and immobilization of biosubstances. *Biotechnol Adv., 1*; *33*(6 Pt 2):1162–1176

Huang, C., Tang, Z., Zhou, Y., Zhou, X., Jin, Y., Li, D., et al., (2012). Magnetic micelles as a potential platform for dual targeted drug delivery in cancer therapy. *International Journal of Pharmaceutics., 429*(1–2), 113–122.

Ito, A., Honda, H., & Kobayashi, T., (2006), Cancer immunotherapy based on intracellular hyperthermia using magnetite nanoparticles, a novel concept of "heat-controlled necrosis" with heat shock protein expression. *Cancer Immunology, Immunotherapy, CII. 55*(3), 320–328.

Jang, D. H., Lee, Y. I., Kim, K. S., Park, E. S., Kang, S. C., Yoon, T. J., et al., (2013). Induced heat property of polyethyleneglycol-coated iron oxide nanoparticles with dispersion stability for hyperthermia. *Journal of Nanoscience and Nanotechnology, 13*(9), 6098–6102.

Jeyadevan, B., (2010). Present status and prospects of magnetite nanoparticles-based hyperthermia. *Journal of the Ceramic Society of Japan., 118*(1378), 391–401.

Jordan, A., Wust, P., Fahling, H., John, W., Hinz, A., & Felix, R., (1993). Inductive heating of ferrimagnetic particles and magnetic fluids, physical evaluation of their potential for hyperthermia. International journal of hyperthermia, the official journal of European Society for Hyperthermic Oncology, *North American Hyperthermia Group. 9*(1), 51–68.

Kaaki, K., Herve-Aubert, K., Chiper, M., Shkilnyy, A., Souce, M., Benoit, R., et al., (2012). Magnetic nanocarriers of doxorubicin coated with poly(ethylene glycol) and folic acid, relation between coating structure, surface properties, colloidal stability, and cancer cell targeting. *Langmuir., 28*(2), 1496–1505.

Kakwere, H., Leal, M. P., Materia, M. E., Curcio, A., Guardia, P., Niculaes, D., et al., (2015). Functionalization of strongly interacting magnetic nanocubes with (thermo)responsive coating and their application in hyperthermia and heat-triggered drug delivery. *ACS Applied Materials and Interfaces, 7*(19), 10132–10145.

Kawai, N., Futakuchi, M., Yoshida, T., Ito, A., Sato, S., Naiki, T., et al., (2008). Effect of heat therapy using magnetic nanoparticles conjugated with cationic liposomes on prostate tumor in bone. *Prostate., 68*(7), 784–792.

Kim, D. H., Lee, S. H., Kim, K. N., Kim, K. M., Shim, I. B., & Lee, Y. K. (2005). Temperature change of various ferrite particles with alternating magnetic field for hyperthermic application. *Journal of Magnetism and Magnetic Materials, 293*(1), 320–327.

Kobayashi, T., (2011). Cancer hyperthermia using magnetic nanoparticles. *Biotechnology Journal, 6*(11), 1342–1347.

Kost, J., Wolfrum, J., & Langer, R., (1987). Magnetically enhanced insulin release in diabetic rats. *Journal of biomedical materials research, 21*(12), 1367–1373.

Lee, G. Y., Qian, W. P., Wang, L., Wang, Y. A., Staley, C. A., Satpathy, M., et al., (2013), Theranostic nanoparticles with controlled release of gemcitabine for targeted therapy and MRI of pancreatic cancer. *ACS Nano, 7*(3), 2078–2089.

Levy, M., Wilhelm, C., Siaugue, J. M., Horner, O., Bacri, J. C., & Gazeau, F., (2008). Magnetically induced hyperthermia, size-dependent heating power of gamma-Fe$_2$O$_3$ nanoparticles. *Journal of Physics Condensed Matter: An Institute of Physics Journal., 20*(20), 204133.

Li, J., Hu, Y., Yang, J., Wei, P., Sun, W., Shen, M., et al., (2015). Hyaluronic acid-modified Fe3O4@Au core/shell nanostars for multimodal imaging and photothermal therapy of tumors. *Biomaterials, 38,* 10–21.

Liu, T. Y., Hu, S. H., Liu, K. H., Shaiu, R. S., Liu, D. M., & Chen, S. Y., (2008). Instantaneous drug delivery of magnetic/thermally sensitive nanospheres by a high-frequency magnetic field. *Langmuir, 24*(23), 13306–13311.

Lopez-Noriega, A., Hastings, C. L., Ozbakir, B., O'Donnell, K. E., O'Brien, F. J., Storm, G., et al., (2014). Hyperthermia-induced drug delivery from thermosensitive liposomes encapsulated in an injectable hydrogel for local chemotherapy. *Advanced Healthcare Materials, 3*(6), 854–859.

Loynachan, C. N., Romero, G., Christiansen, M. G., Chen, R., Ellison, R., O'Malley, T. T., et al., (2015). Targeted Magnetic Nanoparticles for Remote Magnetothermal Disruption of Amyloid-beta Aggregates. *Advanced Healthcare Materials. 4*(14), doi: 10.1002/adhm.201500487.

Lu, Z., Prouty, M. D., Guo, Z., Golub, V. O., Kumar, C. S., & Lvov, Y. M., (2005). Magnetic switch of permeability for polyelectrolyte microcapsules embedded with Co@Au nanoparticles. *Langmuir., 21*(5), 2042–2050.

McGill, S. L., Cuylear, C. L., Adolphi, N. L., Osinski, M., & Smyth, H. D., (2009). Magnetically responsive nanoparticles for drug delivery applications using low magnetic field strengths. *IEEE Trans Nanobioscience, 8*(1), 33–42.

McNerny, K. L., Kim, Y., Laughlin, D. E., & McHenry, M. E., (2010). Chemical synthesis of monodisperse γ-Fe–Ni magnetic nanoparticles with tunable Curie temperatures for self-regulated hyperthermia. *Journal of Applied Physics, 107*(9), 09A312.

Meenach, S. A., Hilt, J. Z., & Anderson, K. W., (2010). Poly(ethylene glycol)-based magnetic hydrogel nanocomposites for hyperthermia cancer therapy. *Acta biomaterialia. 6*(3), 1039–1046.

Meenach, S. A., Shapiro, J. M., Hilt, J. Z., & Anderson, K. W., (2013). Characterization of PEG-iron oxide hydrogel nanocomposites for dual hyperthermia and paclitaxel delivery. *Journal of Biomaterials Science Polymer Edition., 24*(9), 1112–1126.

Mocan, L., Tabaran, F. A., Mocan, T., Bele, C., Orza, A. I., Lucan, C., et al., (2011). Selective ex-vivo photothermal ablation of human pancreatic cancer with albumin functionalized multiwalled carbon nanotubes. *International journal of nanomedicine, 6,* 915–928.

Mohammad, F., & Yusof, N. A., (2014). Doxorubicin-loaded magnetic gold nanoshells for a combination therapy of hyperthermia and drug delivery. *Journal of colloid and interface science. 434,* 89–97.

Moros, M., Pelaz, B., Lopez-Larrubia, P., Garcia-Martin, M. L., & Grazu V, de la Fuente, J. M., (2010), Engineering biofunctional magnetic nanoparticles for biotechnological applications. *Nanoscale, 2*(9), 1746–1755.

N'Guyen, T. T., Duong, H. T., Basuki, J., Montembault, V., Pascual, S., Guibert, C., et al., (2013). Functional iron oxide magnetic nanoparticles with hyperthermia-induced drug release ability by using a combination of orthogonal click reactions. *Angewandte Chemie (International ed in English). 52,* 14152–14156.

Orza, A., Casciano, D., & Biris, A., (2014). Nanomaterials for targeted drug delivery to cancer stem cells. *Drug Metabolism Reviews, 46*(2), 191–206.

Orza, A., Soritau, O., Tomuleasa, C., Olenic, L., Florea, A., Pana, O., et al., (2013). Reversing chemoresistance of malignant glioma stem cells using gold nanoparticles. *International Journal of Nanomedicine, 8*, 689–702.

Patra, S., Roy, E., Karfa, P., Kumar, S., Madhuri, R., & Sharma, P. K., (2015). Dual-responsive polymer coated super paramagnetic nanoparticle for targeted drug delivery and hyperthermia treatment. *ACS Applied Materials & Interfaces, 7*(17), 9235–9246.

Pradhan, P., Giri, J., Rieken, F., Koch, C., Mykhaylyk, O., Doblinger, M., et al., (2010), Targeted temperature sensitive magnetic liposomes for thermo-chemotherapy. *Journal of the Controlled Release Society, 142*(1), 108–121.

Pradhan, P., Giri, J., Samanta, G., Sarma, H. D., Mishra, K. P., Bellare, J., et al., (2007). Comparative evaluation of heating ability and biocompatibility of different ferrite-based magnetic fluids for hyperthermia application. Journal of biomedical materials research Part, B., *Applied Biomaterials, 81*(1), 12–22.

Purushotham, S., & Ramanujan, R. V., (2010). Thermoresponsive magnetic composite nanomaterials for multimodal cancer therapy. *Acta Biomaterialia, 6*(2), 502–510.

Quinto, C. A., Mohindra, P., Tong, S., & Bao, G., (2015), Multifunctional super paramagnetic iron oxide nanoparticles for combined chemotherapy and hyperthermia cancer treatment. *Nanoscale, 7*(29), 12728–12736.

Riviere, C., Wilhelm, C., Cousin, F., Dupuis, V., Gazeau, F., & Perzynski, R., (2007). Internal structure of magnetic endosomes. *The European Physical Journal, E., Soft Matter., 22*(1), 1–10.

Rosensweig, R. E., (2002). Heating magnetic fluid with alternating magnetic field. *Journal of Magnetism and Magnetic Materials, 252*, 370–374.

Sabale, S., Jadhav, V., Khot, V., Zhu, X., Xin, M., & Chen, H., (2015). Super paramagnetic MFe_2O_4 (M = Ni, Co, Zn, Mn) nanoparticles, synthesis, characterization, induction heating and cell viability studies for cancer hyperthermia applications. *Journal of Materials Science Materials in Medicine, 26*(3), 127.

Sadhasivam, S., Savitha, S., Wu, C. J., Lin, F. H., & Stobinski, L., (2015). Carbon encapsulated iron oxide nanoparticles surface engineered with polyethylene glycol-folic acid to induce selective hyperthermia in folate over expressed cancer cells. *International Journal of Pharmaceutics. 480*(1–2), 8–14.

Sadhukha, T., Wiedmann, T. S., & Panyam, J., (2013). Inhalable magnetic nanoparticles for targeted hyperthermia in lung cancer therapy. *Biomaterials, 34*(21), 5163–5171.

Sahu, N. K., Gupta, J., & Bahadur, D., (2015). PEGylated $FePt-Fe_3O_4$ composite nanoassemblies (CNAs), in vitro hyperthermia, drug delivery and generation of reactive oxygen species (ROS). *Dalton Transactions* (Cambridge, England, 2003). *44*(19), 9103–9113.

Satarkar, N. S., & Hilt, J. Z., (2008). Magnetic hydrogel nanocomposites for remote controlled pulsatile drug release. *Journal of the Controlled Release Society, 130*(3), 246–251.

Satpathy, M., Wang, L., Zielinski, R., Qian, W., Lipowska, M., Capala, J., et al., (2014). Active targeting using HER-2-affibody-conjugated nanoparticles enabled sensitive and specific imaging of orthotopic HER-2 positive ovarian tumors. *Small, 10*(3), 544–555.

Saville, S. L., Qi, B., Baker, J., Stone, R., Camley, R. E., Livesey, K. L., et al., (2014). The formation of linear aggregates in magnetic hyperthermia, implications on specific absorption rate and magnetic anisotropy. *Journal of Colloid and Interface Science, 424*, 141–151.

Shah, R. R., Davis, T. P., Glover, A. L., Nikles, D. E., & Brazel, C. S, (2015). Impact of magnetic field parameters and iron oxide nanoparticle properties on heat generation for use in magnetic hyperthermia. *J Magn Magn Mater, 387*, 96–106.

Sharma, R., & Chen, C. J., (2009). Newer nanoparticles in hyperthermia treatment and thermometry. *Journal of Nanoparticle Research., 11*(3), 671–689.

Soares, P. I., Ferreira, I. M., Igreja R. A., & Novo C. M., (2012). Borges, J. P. Application of hyperthermia for cancer treatment, recent patents review. *Recent Patents on Anti-Cancer Drug Discovery, 7*(1), 64–73.

Staruch, R. M., Hynynen, K., & Chopra, R., (2015). Hyperthermia-mediated doxorubicin release from thermosensitive liposomes using MR-HIFU, therapeutic effect in rabbit VΩ2 tumors. International journal of hyperthermia, the official journal of European Society for Hyperthermic Oncology, *North American Hyperthermia Group, 31*(2), 118–133.

Tabatabaei, S. N., Girouard, H., Carret, A. S., & Martel, S., (2015). Remote control of the permeability of the blood-brain barrier by magnetic heating of nanoparticles, A proof of concept for brain drug delivery. *Journal of the Controlled Release Society, 206*, 49–57.

Tao, C., & Zhu, Y., (2014). Magnetic mesoporous silica nanoparticles for potential delivery of chemotherapeutic drugs and hyperthermia. *Dalton Transactions* (Cambridge, England, 2003). *43*(41), 15482–15490.

Thomas, R. G., Moon, M. J., Lee, H., Sasikala, A. R., Kim, C. S., Park, I. K., et al., (2015). Hyaluronic acid conjugated super paramagnetic iron oxide nanoparticle for cancer diagnosis and hyperthermia therapy. *Carbohydrate Polymers. 131*, 439–446.

Urbina, M. C., Zinoveva, S., Miller, T., Sabliov, C. M., Monroe, W. T., & Kumar, C. S. S. R., (2008). Investigation of Magnetic Nanoparticle–Polymer Composites for Multiple-controlled Drug Delivery. *The Journal of Physical Chemistry C., 112*(30), 11102–11108.

Wang, S. Y., Liu, M. C., & Kang, K. A., (2013). Magnetic nanoparticles and thermally responsive polymer for targeted hyperthermia and sustained anti-cancer drug delivery. *Advances in Experimental Medicine and Biology., 765*, 315–321.

Wijaya, A., Brown, K. A., Alper, J. D., & Hamad-Schifferli, K., (2007). Magnetic field heating study of Fe-doped Au nanoparticles. *Journal of Magnetism and Magnetic Materials., 309*(1), 15–19.

Wong, J. E., Gaharwar, A. K., Muller-Schulte, D., Bahadur, D., & Richtering, W., (2008). Dual-stimuli responsive PNiPAM microgel achieved via layer-by-layer assembly, magnetic and thermoresponsive. *Journal of Colloid and Interface Science, 324*(1–2), 47–54.

Wong, J. E., Krishnakumar Gaharwar, A., Müller-Schulte, D., Bahadur, D., & Richtering, W., (2007). Layer-by-layer assembly of a magnetic nanoparticle shell on a thermoresponsive microgel core. *Journal of Magnetism and Magnetic Materials, 311*(1), 219–223.

Zaloga, J., Stapf, M., Nowak, J., Pottler, M., Friedrich, R. P., Tietze, R., et al., (2015). Tangential Flow Ultrafiltration Allows Purification and Concentration of Lauric Acid-/Albumin-Coated Particles for Improved Magnetic Treatment. *Int J Mol Sci., 16*(8), 19291–9307.

Zhang, L., Zhong, X., Wang, L., Chen, H., Wang, Y. A., Yeh, J., et al., (2011). T(1)-weighted ultrashort echo time method for positive contrast imaging of magnetic nanoparticles and cancer cells bound with the targeted nanoparticles. *Journal of Magnetic Resonance Imaging, JMRI., 33*(1), 194–202.

Zhao, D. L., Wang, X. X., Zeng, X. W., Xia, Q. S., & Tang, J. T., (2009). Preparation and inductive heating property of Fe3O4–chitosan composite nanoparticles in an AC magnetic field for localized hyperthermia. *Journal of Alloys and Compounds, 477*(1–2), 739–743.

PART IV

DIVERSE USES OF
NANOBIOMATERIALS

CHAPTER 12

NANO CARRIER SYSTEMS OF UBIDECARENONE (COENZYME Q10) FOR COSMETIC APPLICATIONS

N. K. YADAV,[1] REKHA RAO,[2] O. P. KATARE,[3] and SANJU NANDA[1]

[1]Department of Pharmaceutical Sciences, Maharshi Dayanand University, Rohtak, 124001, India, E-mail: sn_mdu@rediffmail.com

[2]Department of Pharmaceutical Sciences, Guru Jambheshwar University of Science and Technology, Hisar–125001, India

[3]University Institute of Pharmaceutical Sciences, UGC Center of Advanced Studies, Punjab University, Chandigarh, 160–014, India

CONTENTS

ABSTRACT

Coenzyme Q10 (CoQ10) occurs as a natural biomolecule (a compound similar to vitamin) and essential micronutrient present in every cell of our body, chiefly in the mitochondrial membranes. The major role of CoQ10 is to produce energy by acting as a cofactor in electron transport chain. Due to the physiological changes and with advancing age, level of CoQ10 decreases in the body. To restore its concentration in the body and to reverse the conditions arising from its deficiency, CoQ10 needs to be given from outside. Critical role of CoQ10 in cellular bioenergetics lead to its use in neurodegenerative disorders, cardiovascular and neuromuscular diseases. CoQ10, being an established antioxidant has been a part of many skin care products too, especially for anti-aging purposes. In totality, it is widely accepted as a therapeutic agent to treat a number of diseases/disorders, nutraceutical/ food supplement and a cosmetic agent. But it has some limitations too as an entity. CoQ10 is a poorly water soluble, high molecular weight molecule, therefore show poor absorption and thus low bioavailability. Numerous approaches have been explored by the researchers from time to time to overcome these limitations. The present chapter shall discuss the chemistry, source and uses of Ubidecarenone followed by description of various nanosized carrier systems of this biomaterial for its effective and safe delivery through major routes.

12.1   INTRODUCTION

Cosmetics defined under Federal Food, Drug and *Cosmetic* Act (FD and C Act) as "articles intended to be applied to the human body or any part thereof for cleansing, beautifying, promoting attractiveness, or altering the appearance" (Cosmetics and US Law, 2016). However, Federal Food, Drug and *Cosmetic* Act (FD and C Act) does not recognize the term "cosmeceutical," but this word generally denotes cosmetic products in cosmetic industry which behaves drug-like or showing medicinal benefits. Some author also reports word 'cosmeceutical' as products or category which comes between drugs and cosmetics (Fulekar, 2010). Cosmeceutical are also termed as hybrid products that include cosmetics and topical medications. They contain ingredients and their formulations are-based on novel technology that influences physiological and biological function of the skin. These products have an ability to alter and heal skin functions and therefore considered important for cosmetic industries (Li et al., 2011).

Nowadays cosmeceuticals are trending towards the rapidly growing sector of cosmetic industry dedicated for personal care. Technological-based innovations made a drastic impact on the personal cared industry. These products are considered important not only for better absorption, but also for improving stability, acting as targeting agents and prolong release of the actives (De Leeuw et al., 2009; Lohani et al., 2014; Sinico and Fadda, 2009, Vora et al., 1998). Incorporation of novel carrier in cosmetic formulations results in long perfume incense, better sunscreen protection, maintains skin hydration and improved antiaging affect.

Coenzyme Q10 occurs as a natural, vitamin like compound and essential micronutrient occurs in every cell of our body, mainly in mitochondrial membranes. Its natural resources are presented in Figure 12.1. The chemical structure contains benzoquinone moiety and a long side chain containing 10 *trans*-isoprenoid blocks, thus it is referred as coenzyme Q10 (Collins and Jones, 1981). It was first identified in 1940 and afterwards, separated from mitochondria (beef heart) by Frederick Crane in 1957. Besides mitochondria, it is also present in heart, endoplasmic reticulum, kidney, liver, lysosomes, and peroxisomes. Also, it is found in spinach, carrot, broccoli, eggs, beef, fish and meat (Beg et al., 2010).

CoQ10 is also recognized as ubiquinone-10, and acts as co-enzyme in energy-generation pathways of body cells. The major role of CoQ10 is to produce energy by acting in the form of a cofactor in mitochondrial electron transport chain (Kapoor and Kapoor, 2013). The production of energy is stored as Adenosine Triphosphate (ATP) generated from electron transport chain, which governs the cellular functions of the body. It performs a

FIGURE 12.1 Coenzyme Q10 is available naturally in many food items.

vital role in energy generation by transferring electron from Nicotinamide adenine dinucleotide phosphate (NADPH) to cytochrome complex through succinate dehydrogenase. It also perform cofactor role for electron transport enzymes which participate in oxidative phosphorylation.

Chemically CoQ10 is 2,3-dimethoxy-5-methyl-6-decaprenyl-1, 4-ben-zoquinone, which consist P-benzoquinone ring with a side-chain of poly-isoprenoid unit is a highly lipid soluble compound (Figure 12.2). CoQ10 is available as a crystalline powder of yellow to orange color, with melting temperature of 48°C. Due to the presence of multiple isoprene blocks, it is termed as a highly lipophilic substance (Boicelli et al., 1981). It has a large molecular weight (863Da) and is thermo-labile in nature (Borekova et al., 2008). CoQ10 is soluble in ether, very-slightly soluble in dehydrated alcohol and practically insoluble in water.

12.2   SYNTHESIS OF COQ10

CoQ10 biosynthesis in the body is a complex process by multistep pathway comprising mevalonic acid and the precursor is mevalonate. The main step of CoQ10 synthesis includes conversion of phenylalanine or tyrosine into 4-hydroxybenzoate, later 6 to 10 isoprene units are added to produce poly-prenylphenol. Further, an oxygen atom is added at C3 position, and addition of three methyl groups through S-adenosine methionine results in ubiquinol (reduced form of CoQ10), which on reduction gives CoQ10 (Parkhideh, 2008).

Bio-production of ubiquinone at large-scale using microbes has already been validated by some Japanese companies. CoQ10 producing micro-organisms include, yeasts and photosynthetic bacteria, which uses

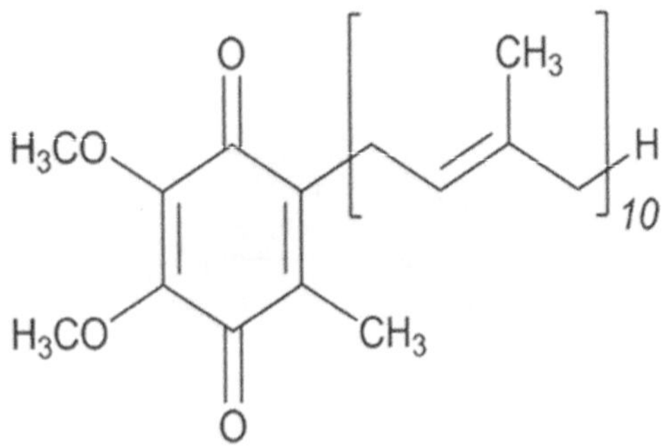

FIGURE 12.2   Structure of Coenzyme Q10.

fermentation-based production of CoQ10 (Yoshida et al., 1998). The most common method of CoQ10 synthesis is through fermentation compared to chemical synthesis. In order to improve the production of CoQ10, genetically modified microorganisms are also used (Okada et al., 1998).

12.3 ROLE OF COENZYME Q10 IN THE BODY

CoQ10 is naturally synthesized in the human body, but due to physiological changes and with advancing age, the amount of CoQ10 is not sufficient to meet the needs of the body. To restore its levels in the body, CoQ10 should be given form outside to reverse the conditions arise from its deficiency. Critical role of CoQ10 in cellular bioenergetics lead its use in neurodegenerative disorders, cardiovascular and neuromuscular diseases (Bhagavan and Chopra, 2006). Neurodegenerative disorders include Parkinson, Alzheimer, Friedreich and Huntington diseases, where many studies showed promising results. Cardiovascular diseases where its effects are well recognized include chronic heart failure, hypertension, myocardial infarction, angina pectoris and atherosclerosis. CoQ10 also exhibits a wide range of pharmacological activities against many chronic ailments like diabetes mellitus, carcinoma, autoimmune diseases, periodontal disease and for maintaining homeostasis (Dhanasekaran and Ren, 2005; Garrido-Maraver et al., 2014).

For its nutraceutical and food supplement use, CoQ10 is easily available over the counter. Its energy production property inside cells lead to its use as a nutraceutical compound. Supplementation of CoQ10 is advisable to persons having physical and mental stress, poor eating habits and infection (Zuelli et al., 2006). Due to large focus of scientific community in past years, CoQ10 became a popular food/nutritional supplement. CoQ10 production decreases in the body with advancing age, thus the required energy can be compensated using daily intake of CoQ10 supplementation. In United States, CoQ10 is third most sold dietary supplement after omega-3 fatty acids and multivitamins. Japan is the only country where it is approved as a drug product for treatment of cardiovascular diseases. Presence of CoQ10 as ubiquinol (reduced form of CoQ10) in human cells acts as a membrane antioxidant. The antioxidant property is attributed to direct interaction with radicals or through ascorbate and tocopherol generation (Crane, 2001, Littarru and Tiano, 2007).

Energy production, membrane antioxidant and a cofactor for enzymatic reactions are the basic functions of CoQ10 which constitute the basis for supporting its use as a *'cosmetic'* agent. It is a well-established antioxidant and already used in many skin care preparations with anti-aging potential (Zhang et al., 2012). The antioxidant property of CoQ10 helps in cell protection against ageing phenomena induced through UV radiations or free radicals. UV radiations and free radicals results in oxidative stress and thus plays a significant role in skin aging. *In vivo* studies suggests, UV-mediated oxidative stress can be effectively reduced using CoQ10 via, collagenase suppression, thiol depletion, phosphotyrosine kinases activation, DNA damage prevention, lipid peroxidation inhibition and also results in the reduction of wrinkle depth (Dhanasekaran and Ren, 2005; Hoppe et al., 1999). Commercial products of CoQ10 are available in various forms, Figure 12.3, represents some of available forms in the market.

12.4   MARKETED PREPARATIONS OF COQ10

CoQ10 is available in the market from years for its pharmaceutical, nutraceutical and cosmeceuticals uses. The most common route of administration for its nutraceutical and pharmaceutical is oral route. CoQ10 is established as a nutraceutical compound and is marketed as a dietary supplement over the counter (OTC) in United States, Europe Union, India and many other countries. However in Japan, it is approved as a drug product for treating heart diseases. Table 12.1 summarizes its marketed formulation for pharmaceutical and nutraceutical/food supplement use.

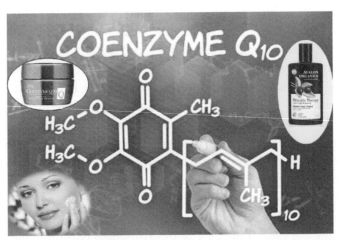

FIGURE 12.3   Commercial cosmetic products of CoQ10 come in various forms.

TABLE 12.1   Marketed CoQ10 Formulations for Oral Use

Region	Approval history	Marketed dosage form (s)/Major brands	
United States	Not approved as drug product, but sold as a dietary food supplement	Capsules (30, 100, 400 mg)	Mega CoQ10 (Twinlab), Ultra CoQ10 (Twinlab)
			Nature's Lab CoQ10
European Union		Soft gel capsules (200, 400 mg)	Beyond CoQ10 (Nature's Plus)
			Q-Sorb (Sundown Naturals)
		Chewable tablets (100 mg)	Smart Q10 CoQ10
		Lozenges (200 mg)	Now CoQ10
		Liquid	Liquid CoQ10
Japan	1974 (for treating heart disease)	Tablets (5, 10 mg)	Neuquinon (Eisai)
		Capsules (5 mg)	Neuquinon (Eisai)
		Granules (1%)	Neuquinon (Eisai)
India	Not approved as drug, sold as nutraceutical	Soft gel Capsules (30, 50, 60, 100, 200, 300 mg)	ATP-30 (VANGAURD), 4UQ10 FORTE (D.R John's)
			COQ (Uni. Medicare), COTEN (Aglowmed)
			MIRAQULE (Alchem), QTEN (Elnova)
		Tablets (100, 300 mg)	COQUEEN (East West), QAIN (Sain)
			ZYME-Q10 (Orchid)

12.4.1   NANOTECHNOLOGY IN COSMETICS/COSMECEUTICALS

Nano carriers and nanoparticles are most widely used and accepted carrier systems in cosmetic and cosmeceutical products. The first cosmetic product containing liposomes was appeared in the market as an anti-aging cream with a brand Capture®, launched by Christian Dior in 1986. After one year, Lancome launched its first cosmetic product containing niosome vesicles, with a brand Niosomes®, into the market in 1987. In early 1990s, Lancome also came up with its successors Niosome Plus® and established as anti-aging cream. Later in 2005, German cosmetics manufacturer Dr. Rimpler GmbH introduced Nano-structured-based cosmetic product, Cutanova Cream NanoRepair Q10 into the market (Li et al., 2011).

CoQ10 is widely accepted ingredients in cosmetic and cosmeceutical products. Various forms of CoQ10 nano formulations are available in the market. The global CoQ10 market was valued at USD 402.4 million in 2015. Grand view research forecast the global market of CoQ10 at USD 8495 million by 2020. Table 12.2, lists nanotechnology or nano-carrier-based marketed cosmetic/cosmeceutical formulations of CoQ10 for improved performances.

12.5　EMERGING TREND OF COENZYME Q10 IN COSMECEUTICALS

In past one decade CoQ10 is established as an important constituent of cosmetic formulations and proved effective in various conditions. CoQ10 is widely used and explored as a promising component for following properties:
- skin antioxidant;
- skin anti-aging;
- photo-induced aging;
- anti-wrinkle;
- nutritional supplement for skin.

TABLE 12.2　Marketed Cosmetic/Cosmeceutical Formulations of COQ10-based on Nanoparticles and Nano-Carriers for Improved Performance

Trade Name	Carrier system	Ingredients	Use
Nano-Lipobelle H-EQ10 cream	Nano-emulsion	CoenzymeQ10, Vitamin E acetate	Anti-aging
Nanorama™-Nano Gold® Mask Pack	Nanoparticles	Nano-gold, amino acid, collagen, coenzyme Q10	Skin tightening, anti-wrinkle
Oleogel®	Nanoparticles	Coenzyme Q10, Vitamin A and E	Anti-aging, antioxidant
Ageless Facelift cream	Liposomes	Coenzyme Q10, Niacinamide	Anti-aging, wrinkle reduction, anti-oxidative
Cutanova Nano Repair Q10 Cream	Nanostructured lipid carriers	Coenzyme Q10	Anti-aging, Revitalising
Platinum Silver Nanocolloid cream	Nanoparticles	Coenzyme Q10 and botanicals	Anti-aging, anti-wrinkle

Antioxidant property of CoQ10 is among its main effects and protect cells against free radical damage. UV radiations and free radicals results in oＸidative stress and plays a significant role in skin aging. UV radiations acts as a trigger and generate reactive oxygen species, a toxic intermediate in skin cells responsible for aging process (Ibbotson et al., 1999). CoQ_{10} is very effective in scavenging of free radicals generated in cellular membranes (Frei et al., 1990). CoQ_{10} combats with UV radiations generated ROS and helps in revivals of aging symptoms. Due to its strong antioxidant property in skin, it is used as an anti-aging agent in skin preparations.

CoQ10 suppress enzyme collagenase, an enzyme responsible for connective tissue damage and thus helps in increase of skin structural proteins and also inhibits lipid peroxidation (Hoppe et al., 1999). The inhibitory function of CoQ10 thus reduces wrinkle depth. The above properties of CoQ10 make it a unique component of cosmetic/cosmeceutical preparations.

12.6 COENZYME Q10 NOVEL NANO CARRIER FORMULATIONS

CoQ10 is a highly lipophilic molecule and thus insoluble in water. Insolubility and large molecular weight of CoQ10 (893 Da), results in poor gastrointestinal absorption and thus show low bioavailability (Chopra et al., 1998). Oral administration of CoQ10 also results in large variation due to its hydrophobic nature. Moreover, its short half-life (2 to 4 h) requires frequent dosing to maintain required blood levels.

New approaches of delivery are aimed at attaining improved therapeutic benefits of CoQ10 as well as its safe and effective delivery. Various novel delivery systems are reported and effective in minimizing the above mentioned drawbacks and results in increase its bioavailability. In recent years, research efforts have been focused towards the development of novel approaches. To entrap full potential of CoQ10, novel drug delivery strategies have been applied by numerous groups of researchers. Figure 12.4 portrays nano-sized carrier systems and their availability in skin preparations. A variety of carrier systems have been explored to overcome pharmaceutical challenges of this molecule. These carriers opened a number of doors to hasten CoQ10 potential. Investigated Novel drug delivery systems till date mainly include nanoparticles, nanostructured lipid carriers (NLC), solid lipid nanoparticles (SLN), liposomes, nano-emulsion, liposomes,

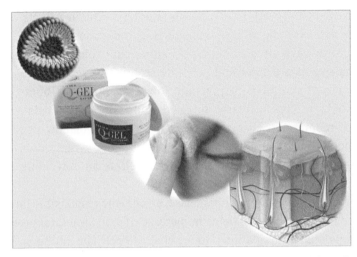

FIGURE 12.4   Nano sized carrier systems are added in skin cosmetic products for better penetration, efficacy and sustained action.

Self-nanoemulsifying drug delivery systems (SNEDDS) and proniosomes. Table 12.3, presents a compilation of some recent studies on novel carrier systems of CoQ10. Various types of nano formulations of CoQ10 are represented in Figure 12.5.

12.6.1   SOLID LIPID NANOPARTICLES (SLN)

SLN are colloidal particles made up of solid biocompatible/biodegradable lipid matrix. They are available in a size ranging between 100–1000 nm. SLNs are colloidal carriers and offer an alternative drug delivery system to liposomes, polymeric nanoparticles and emulsions (Mukherjee et al., 2009). SLN are also termed as lipid emulsions, where the liquid lipid substituted using solid lipid. The key advantages related to these carriers are high drug loading, large surface area, small size, good physical stability and are suitable for pharmaceuticals, nutraceuticals and cosmetic preparations (Joshi and Muller, 2009).

Bunjes et al., developed CoQ10 SLNs using melt-homogenized tripalmitin and reported significant change in physicochemical parameters of SLNs. Characterization of developed SLNs using DSC, XRD, photon correlation spectroscopy and TEM produced reduced crystallization and melting

TABLE 12.3 Selected Recent Studies on Various CoQ10 Novel Delivery Systems with Added Advantages

Delivery System	Remarks	Reference
SLN	Characterization using DSC, photon correlation spectroscopy, XRD and freeze-fracture transmission electron microscopy.	Bunjes et al., 2001
	Improved dermal penetration with anti-wrinkle activity.	Farboud et al., 2011
	Enhanced diffusion of CoQ10 through rat abdominal skin	Korkm et al., 2013
NLC	Better skin penetration and demonstration of antioxidant properties	Yue et al., 2010
	Enhanced skin permeation with increased epidermal CoQ10 accumulation	Chen et al., 2013
	Ultra-small NLC showed enhanced dermal delivery of CoQ10	Schwarz et al., 2013
	Improved antioxidant activity on DPPH scavenging and anti-lipid peroxidation	Nanjwade et al., 2013
SEDDS	Improved oral bioavailability compared to powder formulation.	Kommuru et al., 2001
	Significant increase in Cmax and area under the curve (AUC) compared to powder CoQ10	Balakrishnan et al., 2009
	Demonstrated 5-fold increases in both Cmax and AUC compared with crystalline CoQ10	Onoue et al., 2012
SNEDDS	Developed self nano-emulsified tablet formulation with fast dissolution	Nazzal et al., 2002
	Improved oral bioavailability of CoQ10	Nepal et al., 2010
	Attained maximum liver protection compared to the pure CoQ10	Agrawal et al., 2014
Liposome	Demonstrated production of CoQ10 nanoliposomes at lab and pilot scale.	Xia et al., 2006
	Significant enhancement of CoQ10 accumulation in rat skin for prolonged effect	Lee and Tsai, 2010
	Enhanced permeation through cornea and protection of human lens epithelial cells against oxidative damage	Wang et al., 2011
	Enhanced systemic availability of CoQ10	Shao et al., 2015
Liposomes and SLN	Better anti-oxidative effect of CoQ10 liposomes compared to SLN	Gokce et al., 2012
Polymeric nanoparticles	Superior efficacy compared to commercial formulation	Ankola et al., 2007

TABLE 12.3   (Continued)

Delivery System	Remarks	Reference
	Development of surfactant-free biodegradable nanoparticles of CoQ10	Nehilla et al., 2008
	Improved hepato-protective and anti-inflammatory activity in vivo	Swarnakar et al., 2011
Nano-emulsion	Improved bioavailability compared with conventional formulation	Belhaj et al., 2012
Proniosomes	Developed and optimized CoQ10 proniosomes for better protection in photo-induced aging	Yadav et al., 2016

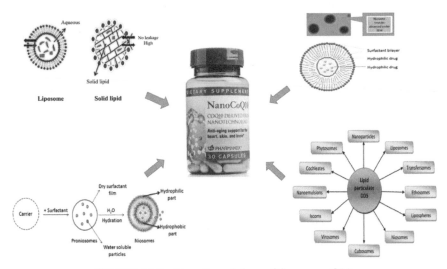

FIGURE 12.5   Nano formulations of Coenzyme Q10.

temperature of particle matrix, whereas conversion of triglyceride into its polymorph was accelerated (Bunjes et al., 2001).

Wissing et al. (2004) incorporated CoQ10 in SLNs and investigated their structure properties using NMR spectroscopy as a principal analytical tool. The spin-lattice relaxation time was measured in rotating frame to characterize the homogenous integration of CoQ10 inside solid lipid matrix. Results demonstrated a large amount of CoQ10 was integrated homogeneously and separate domains on nanometer scale were formed from the remaining small fraction.

Farboud and co-workers (2011) formulated CoQ10-SLNs using lipids (cetylpalmitate or stearic acid), water and surfactants (Tego care 450 or tween 80), using technique of homogenization at high pressure. Developed formulation were characterization in terms of stability, particle size analysis and release studies. Results of in vitro release studies of CoQ10-SLNs cream illustrated biphasic release pattern.

CoQ10 SLNs incorporated into hydrogel was reported by Korkmaz and co-workers (2013), where preparation using high speed homogenization method was done. Characterization of the formulation was performed in the form of rheological properties, *ex vivo* diffusion studies and TEM. Delivery of CoQ10 was successfully demonstrated using diffusion studies using rat abdominal skin. The study results in efficient skin penetration without lowering of its antioxidant properties.

12.6.2　NANOSTRUCTURED LIPID CARRIERS (NLC)

NLCs are drug-delivery systems containing a core matrix made up of both solid and liquid lipids. They are termed as lipid nanoparticles of second generation, developed with the view to counter SLNs' limitations. NLCs symbolise an emerging tool in drug delivery and thus attracted attention in recent years (Fang et al., 2013).

Junyaprasert et al. (2009) prepared CoQ10 nano-emulsion using medium chain triglycerides and NLCs using cetyl palmitate/medium chain triglycerides. Formulations were characterized for viscoelastic and crystal properties for one year at 4, 25 and 40°C. NLC and nano-emulsion formulations showed more than 90% CoQ10 entrapment up to 12 months, however preparation exposed to light showed decreased entrapment. Rheological behavior of NLC and nano-emulsion dispersions results in pseudo-plastic flow, whereas hydrogels of NLC and nano-emulsion showed plastic flow. However, NLCs results in better chemical and physical stability when compared to nano-emulsion formulation.

CoQ10-NLCs prepared using hot high pressure homogenization method was reported by Yue and co-workers. Characterization were performed using freeze-fracture transmission electron microscopy and size. Study of cell viability and morphology demonstrated more effective protection of CoQ10-NLC compared with CoQ10-emulsion in UVA-irradiated fibroblasts. Yue et al. (2010) *concluded that in vivo* studies on CoQ10-NLC displayed enhanced stratum corneum and dermis penetration after topical application.

Chen et al. (2013) prepared CoQ10 NLCs by high-pressure microfluidics technique for epidermal targeting. CoQ10 NLC formulation was optimized using response surface methodology and characterized using TEM, DSC and PXRD technique. Evaluations were also performed using in vitro diffusion studies through animal skin. Concentration of emulsifier and solid lipid in the formulation showed significant role in particle size of NLCs. *In vitro* release study showed biphasic release with 3 h of fast release and prolonged release afterwards. Drug diffusion study conducted in vitro showed 10.11 times higher accumulation of CoQ10 in epidermis compared with CoQ10 emulsion.

The researchers reported that ultra-small size of NLCs shows influence on skin permeation, penetration and physicochemical stability. Another study conducted by Schwarz et al. (2013) formulated ultra-small-NLCs having mean particle size of 80 nm for enhanced dermal delivery of CoQ10. Lesser particle size of CoQ10-NLCs also enhanced skin penetration of CoQ10 along with physiochemical stability.

Nanjwade and co-workers (2013) developed CoQ10-NLCs using solid lipid (glyceryl behenate and glyceryl distearate) and liquid lipid (glyceryl triacetate) by solvent diffusion method. Encapsulation of CoQ10 was improved with addition of liquid lipid into the formulation. Studies concluded higher drug release and CoQ10 encapsulation results in effective antioxidant activity of prepared NLC compared to drug solution. *In vivo* results indicated that CoQ10-NLCs are more advantageous than solution, in terms of antioxidant activity.

12.6.3 SELF-EMULSIFYING DRUG DELIVERY SYSTEMS (SEDDS)

SEDDS are isotropic mixtures of oil and surfactants which usually contains co-surfactants also. Introduction of this mixture into aqueous phase with gentle stirring results in emulsification and produce fine oil-in-water emulsion. Such systems could be prepared from a eutectic interaction between drug and a suitable carrier. This approach for drug delivery is useful in counteracting the formulation problems associated with lower aqueous solubility and poor bioavailability (Singh et al., 2009). SEDDS have been evaluated as a carrier for CoQ10 delivery in many studies. However, as a liquid preparation, SEDDS possesses certain limitations, such as limited choice of dosage forms, stability concerns along with leakage issues, when added into capsules especially of hard gelatin origin. An alternative approach to circumvent the above mentioned limitations is SNEDDS, which involves use of porous solid carriers to adsorb liquid to get high specific surface area. The examples of solid porous carriers includes, amorphous silica (Syloid 244FP) or magnesium aluminometasilicate (Neusilin).

Kommuru et al. (2001) developed and characterized SEDDS of CoQ10. Myvacet 9–45 and CapteX-200 were used as oil phase and polyglycolyzed glycerides as emulsifier, in the formulation. Due to limited solubility of CoQ10 in oil phase drug load was confined to 5.66% (w/w). Significant improvement in bioavailability of CoQ10 was observed where SEDDS results in two-fold bioavailability increase when compared with powder formulation.

Balakrishnan and co-workers (2009) formulated CoQ10 SEDDS for oral administration comprising oil, surfactant and co-surfactant. The formulations used oil phase (Labrafil M 1944 and M 2125), surfactant (Labrasol) and co-surfactant (Lauroglycol FCC and Capryol 90). Characterization of formulations were performed using zeta potential, particle size and drug release behaviour. The pharmacokinetic study in rats using optimized SEDDS formulation showed significant improvement in Cma☐and AUC compared to powder formulation.

In a different approach, Onoue et al., evaluated the plasma concentration of CoQ10 in SEDDS in order to prove possible improvement in pharmacokinetic behavior of molecule after oral administration compared to crystalline CoQ10. Cmax and AUC were 5-fold increase when compared with pure crystalline drug and thus results in oral bioavailability improvements (Onoue et al., 2012).

Study conducted by Nazzal et al. (2002) results in CoQ10 SNEDDS preparation using lemon oil as the oil phase. Lemon oil is used in the formulation due to high solubility of CoQ10 in it. Prepared SNEDDS overcame the shortcomings of the traditional emulsified system, viz. low solubility and drug precipitation with time.

CoQ10 SNEDDS were also prepared by Nepal and co-workers (2010) and evaluated for oral pharmacokinetics of CoQ10 in male rats and plasma concentration was analyzed. The pharmacokinetic parameters, viz. AUC and C_{max} were significantly increased four to five fold, but T_{max} was found reduced, indicating fast drug absorption as compared with powder formulation.

A similar study was performed by Agrawal et al. (2014), where SNEDDS were further formulated in mouth dissolving tablets. The prepared CoQ10

SNEDDS were evaluated for in vitro and in vivo studies and showed enhanced aqueous solubility, dissolution velocity and hepato-protective activity. Results showed a promising improvement in nutraceutical and pharmaceutical values of CoQ10 using SNEDDS.

12.6.4   LIPOSOMES

Liposomes are vesicular delivery systems, where the bilayer vesicles are made up of phospholipid with an aqueous core. Due to amphiphilic nature of the vesicles, both lipophilic (in lipid bilayer) and hydrophilic drugs (in aqueous core) can be encapsulated into liposomes (Verma et al., 2003). Vesicle size in liposomes varies between 20 nm and a few hundred micrometers. Liposomes are widely used in variety of pharmaceutical and cosmeceutical preparations due to their biocompatible, biodegradable, nontoxic, and flexible nature of vesicles.

Xia and co-workers (2006) developed CoQ10-nanoliposomes prepared using ethanol injection and sonication method. Optimization of the formulation was achieved using orthogonal array design. Authors demonstrated the practical usefulness of this formulation technique at pilot scale with the expected encapsulation quality and stability.

Zhang and Wang (2009) investigated the efficacy of CoQ10 liposomes prepared using trimethyl chitosan coating in delaying selenite-induced cataract. Effects of trimethyl chitosan concentration and molecular weight was assessed on CoQ10-loaded liposomes. The study confirmed that trimethyl chitosan could modify pre-corneal retention time and physical properties of liposomes. In addition, these liposomes results in appropriate viscosity and showed better corneal contact.

Lee and Tsai (2010) prepared CoQ10-loaded liposomes composed of soya lecithin and α-tocopherol, by solvent injection method for topical applications. The obtained liposomes showed narrow size distribution and vesicles diameters was less than 200 nm. CoQ10 encapsulation significantly improved its accumulation in rat skin establishing CoQ10 liposomes as a promising formulation for topical application.

Gokce et al. (2012) prepared CoQ10 liposomes using thin film hydration and SLNs using high shear homogenization method. Studies related to CoQ10 formulations biocompatibility/cytotoxicity were performed using cultured cells (human fibroblasts) under o□idativeconditions. Enhanced cell proliferation was shown by the CoQ10-loaded liposomes, however protective effects against ROS accumulation were not reported by CoQ10 loaded SLNs. Authors concluded that CoQ10 liposomes were comparatively more promising than SLNs for effective dermal antioxidant activity.

Chitosan coated TPGS CoQ10 liposomes were reported by Shao and co-workers (2015), along with its optimization and oral bioavailability studies. The system demonstrated good mucoadhesive properties and was found to be more effective in enhancing CoQ10 cellular uptake in Caco-2 cells. Pharmacokinetic study concluded e□tended release profile with 3.4 fold improved systemic exposure of CoQ10 compared to untreated powder.

12.6.5 POLYMERIC NANOPARTICLES

Polymeric nanoparticles are promising colloidal drug carriers, prepared using biodegradable polymers. These are suitable for drug delivery, may be produced by several methods like *in situ* polymerization, spontaneous emulsification-solvent diffusion, emulsification-solvent evaporation and supercritical fluid techniques (Jung and Perrut, 2001). Polymeric nanoparticles were widely explored for enhancing the therapeutic benefits of drugs for oral and parenteral administration. Polymeric nanoparticles have been reported in literature for solubility and stability enhancement of number of drugs. These nanoparticles result in bioavailability improvement and protect the active against first pass effect and enzymatic degradation when administered through oral route (Jain et al., 2011). Over a last decade polymeric nanoparticles are widely investigated as oral delivery vehicles.

Ankola et al. (2007) developed potent CoQ10 loaded PLGA-based nanoparticles by emulsion technique, for oral delivery. The formulation was characterized using particle size, entrapment efficiency, in vitro release and *in situ* uptake studies. The intestinal CoQ10-uptake was evaluated using drug suspension, marketed formulation and developed nanoparticles in male rats. The uptake was found to be greater in developed nanoparticles (79%) as compared to commercial formulation (75%) and suspension (45%). This results in solubility and permeability enhancement using CoQ10 loaded PLGA-based nanoparticles.

Surfactant-free biodegradable PLGA-based CoQ10 nanoparticles were prepared by Nehilla and co-workers (2008) by nanoprecipitation method and characterized for in vitro physicochemical properties. The study mainly focussed on preparation of CoQ10 nanoparticles using surfactant-free nano-precipitation along with novel purification procedure. Prepared nanoparticles acts as a promising sustained CoQ10 delivery for antioxidant purpose.

Swarnakar et al. (2011) designed freeze dried CoQ10 PLGA nanoparticles prepared through scalable emulsion-diffusion-evaporation technique. Developed formulations were characterized for drug crystallinity and stability in simulated gastrointestinal fluids. Further, cellular uptake, sub cellular localization and antio☐idantefficacy of the formulations were investigated. It was concluded from in vitro studies that prepared nanoparticles significantly quenched ROS, with appro☐imately10-times more efficacious than pure CoQ10. Prepared formulation resulted in enhancement of oral bioavailability (4.28 times) over CoQ10 alone. The formulation showed improved hepato-protective and anti- inflammatory activity.

12.6.6   NANO-EMULSION

Nano-emulsion constitutes widely studied colloidal drug delivery system, by virtue of their good thermodynamic stability and sterilization by filtration. These are spontaneously formed oil/water emulsions, with tiny droplet-size, ranging between 10 to 100 nm. Another principal advantage is that a considerable amount of drug can be incorporated, as these have high solubilizing capacity. They consist of lipid, water, surfactant and co-surfactant in an appropriate proportion (Kumar et al., 2016). Till date, nano-emulsions as a carrier for active delivery have been thoroughly studied for multiple research applications, but only few authors have investigated nano-emulsion for CoQ10 delivery.

CoQ10 nano-emulsion was prepared by Belhaj and co-workers (2012) using sonication followed by high pressure homogenization method. The nano-emulsion formulation consisted of salmon oil, salmon lecithin, CoQ10 and water. CoQ10 nano-emulsion demonstrated a superior bioavailability compared to commercial formulation of CoQ10.

Cheuk et al. (2015) encapsulated CoQ10 using octenyl succinic anhydride modified starch, by high pressure homogenization to form nano-emulsion formulation. The resulting nano-emulsion ranged between 200–300 nm particles and zeta potential was reported as 8.4–10.6 mV in absolute. CoQ10 retention from the emulsion was 98.2% and concluded succinate as a main influential factor for zeta potential.

12.6.7   PRONIOSOMES

Proniosomes are pro-vesicular approach-based non-ionic surfactant vesicles that exists in gel or dry powder forms, and have immediate hydration property prior use to produce aqueous niosomal dispersion (Blazek-Welsh and Rhodes, 2001; Vora et al., 1998). Due to potential advantages of PNs over niosomes, latter is increasingly used to enhance drug delivery. Upon simple hydration PNs can be converted to niosomes. In case of topical drug delivery, hydration can be achieved by skin itself. For topical applications, PN gel is the preferred choice. Recently, proniosomes have been widely studied as a novel delivery system for skin penetration enhancement using highly lipophilic drugs.

Yadav et al. (2016) prepared novel proniosomal gel formulation of CoQ10 employing systematic design of experiment (DoE) approach. I-Optimal mixture design was adopted for optimization in systematic manner for proniosomal formulation and evaluation of experimental data was performed against CoQ10 entrapment and in vitro release. Aging was introduced on animal skin using UV radiations and proniosome formulations were applied. The effectiveness of the treatment was evaluated on the basis of biochemical estimation and histopathological studies. By using CoQ10 proniosome gel formulation, levels of superoxide dismutase (SOD), catalase (CA), glutathione (GSH) and total proteins were restored by 81.3%, 72.1%, 74.8%, and 77.1%, respectively, to that of control group. Histopathological studies revealed better protection of skin treated with CoQ10 proniosome gel compared to free CoQ10. Prepared proniosome gel was found to have skin histology closet to normal, hence tolerated by animal skin compare to conventional gel. The study also concluded the usefulness of novel proniosome for effective delivery of drugs.

12.7   CONCLUSION

Cosmetics were basically topical products used for maintaining and improvement of the general appearance of face and other external parts of the body and their prime purpose was cleansing and adornment. But the role of cosmetics has changed in the recent times and the expectations are much more. Microorganisms, chemical toxins, pollutants, extrinsic factors like sunlight, dry weather and other factors like disease, stressful life, poor diet and inadequate sleep are the culprits leading to dull and diseased skin, nails and hair. Cosmetics are now being used in association of active ingredients to check the damage and ageing of skin, hairs, nails, teeth, etc. New developments in drug delivery strategies have opened a new arena in healthcare and personal grooming. Cosmetic products are now being used as vehicles for topical therapeutic formulations. With the increasing interest in youthful, good looking appearance, flawless complexion and blemish free skin, presumably novel carrier systems will continue to establish an indispensable role in the cosmetics industry. An ideal carrier system for cosmetic and cosmeceutical use should be capable to permeate through the barrier skin, modulate the release of active ingredients from the formulation, and maintain an effective concentration in the skin layers. The antioxidant property of CoQ10 imparts versatility to the molecule, owing to which it has created a niche for itself in the cosmetic industry. A biomaterial in origin, it has an added advantage of being safe and useful, both as a health supplement for good skin as well as in the form of skin creams, face masks, etc.

KEYWORDS

- coenzyme Q10
- cosmetics.antioxidants
- nanocarriers
- proniosomes
- ubidecarenone

REFERENCES

Agrawal, A. G., Kumar, A., & Gide, P. S., (2014). Formulation development and *in vivo* hepatoprotective activity of self nanoemulsifying drug delivery system of antioxidant coenzyme Q10. *Arch. Pharm. Res.*, 1–16.

Ankola, D. D., Viswanad, B., Bhardwaj, V., Ramarao, P., & Kumar, M. N., (2007). Development of potent oral nanoparticulate formulation of coenzyme Q10 for treatment of hypertension: can the simple nutritional supplements be used as first line therapeutic agents for prophylaxis/therapy? *Eur. J. Pharm. Bioharm.*, 67(2), 361–369.

Balakrishnan, P., Lee, B. J., Oh, D. H., Kim, J. O., Lee, Y. I., Kim, D. D., et al., (2009). Enhanced oral bioavailability of Coenzyme Q10 by self-emulsifying drug delivery systems. *Int. J. Pharm.*, 374(1), 66–72.

Beg, S., Javed, S., & Kohli, K., (2010). Bioavailability enhancement of coenzyme Q10: An extensive review of patents. *Recent Pat. Drug Deliv. Formul., 4*(3), 245–257.

Belhaj, N., Dupuis, F., Arab-Tehrany, E., Denis, F. M., Paris, C., Lartaud, I., et al., (2012). Formulation, characterization and pharmacokinetic studies of coenzyme Q_{10} PUFA's nanoemulsions. *Eur. J. Pharm. Sci., 47*(2), 305–312.

Bhagavan, H. N., & Chopra, R. K., (2006). Coenzyme Q10: Absorption, tissue uptake, metabolism and pharmacokinetics. *Free Radic. Res., 40*(5), 445–453.

Blazek-Welsh, A. I., & Rhodes, D. G., (2001). SEM imaging predicts quality of niosomes from maltodextrins based proniosomes. *Pharm. Res., 18*(5), 656–661.

Boicelli, C. A., Ramponi, C., Casali, E., & Masotti, L., (1981). Ubiquinones: Stereochemistry and biological implications. *Membr. Biochem., 4*(2), 105–118.

Borekova, M., Hojerova, J., Koprda, V., & Bauerova, K., (2008). Nourishing and health benefits of coenzyme Q10—a review. *Czech J. Food Sci., 26*(4), 229–241.

Bunjes, H., Drechsler, M., Koch, M. H. J., & Westesen, K., (2001). Incorporation of the model drug ubidecarenone into solid lipid nanoparticles. *Pharm. Res., 18*(3), 287–293.

Chen, S., Liu, W., Wan, J., Cheng, X., Gu, C., Zhou, H., et al., (2013). Preparation of Coenzyme Q10 nanostructured lipid carriers for epidermal targeting with high-pressure microfluidics technique. *Drug Dev. Ind. Pharm., 39*, 20–28.

Cheuk, S. Y., Shih, F. F., Champagne, E. T., Daigle, K. W., Patindol, J. A., Mattison, C. P., et al., (2015). Nano-encapsulation of coenzyme Q10 using octenyl succinic anhydride modified starch. *Food Chem., 174*, 585–590.

Chopra, R. K., Goldman, R., Sinatra, S. T., & Bhagavan, H. N., (1998). Relative bioavailability of coenzyme Q10 formulations in human subjects. *Int. J. Vitam. Nutr. Res., 68*(2), 109–113.

Collins, M. D., & Jones, D., (1981). Distribution of isoprenoid quinone structural types in bacteria and their taxonomic implication. *Microbiol. Rev., 45*(2), 316–354.

Cosmetics & US Law. The United States Food and Drug Administration. [Online] 2016. http://www.fda. gov/Cosmetics/GuidanceRegulation/LawsRegulations/ucm2005209.htm.

Crane, F. L., (2001). Biochemical functions of coenzyme Q10. *J. Am. Coll. Nutr., 20*(6), 591–598.

De Leeuw, J., De Vijlder, H., Bjerring, P., & Neumann, H., (2009). Liposomes in dermatology today. *J. Eur. Acad. Dermatol. Venereol., 23*, 505–516.

Dhanasekaran, M., & Ren, J., (2005). The emerging role of Coenzyme Q-10 in aging, neurodegeneration, cardiovascular disease, cancer and diabetes mellitus. *Curr. Neurovasc. Res., 2*(5), 1–13.

Fang, C. L., Al-Suwayeh, S. A., & Fang, J. Y., (2013). Nanostructured lipid carriers (NLCs) for drug delivery and targeting. *Recent Pat. Nanotechnol., 7*(1), 41–55.

Farboud, S. E., Nasrollahi, S. A., & Tabbakhi, Z., (2011). Novel formulation and evaluation of a Q10-loaded solid lipid nanoparticle cream: *in vitro* and *in vivo* studies. *Int. J. Nanomedicine, 6*, 611–617.

Frei, B., Kim, M. C., & Ames, B. N., (1990). Ubiquinol-10 is an effective lipid-soluble antioxidant at physiological concentrations. *Proc. Natl. Acad. Sci. USA, 87*(12), 4879–4883.

Fulekar, M. H., (2010). Nanotechnology: Importance and Application, IK International Publishing House, New Delhi, India.

Garrido-Maraver, J., Cordero, M. D., Oropesa-Avila, M., Vega, A. F., Mata, M., Pavon, A. D., et al., (2014). Coenzyme Q10 therapy. *Mol. Syndromol., 5*(3–4), 187–197.

Gokce, E. H., Korkmaz, E., Tuncay-Tanrıverdi, S., Dellera, E., Sandri, G., Bonferoni, M. C., et al., (2012). Comparative evaluation of coenzyme Q10-loaded liposomes and solid lipid nanoparticles as dermal antioxidant carriers. *Int. J. Nanomedicine, 7*, 5109–5117.

Hoppe, U., Bergemann, J., Diembeck, W., Ennen, J., Gohla, S., Harris, I., et al., (1999). Coenzyme Q10, a cutaneous antioxidant and energizer. *Biofactors, 9*(2–4), 371–378.

Ibbotson, S. H., Moran, M. N., Nash, J. F., & Kochevar, I. E., (1999). The effects of radicals compared with UVB as initiating species for the induction of chronic cutaneous photo-damage. *J. Invest. Dermatol., 112*(6), 933–938.

Jain, A. K., Swarnakar, N. K., Godugu, C., Singh, R. P., & Jain, S., (2011). The effect of the oral administration of polymeric nanoparticles on the efficacy and toxicity of tamoxifen. *Biomaterials, 32*(2), 503–515.

Joshi, M. D., & Muller, R. H., (2009). Lipid nanoparticles for parenteral delivery of actives. *Eur. J. Pharm. Biopharm., 71*(2), 161–172.

Jung, J., & Perrut, M., (2001). Particle design using supercritical fluids: Literature and patent survey. *J. Supercrit. Fluids., 20*(3), 179–219.

Junyaprasert, V. B., Teeranachaideekul, V., Souto, E. B., Boonme, P., & Muller, R. H., (2009). Q10-loaded NLC versus nanoemulsions: stability, rheology and *in vitro* skin permeation. *Int. J. Pharm., 377*(1–2), 207–214.

Kapoor, P., & Kapoor, A. K., (2013). Coenzyme Q10 A novel molecule. *J. I. A. C. M., 14*, 37–45.

Kommuru, T. R., Gurley, B., Khan, M. A., & Reddy, I. K., (2001). Self-emulsifying drug delivery systems (SEDDS) of coenzyme Q10: Formulation development and bioavailability assessment. *Int. J. Pharm., 212*(2), 233–246.

Korkm, E., Gokce, E. H., & Ozer, O., (2013). Development and evaluation of coenzyme Q10 loaded solid lipid nanoparticle hydrogel for enhanced dermal delivery. *Acta Pharm., 63*(4), 517–529.

Kumar, S., Rao, R., Kumar, A., Mahant, S., & Nanda, S., (2016). Novel Carriers for Coenzyme Q10 Delivery. *Curr. Drug deliv.* [Epub ahead of print].

Lee, W. C., Tsai, T. H., Preparation and characterization of liposomal coenzyme Q10 for *in vivo* topical application. *Int. J. Pharm.,* (2010). 395(1–2), 78–83.

Li, D., Wu, Z., Martini, N., & Wen, J., (2011). Advanced carrier systems in cosmetics and cosmeceuticals: A review. *J. Cosmet. Sci., 62*(6), 549–563.

Littarru, G. P., & Tiano, L., (2007). Bioenergetic and antioxidant properties of coenzyme Q10: Recent developments. *Mol. Biotechnol., 37*, 31–37.

Lohani, A., Verma, A., Joshi, H., Yadav, N., & Karki, N., (2014). Nanotechnology-based cosmeceuticals. *ISRN dermatol., 843687.* doi: org/10. 1155/2014/843687.

Mukherjee, S., Ray, S., & Thakur, R. S., (2009). Solid lipid nanoparticles: a modern formulation approach in drug delivery system. *Indian J. Pharm. Sci., 71*(4), 349–358.

Nanjwade, B. K., Kadam, V. T., & Manvi, F. V., (2013). Formulation and characterization of nanostructured lipid carrier of ubiquinone (Coenzyme Q10). *J. Biomed. Nanotechnol., 9*(3), 450–460.

Nazzal, S., Smalyukh, I. I., Lavrentovich, O. D., & Khan, M. A., (2002). Preparation and *in vitro* characterization of a eutectic based semisolid self-nanoemulsifying drug delivery system (SNEDDS) of ubiquinone: mechanism and progress of emulsion formation. *Int. J. Pharm., 235*(1–2), 247–265.

Nehilla, B. J., Bergkvist, M., Popat, K. C., & Desai, T. A., (2008). Purified and surfactant-free coenzyme Q10-loaded biodegradable nanoparticles. *Int. J. Pharm., 348*(1–2), 107–114.

Nepal, P. R., Han, H., & Choi, H. (2010). Enhancement of solubility and dissolution of coenzyme Q_{10} using solid dispersion formulation. *Int. J. Pharm., 383*(1–2), 147–153.

Okada, K., Kainou, T., Matsuda, H., & Kawamukai, M., (1998). Biological significance of the side chain length of ubiquinone in *Saccharomyces cerevisiae*. *FEBS Lett., 431*, 241–244.

Onoue, S., Uchida, A., Kuriyama, K., Nakamura, T., Seto, Y., Kato, M., et al., (2012). Novel solid self-emulsifying drug delivery system of coenzyme Q_{10} with improved photo-chemical and pharmacokinetic behaviors. *Eur. J. Pharm. Sci., 46*(5), 492–499.

Parkhideh, D., (2008). Methods and compositions that enhance bioavailability of coenzyme-Q10. U. S. Patent 7, 438, 903 B2, October 21.

Schwarz, J. C., Baisaeng, N., Hoppel, M., Low, M., Keck, C. M., & Valenta, C., (2013). Ultra-small NLC for improved dermal delivery of coenyzme Q10. *Int. J. Pharm., 447*(1–2), 213–217.

Shao, Y., Yang, L., & Han, H., (2015). TPGS-chitosome as an effective oral delivery system for improving the bioavailability of coenzyme Q10. *Eur. J. Pharm. Biopharm., 89*, 339–346.

Singh, B., Bandopadhyay, S., Kapil, R., Singh, R., & Katare, O., (2009). Self-emulsifying drug delivery systems (SEDDS), formulation development, characterization, and applications. *Crit. Rev. Ther. Drug Carrier Syst., 26*(5), 427–521.

Sinico, C., & Fadda, A. M., (2009). Vesicular carriers for dermal drug delivery. *Expert Opin. Drug Deliv., 8*, 813–825.

Swarnakar, N. K., Jain, A. K., Singh, R. P., Godugu, C., Das, M., & Jain, S., (2011). Oral bioavailability, therapeutic efficacy and reactive oxygen species scavenging properties of coenzyme Q10-loaded polymeric nanoparticles. *Biomaterials, 32*(28), 6860–6874.

Verma, D. D., Verma, S., Blume, G., & Fahr, A., (2003). Particle size of liposomes influences dermal delivery of substances into skin. *Int. J. Pharm., 258*(1–2), 141–151.

Vora, B., Khopade, A. J., & Jain, N. K., (1998). Proniosome based transdermal delivery of levonorgestrel for effective contraception. *J. Control. Release, 54*(2), 149–165.

Wang, S., Zhang, J., Jiang, T., Zheng, L., Wang, Z., Zhang, J., & Yu, P., (2011). Protective effect of Coenzyme Q10 against oxidative damage in human lens epithelial cells by novel ocular drug carriers. *Int. J. Pharm. 403*, 219–229.

Xia, S., Xu, S., & Zhang, X., (2006). Optimization in the preparation of coenzyme Q10 nano-liposomes. *J. Agr. Food Chem., 54*(17), 6358–6366.

Yadav, N. K., Nanda, S., Sharma, G., & Katare, O. P., (2016), Systemically optimized coen-zyme q10-loaded novel proniosomal formulation for treatment of photo-induced aging in mice: Characterization, biocompatibility studies, biochemical estimations and anti-aging evaluation. *J. Drug Target., 24*(3), 257–271. DOI: 10.3109/1061186X.2015.1077845.

Yoshida, H., Kotani, Y., Ochiai, K., & Araki, K., (1998). Production of ubiquinone-10 using bacteria. *J. Gen. Appl. Microbiol., 44*(1), 19–26.

Yue, Y., Zhou, H., Liu, G., Li, Y., Yan, Z., & Duan, M., (2010). The advantages of a novel CoQ10 delivery system in skin photo-protection. *Int. J. Pharm., 392*(1–2), 57–63.

Zhang, J., & Wang, S., (2009). Topical use of coenzyme Q10-loaded liposomes coated with trimethyl chitosan: tolerance, precorneal retention and anti-cataract effect. *Int. J. Pharm., 372*(1–2), 66–75.

Zhang, M., Dang, L., Guo, F., Wang, X., Zhao, W., & Zhao, R., (2012). Coenzyme Q10 enhances dermal elastin expression, inhibits IL-1a production and melanin synthesis in vitro. *Int. J. Cosmet. Sci., 34*, 273–279.

Zuelli, F., Belser, E., Schmid, D., Liechti, C., & Suter, F., (2006). Preparation and properties of Coenzyme Q10 nanoemulsions. *Cosmetic Science Technology*, 40–46.

CHAPTER 13

NANOBASED CNS DELIVERY SYSTEMS

RAHIMEH RASOULI,[1] MAHMOOD ALAEI-BEIRAMI,[2,3]
and FARZANEH ZAAERI[4]

[1]Department of Medical Nanotechnology, International Campus, Tehran University of Medical Sciences, Tehran, Iran, E-mail: r-rasouli@razi.tums.ac.ir

[2]Drug Applied Research Center and Students' Research Committee, Tabriz University of Medical Science, Tabriz, Iran

[3]Faculty of Pharmacy, Tabriz University of Medical Sciences, Tabriz, Iran

[4]Department of Pharmaceutics, Faculty of Pharmacy, Tehran University of Medical Sciences, Tehran, Iran

CONTENTS

13.1   INTRODUCTION

By taking consideration of increasing in the number of *aging populations*, growing trend of central nervous system associated disorders and following obstacles of pharmacotherapy of Central Nervous System (CNS) infections require more portion of research. Improvement of CNS delivery applying new technologies could solve delivery problems of traditional pharmacotherapy and imaging of CNS disorders, for example, reduce drug trafficking and requires high dose of agents which result in toxicity and defect of medical process. Being familiar with physiology, anatomy and mechanism of action of CNS is the basis of improving pharmacodynamical and pharmacokinetical characteristics of drugs and delivery vehicles. Designing efficient delivery systems, possessing required characteristics considering diseases and treatment regimens, could increase treatment success rate. Nanotechnology had been shown as a potent science in designing suitable vehicle to CNS delivery of prognosis, imaging and therapeutical agents, that could authorize most of novel drug delivery systems characteristics, for example, improved drug trafficking and dose optimization, metabolization defense, control release, toxicity control and target oriented delivery. This chapter reviews the achievements of nanoscience used in improving CNS delivery, recently.

13.2   PERSPECTIVES IN CNS DELIVERY AND OBSTACLES

Although novel delivery technologies in drug delivery are appreciable but, could not afford CNS delivery (because of BBB presence) and as a result treatment of CNS diseases like viral or bacterial meningitis, tumor and Alzheimeror Parkinson disease need more researches to reach satisfying outcomes (Jain et al., 2013; Wong et al., 2012; Zhou et al., 2012).

Less availability of CNS by drugs, considering BBB and BCSF presence, solubility problems, In-vivo instaibility, protein binding (PB) and efflu□transporters are case of problems in the view of pharmacokinetice (reaching therapeutic window) (Moghimi et al., 2005; Shah et al., 2013; Wong et al., 2010). Then improvement of target oriented (and site specific) drug delivery systems required to win the battle of CNS problems (Wong et al., 2012). The most important problem in CNS delivery is its' innate defense system against exogenic toxic and undesirable agents, then awareness of these defense systems and mechanism of transporting of agents through these barriers could guide us to our aims (Jain et al., 2013).

13.3 BBB AND BCSF PHYSIOLOGY

BBB and BCSF segregate Brain parenchyma from Blood. These two barriers have firm control on trafficking of agents to CNS. BBB configured from capillary endothelial cells with intercellular tight junctions and possessing so high electrical resistivity, 1500–2000 (1300) Ωcm^2 (Brightman and Reese. 1969; Crone and Christensen. 1981; Crone and Olesen. 1982; Hawkins and Davis. 2005) (compared with other organs having average resistance of 1.8 Ωcm^2, surrounded with basement membrane, astrocytes and pericytes (Figure 13.1). Tight junctions and basal membrane are responsible to the integrity and maintenance of the barrier. Besides pericytes inductive role, they are participated in structure and regulation of BBB and more importantly regeneration of BBB in pathologic conditions (Armulik et al., 2010; Liu et al., 2012). Astrocytes have structural (most abundant glial cells of brain) and homeostasis role in brain like: fuel support (glycogen stoke and release), supporting endothelial cells biochemically, nutrients (e.g., lactate) of nerves, transmitter uptake and release (glutamate transporters for glutamate, ATP, and GABA), synaptic transmission modulation (e.g., in hippocampus act as inhibitors by releasing ATP), extracellular ion balance maintenance, inducing myelinating activity of oligodendrocytes by secreting cytokineleukemia inhibitory factors (LIF), and nervous system repair (circuit repair) by substituting damaged cells by it-self. In addition to these important roles, they have included in hemostasis of CNS by taking part in vasomodulation and BBB structure, also its fundamental role in BBB refers to vasomodulation which affect blood follow of brain (Banks. 2009; Egleton and Davis. 1997; Hamilton et al., 2007; Kusuhara and Sugiyama. 2001; Misra et al., 2003).

Well-designed mechanical barrier besides lack of pinocytosis, efflu□ proteins, high electrical resistance and enzymatic system prevent big molecules, lipophilic agents, hydrophilic agents and miscellaneous, respectively (Gonzalez-Mariscal et al., 2003; Hamilton et al., 2007; Saunders et al., 2008; Yousif et al., 2007). Lipophilic small agents (e.g., hormones) and gases passes BBB by passive diffusion and vice employ catalyzed transport so that glucose, amino acids and nucleosides pass

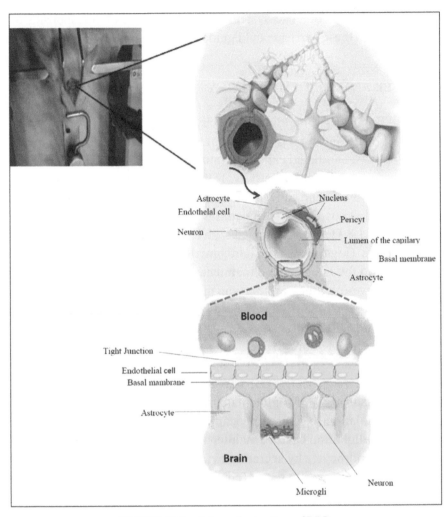

FIGURE 13.1 Schematic structure of BBB.

BBB by carrier mediated transport and bigger molecules as insulin or transferrin employ Receptor-Mediated Transporters. Other agents to pass BBB should have receptor, transporter or be cationic to employee adsorptive transport otherwise it is forbidden to enter CNS. BCSF as a second barrier and defense system of CNS excreted by Choroid Plexus endothelia and control the passage of agents from interstitial fluid to brain parenchyma. Although BCSF possess less area than BBB but its' role in controlling of nutrient and xenobiotic to parenchyma is considerable. Capillary of Choroid Plexus are permeable and have less limitation on water and solute trafficking. Indeed, substantial role of endothelial cells of Choroid Plexus are maintenance of homeostatic composition of CSF (Engelhardt and Sorokin, 2009; Enting et al., 1941–1955; Groothuis and Levy, 1997; Segal, 2000).

13.4 BBB TRAFFICKING

13.4.1 PASSIVE DIFFUSION

Movement of solute across membrane driven by the physicochemical concentration gradient (from high concentration to low concentration) without consumption of (chemical) energy until omission of gradient is termed passive diffusion. Rate of passive diffusion depends on the membrane permeability, solute lipophilicity and molecular weight. Solutes with molecular weight less than 400 Dalton pass BBB using passive diffusion. Most of small molecules are not able to pass BBB considering high molecular weight or preset water solubility (Camenisch et al., 1998; Cohen and Bangham, 1972; van de Waterbeemd et al., 1998). Then it could be concluded that, low concentration, big size and preset water solubility (more hydrogen binding donors) prevent transporting BBB. Because of ATP-Binding caste proteins in BBB which efflux agents to blood, most of agent could not reach effective dose in CNS despite employing passive diffusion. Diffusions could be transcellular (lipophilic agents like ethanol) and paracellular (like cellular Sucrose that limited because of tight junctions).

13.4.2 CATALYZED TRANSPORT

Catalyzed transport (carrier mediated transport and adsorptive or receptor mediated transcytosis) besides diffusion are two ways of transporting from BBB which low molecular weight molecules, cationic molecules, and peptides plus proteins pass BBB by carrier mediated, adsorptive and receptor mediated transport, respectively. Endocytosis could be categorized as bulk (fluid) phase endocytosis (pinocytosis), receptor mediated endocytosis and adsorptive endocytosis. Pinocytosis is non-competitive, unsaturable, temperature and energy dependent and non-dedicated which is done in limited degree in cerebral micro vascular (Pardridge et al.,

1995). Macropinosis as non-dedicated lets macromolecules (big particles) to pass BBB (Amyere et al., 2001).

13.4.2.1 Carrier Mediated Transport

Carrier mediated transport is categorized in two groups. Facilitated by diffusion, which solute pass moving down concentration gradient (downhill) by conformational conversion of specific membrane protein transporter (solute carrier group either facilitative transporters or secondary active transporters also called coupled transports or co-transports which both of them employ entropic energy) following binding solute to it. For example, glucose and nucleosides, cationic amino acids, small peptides, monocarboxylic acid, glutathione, choline, purine, neutral amino acids or their derivatives like melphalan and L-dopa (Cornford et al., 1992; Ganapathy et al., 2009; Pardridge. 2007,2008; Tsuji. 2005; Tsuji and Tamai. 1999; Wade and Katzman. 1975), and active transport by which transfer solute against concentration gradient (uphill) by hydrolyzing ATP like ABC transporters (i.e., MDR, MRPs 1-6, and BCRP) (Crone, 1965; Egleton and Davis, 2005 Zlokovic et al., 1993).

13.4.2.2 Transcytosis

13.4.2.2.1 Adsorptive-Mediated Transcytosis (AMT)

Depended on the interaction of the positive charge of molecule and negative charge of the BBB, there is high capacity for AMT (Bickel et al., 2001) which could be employed in equipping carriers or CNS delivery of cationic molecules (esp. cationic peptides) and cell-penetrating peptides (CPPs), for example Tat-derived peptides and SyN-B vectors (Abbott et al., 2010; Béduneau et al., 2007; Gabathuler, 2010; Pardridge et al., 1993). Toxicity and immunogenicity besides susceptibility of peptide vectors are disadvantages of CPPs (Gabathuler, 2010), which should be improved.

13.4.2.2.2 Receptor Mediated Transcytosis

It is specialized to transport of high weight molecules, vital to brain, by attaching to receptor on the luminal barrier. Different receptors have been evaluated for hormones, growth factors, enzymes and plasma proteins. Transferrin, Insulin, Leptin, insulin-like growth factor I and II, LEPR, low density lipoproteins and transferrin have receptor on BBB (Duffy et al., 1988; Nicholson and Syková, 1998; Pardridge, 2007). These kinds of transporters are being used in designing of CNS oriented colloidal vehicles like nanoparticles and liposomes equipped with receptors' ligand for example transferrin and transferrin-receptor monoclonal antibody conjugated nanoparticles was successful in passing BBB (Boado and Pardridge, 2011; Gabathuler, 2010; Pardridge, 2007, 2008; Wagner et al., 2012). Because of the widely distribution of these receptors in different tissues, applying this mechanism should be evaluated firstly in the view of selectivity (Belfiore et al., 2009; Gao et al., 1993; Lillis et al., 2008).

13.5 CHARACTERISTICS THAT MAKE AGENTS ABLE TO PASS BBB

BBB trafficking is limiting factor of CNS pharmaceutical agents' kinetic, which need more interest in CNS drug designing. Tight junction, electrical resistance, efflux proteins and catalyzing enzymes should be considered in designing agents (Wong et al., 2010, 2012). To pass BBB employing passive diffusion molecular weight of hydrophobic and hydrophilic agents most be less than 400–600 KDa with less than 8–10 hydrogen bands and 150 da, respectively (Banks, 2009; Pardridge, 2001, 2007; Scherrmann, 2002). In addition, since diffusion depends on concentration, high concentration in capillary required to promotes diffusion. Because of this restriction, 98% of small molecules and 100% of large molecules like antibodies even engineering antibodies (around 15–150 KDa) do not pass BBB. This mechanism is responsible to 98% of drugs that do not pass BBB (Garcel et al., 2010; Holman et al., 2011). It could be concluded that small lipophilic molecules with less protein binding (PB) and long serum half-life are good candidates to employee passive diffusion passing BBB. With this regard, lipophilicity, size and ionization degree are magnified physicochemical characteristics which should be considered in designing BBB trafficable agents (Frank et al., 2011).

Passive diffusion could not afford to all of pharmaceuticals and at the other hand noticeable portion of agents or there carriers employee catalyzed transport. For example, carrier mediated transport and receptor mediated transcytosis are goals of carriers or drugs which are conjugated to amino acids, glucose, and nucleosides or bigger molecules like insulin, and transferrin, respectively (Pardridge, 2008; Shah et al., 2013).

Conjugating lipophilic entities, as like methyl or fatty acyl entities, with the aim of increasing lipophilicity could enhance BBB passage although, toxicity following increased distribution to other organs should be analyzed (Joubert et al., 2012; Hansen et al., 1992; Weber et al., 1991). Cationic surface cause to adsorptive transcytosis and could be employed in drug delivery especially Nanoscience, for example, Albumin base nanocarriers. Immunogenicity of Albumin should be considered in long-term usages (Bickel et al., 1994; Gabathuler, 2010; Magga et al., 2010; Pardridge et al., 1990; Sakaeda et al., 2000). Below 200 nm, spherical shape, inert surface ion activities are optimum identities to escape from immune system and phagocytosis which result in elongated blood circulating time of agents, in addition, enhancing colloidal stability applying PEG or surfactant (like polysorbates) is another requirement to elongate half-life (Park et al., 2008; Wong et al., 2010). In addition to increased stability, polysorbates decrease immunogenicity and help to evade from reticuloendothelial system. Variable kinds of agents with different physicochemical and therapeutical identities beside heterogeneous distribution of saturable transporters reveal that characteristics of carriers targeting different parts of Brain should vary (Dalgleish et al., 1984).

Choosing accurate biomaterial and nanocarriers considering transporting system, drug dose, and target characteristics are requirements of carrier designing. Optimizing size, surface charge and hydrophobicity could employee nonspecific passive CNS targeting.

Drugs' kinetic as like as other agents in body is widely affected by efflu□ transporters. Protecting system of CNS, BBB, is fitted by these transporters which limit entrance (distribution) to CNS and facilitate drug elimination from it, in addition equilibrate extracellular-intracellular concentration of exogenous and endogenous agents.

ATP-Binding cassette Proteins (ABC proteins) are one of biggest barrier related proteins superfamily which translocate substrates and metabolites against concentration gradient consuming ATP and over expressed on endothelial and epithelial cells. Because of cellular extrusion of drugs which result in drug resistance, some form of ABC proteins ABCC, called Multi-Drug Resistance Proteins (MRP). ABCB1 [also called MDR1, and P-glycoprotein (P-gp)] (Potschka, 2012), and ABCG2 called as Breast Cancer Resistance

Protein (BCRP), known for first time in breast cancer resistance, are e□tra examples of ABC transporters which are included in extrusion of wide variety of molecules. MRP1–6, MDR1 (P-gp) and ABCG2 are expressed in BBB and should be modulated in advanced CNS drug delivery systems to improve drugs' kinetic. Being hydrophobic, amphipathic, ionic (both cationic and anionic) could be excluded by MDR1 (MRP1) like Antineoplastic agent (Deguchi et al., 2006; Doyle et al., 1998; Gao et al., 2000; Kabanov et al., 2003; Krishna and Mayer, 2000; Löscher and Potschka, 2005; Wong et al., 2010).

It could be concluded that although, lipophilic small molecules with optimized surface activity, low protein binding and long half-life are good candidate to pass BBB and BCSF (Frank et al., 2011) but, their not only characteristics beside toxicity which CNS engineer should consider designing BBB transportable carriers and more challenges required to development of CNS delivery.

13.6 STRATEGIES PROPOSED FOR PASSING THROUGH BBB

13.6.1 INVASIVE

13.6.1.1 BBB Disruption

Temporary paracellular disruption of BBB isone of strategies for CNS delivery which could be done with applying different substrates characterizing hyperosmolar chemical stimulus, like mannitol, urea, or surfactants including Tween 80 or sodium dodecyl sulfate (SDS), that result in uncoupling of tight junctions following shrinking of capillary endothelial cells exerted by water efflux. This phenomenon, which cause to paracellular follow of agents like drugs, tried on animal models and resulted in increased BBB permeability (Fortin et al., 2007). In addition, biological stimulus, like Bradykinin analogues including RMP-7 and some cytokines, induce this mechanism in BBB and tumor locations (Barichello et al., 2011; Bartus et al., 2000; Fortin et al., 2005; Hynynen et al., 2001; Prados et al., 2003; Saija et al., 1997; Sakane et al., 1989). Although these components are toxic, because of being non-dedicated allowing to passage of pathogenic microorganisms, toxins, and plasma proteins, and could not be used for long-term therapies, but could be beneficial in medical crisis or malignant conditions where could be cost effective considering outcome versus disadvantages (Misra et al., 2003).

13.6.1.2   Intracerebral Implantation

Gliadel® wafer, a biodegradable sustain release polifeprosan 20 implant loaded with carmustine, had got FDA approval as delivery vehicle directly to brain tumor. In this method implant is being located in tumor site after removal of tumor, in which drug release gradually following dissolving excipients. Less loading capacity is limitation of this method (Pathan et al., 2009).

13.6.1.3   Convectio4n-Enhanced Delivery (CED)

One of promising technics to brain delivery of chemotherapeutic agents over coming delivery obstacles and improving outcome is CED which had showed its' potential in clinical studies. CED is done by applying several catheters topically in tumor site stereotactically which pharmaceutical formulation conducted to site under concentration and pressure gradient (Debinski and Tatter, 2009). Uniform distribution of drug in tumor site is one of CEDs advantages against implants. Disruption of drug rising from elimination compartments of CNS (Blood and CSF), which could be improved by sustain release formulation, besides practical problems and patient screening are CED's disadvantages (Debinski and Tatter, 2009; Zhou et al., 2012).

13.6.1.4   Focused Ultrasound

Focused ultrasound (FUS) is a kind of BBB disruption in which microbubbles having lipid or albumin membrane with less than 50 μm size in blood vessels receive ultrasound waves and convert acoustic energy to mechanical energy which disrupt BBB (Fan et al., 2014). Erythrocyte extravasation following inertial cavitation is matter of issue of FUS which could be solved by resonant frequency-matched sonication which result stable cavitation that is favor of this method to BBB-Opening. In a study submicron bubbles had delivered chemotherapy agents locally 60 times more than normal state by applying focused ultrasound method (Fan et al., 2014). Safety of FUS is not well studied and is matter of issue (McDannold et al., 2005; Meairs and Alonso, 2007; Vykhodtseva et al., 2008).

13.6.2   NON-INVASIVE

13.6.2.1   Drug Modification

1. Cationization of means (prodrug or carrier) result in interaction with negative charge of barrier, consequently stimulate endocytosis of particles (Alam et al., 2010).
2. Increasing lipophilicity using lipophilic moieties (e.g., methyl, chlorine, fatty acid chains) make agents able to pass lipid bilayer membrane easily. Lipophilic agents are able to distribute to other compartments as like as brain thereupon increase side effects and require high dose of drug which limit this technique (Alam et al., 2010).

13.6.2.2   Cell-mediated Drugs Delivery

Active site targeting of drug, less immunogenicity and toxicity, release management of drug (chronically control), prolonged half-life are advantages of cell-mediated drug delivery. In this method some immune cells (monocytes and macrophages), neutral stem cells, and mesenchymal stem cells is being employed. During inflammation process mononuclear phagocyte (MP) cells called widely through chemotaxis and diapedesis. Because of extravasation and migration ability, these cells pass through BBB and result in its' breakdown. Studies revealed that, because of increased permeability of BBB in some disease associated with inflammation, intravenous application of these cells loaded with active agents could improve CNS delivery (Aboody et al., 2000, 2006; Lee et al., 2009; Najbauer et al., 2007). Less cell capacity and agent degradation inside cell (e.g., inside lysosome of cell) are disadvantages of this method which could be solved by sustain release target delivery formulations requiring less doses of drug (for example, magnetic field could be employed in guiding magnetic nanoparticle to target side.) and loading drug to nanosized formulations resisted to degradation, respectively (Batrakova et al., 2011).

Three methods are common for preparing carrier cells and loading drug to it: (i) loading drug formulation to prepared immunocyte fraction obtained from systemic blood (in-vitro); (ii) Culturing bone marrow biopsy and differentiation of it to immunocyte and then loading formulation to it (in-vitro); (iii) Phagocytosis of immunocyte-oriented (able to targeting immunocyte) nano-formulation injected to body (in-vivo) (Batrakova et al., 2011).

A study showed that formulation of macrophage carrying Indinavir loaded nanoparticle having phospholipid coat was more successful than classic formulations in decreasing HIV count in HIV infected mouse model because of macrophage ability in migration to brain and solving kinetic obstacle of anti-retroviral therapy (ART) in passing CNS (Dou et al., 2007). Neutral and mesenchymal stem cells, have inherent tropism to metastatic tumor and although are able to pass BBB (Aboody et al., 2008; Najbauer et al., 2007) could be applied in CNS delivery of macromolecules unable to pass BBB (including antibodies) which is being used in CNS initial and metastatic tumors' treatment protocols (Balyasnikova et al., 2010; Compte et al., 2009).

13.6.2.3   Olfactory Route (Intranasal or Nose-Brain Pathway)

Drug conduct directly to brain (bypassing BBB) by attaching to olfactory epithelium and passing through olfactory nerves. This method have several advantages including: increased drug bioavailability, shortest distance and most fast adsorption to Brain, and elimination of first pass effect. Safety of this method should be evaluated. Size, solubility and stability of agent affect speed of adsorption so that only one percent of administered agents could pass to brain through this way. Formulation of drug in the form of chitosan and lectin nanoparticles, which have good ability in adhering nasal mucous, could overcome this obstacle. Because of good protection and mucosal retention of intranasal nano-formulations which improve absorption of macromolecules, pharmaceutical researches have a good concentration on Nanotechnology-based nose-brain formulations so that some formulation are in preclinical studies (Illum, 2007).

13.6.2.4   Inhibition of ABC Transporters

Efflux transporters embedded in BBB and BCSF export agents distributed to brain compartment (enhance elimination) [100, 101]. ATP Binding Cassette proteins (ABC transporters) over expressed in endothelial and epithelial membranes are one of biggest superfamily related to membrane, which employ active transport by which transfer solute against concentration gradient (uphill) by hydrolyzing ATP like ABC transporters (Dean et al., 2001).

Synchronous injection of ABC inhibitors with active agents could blockage extruding of agents that is responsible to under therapeutic window concentration of some ABC prone agents and improve outcome. Efflu□ transporter modulation, one of nanocarriers designing goals for drug delivery, is proposed for future research activities (Bickel et al., 2001; Kabanov et al., 2003; Löscher and Potschka, 2005; Potschka, 2012).

13.6.2.5 Exosomes

Exosomes are membrane-enclosed and endosome-derived vesicles bodies (MVBs) with the plasma membrane, the vesicles release in to the e□tracellular milieu, are called Exosomes (Théry et al., 2002; van Niel et al., 2006; Wendler et al., 2013).

Exosomes released from brain tumor cells can escape the blood–brain barrier and have potential to apply in malignant gliomas therapy. Recent finding suggest that e□bsomes-mediated delivery of therapeutic drugs is a new means for drug delivery but more investigations and considerations are required to clarify the safety parameters, immunogenicity, as well as route of delivery across the BBB for future clinical application in CNS diseases (Graner, 2011; Lai and Breakefield, 2012; Lakhal and Wood, 2011; Vlassov et al., 2012).

Brain-targeting exosomes, by fusing neuron-targeting rabies viral glycoprotein (RVG) peptides in the N-terminus of Lamp2b, a murine e□bsomal membrane protein, were loaded with siRNAs then brain-targeting Exosomes were delivered systematically into syngeneic mice, resulting in a significant successful in knock-down of the targeted mRNAs in the brain. This finding suggests brain-targeting E□bsomes have promising potential in RNAi therapy (Lai and Breakefield, 2012; Lai and Breakefield, 2012; Vlassov et al., 2012).

13.6.2.6 Nano-delivery systems

Unique biological and physicochemical characteristics of nanoparticle make them able to pass tissue and cell barriers. Nanoparticles, especially lipid and polymeric ones are widely applied in drug delivery. With the aim of better adaptation of drug carrier with biological systems conception of drug's pharmacokinetic and pharmacodynamic besides conformational variation of BBB in photogenic conditions of CNS and following it drug and carrier engineering for effective targeting and transferring, required. Carriers with high capacity equipped with targeting moieties could be more beneficial (Ewend et al., 2007). Carriers having less than 100 nm are likely to pass BBB without having disruptive characteristic to CNS, then phospholipids, PLGA and some non-ionic surfactant could be good candidate in CNS delivery.

CNS specific Nano-delivery an addition to other Nano-delivery systems advantages (degradation resistant target oriented sustain release catalyze-immune and accurate delivery of low dose (increased bioavailability) plus less systemic toxicity) by transferring BBB solve traditional neuropharmacotherapy regimes defects (e.g., ART) and considering importance and complexity of CNS safety, multifunctional imaging-curing target oriented nanocarriers could elongate life expectancy by curing malignant or incurable disease.

Nanoparticles employ receptor-mediated endocytosis passing BBB, then active targeting (conjugation monoclonal antibodies (mAb), cell penetrating peptides (CPP), and ligand to nanoparticles) could be good solution in improving their BBB transferring. In addition, conjugating target cell oriented ligand or mAb could increase efficacy (Gao et al., 2013; Kawakami et al., 2001).

Although active targeting is likely to be reasonable, but some studies revealed that some not functionalized nanoparticles (e.g., intranasal silica nanoparticle in rodent, gold nanoparticles, and titanium oxide nanoparticles less than 40 nm) have substantial distribution to CNS, which could be considered in CNS drug delivery development (Roy et al., 2003; Wu et al., 2013).

Although targeting is both specific and efficient but not standardized. Safety and selectivity of formulations should be studied. Harmonized composition of carrier system with active ingredient could increase success rate (Fung et al., 1996).

Optimizing of nanoparticles size, surface charge, and hydrophobicity could employ passive targeting which is likely to employ passive diffusion, too (Wong et al., 2010).

13.7 STRATEGIES FOR ACTIVE TARGETING OF NANOPARTICLES

Active targeting apply specialized ligand against receptors which over express in patient (Bae and Park, 2011; Gao et al., 2013; Liu, 2012). Conjugating ligand on nanoparticles could result in the variation of interaction of nanoparticles and serum proteins, cellular compartment and affect blood circulating time In-vitro, then interaction of serum proteins and ligands should be studied and optimized (Fillmore et al., 2011; Florence, 2012). As in all of the systems nanoparticles interact with cellular membrane, surface characteristic of nanoparticles could affect mechanism of internalization and intracellular localization. Nanoparticles should be biocompatible and biodegradable and produce well characterized and do not produce toxic metabolites. Although, data for the applicability of nanoparticles in CNS delivery are hopeful, but there application is in experimental phase mostly, new methods for improving specificity, safety and efficiency should be tested.

Two methods are used for brain targeted drug delivery: (i) Direct method: in which nanoparticle link to a substrate of endothelial cells receptors, as an instance: nano particle linked to ApoE which is a substrate of LDL receptors on brain's endothelial cell; and (ii) Indirect method: so that nanoparticle link to an agent which could attach to brain endothelial cells 'substrate in body as an instance: nanoparticles linked to polysorbates (which is a non-ionic surfactant) could attached to Apo E in body (Göppert and Müller, 2005; Kerwin, 2008; Kerwin, 2008; Michaelis et al., 2006).

Monoclonal antibody, CPP, receptors' substrates are used as targeting agent on the nanoparticle surface for specific drugs delivery. As an example, OX26 (anti transferrin antibody), which are linked to liposome are as targeting agents for brain drug delivery. As an another example, liposomes which are conjugated with folate or transferrin, or are able directly or indirectly attach to Apo-A or ApoE could increase CNS delivery significantly (Huwyler et al., 1997; Liu et al., 2008; Michaelis et al., 2006; Wong et al., 2010). Since in all systems nanoparticles interact with cell membrane, surface characteristics could affect internalization and intracellular localization mechanism (Murakami et al., 1999). Different ligands have been applied in CNS delivery which summarized in Table 13.1.

TABLE 13.1 Different Ligands Have Been Applied in CNS Delivery

Molecular Imaging	Molecular Prob	Strengths	Weakness
	Antibody	• High Specificity • High affinity • Low antigenicity and acceptable toxicity and high specificity to targets. • Many clinically approved antibodies available for labeling	• Immunogenicity • High cost • Long blood half-life decreases specificity of signal.
	Small molecule Amino acid	• Intracellular imaging • High specificity, rapid clearance. Intra-cellular targets are available for imaging	• Non-specific binding • Fluorochromes and their comparable size to small molecules may affect pharmacokinetics and biodistribution of the resulting labeled ligands
	Folic acid Sugar moieties		
	Mannose/ glucose/ galactose		
	Peptide	• Low size • Low immunogenicity • Low cost • High specificity to targets, easy for synthesis and feasible for conjugation with contrast agents, rapid clearance times.	• Low affinity • Degradation in vivo • Many peptides have brief serum half-lives, usually caused by degradation or excretion

TABLE 13.1 (Continued)

Molecular Imaging	Molecular Prob	Strengths	Weakness
	Aptamer	• Low size • Low immunogenicity • High affinity	• High cost • Undefined toxicity
	Affibody	• Low size • Low cost • These structures retain high binding affinity and specificity. Clearance times well-suited for imaging	• Low detection ability due to lack of FC fragment for 2' staining • More complex to formulate compared to whole antibodies

13.7.1 UTILIZING MONOCLONAL ANTIBODIES

In recent years nanotechnology researches concentrated on developing new moites able to absorb on surface of nanoparticles. The aim of this phenomenon is to improve the ability in passing BBB. Recent researches on functionalizing nano systems, investigated different proteins which express in the surface of tumor cells. Clinical evaluation of EGFR's, over express in tumor cells, targeting ligand showed that functionalized moite are able to recognized proteins over expressed in both tumor cells and surrounding neovascular, which is interesting for researchers (Mickler et al., 2012). Integrin receptor expression in both neovascular and tumor cell is more favorable for targeting both BBB and tumor. Transferrin receptor (Tfr) is being expressed in BBB, could be used for targeting, too (Fishman et al., 1987). LRP 1 is another an example which is being over expressed in tumor (glioma) cells and BBB. A study recently revealed that Angiopep-2, ligand of LRP1, was able to target brain tumor (Xin et al., 2011).

Employing several BBB's endogenous receptors, including: TfR, Insulin receptor (InsR), Glutamate receptor (GLR), LDL receptor, and LRP1 receptor, in recent studies, has been evaluated for brain targeting, reveal that receptors could be favor target for delivery and targeting of brain tumors (Rip et al., 2009). Because of the ability of anigiopep-2 in passing BBB, most companies are trying to use this peptide in brain delivery so that ANG1005 (Paclitaxel + angiopep-2) was in phase II clinical trial in 2001, and patients receiving this drug has no any antibody or immune response against that (Kurzrock et al., 2012).

13.7.2   UTILIZING CELL-PENETRATING PEPTIDES (CPP)

CPPs are amphipathic peptides possessing positive charge which have a noticeable tendency to interact with biological barriers. Tat, a HIV peptide composed of 6 arginine and 2 Lysine, possess positive charge which interact with negative charge of cell surface and result in adsorptive mediated transcytosis and increased BBB transporting.

Studies about applying Tat in HIV therapy has been showed that completely tagged or scraped Tat to nanoparticles could have good interaction with vascular endothelial cells and result in increased brain bioavailability of ART agents comparing Tat free nanoparticles. Penetration antenapedia is another CPP which has been studied. It should be considered that long time risk to icity, immune response, and specificity of targeting of CPPs are cases of question and study (Kim and Sharp, 2001; Kravcik et al., 1999; New et al., 1997; Sabatier et al., 1991; Vives, 2005; Wei et al., 2009).

13.7.3   UTILIZING LIGANDS

Since specific targeting is not cost-effective mostly and requires high quality monoclonal antibodies in clinical uses, more efforts is being made to employ carrier mediated transport for crossing agents (like induction of BBB transport by Glut1, mannose) or nanoparticles conjugated to ligands (like Apolipoproteins, transferrin, and folate) to induce receptor mediated transcytosis. Although mAbs are more specific but, ligand have less immune response because of endogenous characteristic (Costantino et al., 2009; Wong et al., 2012).

13.8   NANOPARTICLES EMPLOYED IN CNS DELIVERY

13.8.1   POLYMERIC NANOPARTICLES

Polymeric   nanoparticle   especially   Polybutylcyanoarylate   (PBCA) nanoparticles coated with polysorbates 80 is widely being used in CNS drug delivery (Kreuter et al., 2003; Rempe et al., 2011). PBCA have good BBB permeability since inducing both receptor mediated endocytosis, with adsorbing lipoproteins on their surface as a result employing LDL receptors in microvascular endothelial, and temporary BBB disruption (Rempe et al., 2011). Acrylic nanoparticle pass BBB by both paracellular and transcellular pathway although, paracellular possess less contribution than transcellular. Acrylic nanoparticle are biodegradable and have less toxic characteristics because undergo fast enzymatic hydrolization having water soluble byproducts, type 1 alcohol, buthanol, and poly cyanoacrylate, which eliminate rapidly and do not concentrate in CNS at the other hand, formaldehyde as a toxic byproduct of this polymer could compensate its beneficial, partially. Although, literature refer to more review about this polymer toxicity but, as byproducts of polyesters are water and CO_2 which are not toxic and just acidifying environment slightly, PLA and PLGA as polyester have got certificate to be used in clinic (Dechy-Cabaret et al., 2004). Chitosan is another biodegradable polymer which chitosan-based nanoparticles are in preclinical phase (Costantino and Boraschi, 2012).

Loperamide, methotrexate, doxorubicin, and temozolamine are the examples which have been delivered to CNS using PBCA (Gao and Jiang, 2006; Tian et al., 2011).

Polymeric nanoparticles with different ligand have been applied in CNS delivery which summarized in Table 13.2.

13.8.2 LIPOSOMES

Liposomes, bilayer amphiphilic vesicle with unilamellar or multilamellar (Webb et al., 2007) structure with low toxicity, could carry both hydrophilic and hydrophobic agents so that, these properties make this colloidal systems promising in CNS drug delivery improvement (Costantino et al., 2009). Rapid systematic elimination and undergoing metabolic decomposition of phospholipids, short shelf-life (several months), and being unable to be sustain release are disadvantages of liposomes (Webb et al., 2007; Bertrand et al., 2010). Pegylationis solution which increases stability and circulating time of liposome (Bertrand et al., 2010). Recent studies on patients with primary brain tumor have showed that Glutathione pegylationliposomal doxorubicin could be safe and effective in these patients which guided this product to phase 3 clinical trial. It is expected that more lipid formulation will be used in CNS delivery in future (Costantino and Boraschi, 2012).

TABLE 13.2 Polymeric Nanoparticles with Different Ligand Have Been Applied in CNS Delivery

Ligand	Carrier	References
Monoclonal Antibodies Against the Transferrin Receptor (TR)	Human serum albumin (HSA) nanoparticles	Ulbrich et al., 2009; Chang et al. 2009
Low-density lipoprotein receptor	PLGA	Kim et al., 2007
Monoclonal Antibodies against the Insulin Receptor (IR)	Human serum albumin (HSA) nanoparticles	Ulbrich et al., 2011
Low-density Lipoprotein Receptor-related Protein (LRP)-1	Poly (methoxypolyethyleneglycol cyanoacrylate-co hexadecylcyanoacrylate) (PEG-PHDCA) nanoparticles	Kim et al., 2007
Apolipoprotein A, B, E	Albumin nanoparticles	Kreuter et al., 2007
Melanotransferrin	Albumin nanoparticles	
Poly (ethylene glycol)-co-poly (ε-caprolactone) nanoparticles Paclitaxel	Angiopep family of Peptides Angiopep2	Xin et al., 2011; Regina et al., 2008; Thomas et al., 2009
Single domain antibody	Phospholipids unilamellar vesicles (ULVEs)	Katsaras, 2014
Leptin	Liposomes	Tosi et al., 2012
SynB1 protein transduction domains (PTDs)	Dalargin	Rousselle et al., 2003
Poly-arginines	Albumin	Westergren and Johannson, 1993
Rabies virus glycoprotein (RVG) peptide	Chitosan	Gao et al., 2014
Tat peptide	Solid lipid Nanoparticle (SLN) PLGA	Choi et al., 2014; Nam et al., 2002
Diphtheria toxin receptor (DTR)	Poly Butyl Cyano Acrylate (PBCA)	Kuo and Chung, 2012
Mannose	Liposomes	Mora et al., 2002
Folic acid	Liposomes	Anderson et al., 2001; Lee and Low, 1995; Wang et al., 1995
Thiamine	Solid lipid Np	Lockman et al., 2003
Simil-opioid peptide	PLGA, albumin-ApoE, albuminApoA-I/ApoB100	Costantino et al., 2005
Glutathione	PLGA-PEG	Geldenhuys et al., 2011

13.8.3 SOLID LIPID NANOPARTICLES (SLN)

Solid lipid nanoparticles (SLNs) are submicron solid lipid matrix at room temperature and body containing rigid and solid hydrophobic lipid core surrounded by phospholipid layer and stabled with surfactant. Pegylation SLNs pass BBB and improve CNS delivery (MuÈller et al., 2000), inherent low toxicity, physiological stability, protection of sensitive drugs against decomposition, able to control drug release, high loading efficacy, easy preparation process, and having potential to loading both lipophilic and hydrophilic drug or diagnostic agent are advantages of SLNs which attract attention of drug delivery researchers (Wong et al., 2007).

13.8.4 NANO-EMULATIONS AND NANO-SUSPENSIONS

Nano-emulsion, which could be oil in water or water in oil, could carry hydrophobic drugs especially imaging agent in oil phase and pass biological barriers, and also could be targeting and overcome to efflux pumps by conjugating to ligands on its surface (Ganta et al., 2010; Sarker, 2005). Nano-suspension contains homogenized solid drug dispersed in stable surfactant, which large active surface make these potential to increase drug bioavailability in systematic circulating system and brain (Wong et al., 2012).

13.8.5 DENDRIMERS

Dendrimers are macromolecules consist of repeated unit, like a tree, could carry drug through direct non-covalent band or by encapsulation. Because of their highly branched construction and cross-linker profile, their decomposition is not well known, there are concerns about their safety and this subject should be considered (Costantino and Boraschi, 2012).

13.9 CONCLUSIONS AND CONSIDERATIONS

Cancer is the most important cause of death all over the world. Brain cancers have significant stage in man death statistics. Glioma, as a brain cancer, is being treated with different technics include mix of surgery, radiotherapy, systematic chemotherapy and photodynamic therapy, despite all improving advances, patients survive for several months, only (Stupp et al., 2006). Cancer is not unique in CNS problems so, that increasing trend of neurodegenerative disease, following increasing life expectancy besides aging of HIV infected patients, and CNS infections, especially hidden HIV in CNS acting like Trojan horse, is being considered in CNS drug delivery studies. BBB is first of the important obstacles in CNS therapy which could be leaved behind by potential nanosystems or automated drugs (Guarnieri et al., 2014).

Nanoparticles showed a good potential in drug delivery to various compartments of body like CNS (Guarnieri et al., 2014). Among variety of proposed both invasive and noninvasive CNS drug delivery strategies, nanoparticles are just means which are being used broadly and successfully, nowadays. Disability of other strategies to pass tight junction BBB and deliver desirable drug level to CNS resulted to this phenomenon. It should be considered that designing of effective delivery systems require good knowledge of basic medical sciences (anatomy, physiology, biochemistry, …), pathology, pharmacodynamic, and pharmacokinetic besides special profession on chemical engineering, physics, and mathematic (Alam et al., 2010). There are a little documented statistics about CNS tumors in Iran. A ten years study directed by National Cancer Registry (NRC) of Iran during 2000–2010, reported that rate of malignant, benign lesions, and total primary brain tumors per 1000 person over one year was 2.74, 2.95, and 5.69, respectively (Jazayeri et al., 2013). Target oriented nanodelivery systems provided a hope to man in increasing life expectancy of CNS disorders patients (Gao et al., 2013). Although size, shape, surface charge and coating are factors which affect nanoparticles BBB transferring but, there are less information about effect of size, surfactant, and zeta potential (Schöler et al., 2001). Designing effective delivery system to CNS require optimized harmony of size, surface charge, chemical composition, and route of delivery. Study of toxicity and immune response of nanomaterial could guide increasing trend of usage of nanotechnology in safe drug delivery (De and Borm, 2008; Gao et al., 2013). Ligand conjugating to nanoparticles could affect protein binding of them as a result serum-circulating time and In-vivo targeting could be altered. More studies on interaction of serum proteins and targeting ligands required to optimizing targeting systems (Gao et al., 2013). Active targeting of nanoparticles by using conjugating technology of cell surface marker's ligand enhance delivery efficacy to tumor site while its' toxic effect to other compartments decrease (Hernández-Pedro et al., 2013). In addition, pegylation decrease activation of scavenging by complement, opsonization, and phagocyte by reticuloendothelial system (RES), which used to enhancing serum circulating time besides managing toxicity (Moghimi and Szebeni, 2003). Transcytosis by dedicated receptor on BBB is more acceptable mechanism to absorption of nanoparticles by CNS. Lipoproteins, insulin, and transferrin receptor are being employed to this technic. Basal membrane and pericytes are abnormal in tumor site, mostly. In other hand, BBB are defective in gliomas (Hwang and Kim, 2014). Employing insulin receptors had showed that could increase absorption of gold nanoparticles tenfold more than other receptors (Shilo et al., 2014). Administration route of drug alter its' fate (Knudsen et al., 2014; Prilloff et al., 2010). Although, most of nanoparticles' biodistribution evaluated but, more researches are needed analyzing nanoparticles transports from barriers, for this aim, MRI has a good potential for evaluation of nanoparticles transport from barriers of CNS and magnetic nanoparticles are good choice for In-vivo tracking of nanoparticles in brain (Knudsen et al., 2014; Prilloff et al., 2010). Single-Photon Emission computerized tomography (SPECT), Positron Emission tomography (PET), and Computerized Tomography (CT) are another extra scanning technics which could be applied in CNS pharmacokinetic in addition to MRI (Su et al., 2011).

Basic considerations in novel target oriented improved deliveries: Carrying (carriers) – loading efficacy, carrier size (bellow 100 nm), safety, biodegradability and biocompatibility should be considered. Targeting scape (although ligands or antibodies are being used for logical accurate targeting, but using unique identities of target size could be resulted in different targeting techniques, for example, tumor's gently high temperature are base of sol–gel targeting carriers. In addition, ex-vivo magnetic controllable carriers are independent example of these two targeting types) from biological metabolization (although prodrug which is stable against aberrant degradation and susceptible around target site could increase potency and prevent systemic toxicity, but prodrug toxicity is a matter of consideration). Kinetically improved formulizations (sustain released, multidrug systems, target oriented release, easy to transport) multifunctional systems (imaging-curing which decrease burden deliveries following toxicity and costs) Glutathione pegylated liposomal doxorubicin hydrochloride (2B3–101) and ANG-1005 are two FDA approved successful brain-targeted delivery systems which give more to brain delivery researchers to stand on their interests and work so hard with hope to future (Costantino and Boraschi, 2012; Del Burgo et al., 2013; Kreuter, 2013; Wilson, 2009).

KEYWORDS

- **blood brain barrier**
- **CNS**
- **delivery system**
- **invasive delivery system**
- **nanoparticles**
- **nanotechnology**
- **non-invasive delivery system**

REFERENCES

Abbott, N. J., et al., (2010). Structure and function of the blood–brain barrier. *Neurobiology of Disease, 37*(1), 13–25.

Aboody, K. S., et al., (2000). Neural stem cells display extensive tropism for pathology in adult brain: Evidence from intracranial gliomas. *Proceedings of the National Academy of Sciences, 97*(23), 12846–12851.

Aboody, K. S., et al., (2006). Targeting of melanoma brain metastases using engineered neural stem/progenitor cells. *Neuro-Oncology, 8*(2), 119–126.

Aboody, K., J. Najbauer, & Danks, M. (2008). Stem and progenitor cell-mediated tumor selective gene therapy. *Gene Therapy, 15*(10), 739–752.

Alam, M. I., et al., (2010). Strategy for effective brain drug delivery. *European Journal of Pharmaceutical Sciences, 40*(5), 385–403.

Amanlou, M., et al., (2011). Gd 3+ DTPA-DG: Novel nanosized dual anticancer and molecular imaging agent. *Int J Nanomedicine, 6*, 747–763.

Amyere, M., et al., (2001). Origin, originality, functions, subversions and molecular signalling of macropinocytosis. *International Journal of Medical Microbiology, 291*(6), 487–494.

Anderson, K. E., et al., (2001). Formulation and evaluation of a folic acid receptor-targeted oral vancomycin liposomal dosage form. *Pharmaceutical Research, 18*(3), 316–322.

Armulik, A., et al., (2010). Pericytes regulate the blood-brain barrier. *Nature, 468*(7323), 557–561.

Bae, Y. H., & Park, K., (2011). Targeted drug delivery to tumors: myths, reality and possibility. *Journal of Controlled Release, 153*(3), 198.

Balyasnikova, I. V., et al., (2010). Mesenchymal stem cells modified with a single-chain antibody against EGFRvIII successfully inhibit the growth of human xenograft malignant glioma. *PloS One, 5*(3), e9750.

Banks, W. A., (2009). Characteristics of compounds that cross the blood-brain barrier. *BMC Neurology, 9*(Suppl 1), S3.

Barichello, T., et al., (2011). A kinetic study of the cytokine/chemokines levels and disruption of blood-brain barrier in infant rats after pneumococcal meningitis. *Journal of Neuroimmunology, 233*(1), 12–17.

Bartus, R. T., et al., (2000). Intravenous cereport (RMP-7) modifies topographic uptake profile of carboplatin within rat glioma and brain surrounding tumor, elevates platinum levels, and enhances survival. *Journal of Pharmacology and Experimental Therapeutics, 293*(3), 903–911.

Batrakova, E. V., Gendelman, H. E. & Kabanov, A. V., (2011). Cell-mediated drug delivery. *Expert Opinion on Drug Delivery, 8*(4), 415–433.

Béduneau, A., Saulnier, P., & Benoit, J. P., (2007). Active targeting of brain tumors using nanocarriers. *Biomaterials, 28*(33), 4947–4967.

Belfiore, A., et al., (2009). Insulin receptor isoforms and insulin receptor/insulin-like growth factor receptor hybrids in physiology and disease. *Endocrine Reviews, 30*(6), 586–623.

Bertrand, N., Simard, P. &. Leroux J. C., (2010). Serum-stable, long-circulating, pH-sensitive PEGylated liposomes, *in Liposomes,* Springer, p. 545–558.

Bickel, U., et al., (1994). Development and *in vitro* characterization of a cationized monoclonal antibody against βA4 protein: A potential probe for Alzheimer's disease. *Bioconjugate Chemistry, 5*(2), 119–125.

Bickel, U., Yoshikawa, T., & Pardridge, W. M., (2001). Delivery of peptides and proteins through the blood–brain barrier. *Advanced Drug Delivery Reviews, 46*(1), 247–279.

Boado, R. J., & Pardridge, W. M., (2011). The Trojan horse liposome technology for nonviral gene transfer across the blood-brain barrier. *Journal of Drug Delivery,* 12–25

Brightman, M., & Reese, T., (1969). Junctions between intimately apposed cell membranes in the vertebrate brain. *The Journal of Cell Biology, 40*(3), 648–677.

Camenisch, G., et al., (1998). Estimation of permeability by passive diffusion through Caco-2 cell monolayers using the drugs' lipophilicity and molecular weight. *European Journal of Pharmaceutical Sciences, 6*(4), 313–319.

Chang, J., et al., (2009). Characterization of endocytosis of transferrin-coated PLGA nanoparticles by the blood–brain barrier. *International Journal of Pharmaceutics, 379*(2), 285–292.

Choi, Y. S., et al., (2014). Nanoparticles for gene delivery: therapeutic and toxic effects. *Molecular & Cellular Toxicology, 10*(1), 1–8.

Cohen, B., & Bangham, A., (1972). Diffusion of small non-electrolytes across liposome membranes. *Nature, 236,* 1731–1734.

Compte, M., et al., (2009). Tumor immunotherapy using gene modified human mesenchymal stem cells loaded into synthetic extracellular matrix scaffolds. *Stem Cells, 27*(3), 753–760.

Cornford, E. M., et al., (1992). Melphalan penetration of the blood-brain barrier via the neutral amino acid transporter in tumor-bearing brain. *Cancer Research, 52*(1), 138–143.

Costantino, L., &. Boraschi, D., (2012). Is there a clinical future for polymeric nanoparticles as brain-targeting drug delivery agents?. *Drug Discovery Today, 17*(7), 367–378.

Costantino, L., et al., (2005). Peptide-derivatized biodegradable nanoparticles able to cross the blood–brain barrier. *Journal of Controlled Release, 108*(1), 84–96.

Costantino, L., et al., (2009). Colloidal systems for CNS drug delivery. *Progress in Brain Research, 180,* 35–69.

Crone, C., & Olesen, S., (1982). Electrical resistance of brain microvascular endothelium. *Brain Research, 241*(1), 49–55.

Crone, C., &. Christensen, O., (1981). Electrical resistance of a capillary endothelium. *The Journal of General Physiology, 77*(4), 349–371.

Crone, C., (1965). Facilitated transfer of glucose from blood into brain tissue. *The Journal of Physiology, 181*(1), 103.

Dalgleish, A. G., et al., (1984). The CD4 (T4) antigen is an essential component of the receptor for the AIDS retrovirus. *Science 232,* 1123–1127.

De Jong, W. H., & Borm, P. J., (2008). Drug delivery and nanoparticles: applications and hazards. *International Journal of Nanomedicine, 3*(2), 133.

Dean, M., Hamon, Y. & Chimini G., (2001). The human ATP-binding cassette (ABC) transporter superfamily. *Journal of Lipid Research, 42*(7), 1007–1017.

Debinski, W., &. Tatter, S. B (2009). Convection-enhanced delivery for the treatment of brain tumors. *Expert Rev Neurother, 9*(10), 1519–1527.

Dechy-Cabaret, O., Martin-Vaca, B, &. Bourissou, D., (2004). Controlled ring-opening polymerization of lactide and glycolide. *Chemical Reviews, 104*(12), 6147–6176.

Deguchi, T., et al., (2006). Involvement of organic anion transporters in the efflux of uremic toxins across the blood–brain barrier. *Journal of Neurochemistry, 96*(4), 1051–1059.

del Burgo, L. S., et al., (2013). Nanotherapeutic approaches for brain cancer management. *Nanomedicine: Nanotechnology, Biology and Medicine, 10*(5), e905–e919.

Dou, H., et al., (2007). Laboratory investigations for the morphologic, pharmacokinetic, and anti-retroviral properties of indinavir nanoparticles in human monocyte-derived macrophages. *Virology, 358*(1), 148–158.

Doyle, L. A., et al., (1998). A multidrug resistance transporter from human MCF-7 breast cancer cells. *Proceedings of the National Academy of Sciences, 95*(26), 15665–15670.

Duffy, K. R., Pardridge, W. M., & Rosenfeld, R. G., (1988). Human blood-brain barrier insulin-like growth factor receptor. *Metabolism, 37*(2), 136–140.

Egleton, R. D., & Davis, T. P., (2005). Development of neuropeptide drugs that cross the blood-brain barrier. *NeuroRx, 2*(1), 44–53.

Egleton, R. D., & Davis, T. P., (1997). Bioavailability and transport of peptides and peptide drugs into the brain. *Peptides, 18*(9), 1431–1439.

Engelhardt, B., & Sorokin L., (2009). The blood–brain and the blood–cerebrospinal fluid barriers: Function and dysfunction. in Seminars in immunopathology. Springer.

Enting, R. H., et al., (1998). Antiretroviral drugs and the central nervous system. *Aids, 12*(15), 1941–1955.

Ewend, M. G., et al., (2007). Treatment of single brain metastasis with resection, intracavity carmustine polymer wafers, and radiation therapy is safe and provides excellent local control. *Clinical Cancer Research, 13*(12), 3637–3641.

Fan, C. H., et al., (2014). Submicron-bubble-enhanced focused ultrasound for blood-brain barrier disruption and improved CNS drug delivery. *PLoS One, 9*(5), p. e96327.

Fillmore, H. L., et al., (2011). Conjugation of functionalized gadolinium metallofullerenes with IL-13 peptides for targeting and imaging glial tumors. *Nanomedicine, 6*(3), 449–458.

Fishman, J., et al., (1987). Receptor-mediated transcytosis of transferrin across the blood-brain barrier. *Journal of Neuroscience Research, 18*(2), 299–304.

Florence, A. T., (2012). "Targeting" nanoparticles: The constraints of physical laws and physical barriers. *Journal of Controlled Release, 164*(2), 115–124.

Fortin, D., et al., (2005). Enhanced chemotherapy delivery by intraarterial infusion and blood-brain barrier disruption in malignant brain tumors. *Cancer, 103*(12), 2606–2615.

Fortin, D., et al., (2007). Enhanced chemotherapy delivery by intraarterial infusion and blood-brain barrier disruption in the treatment of cerebral metastasis. *Cancer, 109*(4), 751–760.

Frank, R. T., Aboody, K. S., & Najbauer, J., (2011). Strategies for enhancing antibody delivery to the brain. *Biochimica et Biophysica Acta (BBA)-Reviews on Cancer, 1816*(2), 191–198.

Fung, L. K., et al., (1996). Chemotherapeutic Drugs Released from Polymers: Distribution of 1, 3-bis (2-chloroethyl)-l-nitrosourea in the Rat Brain. *Pharmaceutical Research, 13*(5), 671–682.

Gabathuler, R., (2010). Approaches to transport therapeutic drugs across the blood–brain barrier to treat brain diseases. *Neurobiology of Disease, 37*(1), 48–57.

Ganapathy, V., Thangaraju, M., & Prasad, P. D., (2009). Nutrient transporters in cancer: Relevance to Warburg hypothesis and beyond. *Pharmacology & Therapeutics, 121*(1), 29–40.

Ganta, S., et al., (2010). A review of multifunctional nanoemulsion systems to overcome oral and CNS drug delivery barriers. *Molecular Membrane Biology, 27*(7), 260–273.

Gao, B., et al., (2000). Organic anion-transporting polypeptides mediate transport of opioid peptides across blood-brain barrier. *Journal of Pharmacology and Experimental Therapeutics, 294*(1), 73–79.

Gao, H., et al., (2013). Ligand modified nanoparticles increases cell uptake, alters endocytosis and elevates glioma distribution and internalization. *Scientific Reports, 3*.

Gao, K., & Jiang, X., (2006). Influence of particle size on transport of methotrexate across blood brain barrier by polysorbate 80-coated polybutylcyanoacrylate nanoparticles. *International Journal of Pharmaceutics, 310*(1), 213–219.

Gao, L., et al., (1993). Direct *in vivo* gene transfer to airway epithelium employing adenovirus-polylysine-DNA complexes. *Human Gene Therapy, 4*(1), 17–24.

Gao, Y., et al., (2014). RVG-Peptide-Linked Trimethylated Chitosan for Delivery of siRNA to the Brain. *Biomacromolecules, 15*(3), 1010–1018.

Garcel, A., et al., (2010). *In vitro* blood brain barrier models as a screening tool for colloidal drug delivery systems and other nanosystems. *International Journal of Biomedical Nanoscience and Nanotechnology, 1*(2), 133–163.

Geldenhuys, W., et al., (2011). Brain-targeted delivery of paclitaxel using glutathione-coated nanoparticles for brain cancers. *Journal of Drug Targeting, 19*(9), 837–845.

Gonzalez-Mariscal, L., et al., (2003). Tight junction proteins. *Progress in Biophysics and Molecular Biology, 81*(1), 1–44.

Göppert, T. M., & Müller, R. H., (2005). Polysorbate-stabilized solid lipid nanoparticles as colloidal carriers for intravenous targeting of drugs to the brain: Comparison of plasma protein adsorption patterns. *Journal of Drug Targeting, 13*(3), 179–187.

Graff, C. L., & Pollack, G. M., (2004). (Section B: Integrated Function of Drug Transporters *In Vivo*) Drug Transport at the Blood-Brain Barrier and the Choroid Plexus. *Current Drug Metabolism, 5*(1), 95–108.

Graner, M. W., (2011). Brain tumor exosomes and microvesicles: pleiotropic effects from tiny cellular surrogates. *Molecular Targets of CNS Tumors, 43*–78.

Groothuis, D. R., & Levy, R. M., (1997). The entry of antiviral and antiretroviral drugs into the central nervous system. *Journal of Neurovirology, 3*(6), 387–400.

Guarnieri, D., Muscetti, O. & Netti, P. A., (2014), A method for evaluating nanoparticle transport through the blood–brain barrier *in vitro*, in: *Drug Delivery System, Springer*, 185–199.

Hamilton, R., Foss, A., & Leach, L., (2007). Establishment of a human in vitro model of the outer blood–retinal barrier. *Journal of Anatomy, 211*(6), 707–716.

Hansen, Jr, D. W., et al., (1992). Systemic analgesic activity and. delta. opioid selectivity in [2, 6-dimethyl-Tyr1, D-Pen2, D-Pen5] enkephalin. *Journal of Medicinal Chemistry*, *35*(4), 684–687.

Hawkins, B. T., & Davis, T. P., (2005). The blood-brain barrier/neurovascular unit in health and disease. *Pharmacological Reviews*, *57*(2), 173–185.

Hernández-Pedro, N. Y., et al., (2013). Application of Nanoparticles on Diagnosis and Therapy in Gliomas. *BioMed Research International.* Apr 18;2013.

Holman, D. W., Klein, R. S., & Ransohoff, R. M., (2011). The blood–brain barrier, chemokines and multiple sclerosis. *Biochimica et Biophysica Acta (BBA)-Molecular Basis of Disease, 1812*(2), 220–230.

Huwyler, J., Yang, J., & Pardridge, W. M., (1997). Receptor mediated delivery of daunomycin using immunoliposomes: pharmacokinetics and tissue distribution in the rat. *Journal of Pharmacology and Experimental Therapeutics, 282*(3), 1541–1546.

Hwang, S. R., &. Kim, K., (2014). Nano-enabled delivery systems across the blood–brain barrier. *Archives of Pharmacal Research, 37*(1), 24–30.

Hynynen, K., et al., (2001). Noninvasive MR Imaging–guided Focal Opening of the Blood-Brain Barrier in Rabbits 1. *Radiology, 220*(3), 640–646.

Illum, L., (2007). Nanoparticulate systems for nasal delivery of drugs: A real improvement over simple systems?. *Journal of Pharmaceutical Sciences, 96*(3), 473–483.

Jain, A., Soni, V., & Jain, S. K. Receptor Mediated Pegylated Nanoparticles Bearing Paclitaxel For Brain Tumors, International Joint Conference on Pharmacology and Pharmaceutical Technology. New Delhi, 6 October 2013.

Jazayeri, S. B., et al., (2013). Epidemiology of primary CNS tumors in Iran: A systematic. *Asian Pacific Journal of Cancer Prevention, 14*(6), 3979–3985.

Joubert, J., et al., (2012). Polycyclic cage structures as lipophilic scaffolds for neuroactive drugs. *Chem. Med. Chem., 7*(3), p. 375–384.

Juillerat-Jeanneret, L., (2008). The targeted delivery of cancer drugs across the blood–brain barrier: chemical modifications of drugs or drug-nanoparticles?. *Drug Discovery Today, 13*(23), 1099–1106.

Kabanov, A. V., Batrakova, E. V., & Miller, D. W., (2003). Pluronic® block copolymers as modulators of drug efflux transporter activity in the blood–brain barrier. *Advanced Drug Delivery Reviews, 55*(1), 151–164.

Katsaras, J., (2014). Self-assembled lipid nanoparticles for imaging disease. In: Biomedical Science and Engineering Center Conference (BSEC), 2014 Annual Oak Ridge National Laboratory. IEEE.

Kawakami, M., et al., (2001). Mutation and functional analysis of IL-13 receptors in human malignant glioma cells. *Oncology Research Featuring Preclinical and Clinical Cancer Therapeutics, 12*(11–12), 11–12.

Kerwin, B. A., (2008). Polysorbates 20 and 80 used in the formulation of protein biotherapeutics: structure and degradation pathways. *Journal of Pharmaceutical Sciences, 97*(8), 2924–2935.

Kim, H. R., et al., (2007). Low-density lipoprotein receptor-mediated endocytosis of PEGylated nanoparticles in rat brain endothelial cells. *Cellular and Molecular Life Sciences, 64*(3), 356–364.

Kim, J. B., & Sharp, P. A., (2001). Positive transcription elongation factor B phosphorylates hSPT5 and RNA polymerase II carboxyl-terminal domain independently of cyclin-dependent kinase-activating kinase. *Journal of Biological Chemistry, 276*(15), 12317–12323.

Knudsen, K. B., et al., (2014). Biodistribution of rhodamine B fluorescence-labeled cationic nanoparticles in rats. *Journal of Nanoparticle Research, 16*(2), 1–11.

Kravcik, S., et al., (1999). Cerebrospinal fluid HIV RNA and drug levels with combination ritonavir and saquinavir. *JAIDS Journal of Acquired Immune Deficiency Syndromes, 21*(5), 371–375.

Kreuter, J., (2013). Drug delivery to the central nervous system by polymeric nanoparticles: What do we know?. *Advanced Drug Delivery Reviews. 71*, pp. 2–14.

Kreuter, J., et al., (2003). Direct evidence that polysorbate-80-coated poly (butylcyanoacrylate) nanoparticles deliver drugs to the CNS via specific mechanisms requiring prior binding of drug to the nanoparticles. *Pharmaceutical Research, 20*(3), 409–416.

Kreuter, J., et al., (2007). Covalent attachment of apolipoprotein AI and apolipoprotein B-100 to albumin nanoparticles enables drug transport into the brain. *Journal of Controlled Release, 118*(1), p. 54–58.

Krishna, R., & Mayer, L. D., (2000). Multidrug resistance (MDR) in cancer: mechanisms, reversal using modulators of MDR and the role of MDR modulators in influencing the pharmacokinetics of anticancer drugs. *European Journal of Pharmaceutical Sciences, 11*(4), 265–283.

Kuo, Y. C., & Chung, C. Y., (2012). Transcytosis of CRM197-grafted polybutylcyanoacrylate nanoparticles for delivering zidovudine across human brain-microvascular endothelial cells. *Colloids and Surfaces B: Biointerfaces, 91*, 242–249.

Kurzrock, R., et al., (2012). Safety, pharmacokinetics, and activity of GRN1005, a novel conjugate of angiopep-2, a peptide facilitating brain penetration, and paclitaxel, in patients with advanced solid tumors. *Molecular Cancer Therapeutics, 11*(2), 308–316.

Kusuhara, H., & Sugiyama, Y., (2001). Efflux transport systems for drugs at the blood–brain barrier and blood–cerebrospinal fluid barrier (Part 1). *Drug Discovery Today, 6*(3), 150–156.

Lai, C. P. K., &. Breakefield X. O, (2012). Role of exosomes/microvesicles in the nervous system and use in emerging therapies. *Frontiers in physiology, 3.*, 228.

Lakhal, S., & Wood, M. J., (2011). Exosome nanotechnology: An emerging paradigm shift in drug delivery. *Bioessays, 33*(10), 737–741.

Lee, D. H., et al., (2009). Targeting rat brainstem glioma using human neural stem cells and human mesenchymal stem cells. *Clinical Cancer Research, 15*(15), 4925–4934.

Lee, R. J., & P. S. Low, (1995). Folate-mediated tumor cell targeting of liposome-entrapped doxorubicin *in vitro. Biochimica et Biophysica Acta (BBA)-Biomembranes, 1233*(2), 134–144.

Lillis, A. P., et al., (2008). LDL receptor-related protein 1: Unique tissue-specific functions revealed by selective gene knockout studies. *Physiological Reviews, 88*(3), 887–918.

Liu, L., et al., (2008). Biologically active core/shell nanoparticles self-assembled from cholesterol-terminated PEG–TAT for drug delivery across the blood–brain barrier. *Biomaterials, 29*(10), 1509–1517.

Liu, S., et al., (2012). The role of pericytes in blood-brain barrier function and stroke. *Curr Pharm Des,. 18*(25), 3653–3662.

Liu, Y., & Lu, W., (2012). Recent advances in brain tumor-targeted nano-drug delivery systems. *Expert Opinion on Drug Delivery, 9*(6), 671–686.

Lockman, P. R., et al., (2003). Brain uptake of thiamine-coated nanoparticles. *Journal of Controlled Release, 93*(3), 271–282.

Löscher, W., & Potschka, H., (2005). Blood-brain barrier active efflux transporters: ATP-binding cassette gene family. *NeuroRx, 2*(1), 86–98.

Magga, J., et al., (2010). Human intravenous immunoglobulin provides protection against Ab toxicity by multiple mechanisms in a mouse model of Alzheimer's disease. *J. Neuroinflammation*, 7, 90.

McDannold, N., et al., (2005). MRI-guided targeted blood-brain barrier disruption with focused ultrasound: histological findings in rabbits. *Ultrasound in Medicine & Biology*, 31(11), 1527–1537.

Meairs, S., & A. Alonso, (2007). Ultrasound, microbubbles and the blood–brain barrier. *Progress in Biophysics and Molecular Biology*, 93(1), 354–362.

Mehravi, B., et al., (2014). Breast cancer cells imaging by targeting methionine transporters with gadolinium-based nanoprobe. *Molecular Imaging and Biology*, 16(4), 519–528

Michaelis, K., et al., (2006). Covalent linkage of apolipoprotein e to albumin nanoparticles strongly enhances drug transport into the brain. *Journal of Pharmacology and Experimental Therapeutics*, 317(3), 1246–1253.

Mickler, F. M., et al., (2012). Tuning nanoparticle uptake: live-cell imaging reveals two distinct endocytosis mechanisms mediated by natural and artificial EGFR targeting ligand. *Nano letters*, 12(7), 3417–3423.

Misra, A., et al., (2003). Drug delivery to the central nervous system: A review. *J Pharm Pharm Sci*, 6(2), 252–273.

Moghimi, S. M., Hunter, A. C., & Murray, J. C., (2005). Nanomedicine: Current status and future prospects. *The FASEB Journal*, 19(3), 311–330.

Moghimi, S., & J. Szebeni, (2003). Stealth liposomes and long circulating nanoparticles: critical issues in pharmacokinetics, opsonization and protein-binding properties. *Progress in Lipid Research*, 42(6), 463–478.

Mohammadi, E., et al., (2014). Cellular uptake, imaging and pathotoxicological studies of a novel Gd [iii]–DO3A-butrol nano-formulation. *RSC Advances*, 4(86), 45984–45994.

Mora, M., et al., (2002). Design and characterization of liposomes containing long-chain N-acylPEs for brain delivery: penetration of liposomes incorporating GM1 into the rat brain. *Pharmaceutical Research*, 19(10), 1430–1438.

MuÈller, R. H.,. MaÈder, K, &. Gohla, S, (2000). Solid lipid nanoparticles (SLN) for controlled drug delivery–a review of the state of the art. *European Journal of Pharmaceutics and Biopharmaceutics*, 50(1), 161–177.

Murakami, H., et al., (1999). Preparation of poly (DL-lactide-co-glycolide) nanoparticles by modified spontaneous emulsification solvent diffusion method. *International Journal of Pharmaceutics*, 187(2), 143–152.

Najbauer, J., et al., (2007). Neural stem cell-mediated therapy of primary and metastatic solid tumors. Progress in gene therapy, autologous and cancer stem cell gene therapy. World Scientific: Singapore, 335–372.

Nam, Y. S., et al., (2002). Intracellular drug delivery using poly (d, l-lactide-co-glycolide) nano-particles derivatized with a peptide from a transcriptional activator protein of HIV-1. *Biotechnology letters*, 24(24), 2093–2098.

New, D. R., et al., (1997). Human immunodeficiency virus type 1 Tat protein induces death by apoptosis in primary human neuron cultures. *Journal of Neurovirology*, 3(2), 168–173.

Nicholson, C., & Syková, E., (1998). Extracellular space structure revealed by diffusion analysis. *Trends in Neurosciences*, 21(5), 207–215.

Olivier, J. C., (2005). Drug transport to brain with targeted nanoparticles. *NeuroRx*, 2(1), 108–119.

Pardridge, W. M., (2001). Crossing the blood–brain barrier: Are we getting it right?. *Drug Discovery Today*, 6(1), 1–2.

Pardridge, W. M., (2007). Blood–brain barrier delivery. *Drug Discovery Today, 12*(1), 54–61.

Pardridge, W. M., (2008). Re-engineering biopharmaceuticals for delivery to brain with molecular Trojan horses. *Bioconjugate Chemistry, 19*(7), 1327–1338.

Pardridge, W. M., Boado, R. J., & Kang, Y. S., (1995). Vector-mediated delivery of a polyamide (" peptide") nucleic acid analogue through the blood-brain barrier in vivo. *Proceedings of the National Academy of Sciences, 92*(12), 5592–5596.

Pardridge, W. M., et al., (1990). Evaluation of cationized rat albumin as a potential blood-brain barrier drug transport vector. *Journal of Pharmacology and Experimental Therapeutics, 255*(2), 893–899.

Pardridge, W., et al., (1993). Protamine-mediated transport of albumin into brain and other organs of the rat. Binding and endocytosis of protamine-albumin complex by microvascular endothelium. *Journal of Clinical Investigation, 92*(5), 2224.

Park, J. H., et al., (2008). Polymeric nanomedicine for cancer therapy. *Progress in Polymer Science, 33*(1), 113–137.

Pathan, S. A., et al., (2009). CNS drug delivery systems: novel approaches. *Recent Patents on Drug Delivery & Formulation, 3*(1), 71–89.

Potschka, H., (2012). Role of CNS efflux drug transporters in antiepileptic drug delivery: Overcoming CNS efflux drug transport. *Advanced Drug Delivery Reviews, 64*(10), 943–952.

Prados, M. D., et al., (2003). A randomized, double-blind, placebo-controlled, phase 2 study of RMP-7 in combination with carboplatin administered intravenously for the treatment of recurrent malignant glioma. *Neuro-Oncology, 5*(2), 96–103.

Prilloff, S., et al., (2010). *In vivo* confocal neuroimaging (ICON), non-invasive, functional imaging of the mammalian CNS with cellular resolution. *European Journal of Neuroscience, 31*(3), 521–528.

Regina, A., et al., (2008). Antitumour activity of ANG1005, a conjugate between paclitaxel and the new brain delivery vector Angiopep-2. *British Journal of Pharmacology, 155*(2), 185–197.

Rempe, R., et al., (2011). Transport of Poly (*n*-butylcyano-acrylate) nanoparticles across the blood–brain barrier *in vitro* and their influence on barrier integrity. *Biochemical and Biophysical Research Communications, 406*(1), 64–69.

Rip, J., Schenk, G., & De Boer, A., (2009). Differential receptor-mediated drug targeting to the diseased brain. *Expert Opinion on Drug Delivery, 6*(3), 227–237.

Rousselle, C., et al., (2003). Improved brain uptake and pharmacological activity of dalargin using a peptide-vector-mediated strategy. *Journal of Pharmacology and Experimental Therapeutics, 306*(1), 371–376.

Roy, I., et al., (2003). Calcium phosphate nanoparticles as novel non-viral vectors for targeted gene delivery. *International Journal of Pharmaceutics, 250*(1), 25–33.

Sabatier, J., et al., (1991). Evidence for neurotoxic activity of tat from human immunodeficiency virus type 1. *Journal of Virology, 65*(2), 961–967.

Saija, A., et al., (1997). Changes in the permeability of the blood-brain barrier following sodium dodecyl sulphate administration in the rat. *Experimental Brain Research, 115*(3), 546–551.

Sakaeda, T., et al., (2000). Enhancement of transport of D-melphalan analogue by conjugation with L-glutamate across bovine brain microvessel endothelial cell monolayers. *Journal of Drug Targeting, 8*(3), 195–204.

Sakane, T., et al., (1989). The effect of polysorbate 80 on brain uptake and analgesic effect of D-kyotorphin. *International Journal of Pharmaceutics, 57*(1), 77–83.

Sarker, D. K., (2005). Engineering of nanoemulsions for drug delivery. *Current Drug Delivery, 2*(4), 297–310.

Saunders, N. R., et al., (2008). Barriers in the brain: a renaissance?. *Trends in Neurosciences*, *31*(6), 279–286.

Scherrmann, J. M., (2002). Drug delivery to brain via the blood–brain barrier. *Vascular Pharmacology, 38*(6), 349–354.

Schöler, N., et al., (2001). Surfactant, but not the size of solid lipid nanoparticles (SLN) influences viability and cytokine production of macrophages. *International Journal of Pharmaceutics, 221*(1), 57–67.

Segal, M. B., (2000). The choroid plexuses and the barriers between the blood and the cerebrospinal fluid. *Cellular and Molecular Neurobiology, 20*(2), 183–196.

Shah, L., Yadav, S., & Amiji, M., (2013). Nanotechnology for CNS delivery of bio-therapeutic agents. *Drug Delivery and Translational Research, 4*, 336–351.

Shilo, M., et al., (2014). Transport of nanoparticles through the blood–brain barrier for imaging and therapeutic applications. *Nanoscale, 6*(4), 2146–2152.

Stupp, R., et al., (2006). Changing paradigms an update on the multidisciplinary management of malignant glioma. *The Oncologist, 11*(2), 165–180.

Su, X., et al., (2011). Magnetic nanoparticles in brain disease diagnosis and targeting drug delivery. *Current Nanoscience, 7*(1), 37–46.

Théry, C., Zitvogel, L., & Amigorena, S. (2002). Exosomes: Composition, biogenesis and function. *Nature Reviews Immunology, 2*(8), 569–579.

Thomas, F. C., et al., (2009). Uptake of ANG1005, a novel paclitaxel derivative, through the blood-brain barrier into brain and experimental brain metastases of breast cancer. *Pharmaceutical Research, 26*(11), 2486–2494.

Thompson, A., et al., (1992). Serial gadolinium-enhanced MRI in relapsing/remitting multiple sclerosis of varying disease duration. *Neurology, 42*(1), 60–60.

Tian, X. H., et al., (2011). Enhanced brain targeting of temozolomide in polysorbate-80 coated polybutylcyanoacrylate nanoparticles. *Int. J. Nanomedicine, 6*, 445–452.

Tosi, G., et al., (2012). Can leptin-derived sequence-modified nanoparticles be suitable tools for brain delivery? *Nanomedicine, 7*(3), 365–382.

Tsuji, A., & Tamai, I., (1999). Carrier-mediated or specialized transport of drugs across the blood–brain barrier. *Advanced Drug Delivery Reviews, 36*(2), 277–290.

Tsuji, A., (2005). Small molecular drug transfer across the blood-brain barrier via carrier-mediated transport systems. *NeuroRx, 2*(1), 54–62.

Ulbrich, K., et al., (2009). Transferrin-and transferrin-receptor-antibody-modified nanoparticles enable drug delivery across the blood–brain barrier (BBB). *European Journal of Pharmaceutics and Biopharmaceutics, 71*(2), 251–256.

Ulbrich, K., Knobloch, T., & Kreuter, J., (2011). Targeting the insulin receptor: nanoparticles for drug delivery across the blood-brain barrier (BBB). *Journal of Drug Targeting, 19*(2), 125–132.

van de Waterbeemd, H., et al., (1998). Estimation of blood-brain barrier crossing of drugs using molecular size and shape, and H-bonding descriptors. *Journal of Drug Targeting, 6*(2), 151–165.

van Niel, G., et al., (2006). Exosomes: A common pathway for a specialized function. *Journal of Biochemistry, 140*(1), 13–21.

Vives, E., (2005). Present and future of cell-penetrating peptide mediated delivery systems: Is the Trojan horse too wild to go only to Troy?. *Journal of Controlled Release, 109*(1), 77–85.

Vlassov, A. V., et al., (2012). E□osomes: Current knowledge of their composition, biological functions, and diagnostic and therapeutic potentials. *Biochimica et Biophysica Acta (BBA)-General Subjects, 1820*(7), 940–948.

Vykhodtseva, N., McDannold, N., & Hynynen, K., (2008). Progress and problems in the application of focused ultrasound for blood–brain barrier disruption. *Ultrasonics, 48*(4), 279–296.

Wade, L. A., & Katzman, R., (1975). Rat brain regional uptake and decarboxylation of L-DOPA following carotid injection. *Am J Physiol., 228*(2), 352–359.

Wagner, S., et al., (2012). Uptake mechanism of ApoE-modified nanoparticles on brain capillary endothelial cells as a blood-brain barrier model. *PloS One, 7*(3), e32568.

Wang, S., et al., (1995). Delivery of antisense oligodeoxyribonucleotides against the human epidermal growth factor receptor into cultured KB cells with liposomes conjugated to folate via polyethylene glycol. *Proceedings of the National Academy of Sciences, 92*(8), 3318–3322.

Webb, M. S., et al., (2007). Liposomal drug delivery: recent patents and emerging opportunities. *Recent Patents on Drug Delivery & Formulation, 1*(3), 185–194.

Weber, S. J., et al., (1991). Distribution and analgesia of [3H][D-Pen2, D-Pen5] enkephalin and two halogenated analogs after intravenous administration. *Journal of Pharmacology and Experimental Therapeutics, 259*(3), 1109–1117.

Wei, B., et al., (2009). Development of an antisense RNA delivery system using conjugates of the MS2 bacteriophage capsids and HIV-1 TAT cell penetrating peptide. *Biomedicine & Pharmacotherapy, 63*(4), 313–318.

Wendler, F., Bota-Rabassedas, N., & Franch-Marro, X., (2013). Cancer becomes wasteful: Emerging roles of exosomes† in cell-fate determination. *Journal of Extracellular Vesicles, 2.*

Westergren, I., & Johansson, B. B., (1993). Altering the blood-brain barrier in the rat by intracarotid infusion of polycations: A comparison between protamine, poly-L-lysine and poly-L-arginine. *Acta Physiologica Scandinavica, 149*(1), 99–104.

Wilson, B., (2009). Brain targeting PBCA nanoparticles and the blood–brain barrier. *Nanomedicine, 4*(5), 499–502.

Wohlfart, S., et al., (2011). Kinetics of transport of doxorubicin bound to nanoparticles across the blood–brain barrier. *Journal of Controlled Release, 154*(1), 103–107.

Wong, H. L., et al., (2007). Chemotherapy with anticancer drugs encapsulated in solid lipid nanoparticles. *Advanced Drug Delivery Reviews, 59*(6), 491–504.

Wong, H. L., et al., (2010). Nanotechnology applications for improved delivery of antiretroviral drugs to the brain. *Advanced Drug Delivery Reviews, 62*(4), 503–517.

Wong, H. L., Wu, X. Y. & Bendayan, R., (2012). Nanotechnological advances for the delivery of CNS therapeutics. *Advanced Drug Delivery Reviews. 64*(7), 686–700.

Wu, S. H., Mou, C. Y., & Lin, H. P., (2013). Synthesis of mesoporous silica nanoparticles. *Chemical Society Reviews, 42*(9), 3862–3875.

Xin, H., et al., (2011). Angiopep-conjugated poly (ethylene glycol)-co-poly (ε-caprolactone) nanoparticles as dual-targeting drug delivery system for brain glioma. *Biomaterials, 32*(18), 4293–4305.

Yousif, S., et al., (2007). Expression of drug transporters at the blood–brain barrier using an optimized isolated rat brain microvessel strategy. *Brain Research, 1134,* 1–11.

Zhou, J., et al., (2012). Novel delivery strategies for glioblastoma. *Cancer Journal (Sudbury, Mass), 18*(1), 89.

Zlokovic, B. V., et al., (1993). Differential e□pression of Na, K-ATPase alpha and beta subunit isoforms at the blood-brain barrier and the choroid plexus. *Journal of Biological Chemistry, 268*(11), 8019–8025.

CHAPTER 14

NANOBIOMATERIAL FOR NON-VIRAL GENE DELIVERY

KIRTI RANI SHARMA

Amity Institute of Biotechnology, Amity University Uttar Pradesh, Noida, Sector 125, Noida–201303 (UP), India, E-mail: krsharma@amity.edu, Kirtisharma2k@rediffmail.com

CONTENTS

14.1   INTRODUCTION

Nanomaterials are known as chemically or biologically engineered materials with having dimension in thenano-range of 1–100 nm. Particles having "nano" dimension have been shown to exhibit various novel properties, for example, enhanced reactivity, greater sensing capability and improved mechanical strength. Various conventional and improved methods have also been found to offer easy, safe, fast, efficient, and cost effective nanotechniques to synthesis the desired bionanomaterials (Arivalagan et al., 2011; Salata et al., 2004). Previously, nano-sized formulations and dispersions had been proposed for drug delivery in major and various established pharmaceutical companies such as colloidal silver nanoparticles, titania nanoparticles, carbon nanotubes, nanoscaffold and silver nanoparticles and other metallo nanoparticles in tissue engineering, protein detection, cancer therapy, multicolor optical coding for biological assay at genome level (Salata et al., 2004). These days, nonviral gene therapy is well-known and promising therapeutic modality for the treatment of genetic, metabolic and neurodegenerative disorders. Nonviral approaches has been found to be a excellent and safe alternative gene transfer vehicles to other popular viral vectors due to having significant favorable properties such as lack of immunogenicity, low toxicity, and potential for tissue specificity and targeted drug delivery. Hence, these nonviral driven clinical and genetic approaches have also been tested in preclinical studies and human clinical trials over the last decades (Al-Dosari et al., 2009).

14.2   NANOBIOMATERIAL AS NONVIRAL GENE DELIVERY

At present, no FDA (Food and Drug Administration) approved gene medicinal products are available to be used as direct nonviral gene therapy. Therefore, many improved procedures are still under innovative considerations to enhance the potency, safety, and efficiency of chosen plasmid DNA vectors to get safe vector design itself in terms of loading of desired protein and antibiotic resistance genes in the host for their site specific targeted delivery (Mairhofer et al., 2008). Viral gene delivery system might cause pathogenesis in host cell during the gene delivery. So, designing of new multifunctional nano devices and nanobiomaterials are become more promising and safer choice for gene delivery over viral vector delivery such as use of E. coli, Lentil virus, Adenovirus, Herpes virus and Retro viruses for gene delivery. Use of non-viral gene vectors was also proposed for neuronal transfection via siRNA delivery mediated and still under experimental trials for their further improvements (Gao et al., 2007; Perez-Martinez et al., 2011). Recently, non-viral gene vectors, for example, lipoplexes, liposomes, polyplexes, and nanoparticles have also been experimented to relocate them as therapeutic DNA-based nano-medicine into the target cell that can facilitate development of nucleic acids-based nanotherapy (Pathak et al., 2009). Advanced nanotechnology has been stimulated to develop the multifunctional biomaterials for tissue engineering applications having synergistic interactions between nanomaterials and stem cell engineering to offer various possibilities to address some of the daunting challenges in regenerative medicine, such as control-ling trigger differentiation, immune reactions, limited supply of stem cells, and engineering complex tissue structures. These new advanced nanomaterial–stem cell interactions will facilitate improved biomaterial design for a range of biomedical and biotechnological applications (Kerativitayanan et al., 2015). Non-viral vectors-based gene therapy using cationic lipids or polymers, gold Nanoparticles, magnetic nanoparticles, quantum dots, silica nanoparticles, fullerenes, carbon nanotubes was found to have promising safe delivery potential as compared to viral gene delivery. nanoparticulate approaches like Cationic lipids, Cationic polymers, Supramolecular sys-tems (Katragadda et al., 2012; Ramamoorth et al., 2015). Use of nanobio-materials using cationic lipids, polyethylenimine derivatives, dendrimers, carbon nanotubes and combination of carbon-made nanoparticles with dendrimers was also proposed for targeted delivery system as new innova-tive nanoprocedure (Posadas et al., 2010). Recently, the delivery of nucleic acid was performed by using nano-carriers including calcium phosphates, lipids, and cationic polymers, polyplexes, lipoplexes, lipopolyplexes, chi-tosan, polyethylenimine, polyamidoamine dendrimers, and poly (lactide-co-glycolide) (Germershaus et al., 2015; Jin et al., 2014). Nanobiomaterials were loaded with desired delivery materials to be used as non-viral gene delivery using polyethylenimine-coated magnetic nanoparticlesand cell-penetrating peptides (CPPs) and used their impregnation into host genome. And, it was studied for efficient gene/ nucleic acid delivery in to the host cell (Kami et al., 2011; Huang et al., 2015). Nanoparticles gene carrier were developed with ease due to having their advantages such as low toxicity and immunogenicity, good biocompatibility, easy to produce, and a prom-ising perspective for application in tumor gene therapy to improve transfec-tion efficiency in host cell as nonviral gene vector (Blasiack et al., 2013). Currently, nanoparticles gene carriers have also been scale-up to encour-age the investigation of other potential nanoscaffolds exogenous DNA into targeted tissue as nanobiomaterials driven nonviral vector delivery or therapy especially used in treating genetic disorders such as severe combined immunodeficiency, cystic fibrosis and Parkinson's disease (Mintzer et al., 2009). Nonviral Gene therapy has been recently reported a promising therapeutic tool for cancer treatment as nanotechnology-based safe and effective delivery method. Genes are wrapped up in prepared nanoparticles which could be taken up easily by cancer or tumor cells, not to their healthy neighboring cells to investigate the improved anticancer efficacy suggesting that nanomedicine provides novel opportunities to safely gene deliver method to treat cancers or tumors (Suna et al., 2014). Significant advances were also achieved in both regenerative medicine and nanomedicine offering the significant use of tissue and organ specific regenerative treatments to develop functionalized bioactive scaffolds and nanoparticles that promote cell proliferation, migration and differentiation to open up their presumed expectations in clinical applications (Pean et al., 2012). Nonviral gene delivery systems have been increasingly proposed as a safer alternative to viral vehicles by using synthesized water-soluble chitin by aminoalkylating onto chitin at the C-6 position, Aminoethyl-chitin (AEC) complexed with DNA and AEC/DNA nanoparticles and further tested for their transfection efficiency in host genome. The transfection efficiency of AEC/DNA nanoparticles was investigated in a human embryonic kidney cell line (HEK293), which was showed the better and improved transfection efficiency of AEC/DNA nanoparticles as compare to naked DNA (Je et al., 2006).

14.3 GOLD NANOPARTICLES (AUNPS) AS NONVIRAL GENE DELIVERY VEHICLE

Gold nanoparticles were found to be an attractive nanoscaffold as nontoxic delivery vehicle of nucleic acids exhibiting their exploitable properties such as biocompatibility, tunable size and easy functionalization. Their use as targeting nucleic acid can be achieved by decorating the surface of these nanobiovehicles with desired specific antibodies. This fabricated system was further targeted to the disease cells and their grafting with fabrication of active functional groups (e.g., polyethylene glycol and zwitterionic entities) on their surface that lead to increase plasma protein adsorption followed by improving the pharmacokinetics and evading immune surveillance (Ding et al., 2014).

14.3.1 AUNPS AS NUCLEIC ACID DELIVERY VEHICLE

Key features of oligonucleotide and small interfering RNA-modified gold nanoparticle conjugates were found to be more promising intracellular gene regulation agents over various conventional agents. It was also reported that gold nanoparticles were stably functionalized with covalently attached oligonucleotides to activate immune-related genes and pathways in human peripheral blood mononuclear cells, but not an immortalized lineage-restricted cell line. These unique observations have strong implications for the application of oligonucleotide-modified gold nanoparticle conjugates in translational research and to interpret further development of their applications in therapeutics and gene delivery systems (Kim et al., 2011). In the last decade, the potential of double-stranded RNAs to interfere with gene expression has been studied to carry out new therapeutic implication in history of medicinal research. Since small interfering RNA (siRNAs, 21 base pair double-stranded RNA) was found to more efficient to elicit RNA interference (RNAi) from the administration site to the target cell that provide evidence of RNAi triggering, specifically silencing c-myc protooncogene, via the synthesis of a library of novel multifunctional gold nanoparticles (AuNPs). The efficiency of the AuNPs is demonstrated its hierarchical approach to drive chemical and biological approaches led to a safe, nonpathogenic, self-tracking, and universally valid nanocarrier that could be exploited for therapeutic RNAi (Conde et al., 2012). Monolayer-functionalized gold nanoparticles were also providing attractive nonviral nanobiovehicles for pharmaceutical delivery approaches because of their small size and site-specific delivery (Han et al., 2007).

14.3.2 AUNPS USED IN MULTIFUNCTIONAL TUMOR GENE DELIVERY

Novel therapeutic strategies are desired to achieve tumor eradication recently using gold nanoparticles (AuNPs) against hyperthermic tumor cell ablation. Same as that by keeping that vision, adenoviral (Ad) vectors coupled with AuNP-mediated hyperthermia have been used for targeting, imaging, and cancer therapy. It was tested for the capability of these AuNP coupled Ad vectors for hyperthermic tumor cell ablation and found to be quite suitable with laser induced hyperthermic tumor cell killing therapy. In addition, it was observed that AuNPs outside and inside the cell contribute differentially towards hyperthermia induction. However, it may be needed more suitable nanotechnological advances to realize the exact potential of the multifunctional AuNP-coupled Ad vector system for simultaneous targeting, imaging, and combined hyperthermia and tumor gene therapy (Saini et al., 2014). Gold nanoparticles are also fabricated by grafting various compatible biocompatible polymers and natural or synthethic biomolecules to be used as potential nonviral gene delivery vehicle for cancer gene therapy being a novel nano-avenue-based on engineering multifunctional smart delivery systems (Bahadur et al., 2014).

14.3.3 AUNPS AS NANOTHERANOSTIC TOOL

Nanotheranostics had couple of advantages of nanotechnology-based systems in order to diagnose and treat a specific disease at an early stage, to direct a suitable therapy toward the target tissue-based on the molecular profile of the altered phenotype that subsequently facilitate monitoring and treatment of targeted disease. Hence, this can be called a advanced nonviral driven tailored therapeutic strategy when used gold nanoparticles having unique optical, their ease of surface modification and high surface-to-volume ratio that enable to reduce the off-target effects associated with universal treatments to improve the safety profile of a given gene therapy (Sun et al., 2013). Now with the development of nanobiotechnology, the applications of nano-magnetic particles and gold nanoparticles in biomedical field have increasingly become one of the hot avenue spots lead to elicit more clear advantageous cell targeting transfection efficiency (Vinhas et al., 2015; Guo et al., 2010). The study-based on the process of magnetofection was showed that it did not change the mechanism of gene complexes' endocytosis. It is accelerated gene complexes sedimentation to the cell surface to promote the phagocytosis of gene complexes efficiently, to enhance the fast the gene transfection even carrying a plasmid DNA harboring target genes or a synthesized small interference RNAs (siRNA) (Jain et al., 2012). Charge-reversal functional gold nanoparticles were also prepared by using layer-by-layer technique to deliver small interfering RNA (siRNA) and plasmid DNA into cancer cells rapidly. The expression efficiency of enhanced green fluorescent protein (EGFP) was also improved by adjuvant transfection with charge-reversal functional gold nanoparticles that had much lower toxicity to the cell proliferation. Lamin A/C, an important nuclear envelope protein, was effectively silenced by lamin A/C-siRNA and delivered by charge-reversal functional gold nanoparticles with its knockdown efficiency was observed to be better than that of commercial Lipofectamine 2000. As well as, Confocal laser scanning microscopic images was done which was indicated that there was more cy5-siRNA distributed throughout the cytoplasm for cyanine 5-siRNA/polyethyleneimine/cis-aconitic anhydride-functionalized poly (allylamine)/ polyethyleneimine/11-mercaptoundecanoic acid-gold nanoparticle (cy5-siRNA/PEI/PAH-Cit/PEI/MUA-AuNP) complexes. And, these novel findings were reported to demonstrate the feasibility of this procedure using of charge-reversal functional gold nanoparticles to improve the nucleic acid delivery efficiency (Jain et al., 2012; Phillips et al., 2012).

14.3.4 AUNPS AS NANOTHERANOSTIC NONVIRAL GENE TOOL USED IN CORNEAL SURGERY

Gold nanoparticles are also found to be stabilized in naturally occurring non-toxic gum arabic (GA-AuNP) which were developed recently at the University of Missouri-Columbia. These synthesized AuNPs were found to be excellent nanoscale in range of 16 nm to 20 nm and subjected to binding with DNA that has shown their exceptional in vivo stability, uptake and clearance in pigs and rodents. The aim of this study was to evaluate GA-AuNPs cytotoxicity to the cornea, to examine the effects of GA-AuNPs on corneal morphology and function that determine their suitability for corneal gene therapy. Donor human corneas and cultured human corneal fibroblasts (HSF) were maintained at 37°C with 5% CO_2 and exposed to GA-AuNP (0–60 µg/ml) for time variation ranging from 0–72 hours, Trypan blue exclusion, MTT, MultiTox-Fluor, glutathione and TUNEL assays to evaluate cytotoxicity. Immunocytochemistry, bright/fluorescent microscopy and transmission electron microscopy (TEM) were also done to analyze morphological changes, relative populations of live and dead cells and/or intracellular trafficking of GA-AuNP in the human cornea and HSF (human embryonic stem cells). Along with, inductive coupled plasma enhanced absorption spectroscopy (ICP-EAS) and neutron activation atomic absorption (NAA) were also carried out to determine cellular uptake of gold in human cornea in vitro and ex vivo. As a result, significant uptake of gold was detected in HSF at 4 hours (22.5 ± 0.85–30.85 ± 1.20 ng/ml) by ICP-EAS and in ex-vivo human cornea at 4 hours (2.3–22.5 ppm) and 24 hours (6.6–47.0 ppm) by the NAA analysis. TEM analysis was also demonstrated that GA-AuNPs can easily enter the human corneal epithelial, stromal and endothelial cells freely and confirmed the presence of GA-AuNPs in the cytoplasm and GA-AuNPs did not any kind of cause damage to the vital cell organelles (Phillips et al., 2012). A study considering corneal transplant was also conducted by using adeno-associated virus (AAV) and gold nanoparticles stabilized in polyethyleneimine (GNP-PEI) vectors for delivering therapeutic genes into human corneal endothelial cells as an effective and attractive gene therapy approach to treat/cure corneal endothelial diseases. HPV16-E6/E7 was observed to be transformed human corneal endothelial (HCN) cultures and incubated with AAV6, AAV8, AAV9 or GNP-PEI vector expressing reporter gene under control of Rous sarcoma virus or CMV + chicken beta actin promoter for 6 hours. The AAV titer had reported to have 109 genomic copies/µl and GNP-PEI transfection solution had 5µg DNA with PEI nitrogen to DNA phosphate ratio180. The amount of transgene delivery was studied with the quantification done with cytochemistry, immunostaining and/or real-time polymerase chain reactions. As well as, toxicity was also tested by Trypan blue assay and morphological examinations. These finding was lead to confirm the significant gene delivery into human corneal endothelial cells was done with tested AAV and GNP-PEI vectors. Quantification of experimental data revealed 0.6–4.4% transgene delivery by AAV vectors into HCN. AAV6 vector showed highest transduction among tested AAV vectors as well as contrary to AAV vectors GNP-PEI vector was showed remarkably high transgene delivery into HCN (Sinha et al., 2008).

14.3.5   AUNPS-NANOSCAFFOLDAS NONVIRAL GENE DELIVERY

Gold nanoparticles was also reported an attractive and applicable scaffold for delivery of nucleic acids when used as covalent and noncovalent gold nanoparticle conjugates for applications in gene delivery and RNA-interference technologies including their effective endosomal entrapment/escape and active delivery expression potential of nucleic acids in the host genome (Mieszawska et al., 2013). The surface of gold nanoparticles can easily be modified according to the desired binding ligands for targeting, drugs or biocompatible coatings and addition of other imaging tags/labels to study their vehicle trafficking in the host cell. As well as, AuNPs can be easily incorporated into polymeric nanoparticles or liposomes that have been used in various biomedical application due to their amenability of synthesis and multisystem functionalization (Tiwari et al., 2011; Yah et al., 2013).

14.3.6 AUNPS-DENDRIMER (AUDENPS) AS NONVIRAL GENE DELIVERY

The use of partially acetylated dendrimer-entrapped gold nanoparticles (AuDENPs) was also designed and studied for gene delivery applications. Partially acetylated generation 5 poly (amidoamine) dendrimers were used as templates to synthesize AuDENPs and characterized via using complex two different pDNAs encoding luciferase (Luc) and enhanced green fluorescent protein (EGFP), respectively for gene transfection studies which further characterized by gel retardation assay, dynamic light scattering and zeta potential measurements. It was reported that all acetylated AuDENPs are able to effectively compact the pDNA and transfect genes to the model cell line with high efficiency comparable to the AuDENPs without acetylation. It was also proved that partially acetylated AuDENPs were less toxic than that of non-acetylated AuDENPs by cell viability assay. Hence, these developed partially acetylated AuDENPs may serve as key nonviral gene delivery nano-bio-vehicle for safe gene delivery applications with desired gene transfection efficiency (Hou et al., 2015; Hossain et al., 2015).

14.3.7 CATIONIC AUNPS AS AUDENPS FOR GENE DELIVERY

Cationic gold nanoparticles were synthesized by carrying out possible modification done with 2-aminoethanethiol, 8-amino-1-octanethiol, and 11-amino-1-undecanethiol by $NaBH_4$ reduction of $HAuCl_4$ in the presence of thiols in water or a water/ethanol mixture solvent. As a result, using of high concentration of $HAuCl_4$ solution was lead to induce the synthesis of gold nanoparticles. When thiols were used for further fabrications, changed surface charge of the gold nanoparticles was shift from negative to positive that stabilized the gold nanoparticles. These synthesized cationic gold nanoparticles were observed to promote the gene transfection with high efficacy (Chen et al., 2008). Di-sulfide linked polyethylenimine coated gold nanoparticles (ssPEI-GNPs) were also prepared previously which are observed in the size of 20 nm size and subjected to deliver the genes to target site. Coating of ssPEI onto prepared gold nanoparticles was observed for their stability by having weight ratio of 1:3, the 19 ± 1.14 nm of average hydrodynamic diameter of the ssPEI-GNPs and 41 ± 1.23 mV of zeta potential value. Gene expression efficacy of the nanoparticles were confirmed by fluorescent microscopy and luciferase assay which demonstrated the transgene delivery capability of the ssPEI-GNPs (Uthaman et al., 2015).

14.3.8 ULTRA-SMALL AUNPS AS NONVIRAL GENE DELIVER TOOL

A novel current clinical study was carried out the size-dependent gene tranfection potential of ultra-small gold nanoparticles (upto 2 nm in diameter) and further studied for their potential application of intranucleus delivery nonviral gene delivery vehicle or nanotherapetic tool. Their intracellular distribution was studied in MCF-7 breast cancer cells and it was investigated the possibility of using ultra-small 2 nm nanoparticles being as nano-nonviral gene carriers for nuclear delivery of a tripleX-forming oligonucleotide (TFO) that binds to the c-myc promoter. It was also found that on comparing with free TFO, the nanoparticle-conjugated TFO was more active at reducing c-myc RNA and c-myc protein that resulted in reduced cell viability (Huo et al., 2014).

14.4 SILVER NANOPARTICLES (AGNPS) AS NONVIRAL GENE DELIVERY

Silver nanoparticles (AgNPs) exhibit a consistent amount of versatile properties which endorse their vast spectrum of applications in bio-nano-medicine. Researchers found that AgNPs have high antimicrobial efficacy against many pathogenic bacteria species including *Escherichia coli, Neisseria gonorrhea, Chlamydia trachomatis* and other viruses. In biomedical engineering, silver nanoparticles are found to be considered potent and ideal gene delivery systems for tissue regeneration and used as safe biosensors for nonviral gene therapeutic tool (Marin et al., 2015).

14.4.1 AGNPS AS NANOTHERANOSTIC TOOL

Targeted delivery and controlled release of oligonucleotide therapeutics in vivo were studied by using silver nanoparticle (AgNPs) tagged with photolabile nucleic acid conjugates to performing the inducible gene silencing. There were used for the delivery of therapeutic agents such as antisense oligonucleotides (Brown et al., 2011) Biofunctionalized stable AgNPs with good DNA binding ability were studied for efficient transfection and minimal toxicity to be used in biomedical applications especially in wound healing and cancer gene therapy. One-pot facile green synthesis of polyethylene glycol (PEG) was stabilized with chitosan-g-polyacrylamide modified AgNPs. To enhance the efficiency of gene transfection, the Arg–Gly–Asp–Ser (RGDS) peptide was immobilized on the silver nanoparticles to increase the transfection efficiency of AgNPs in host genome (Sarkar et al., 2015). Montmorillonite, type of nanoparticle was also conjugated with silver nanoparticles and further stabilized with montmorillonite, starch, citrate, polylysine and multiwalled carbon nanotubes that were used for binding with plasmid pcDNA-GFP to carry out desired a gene delivery. And, it was found that this AbNPs conjugated nonviral gene delivery vector had potential to revolutionize the area of biosensing, imaging, diagnosis and gene therapy (Sironmani et al., 2015). Silver nanoparticles (Ag NPs) are also investigated their role in molecular mechanism and cytotoxic effects of AgNPs mediated cytotoxicity in both cancer and non-cancer in vitro for their further therapeutic use in conjunction with conventional gene therapy. The synergistic effect of Ag NPs was also experimented on the uracil phosphoribosyl transferase expression system sensitized the cells more towards treatment with the drug 5-fluorouracil to induce the apoptotic pathway in the host disease cell that can be considered new chemosensitization strategy for future application in gene therapy (Cardenas et al., 2006).

14.4.2 AGNPS AS OILIGONUCLEOTIDE-THIOL CONJUGATED NONVIRAL GENE DELIVERY VEHICLE

Recent clinical study was performed to know the possible nonspecific interactions of thiol–ssDNA and dsDNA macromolecules with gold nanoparticles by using dynamic light scattering and cryogenic transmission electron measure the nano-visualization and functionalization of gold nanoparticles with thiol–ssDNA and nonthiolated dsDNA (Gopinath et al., 2008). As well as, poly adenine (polyA) DNA functionalized gold nanoparticles (AuNPs) were fabricated with high density of DNA attachment and high hybridization ability similar to those of its thiolated counterpart. This nanoconjugate was used polyadenine as an anchoring block for binding with the AuNPs surface to facilitate the appended recognition block a better upright conformation for hybridization that demonstrate its great potential to be a tunable plasmonic biosensor. Sensitive fluorescence turN-on strategy was used to investigate and quantified the dissociation of polyA DNA on gold nanoparticles in diverse experimental conditions to maintain a good stability of this prepared polyA DNA–AuNPs nanoconjugates (Lu et al., 2015; Hill et al., 2009).

14.4.3 AGNPS AS PHOTOACTIVATED ANTISENSE MEDIATED NONVIRAL GENE DELIVERY VEHICLE

The unique photophysical properties of metal nanoparticles such as silver nanoparticles were characterized for their clinical contribution as potential photoactivated drug delivery vectors. For this study, synthesis and characterization of 60–80 nm silver nanoparticles (SNPs) were done with thiol-terminated photolabile DNA oligonucleotides. In vitro study, fluorescent confocal microscopy of treated cell cultures was performed that showed efficient UV-wavelength photoactivation of surface-tethered caged ISIS2302 antisense oligonucleotides possessing internal photocleavable linkers to perform efficient cellular uptake as compared to commercial transfection vectors. Their potential as multicomponent delivery agents was also studied for oligonucleotide therapeutics through regulation of ICAM-1 (Intracellular Adhesion Molecule-1) silencing. Following results were suggested for better achieving light-triggered, spatiotemporally controlled gene silencing via nontoxic silver nanocarriers that provide as tailorable strategy for nanomedicine, gene expression studies and genetic therapies (Brown et al., 2013). Photoluminescent (PL) graphene quantum dots (GQDs) with large surface area and superior mechanical flexibility were exhibited to fascinate the optical and electronic properties and possess their great promising applications in biomedical engineering. For this study, a multifunctional nanocomposite of poly-L-lactide (PLA) and polyethylene glycol (PEG)-grafted GQDs (f-GQDs) was proposed for simultaneous intracellular microRNAs (miRNAs) imaging analysis and subjected to combined gene delivery for enhanced therapeutic efficiency. The functionalization of GQDs with PEG and PLA was found to be imparted the nanocomposite with super physiological stability and stable photoluminescence was established over a broad pH range, which is vital for cell imaging. These conducted experiments were demonstrate the f-GQDs excellent biocompatibility, lower cytotoxicity and protective properties in HeLa cell line. It was observed the f-GQDs effectively delivered the desired miRNA probe for intracellular miRNA imaging analysis and regulation. Large surface of GQDs was notably found to be capable of simultaneous adsorption of agents targeting miRNA- 21 and survivin, respectively. The combined conjugation of miRNA-21-targeting and survivin-targeting agents were also better induced with improved inhibition of cancer cell growth and more induced apoptosis of cancer cells, compared with conjugation of agents targeting miRNA-21 or survivin alone. Hence, these recent findings are highlighted the key role of the highly flexible multifunctional nanocomposite in biomedical application of intracellular molecules interpretations and clinical gene therapeutics (Dong et al., 2015).

14.5 IRON/MAGNETO NANOPARTICLES AS NONVIRAL GENE DELIVERY VEHICLE

In the recent survey, magnetic iron nanoparticles along with various chemically modified biopolymers were studied for gene therapy with the ease of their synthesis and bioconjugation processes. Super paramagnetic nanoparticles are found to be more promising nonviral gene delivery tool as somatic cell nuclear transfer technique that performed by magnetofection using magnetic Fe^3O^4 nanoparticles as gene carriers. Surface modification was done by polyethylenimine and prepared the spherical magnetic Fe^3O^4 nanoparticles showed strong binding affinity for DNA plasmids expressing the genes encoding a green (DNAGFP) or red (DNADsRed) fluorescent protein. Atomic force microscopy an analysis was also confirmed binding of the spherical magnetic nanoparticles to stretched DNA strands up to several hundred nanometers in length. As a result, stable and efficient co-expression of GFP and DsRed in porcine kidney PK-15 cells was also achieved by magnetofection that demonstrate the potential application of magnetic nanoparticles as an attractive nonviral gene delivery system for animal genetics and breeding studies (Mulens et al., 2013; Wang et al., 2014). Gold/iron-oxide magnetic Nanoparticles (GoldMAN) were also observed to study for their useful magnetic properties to various biomolecules to induce intracellular transduction. This proposed method had great potential for application of the adenovirus gene delivery vector (Ad), widely used for in vitro/in vivo gene transfer, to Ad-resistant cells. It was also noted that Ad was easily conjugated on to GoldMAN and the Ad/GoldMAN complex and it was introduced into the cell by the magnetic field which increased gene expression over 1000 times that of Ad alone. The penetration of GoldMAN was noticed fast in plasma membrane and independent of the cell-surface virus receptors in endocytosis pathway. This mechanism may be fruitful to contribute the proposed improvement in the gene expression efficiency of Ad. Hence, this technology can be a useful tool for extending Ad tropism and enhancing transduction efficiency by using Gold/iron-oxide magnetic nanoparticles. New developed magnetic nucleic acid delivery system was also under new innovative strategy composed of iron oxide nanoparticles and cationic lipid-like materials termed lipidoids. Coated nanoparticles are found to be capable of delivering DNA and siRNA to cells in host cell culture. The iron oxide nanoparticle delivery platform might be offered the potential magnetic targeting system to provide opportunity for effective gene therapy with MRI imaging and magnetic hyperthermia (Kamei et al., 2009).

Magnetic nanoparticle-based gene transfection has been shown to be effective nonviral nanodevice in both viral vectors and with non-viral gene delivery. Therapeutic or reporter genes are attached to magnetic nanoparticles which are then focused to the target site/cells via high-field/high-gradient to perform rapid in-vitro transfection and compares well with cationic lipid-based reagents, producing good overall precise transfection levels and shorter transfection times. In order to improve the overall transfection levels, a novel oscillating magnet array system was developed to add lateral motion to the particle/gene complex in order to promote transfection levels in human airway epithelial cells compared to both tested static field techniques and the cationic lipids. Hence, it was led to demonstrate their advantages of magnetofection with rapid transfection times and requiring lower levels of DNA than cationic lipid-based transfection agents as more potential non-viral gene delivery vehicle both in vitro and in vivo (Jiang et al., 2013; McBain et al., 2008). Polyethyleneimine (PEI) coated iron oxide magnetic nanoparticles (MNPs) were used as gene carriers for binding and condensing with plasmid DNA e⬜pressing enhanced green fluorescent protein (EGFP). Their characterization was done with transmission electron microscopy (TEM) and atomic force microscopy (AFM). And, it was confirmed that EGFP gene was successfully e⬜pressed under mediation of an e⬜ternal magnetic field to get efficient targeting of EGFP gene into host genome that can be used treatment of many diseases such as tumor and orthopedic disease (McBain et al., 2008; Wang et al., 2014). Super paramagnetic iron oxide nanoparticle (NP) was reported to enable magnetic resonance imaging when coated with a novel copolymer (CP-PEI) comprised of short chain polyethylenimine (PEI) and poly (ethylene glycol) (PEG) grafted to the natural polysaccharide, chitosan (CP), which allows efficient loading and protection of the nucleic acids. Significantly, NP-cP-PEI was demonstrates an innocuous to⬜ic profile and a high level of expression of the delivered plasmid DNA in a C6 xenograft mouse model. It was led to make it more potential nonviral nanocandidate for safe in-vivo delivery of DNA used in targeted gene therapy (Kamau et al., 2005). In another study, efficacy of a nanocarrier (polyethyleneimine [PEI]-super paramagnetic iron oxide nanoparticle [SPIO]), was examined which are composed of a core of iron oxide and a shell of PEI, in the systemic delivery of therapeutic siRNA to experimental arthritic joints.

PEI-SPIO/siRNA nanoparticles were synthesized and characterized in vitro and administered intravenously to arthritic rats to analyze cellular uptake, tissue distribution and the therapeutic effect of a siRNA against the IL-2/-15 receptor β chain (IL-2/IL-15Rβ). These e⬜amined findings were confirmed that PEI-SPIOs loaded with siRNA was displayed negligible cytotoxicity, improved siRNA stability, efficient uptake by macrophages and the ability to induce specific gene silencing in vitro. PEI-SPIO-delivered siRNA was accumulated easily in inflamed joints and was efficiently taken up by joint macrophages and T cells. Although, IL-2/IL-15Rβ siRNA-loaded PEI-SPIOs alone were efficacious in the treatment of studied e⬜perimental arthritis when given in combination therapy with both PEI-SPIO/IL-2/IL-15Rβ siRNA and a magnetic field displayed an additive anti-inflammatory effect especially in rheumatoid arthirits (Kievit et al., 2009). Hence, these theranostic magnetic nanoparticles were studies from last many decades and further considerable fabrications were performed for their encapsulation or coated them with polymers such as polysaccharides to make them more efficient to deliver various drugs and therapeutic genes in host cell (Duan et al., 2013). Optimization of silica-iron oxide magnetic nanoparticles with surface phosphonate groups was also carried out to decorate them with 25-kD branched polyethylenimine

(PEI) for gene delivery. Moderate PEI-decoration of MNPs was resulted in charge reversal and destabilization. Surface decoration of the silica-iron oxide nanoparticles with a PEI-to-iron was done in w/w ratio of 10–12% that leads to allows for efficient viral gene delivery and labeled cell detection by MRI (Uthaman et al., 2015). Intrinsic superparamagnetism of iron nanoparticles was also enabled noninvasive magnetic resonance imaging (MRI) and their biodegradability which make it more viable for effective in-vivo delivery.

A therapeutic super paramagnetic iron oxide nanoparticle (SPION) typically is consisted of three primary components: an iron oxide nanoparticle core that serves as both a carrier for therapeutics and contrasting tool for MRI, a coating on the iron oxide nanoparticle that promotes favorable interactions between the SPION and host cell system and a therapeutic payload that performs the designated function in-vivo. That design may also include a targeting ligand that recognizes the receptors over-expressed on the exterior surface of cancer cells. Acquired strategies were planned to bypass the physiological barriers such as liver, kidneys, and spleen, involve tuning the overall size and surface chemistry of the SPION to maximize blood half-life and facilitate the observable navigation in the host body. The payload can be genes, proteins, chemotherapy drugs, or a combination of these other conjugated molecules coated with silica or other designed polymer. These critical design parameters have been optimized and these nanoparticles, combined with imaging modalities, can serve as truly multifunctional theranostic nonviral agents to targeted the desired or site specific gene delivery into host genome (Mykhaylyk et al., 2012; Kievit et al., 2011).

14.6 OTHERS NEWLY PROPOSED BIONANOPARTICLES USED AS NONVIRAL GENE DELIVERY SYSTEM

Recently, a number of investigations were carried out on various non-toxic, non-allergic and biocompatible biomolecules such as chitsoan, glycanoparticles and other fabricated polymers. These biopolymer-based prepared bionanoparticles were also described effective and more potent nonviral gene delivery vehicle.

14.6.1 CHITOSAN–DNA-FAP-B NANOPARTICLES AS NONVIRAL GENE DELIVERY VEHICLE

Chitosan–DNA-FAP-B nanoparticles were found to be novel non-viral vectors for specific gene delivery to the lung epithelial cells. Their stability studies were showed that chitosan-DNA-FAP-B nanoparticles were stable for upon preservation at −20°C having constant transfection efficacy. It was proved that these polymer derived nanoparticles can be proved key nonviral gene delivery vehicles (Mohmmadi et al., 2012; Pack et al., 2005). Hence, many research groups and active international clinical organizations have dedicated considerable efforts have been streamlined to improve the efficiency of nonviral gene delivery systems composed of DNA and safe, biocompatible, non-toxic and cost effective nanomaterials such as lipids, polymers, peptides, dendrimers and surfactants to carry out high levels of transfection, biocompatibility and tissue-targeting ability (Donkuru et al., 2015). Currently, the block copolymer was prepared that-based on the biocompatible polymer poly (2-methyl-2-oxazoline) (PMOXA) combined with the biocleavable peptide block poly (aspartic acid) (PASP) and finally modified with diethylenetriamine (DET). PMOXA-b-PASP(DET). The polymer–peptide hybrid system was investigated for efficient transfection of HEK293 and HeLa cells with GFP pDNA in-vitro (Witzigmann et al., 2015).

14.6.2 POLYMERIC/PLOYCATIONIC FABRICATED NANOPARTICLES AS NONVIRAL GENE DELIVERY VEHICLE

The potential non-viral gene delivery was also performed with polymeric nanoparticles composed of poly ([beta]-amino esters) (PBAEs) and DNA that can be formulated to be stable in the presence of serum proteins and have high gene delivery without toxicity to human primary cells. The biophysical properties of PBAE/DNA nanoparticles had good correlation to transfection efficacy when tested in the appropriate media conditions. As well as, it is showed that effective electrostatic interactions can drive peptide coating of nanoparticles and enable ligand-specific gene delivery to have their biphasic efficacy relationships. Even the small modifications to the termini of a polymer can significantly increase its in-vivo activity and demonstrate potential utility of these polymers in the fields of cancer therapy, genetic vaccines, and stem cell engineering. And this proposed nonviral carrier systems have many attractive properties over a virus including high safety, low immunogenicity, high nucleic acid cargo capacity and ease in manufacture. Hence, a potential coating method for targeted delivery and flexibility for future delivery design improvements was developed (Green, 2007). So, non-viral gene delivery using polymeric nanoparticles has emerged as an attractive approach for gene therapy to treat various genetic diseases, neuro-degenerative disorders, heart disease, kidney diseases in the field of regenerative medicine. Examples of biomaterials nanoassemblies which have been proposed to form nanoscale polycationic gene delivery nanoparticles include polylysine, polyphosphoesters, poly (amidoamines) and polyethylenimine (PEI) which are studied to carry out earlier proposed nucleic acid delivery, human retinal endothelial cells (HRECs), mouse mammary epithelial cells, human brain cancer cells and macrovascular (human umbilical vein, HUVECs) endothelial cells (Shmueli et al., 2013). Over the past two decades, many natural and synthetic polymer-based micro and nanocarriers, had been developed for their promising application in various types of tissue regeneration, including bone, cartilage, nerve, blood vessels and skin. The development of suitable polymers scaffold was also proposed to designs to aid the repair of specific diseased cell types and to find out its important potentials in tissue restoration.

Fibrinogen (Fbg)- and fibrin (Fbn)-based micro and nanostructures can provide suitable natural matrix environments and had been effectively used in tissue engineering and regenerative medicine (Rajangam and An, 2013). A current study of nanotherapeutics was also studied for the use of cationized gelatin nanoparticles as biodegradable and low cell toxic alternative carrier to existing DNA delivery systems. These native gelatin nanoparticles were synthesized by using a two step desolvation method and trigger to bind with DNA by electrostatic interactions onto the surface of the particles via the quaternary amine cholamine that was covalently coupled to the particles. The modified bionanoparticles were loaded with different amounts of plasmid and compared to polyethyleneimine-DNA complexes (PEI polyplexes) as gold standard. Transfection ability of the loaded nanoparticles was carried out to test the B16 F10 cell and their cell toxicity was monitored. In this proposed experiment, exponential increase of gene expression was observed with a certain delay after transfection. In contrast to PEI polyplexes, cationized gelatin nanoparticles almost did not show this kind of significant cytotoxic effects during the proposed delivery. These observations are confirmed that cationized gelatin nanoparticles are more potent carrier for non-viral gene delivery (Zwiorek et al., 2004). Various lipids, peptides, cationic polymers and certain inorganic nanomaterials have been also reported as gene delivery vectors being effective multifunctional bionanocarriers (Xian and Lee, 2012).

14.6.3 FABRICATED POLYMERIC/METALLIC/NON-METALLIC/LIPOSOMIC/QUANTAM DOTS COUPLED HYBRID BIONANOSCAFFOLDS AS NONVIRAL GENE DELIVERY VEHICLE

Recently, gene delivery scaffolds-based on DNA plasmid condensation with colloidal gold/cationic polymer was also developed via electrostatic interaction yielding gold/polyethyleneimine (PEI), gold/chitosan and gold/chitosan/PEI complexes. Luciferase-encoding plasmid DNA was subsequently added and adsorbed on the prepared scaffolds to be used as a non-viral gene delivery carrier. Confocal fluorescent microscopy was carried out to verify the presence of DNA in the cell using gold nano-scaffold as a carrier. Transfection efficiency assay using A549 and HeLa human cell lines was performed that reveal gold/polymer-based nano-scaffolds provided transfection efficiency approximately 10 times higher than polymeric-based gene carriers (Tencomnao et al., 2011; Nitta and Numata, 2013). Native glucagon like peptide 1 (GLP-1), an incretin hormone that regulates blood glucose level post-prandially, is used for the treatment of type 2 diabetes mellitus. In this study, to exploit the function of GLP-1, the glucosamine-based polymer chitosan was used as a cationic polymer-based in vitro delivery system for GLP-1, DPP-IV resistant GLP-1 analogues and siRNA targeting DPP-IV mRNA. It was reported that all chitosan–DPP-IV siRNA nanocomplexes were capable of DPP-IV silencing and carry out effective abrogating enzymatic activity of DPP-IV in media of silenced cells, and with no apparent cytotoxicity. These findings are lead to confirm the versatility of these specific formulations to deliver plasmid DNA and siRNA render their use as a combined in vivo therapy for the control of type 2 diabetes (Jean et al., 2012).

Current investigations on effective nonviral gene delivery are still under experimentations and innovative strategies to focused on developing more safe and potent nonviral gene carriers that can be effectively protect naked DNA, RNA and siRNA that can be rapidly degraded by various metabolizing enzymes present in the blood and easily traceable during the delivery in host cell with good ease of gene trafficking at the targeted site (Choi et al., 2014). The nonviral bionanodelivery systems can significantly improved to enhance their exploitable biopharmaceutical features, pharmacokinetic properties and therapeutic efficacy of entrapped drugs and desired gene or proteins including branched polyethylenimine (PEI) as a cationic polymer that contains primary, secondary and tertiary amino groups having high density of amines. This improvised strategy was reported to make it most promising cationic nonviral nanovectors for gene delivery as PEI-based polymeric nanoparticle system, PEI/silica nanoparticle systems and PEI/metal nanoparticle systems (Wang et al., 2015; Vasir and Labhasetwar, 2006). Fabricated hybrid carbon-nanotubes were also popular to be used as DNA loaded nonviral gene delivery vehicle from last decade and they are also described more multifunctional gene or drug delivery vehicle (Bates and Kostarelos, 2013). Liposomes conjugated polymer/metal nanoparticles driven gene delivery vehicle were also studied for improved tranfection efficiency in prescribed and controlled delivery doses due to having high toxicity reported in some designed experiments (Liu et al., 2011). Gene-associated drugs can be loaded with ease within a hybrid Quantum Dots (QD) core or attached to the surface of these nanoparticles via direct conjugation or electrostatic complexation (Bijju et al., 2010; Probst et al., 2013).

Newly developed novel nanomaterials and their use in biomedicine has been going proved a potential and automotive tool which includes iron, carbon, gadolinium, gold, silicon mediated nanovectors like nanotubes, nanorods, dendrimers, nanospheres, nanoantennas or nanowires. These are used for the targeted drug and gene delivery nonviral vehicle stem cell therapy and cancer therapy too. Prepared nanoparticles can be used to target antigens or bio-markers or gene or specific desired protein that are highly specific to cancer cells or diseased cell that were subjected to antigen binding peptide ligands binding to the nanostructure. This bionanoactive complex may play potential role of effective drug delivery systems and to develop hybrid nanoscaffolds and nanowires to be used in nerve regeneration therapy too (Rani et al., 2014a). New landmark payload of nanomedicine was also noticed named, vertical silicon nanowires (NWs) that can be better tool to perform the targeted gene-specific manipulation of diverse murine and human immune cells with negligible toxicity. This NW-mediated gene silencing was subjected to investigative the role of the Wnt signaling pathway in chronic lymphocytic leukemia (CLL). Remarkable CLL-B cells from different patients were interpreted for their observed results and it was exhibited the tremendous heterogeneity in their response to the knockdown of a single gene, *LEF1*. Overall, these findings were led to suggest functional classifications of potentially markers for the selection of patient-specific therapies in CLL and highlighted their opportunities in nanomedicine (Shalek et al., 2012). While polymer, lipid conjugated nanowires-based delivery system was proved quite active nonviral gene delivery tool and there are still further under new improvements that are going to be carried out by varying electroporation techniques continue to be make them more refined to desired substrate mediated delivery in host cell (Blow, 2009).

14.6.4 FABRICATED HYBRID ALBUMIN BIONANOPARTICLES AS NONVIRAL GENE DELIVERY VEHICLE

There has been a great interest in application of nanoparticles as biomaterials for delivery of therapeutic molecules such as drugs and genes used for tissue engineering. Now these days, number of improved fabrications of albumin or protein-based nanoparticles were under various investigations to being consisting of protein molecules such as protein, silk, collagen, gelatin, β-casein, zein and albumin, protein-mimicked polypeptides along with polysaccharides such as chitosan, alginate, pullulan, starch and heparin. Recently, various enzyme bound biochemically active bovine serum albumin and egg albumin nanoparticles were synthesized by emulsification and desolvation process to improve their loading efficacy of the enzyme to study their controlled delivery in the delivery system with variable concentration of proteolytic enzyme (Rani et al., 2015b, 2015c, 2015d; Rani, 2015e; Rani and Chauhan, 2015f, 2014g; Sharma, 2012; Rani, 2015h). Further effects of the selected chemically modifications, bioconjugation processes and the fabrication processes on the characteristics of the nanoparticles were studied that can be helpful to investigate the application of formulated bionanomaterials as nonviral delivery carriers of therapeutic drugs/genes/biomaterials used for various tissue engineering techniques (Lohcharoenkal et al., 2014).

Moreover, due to their defined structure, protein-based bionanoparticles are offered various possibilities for safe and cost-effective surface modifications including covalent attachment of drugs and targeting ligands to be considered as effective site specific nonviral targted loading vehicle used in cancer gene therapy (Chen et al., 2006; Salatin et al., 2015). The genetic treatment of neurodegenerative diseases can be summed up in safe and considerable clinical extent by using albumin hybrid nanomaterials to deliver the desired therapeutic material in relevant concentrations into the brain as compared to viral vectors that sometimes, elicits the risk of immune and inflammatory responses in to host cell. Their ability to cross the blood–brain barrier (BBB) and deliver the desired drugs can be proved safer and cost effective alternative nanotherapeutic approach to treat various neurodegenerative disorders. And, luciferase control vector pGL3 was considered effective reporter plasmid encoding for the firefly luciferase protein, linear polyethylenimine (22 kDa) as endosomolytic agent for enhancing the cells transfection in diseased neuronal cells. Observed studies were interpreted for their particle characteristics, their cellular uptake into aforementioned cell lines and on sub-cellular localization and transfection efficiency in the cerebellar cells to get improved feasibility of nanoparticle-based gene delivery (Langiu et al., 2014). Small interfering RNA (siRNA) targeted therapeutics (STT) were also performed to depict their compelling alternative to tradition medications for treatment of genetic diseases by providing a means to silence the e□pressionof specific aberrant proteins through interference at the expression level. Since serum proteins has been considered as safe, biocompative nonviral delivery vehicles for targeting tumors and further, investigated for their effects of incorporation of human serum albumin (HSA) in branched polyethylenimine (bPEI)-siRNA polyplexes in their functionalized internalization in epithelial and endothelial cells to study their silencing efficiency.

Furthermore, the uptake of the HSA-bPEI-siRNA ternary polyplexes was observed that the occurrence of caveolae-mediated endocytosis that providing the evidence for a clear role for HSA in polyplex internalization to explore the role of serum proteins in delivery of siRNA in host genome (Karimi et al., 2013). However, previous studies had showed that systemic delivery of cationic gene vectors mediates specific and efficient transfection within the lung which resulted due to interaction of the vectors with serum proteins-based on these previous observations, a novel and charge-density-controllable siRNA nanohybrid nonviral delivery system is developed to treat lung metastatic cancer by using cationic bovine serum albumin (CBSA) as the gene vector. By surface modification of BSA, CBSA with different isoelectric points (pI) were synthesized and the optimal their cationization degree for considerable siRNA binding and delivery ability as well as decreasing the extent of cell toxicity. And, it was noted that CBSA can form stable nanosized particles with siRNA and protect siRNA from degradation during the delivery into host cell to reach at targeted site with desired concentration of loaded payloads. CBSA was found to have excellent abilities to intracellularly deliver siRNA and mediate significant accumulation in the lung. When Bcl2-specific siRNA is introduced to this proposed system,

CBSA/siRNA nanoparticles were found to be exhibited an efficient gene-silencing effect that induces notable cancer cell apoptosis and subsequently inhibits the tumor growth in a B16 lung metastasis model. Hence, these results were depicted the role of CBSA-based self-assembled nanoparticles that can be a promising nonviral vehicle delivery strategy using fabricated albumin nanoparticles used for siRNA delivery for targeting diseased cell in lung and metastatic cancer therapy (Han et al., 2014).

ACKNOWLEDGMENT

This chapter is dedicated to Blessing of Almighty and my parents.

KEYWORDS

- **dendrimers**
- **gene delivery**
- **liposomes**
- **magnetic nanoparticles**
- **micelles**
- **natural polymers**
- **stem cells**

REFERENCES

Al-Dosari, M. S., & Ao, G. X., (2009). Nonviral Gene Delivery: Principle, Limitations, and Recent Progress. *AAPS J.*, *11(4)*, 671–681.

Arivalagan, K., Ravichandran, S., Rangasamy, K., & Karthikeyan, E., (2011). Nanomaterials and its potential applications. *Int J Chem Tech Res.*, *3(2)*, 534–538.

Bahadur, K. C., Thapa, B., & Bhattarai, N., (2014). Gold nanoparticle-based gene delivery: promises and challenges. *Nanotechnol Rev.*, *3(3)*, 269–280.

Bates, K., & Kostarelos, K., (2013). Carbon nanotubes as vectors for gene therapy: past achievements, present challenges and future goals. *Advn Drug Delivery Res.*, *65(15)*, 2023–2033. DOI: 10.1016/j.addr. 2013.10.003.

Biju, V., Mundayoor, S., Omkumar, R. V., Anas, A., & Ishikawa, M., (2010). Bioconjugated quantum dots for cancer research: present status, prospects and remaining issues. *Biotechnol Adv.*, *27*, 27.

Blasiack, B., van Veggeal, F. C. J. M., & Tomanek, B., (2013). Applications of Nanoparticles for MRI Cancer Diagnosis and Therapy. *J. Nanomaterials*. 1–12. http://dx.doi.org/10.1155/2013/148578.

Blow, N., (2009). Journeys across the membrane. *Nature Methods*, 6, 305–309. DOI: 10.1038/nmeth0409–0305.

Brown, P. K., Qureshi, A. T., Hayes, D. J., & Monroe, W. T., (2011). Targeted Gene Silencing With Light and a Silver Nanoparticle Antisense Delivery System. *ASME.*, 573–574.

Brown, P. K., Qureshi, A. T., Moll, A. N., Hayes, D. J., & Monroe, W. T., (2013). Silver nanoscale antisense drug delivery system for photoactivated gene silencing. *ACS Nano., 7(4)*, 2948–2959.

Cardenas, M., Barauskas, J., Schillen, K., Jennifer, L., Brennan, J. L., Brust, M., & Nylander, T., (2006). Thiol-Specific and Nonspecific Interactions between DNA and Gold Nanoparticles. *Langmuir, 22(7)*, 3294–3299. DOI: 10.1021/la0530438.

Chen, G., Takezawa, M., Kawazoe, N., & Tateishi, T., (2008). Preparation of Cationic Gold Nanoparticles for Gene Delivery. *Open Biotechnol J., 2*, 152–156.

Chen. L., Remondetto, G. E., & Subirade, M., (2006). Food proteinbased materials as nutraceutical delivery systems. *Trends Food Sci Technol., 17*(5), 272–283.

Choi, Y. S., Lee, M. Y., & David, A. E., (2014). Nanoparticles for gene delivery: therapeutic and toxic effects. *Mol Cellular Toxicol., 10*(1), 1–8.

Conde, J., Ambrosone, A., Sanz, V., Hernandez, Y., Marchesano, V., Tian, F., et al., (2012). Design of Multifunctional Gold Nanoparticles for In Vitro and In Vivo Gene Silencing. *ACS Nano., 6*(9), 8316–8324.

Ding, Y., Jiang, Z, , Saha, K., Kim, C. S., Kim, S. T., Landis, R. F., & Rotello, V. M., (2014). Gold Nanoparticles for Nucleic Acid Delivery. *Mol Therapy., 22*(6), 1075–1083. DOI: 10.1038/mt.2014.30.

Dong, H., Dai, W., Ju, H., Lu, H., Wang, S., Xu, L., et al., (2015). Multifunctional Poly(L-lactide)-Polyethylene Glycol-Grafted Graphene Quantum Dots for Intracellular MicroRNA Imaging and Combined Specific-Gene-Targeting Agents Delivery for Improved Therapeutics. *ACS Appl. Mater. Interfaces., 7*, 11015–11023. DOI: 10.1021/acsami.5b02803.

Donkuru, M., Badea, I., Wetting, S., Verrall, S., Elsabahy, M., & Foldvari, M., (2015). Advancing nonviral gene delivery: lipid- and surfactant-based nanoparticle design strategies. *Nanomedicine, 5*(7), 1103–1127.

Duan, J., Dong, J., Zhang, T., Su, Z., Ding, J., Zhang, Y., & Mao, X., (2013). Polyethyleneimine-functionalized iron oxide nanoparticles for systemic siRNA delivery in experimental arthritis. *Nanomedicine, 9(6)*, 789–801.

Gao, Y., Gu, W., Chen, L., Xu, Z., & Li, Y., (2007). A multifunctional nano device as non-viral vector for gene delivery: In vitro characteristics and transfection. *J Controlled Release, 118*(3), 381–388.

Germershaus, O., & Nultsch, K., (2015). Localized, non-viral delivery of nucleic acids: Opportunities, challenges and current strategies. *Asian J Pharma Sci., 10*(3), 159–175.

Gopinath, P., Gogoi, S. K., Chattopadhyay, A., & Ghosh, S. S., (2008). Implications of silver nanoparticle induced cell apoptosis for in vitro gene therapy. *Nanotechnol., 19*(7), 075104.

Green, J. J., (2007). Enhanced polymeric nanoparticles for gene delivery. *MIT*, http://hdl.handle.net/1721.1/44299.

Guo, S., Huang, Y., Jiang, Q., Sun, Y., Deng, L., Liang, Z., et al., (2010). Enhanced Gene Delivery and siRNA Silencing by Gold Nanoparticles Coated with Charge-Reversal Polyelectrolyte. *ACS Nano., 4*(9), 5505–5511.

Han, G., Ghosh, P., De, M., & Rotell, V. M., (2007). Drug and gene delivery using gold nanoparticles. *NanoBiotechnol., 3*(1), 40–45.

Han, J., Wang, Q., Zhang, Z., Gong, T., & Sun, X., (2014). Cationic bovine serum albumin based self-assembled nanoparticles as siRNA delivery vector for treating lung metastatic cancer. *Small. 10,* 524–535. DOI: 10.1002/smll.201301992.

Hill, H. D., Millstone, J. E., Banholzer, M. J., & Mirkin, C. A., (2009). The role radius of curvature plays in thiolated oligonucleotide loading on gold nanoparticles. *ACS Nano., 3,* 418–424.

Hossain, U., & Kojima, C., (2015). Dendrimers for theranostic applications. *Biomol Concepts. 6*(3), 205–217. DOI: 10.1515/bmc-2015-0012.

Hou, W., Wen, S., Guo, R., Wang, S., & Shi, X., (2015). Partially Acetylated Dendrimer-Entrapped Gold Nanoparticles with Reduced Cytotoxicity for Gene Delivery Applications. *J Nanosci Nanotechnol., 15*(6), 4094–4105.

Huang, Y., Lee, H., Tolliver, L. M., & Aronstam, R. S., (2015). Delivery of Nucleic Acids and Nanomaterials by Cell-Penetrating Peptides: Opportunities and Challenges. *BioMed Res Int., 16,* Article ID 834079.

Huo, S., Jin, S., Ma, X., Xue, X., Yang, K., Kumar, A., et al., (2014). Ultrasmall Gold Nanoparticles as Carriers for Nucleus-Based Gene Therapy Due to Size-Dependent Nuclear Entry. *ACS Nano., 8(6),* 5852–5862.

Jain, S., Hirst, D. G., & O'Sullivan, J. M., (2012). Gold nanoparticles as novel agents for cancer therapy. *Br J Radiol., 85*(1010), 101–113.

Je, J. Y., Cho, Y. S., & Kim, S. K., (2006). Characterization of (aminoethyl)chitin/DNA nanoparticle for gene delivery. *Biomacromol., 7*(12), 3448–3451.

Jean, M., Alameh, M., Chang, C. Y., Thibault, M., Lavertu, M., Darras, V., et al., (2012). Chitosan-based nanoparticles for GLP-1 gene delivery and DPP-IV gene silencing of in vitro cell lines relevant to type 2 diabetes. *Conference Nano. Quebec., 45*(1–2), 138–149.

Jiang, S., Eltoukhy, A. A., Love, K. T., Langer, R., & Anderson, D. G., (2013). Lipidoid-Coated Iron Oxide Nanoparticles for Efficient DNA and siRNA delivery. *Nano Lett., 13*(3), 1059–1064.

Jin, L., Zeng, X, Liu, M., Deng, Y & He, N., (2014). Current Progress in Gene Delivery Technology Based on Chemical Methods and Nano-carriers. *Theranostics, 4*(3), 240–255.

Kamau, S. W., Schulze, K., Steitz, B., Petri-Fink, A., Hassa, P. O., Hottiger, M. O., et al., (2005). Superparamagnetic iron oxide nanoparticles (SPIONs) as non-viral vectors for gene delivery in vitro. *European Cells and Mater., 10*(5).

Kamei, K., Mukai, Y., Kojima, H., Yoshikawa, T., Yoshikawa, M., Kiyohara, G., et al., (2009). Direct cell entry of gold/iron-oxide magnetic nanoparticles in adenovirus mediated gene delivery. *Biomaterials, 30*(9), 1809–1814.

Kami, D., Takeda, S., Itakura, Y., Gojo, S., Watanabe, M., & Toyoda, M., (2011). Application of Magnetic Nanoparticles to Gene Delivery. *Int. J. Mol. Sci., 12,* 3705–3722.

Karimi, M., Avci, P., Mobasseri, R., Hamblin, M., & Naderi-Manesh, H., (2013). The novel albumin–chitosan core–shell nanoparticles for gene delivery: preparation, optimization and cell uptake investigation. *J Nanopart Res 15*(5), 1–14.

Katragadda, C. S., Choudhury, P. K., & Murthy, P. N., (2010). Nanoparticles as Non-Viral Gene Delivery Vectors. *Indian J. Pharm. Educ. Res., 44*(2), 109–120.

Kerativitayanan, P., Carrow, J. K., & Gaharwar, A. K., (2015). Nanomaterials for Engineering Stem Cell Responses. *Advan Healthcare Materials, 4,* 1600–1627.

Kievit, F. M., & Zhang, M., (2011). Surface engineering of iron oxide nanoparticles for targeted cancer therapy. *Acc Chem Res., 44*(10), 853–862.

Kievit, F. M., Veiseh, O., Bhattarai, N., Fang, C., Gunn, J. W., Lee, D., Ellenbogen, R. G., Olson, J. M., et al., (2009). PEI-PEG-Chitosan Copolymer Coated Iron Oxide Nanoparticles for Safe Gene Delivery: synthesis, complexation, and transfection. *Adv Funct Mater*, *19*(14), 2244–2251.

Kim, E. Y., Schulz, R., Swantek, P., Kunstman, K., Malim, M. H., & Wolinsky, S. M., (2011). Gold nanoparticle-mediated gene delivery induces widespread changes in the expression of innate immunity genes. *Gene Therapy.*, 1–7.

Langiu, M., Dadparvar, M., Kreuter, J., & Ruonala, M. O., (2014). Human Serum Albumin-Based Nanoparticle-Mediated In Vitro Gene Delivery. *PLOSone*, DOI: 10.1371/journal.pone.0107603.

Liu, C., & Zhang, N., (2011). Nanoparticles in gene therapy: principles, prospects, and challenges. *Prog Mol Biol Transl Sci.*, *104*, 509–562.

Lohcharoenkal, W., Wang, L., Chen, Y. C., & Rojanasakul, Y., (2014). Protein Nanoparticles as Drug Delivery Carriers for Cancer Therapy. *BioMed Res Int.*, 1–12. http://dx.doi.org/10.1155/2014/180549.

Lu, W., Wang, L., Li, L., Zhao, Y., Zhou, Z., Shi, J., Zuo, X., & Pan, D., (2015). Quantitative investigation of the poly-adenine DNA dissociation from the surface of gold nanoparticles. *Scientific Reports*, *5*, 10158. DOI: 10.1038/serp10158.

Mairhofer, J., & Grabherr, R., (2008). Rational Vector Design for Efficient Non-viral Gene Delivery: Challenges Facing the Use of Plasmid DNA. *Mol Biotechnol.*, *39*, 97–104.

Marin, S., Vlasceanu, G. M., Tiplea, R. E., Bucur, I. R., Lemnaru, M., Marin, M. M., & Grumezescu, A. M., (2015). Applications and toxicity of silver nanoparticles: a recent review. *Curr Top MedChem.*, *15*(16), 1596–1604.

McBain, S. C., Griesenbach, U., Xenariou, S., Keramane, A., Batich, C. D., Alton, E. W. F. W., & Dobson, J., (2008). Magnetic nanoparticles as gene delivery agents: enhanced transfection in the presence of oscillating magnet arrays. *Nanotechnol.*, *19*(40).

McBain, S. C., Yiu, H. H. P., & Dobson, J., (2008). Magnetic nanoparticles for gene and drug delivery. *Int J Nanomedicine*, *3*(2), 169–180.

Mieszawska, A. J., Mulder, W. J. M., Fayad, Z. A., & Cormode, D. P., (2013). Multifunctional Gold Nanoparticles for Diagnosis and Therapy of Disease. *Mol. Pharm.*, *10*(3), 831–847.

Mintzer, M. A., & Simanek, E. E., (2009). Nonviral Vectors for Gene Delivery. *Chem. Rev.*, *109*, 259–302.

Mohmmadi, Z., Dorkoosh, F. A., Hosseinkhani, S., Amini, A., Rahimi, A. A., et al., (2012). Stability studies of chitosan-DNA-FAP-B nanoparticles for gene delivery to lung epithelial cells. *Acta Pharm.*, *62*, 83–92.

Mulens, V., Morales, M. P., & Barber, D. F., (2013). Development of Magnetic Nanoparticles for Cancer Gene Therapy: A Comprehensive Review. *ISRN Nanomaterials*, Article ID 646284.

Mykhaylyk, O., Sobisch, T., Almstatter, I., Antequera, Y., Brandt, S., & Anton, M., (2012). Silica-Iron Oxide Magnetic Nanoparticles Modified for Gene Delivery: A Search for Optimum and Quantitative Criteria. *Pharma Res.*, *29*(5), 1344–1365.

Nicoli, E., Syga, M. I., Bosetti, M., & Shastri, V. P., (2015). Enhanced Gene Silencing through Human Serum Albumin-Mediated Delivery of Polyethylenimine-siRNA Polyplexes. *PLOSone*, DOI: 10.1371/journal.pone.0122581.

Nitta, S. K & Numata, K., (2013). Biopolymer-Based Nanoparticles for Drug/Gene Delivery and Tissue Engineering. *Int J Mol Sci.*, *14*, 1629–1654.

Pack, D. W., Hoffman, A. S., Pun, S., & Stayton, P. S., (2005). Design and development of polymers for gene delivery. *Nat Rev Drug Discovery*, *4*, 581–593.

Pathak, A., Patnaik, S., & Gupta, K. C., (2009). Recent trends in non-viral vector-mediated gene delivery. *Biotechnol. J., 4,* 1559–1572.

Pean, M., Gracia, M. A., Ruiz, E. L., Bustamante, M., Jimenez, M., 4, Madeddu, R. & Marchal, J. A., (2012). Functionalized Nanostructures with Application in Regenerative Medicine. *Int J Mol Sci., 13,* 3847–3886, DOI: 10.3390/ijms13033847.

Perez-Martínez, F. C., Guerra, J., Posadas, I., & Cena, V., (2011). Barriers to Non-Viral Vector-Mediated Gene Delivery in the Nervous System. *Pharma Res., 28,* 1843–1858.

Phillips, D. L., Sharma, A., Tovey, J. C. K., Ghosh, A., Klibanov, A. M., J. W. Cowden, J, W., et al., (2010). Gold Nanoparticles: A Potent Vector for Non-Viral Corneal Endothelial Gene Therapy. *Investigative Ophthalmol Visual Sci., 51,* 433.

Posadas, I., Guerra, F. J., & Cena, V., (2010). Nonviral Vectors for the Delivery of Small Interfering RNAs to the CNS. *Nanomedicine, 5*(8), 1219–1236.

Probst, C. E., Zrazhevskiy, P., Bagalkot, V., & Gao, X., (2013). Quantum dots as a platform for nanoparticle drug delivery vehicle design. *Adv Drug Deliv Rev., 65*(5), 703–718.

Rajangam, T., & An, S. S. A., (2013). Fibrinogen and fibrin based micro and nano scaffolds incorporated with drugs, proteins, cells and genes for therapeutic biomedical applications. *Int J Nanomedicine, 8,* 3641–3662.

Ramamoorth, M., & Narvekar, A. (2015). Non-Viral Vectors in Gene Therapy- An Overview. *J Clin Diagn Res., 9*(1), 01–06.

Rani, K., Chauhan, C., & Kaur, H., (2014a). Potential and automotive applications of nanomaterials in combating cancer and stem cell therapy: An informative overview on nanotherapeutics. *J Nanotech and Smart Materials, 1,* 1–6.

Rani, K., Goyal, S., & Chauhan, C., (2015b). Novel approach of alkaline protease mediated biodegradation analysis of mustard oil driven emulsified bovine serum albumin nanospheres for controlled release of entrapped *Pennisetum glaucum* (Pearl Millet) amylase. *American J Advn Drug Delivery, 3*(2), 135–148.

Rani, K., Gupta, C., & Chauhan, C., (2015c). Biodegradation of almond oil driven bovine serum albumin nanoparticles for controlled release of encapsulated Pearl millet amylase. *American J Phytomedicine Clin Therapeutics, 3*(3), 222–230.

Rani, K., & Kant, S., (2015d). Alkaline Protease Mediated Bioproteolysis of Jasmine Oil Activated Pennisetum glaucum Amylase Loaded BSA Nanoparticles for Release of Encapsulated Amylase. *Int J Chem Sci and Appl., 6*(2), 56–63.

Rani, K., (2015e). Novel Biodegradation Analysis of Olive Oil Driven Emulsified Bovine Serum Albumin Nanopreparation with Alkaline Protease for Controlled Release of Encapsulated Pennisetum glaucum Amylase. *J Chem Chemical Sci., 5*(6), 341–350.

Rani, K., & Chauhan, C., (2015f). Preparation of Cicer Artienium Amylase Loaded BSA Nanoparticles and Their Bioproteolysis to be used as Detergent Additive. *Bioengg and Biosci., 3*(5), 72–82.

Rani, K., & Chauhan, C., (2014g). Biodegradation of Cicer Arietinum Amylase loaded Coconut oil driven Emulsified Bovine Serum Albumin Nanoparticles and their application in Washing Detergents as Eco-Friendly Bio-Active Addictive. *World J Pharm Pharmaceutical Sci., 3(12),* 924–936.

Rani, K., (2015h). Applicative biodegradation study of egg albumin nanospheres by alkaline protease for release of encapsulated cicer arietinum amylase in washing as bio-active detergent additive. *World J Pharmaceutical Res, 4(1),* 1–13.

Saini, V., Martyshki, D. V., Towner, V. D., Mirov, S. B., & Everts, M., (2014). Limitations of Adenoviral Vector-Mediated Delivery of Gold Nanoparticles to Tumors for Hyperthermia Induction. *Open Nanomed J., 2.*

Salata, V., (2004). Applications of nanoparticles in biology and medicine. *J BioNanotechnol.*, *2*, 3, DOI: 10.1186/1477-3155-2-3.

Salatin, S., Jelvehgari, M., Maleki-Dizaj, S., & Adibkia, K., (2015). A sight on protein-based nanoparticles as drug/gene delivery systems. *Therapeutic Delivery, 6*(8), 1017–1029.

Sarkar, K., Banerjee, S. L., Kundu, P. P., Madras, G., & Chatterjee, K., (2015). Biofunction-alized surface-modified silver nanoparticles for gene delivery. *J. Mater. Chem. B., 3,* 5266–5276.

Shalek, A. K., Gaublomme, T., Wang, L., Yosef, N., Chevrier, N., Andersen, N. S., et al., (2012). Nanowire-Mediated Delivery Enables Functional Interrogation of Primary Immune Cells: Application to the Analysis of Chronic Lymphocytic Leukemia. *Nano Lett., 12*(12), 6498–6504. DOI: 10.1021/nl3042917.

Sharma, K. R., (2012). Preparation of emulsified encapsulated nanoparticles of bovine serum albumin of bound glucose oxidase and their application in soft drinks/non-alcoholic beverages. *Biotechnol and Biomaterials, 2*(2), 1–4.

Shmueli, R. B., Bhise, N. S., & Green, J. J., (2013). Evaluation of Polymeric Gene Delivery Nanoparticles by Nanoparticle Tracking Analysis and High-throughput Flow Cytom-etry. *J. Vis. Exp., 73,* e50176.

Sinha, S., McKnight, D., Katti, K. V., Kannan, R., Robertson, D. J., Cowden, J. W., et al., (2008). Gold Nanoparticles Stabilized in Gum Arabic for Corneal Gene Therapy. *Inves-tigative Ophthalmol Visual Sci., 49,* 4787.

Sironmani, T. A., (2015). Comparison of nanocarriers for gene delivery and nanosensing using montmorillonite, silver nanoparticles and multiwalled carbon nanotubes. *Applied Clay Sci., 103,* 55–61.

Sun, N. F., Liu, Z. A., & Hu, S. Y., (2013). The Application of Nanoparticles for Gene Vector in Tumor Gene Therapy. *J Tumor, 1*(5), 24–27.

Suna, N. F., Liua, Z., Huanga, W., Tiana, Ai-Ling. & Hua, San-Yua., (2014). The research of nanoparticles as gene vector for tumor gene therapy. *Critical Rev Oncol/Hemat, 89*(3), 352–357.

Tencomnao, T., Apijaraskul, A., Rakkhithawatthana, V., Chaleawlert-umpon, S., Pimpa, N., Sajomsang, W, et al., (2011). Gold/cationic polymer nano-scaffolds mediated transfec-tion for non-viral gene delivery system. *Carbohydrate Polymers, 84*(10), 216–222.

Tiwari, P. M., Vig, K., Dennis, V. A., & Singh, S. R., (2011). Functionalized Gold Nanopar-ticles and Their Biomedical Applications. *Nanomaterials., 1,* 31–63.

Uthaman, S., Lee, S. J., Cherukula, K., Cho, C., & Park, I., (2015). Polysaccharide-Coated Magnetic Nanoparticles for Imaging and Gene Therapy. *BioMed Res Int.*, Article ID 959175.

Uthaman, S., Moon, M. J., Lee, D., Kim, W. J., & Park, I. K., (2015). Di-Sulfide Linked Polyethylenimine Coated Gold Nanoparticles as a Non-Viral Gene Delivery Agent in NIH-3T3 Mouse Embryonic Fibroblast. *J. Nanosci Nanotechol., 15*(10), 7895–7899.

Vasir, J. K., & Labhasetwar, V., (2006). Polymeric nanoparticles for gene delivery. *Expert Opin Drug Deliv., 3*(3), 325–344.

Vinhas, R., Cordeiro, M., Carlos, F. F., Mendo, S., Fernandes, A. R., Figueiredo, S., et al., (2015). Gold nanoparticle-based theranostics: disease diagnostics and treatment using a single nanomaterial. *Nanobiosensors Dis Diag., 4,* 11—23.

Wang, X., Niu, D., Hu, C., & Li, P., (2015). Polyethyleneimine-Based Nanocarriers for Gene Delivery. *Curr Pharma Design, 21*(42), 6140–6156.

Wang, Y., Cui, H., Li, K., Sun, C., Du, W., Cui, J., et al., (2014). Magnetic Nanoparticle-Based Multiple-Gene Delivery System for Transfection of Porcine Kidney Cells. *Plos One*, *9*(8), e106612.

Wang, Y., Cui, H., Yang, Y., Zhao, X., Sun, C., Chen, W., et al., (2014). Mechanism Study of Gene Delivery and Expression in PK-15 Cells Using Magnetic Iron Oxide Nanoparticles as Gene Carriers. *Nano Life*, *4*(4), 1441018–1441026.

Witzigmann, D., Wu, D., Schenk, S. H., Balasubramanian, V., Meier, W., & Huwyler, J., (2015). Biocompatible Polymer–Peptide Hybrid-Based DNA Nanoparticles for Gene Delivery. *ACS Appl. Mater. Interfaces*, *7*(19), 10446–10456.

Xian, J. L., & Lee, T., (2012). Gene Delivery by Functional Inorganic Nanocarriers. *Recent Patents on DNA & Gene Sequences*. *6*(2), 108–114.

Yah, C. S., (2013). The toxicity of Gold Nanoparticles in relation to their physiochemical properties. *Biomed Res.*, *24*(3), 400–413.

Zwiorek, K., Kloeckner, J., Wagner, E., & Coester, C., (2004). Gelatin nanoparticles as a new and simple gene delivery system. *J Pharm Pharmaceut Sci.*, *7*(4), 22–28.

INDEX

Printed and bound by CPI Group (UK) Ltd, Croydon, CR0 4YY

23/10/2024

01777701-0003